COASTAL ENGINEERING

– For Engineers and Practitioners –

Yoshimichi YAMAMOTO
Ca Thanh VU
Harshinie KARUNARATHNA

Rikohtosho

Author Biographies

Editor and Author: Yoshimichi Yamamoto

Dr. of Engineering (Coastal Engineering)

Pro. Engineer in Japan (Civil Engineering)

Exec. Pro. Civil Engineer (Disaster Prevention) of JSCE

Fellow member of JSCE

Former Prof. of Tokai University, Japan

Author: Ca Thanh Vu

PhD. (Biological and Environmental Sciences)

Associate Professor, Principal Lecturer

Ha Noi University of Natural Resources and Environment, Viet Nam

Former Director General, Viet Nam Institute of Seas and Islands

Author: Harshinie Karunarathna

PhD. (Coastal Engineering)

Professor in Coastal Engineering

Swansea University, United Kingdom

COASTAL ENGINEERING

- For Engineers and Practitioners -

copy right © 2025 Yoshimichi YAMAMOTO, Ca Thanh VU, Harshinie KARUNARATHNA

rikohtosho Co.

All right reserved.

ISBN:978-4-8446-0974-2

Printed in Japan

All right reserved. No part of this publication may be reproduced, stored in a retrieval system, or transmitted in any form or by any means, electronic, mechanical, photocopying, recording, or otherwise without either the prior written permission of authors or a license permitting restricted copying issued by the aforementioned.

PREFACE

Although records and monuments of harbour and shipping technology have survived since B.C., coastal engineering was not systematised as a discipline until the 20th century. The study of tidal and wave theory in coastal engineering is relatively old, Newton published the equilibrium tidal theory in his Principia in 1687, but it could not explain the actual tidal motion well, and Kelvin (W. Thomson) published the dynamical tidal theory in 1879 as a resonance phenomenon of long waves generated by tidal force based on the equilibrium tidal theory. For waves, Gerstner first published his trochoidal wave theory in 1802, Airy published his small-amplitude wave theory in 1845, which is still widely used today, and Stokes published his finite-amplitude wave theory in 1847. Further theoretical support for the solitary wave was published by Boussinesq in 1871 and Rayleigh in 1876, and the theory of waves in extremely shallow water (the cnoidal wave theory) was published by Korteweg and de Vries in 1895. Thus, the basic theories of waves were almost complete by the end of the 19th century.

In the 1900s, Gaillard published the hydrodynamic pressure formula in 1905, Hiroi the wave pressure equation after wave breaking in 1919, Sainflau the wave pressure equation for standing waves in 1928 and Bagnold the pressure generation mechanism of impact waves in 1939. In the 1930s, research into coastal erosion began in earnest at the Beach Erosion Board of the US Army Corps of Engineers and elsewhere, and the theory of wave generation, development, and propagation also began to be studied in earnest, with Svedrup and Munk developing the Significant Wave Method in 1942 and publishing it in 1946 and 1947.

In 1950, O'Brien of the University of California, Berkeley, decided that there was a need for a forum for comprehensive discussion of coastal issues and organised the first Coastal Engineering Conference in Long Beach, California, where he gave the name "Coastal Engineering" to engineering related to the coast. The conference was renamed the International Conference on Coastal Engineering and is still held every two years.

Coastal engineering research since the 1950s has been dominated by Bretschneider's improvements to the significant wave method, Pierson and others' publication of the spectral method, Minikin's impact wave pressure equation, Morison's wave force equation on piers, Hudson's rubble stability equation, and Sato and Saville's proposed methods for assessing run-up heights. In the 1960s, research on spectral analysis progressed, Longuet-Higgins and Stewart published the radiation stress theory associated with nearshore currents, and research on sand drift, such as the critical water depth equation for sediment movement and the sediment transport rate formula, began to gain momentum. From 1970 onwards, in addition to the proposal of various evaluation formulae such as the wave force calculation formula by Goda et al, research on numerical simulation models for wave generation, development, and propagation, wave run-up and overtopping, nearshore currents, storm surges, wind-driven currents, and tsunamis, wave forces on various structures, topographical changes caused by waves and currents, etc., progressed rapidly with the rapid development of computer processing power.

A wide variety of technical books summarising the results of these studies have been published by experts in many countries (e.g. introductory books with gentle explanations of the main issues of coastal engineering, dictionary-like books covering a wide range of coastal engineering, technical books with detailed explanations of some specific areas, etc.). Under these circumstances, we decided to publish a handy technical book for engineers that summarises the main points in a concise manner and covers as wide a range of coastal engineering as possible.

The features of this book are as follows:

(1) We have tried to include the latest research as much as possible.

1) Introduction to the main wave estimation methods and existing open sources (including URLs for download).

2) Introduction to methods for estimating wave run-up heights and overtopping rates, considering wave grouping effects.

3) Reports on damage patterns based on field surveys of the 2004 Indian Ocean tsunami and the 2011 Tohoku tsunami.

4) Presentation of all evaluation methods for predicting beach topographic changes due to large waves and existing open sources (including URLs for download).

5) Introduction to the mechanism, prediction models and countermeasures for the outflow of backfill materials from coastal dikes and seawalls.

6) Introduction to the main causes and main countermeasures of coastal erosion.

7) Introduction of appropriate formulas for calculation of tsunami force and drift impact force.

8) Introduction of numerical models of topographic change of a tsunami inundation area and simple evaluation diagrams of building collapse.

9) Introduction of countermeasures against microplastics and other drifting debris.

10) Introduction to the main types of wave power generation and examples.

(2) A set of programs and manuals for the following numerical models for predicting wave-induced topographic changes will be provided

1) Numerical prediction model of shoreline changes

This numerical model can be used to predict shoreline change over a wide area and over a long period of time, considering the influence of structures on a straight beach.

2) Numerical prediction model of beach change due to large waves

This numerical model can be used to predict the horizontal two-dimensional beach profile change due to irregular waves under the installation of impermeable and permeable structures, including wave run-up areas.

3) Numerical prediction model of topographic changes due to tsunamis.

This numerical model can be used to predict the horizontal two-dimensional topographic change due to tsunamis, including inundated areas on land.

We hope that interested readers will find this book "Coastal Engineering for Engineers and Practitioners" useful.

Finally, we would like to express our appreciation to the staff of Rikohtosho Co. for their efforts in publishing this book.

April 2025

Yoshimichi YAMAMOTO

CONTENTS

1. STATISTICAL PROPERTIES AND GENERATION MECHANISMS OF WAVES ——————————— 1

1.1 Statistical Properties of Waves /2

  1) Basic Representation and Synthesis of Regular Waves /2

  2) Irregular Waves and Wave Grouping /4

  3) Wave-by-wave Analysis Method /4

  4) Spectral Analysis Method /7

1.2 Generation Mechanism of Sea Waves /16

  1) Basis of Wave Generation Theory /16

  2) Wave Forecasting Models /19

2. WAVE THEORIES AND, PROPAGATION AND DEFORMATION —————————————— 39

2.1 Small Amplitude Wave Theory /40

  1) Assumptions for Derivation /40

  2) Derivation of the Water Particle Velocity Equations /41

  3) Equations for Wavelength and Wave Celerity /43

  4) Acceleration and Orbit of Water Particles /45

  5) Method for Measuring Hydrodynamic Pressure and Wave Height /46

  6) Wave Group Velocity /48

  7) Wave Energy and Power /49

  8) Standing Waves /51

  9) Partial Standing Waves /52

2.2 Finite Amplitude Wave Theory /54

  1) Characteristics and Applicable Range of Finite Amplitude Wave Theory /54

  2) Stokes Wave Theory /55

  3) Cnoidal Wave Theory /57

2.3 Wave Propagation and Deformation /59

  1) Shallow Water Deformation and Refraction Deformation of Waves /59

  2) Diffraction Deformation of Waves /62

  3) Transmission and Reflection of Waves /65

  4) Wave Breaking /72

  5) Wave Height Reduction Due to Seabed Friction and Other Factors /82

2.4 Numerical Models for Calculating Wave Fields /86

  1) Models Based on Energy Conservation Law /86

2) Models Based on the Equation of Motion or the Mild Slope Equation　/86

3. WAVE RUN-UP, WAVE OVERTOPPING AND WAVE FORCES ————————————— 95

3.1 Wave Run-up and Wave Overtopping　/96

1) Influence of Wave Irregularity on Wave Run-up　/96

2) Long-period Gravity Waves Due to Wave Grouping　/98

3) Calculation of Wave Run-up Height　/103

4) Calculation of Wave Overtopping Rate　/113

3.2 Wave Forces　/124

1) Wave Forces Acting on an Upright Walls　/124

2) Stability of Stones or Blocks Covering a Sloping Breakwater　/137

3) Wave Forces Acting on a Pillar Structure　/143

4. CURRENTS IN THE SEA ————————————————————————————— 151

4.1 Nearshore Currents　/152

1) Radiation Stress　/152

2) Average Water Level Inside and Outside the Breaking Zone　/154

3) Planar Structure of Nearshore Currents　/155

4) Numerical Simulation Models of Nearshore Currents　/160

4.2 Drift Currents　/162

4.3 Ocean Currents　/165

4.4 Tidal Currents　/167

1) Tides　/167

2) Harmonic Analysis of Tides　/169

3) Prediction of Tidal Currents　/169

5. STORM SURGES AND TSUNAMIS ————————————————————————— 173

5.1 Storm Surges　/174

1) Generation of Storm Surges　/174

2) Reality of Storm Surge Disasters　/176

3) Methods for Calculating Tide Levels during Storm Surges　/177

4) Secondary Undulation and Continental Shelf Waves　/180

5.2 Tsunamis　/182

1) Generation of Tsunamis　/182

2) Reality of Tsunami Disasters　/185

3) Evaluation Methods for Various Tsunami Parameters　/187

4) Tsunami Forces /195

5) Tsunami Debris Impact Forces /200

6. COASTAL TOPOGRAPHIC CHANGE —————————————————— 209

6.1 Beach Profile Change /210

1) Beach Cross-section Topography /210

2) Beach Cross-sectional Topographic Change /211

3) Methods for Setting Beach Stability Cross-sections /215

6.2 Drifting Sand /220

1) Patterns of Sediment Transport /220

2) Critical Depth for Sediment Movement /220

3) Equations for Calculating Bed Load Transport Rates /221

4) Equations for Calculating Suspended Load Transport /225

5) Continuity Equation of Drifting Sand /226

6.3 Coastal Topographic Change Prediction Methods /228

1) Line Models and Beach Profile Models Based on Wave Parameters /229

2) Bathymetric Change Models Based on Waves and Nearshore Currents /236

6.4 Coastal Erosion /244

1) Causes of Coastal Erosion /244

2) Basic Approach to Coastal Erosion Control /249

6.5 Scour and Sand Outflow /252

1) Scour Due to Large Waves /252

2) Outflow (Suction) of Backfill Materials /255

3) Scour Due to Tsunamis /265

6.6 Wind-blown Sand /269

1) Critical Friction Velocity of Wind-blown Sand /269

2) Equations for Calculating Friction Velocities and Wind-blown Sand Rates /269

7. COASTAL PROTECTION AND VARIOUS OTHER STRUCTURES ———————————— 281

7.1 Coastal Structures /282

1) Facilities on Coasts /282

2) Facilities on Ports and Harbours /286

7.2 Measures against Large Waves /289

1) Occurrence of Large Waves /289

2) Characteristics of Large Wave Damage /289

3) Concept of High Wave Protection /291

4) Measures against Wave Forces　/293

5) Wave Overtopping Measures　/297

6) Measures against Coastal Erosion and Sedimentation in Harbours　/298

7) Measures against Scour and Sand Outflow　/306

8) Measures against Splash and Wind-blown Sand　/307

7.3 Storm Surge Protection　/309

1) Characteristics of Storm Surge Damage　/309

2) Measures against Storm Surges　/310

7.4 Tsunami Protection　/313

1) Characteristics of Tsunami Damage　/313

2) Tsunami Protection Measures　/317

7.5 Performance-based Design and Reliability-based Design　/329

1) Performance-based Design　/329

2) Reliability-based Structural Design　/329

8. ENVIRONMENTAL PROTECTION AND WAVE POWER GENERATION ———————————— 341

8.1 Protection of Water Quality in Marine Areas　/342

1) Actual Situation of Water Quality Deterioration　/342

2) Actions for Water Quality Protection　/349

8.2 Marine Litter　/352

1) Actual Situation of Drifted Litter　/352

2) Measures against Drifted Litter　/354

3) Countermeasures against the Smell of Rotting Seaweed and Seagrass　/355

8.3 Widespread Use of Wave Power　/360

1) Renewable Energy　/360

2) Examples of Wave Power Generation Methods　/363

3) A Thought for the Spread of Wave Power and Small-scale Hydropower　/372

Appendix　MANUALS OF NUMERICAL PREDICTION MODELS ———————————————— 387

1. Numerical Model for Shoreline Change Due to Waves　/388

2. Numerical Model for Beach Change Due to High Waves　/410

3. Numerical Model for Coastal Topographic Change Due to Tsunamis　/434

SYMBOL LIST ——— 455

TECHNICAL TERM LIST ————————————————————————————————— 471

Chapter:1

STATISTICAL PROPERTIES AND GENERATION MECHANISMS OF WAVES

1. STATISTICAL PROPERTIES AND GENERATION MECHANISMS OF WAVES

Ocean waves include **gravity waves** caused by wind, **tsunamis** with a period of several minutes or more caused by a strong undersea earthquake, **storm surges** with a period of several hours caused by atmospheric depressions, and **tidal waves** with a period of half a day or one day caused by the gravitational forces of the moon and the sun.

This chapter explains the statistical properties, theory of occurrence, and prediction methods of ocean waves. Incidentally, waves are classified into **ripples** with periods of less than 0.1 second, **wind waves** and **swell** with periods of 0.1 to 30 seconds, and **long-period gravity waves** or **infragravity waves**, which have periods of 30 seconds to several minutes.

1.1 Statistical Properties of Waves
1) Basic Representation and Synthesis of Regular Waves
(1) Basic Representation of Regular Waves

In the case of the regular wave shown in **Fig. 1.1**, the length from the first crest to the next crest is defined as the **wavelength** (L) and the vertical distance from the crest to the trough as the **wave height** (H). The following cosine function can express the water surface elevation (η) displacement of a water wave from the still water surface.

$$\eta = \frac{H}{2}\cos\left(\frac{2\pi}{L}x\right) = \frac{H}{2}\cos(\kappa x) \tag{1.1.1}$$

where x is the horizontal coordinate, and $\kappa = 2\pi/L$ is called the **wave number**.

Furthermore, in **Fig. 1.1**, the time between the passage of one wave and the arrival of the next wave, i.e. the time taken by one wavelength to pass, is defined as the **wave period** (T), and the following equation can express a cosine wave travelling in the x direction.

$$\eta = \frac{H}{2}\cos\left(\frac{2\pi}{L}x - \frac{2\pi}{T}t\right) = \frac{H}{2}\cos(\kappa x - \omega t) = \frac{H}{2}\cos\kappa(x - Ct) \tag{1.1.2}$$

where t is the time, $\omega = 2\pi/T$ is the **angular frequency,** and $C = L/T = \omega/\kappa$ is called the **phase velocity** or the **wave celerity**.

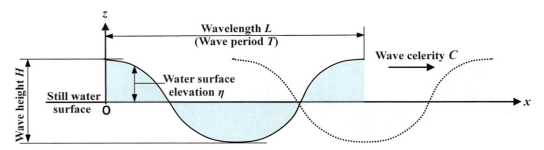

Figure 1.1 Definition of wave parameters

1.1 Statistical Properties of Waves

(2) Synthesis of Regular Waves

For simplicity, the wave number is hidden and a composite of two regular waves with respective wave heights of H_1 and H_2, and periods of T_1 and T_2 is considered. In this case, the water surface elevation from the still water surface η is expressed with Eq. (1.1.3).

$$\eta = \frac{H_1}{2}\cos\frac{2\pi}{T_1}t + \frac{H_2}{2}\cos\frac{2\pi}{T_2}t = (H_+ + H_-)\cos(T_+ + T_-) + (H_+ - H_-)\cos(T_+ - T_-)$$
$$= H_+\left[\cos(T_+ + T_-) + \cos(T_+ - T_-)\right] + H_-\left[\cos(T_+ + T_-) - \cos(T_+ - T_-)\right] \quad (1.1.3)$$

Here, $H_+ = \dfrac{H_1/2 + H_2/2}{2}$, $H_- = \dfrac{H_1/2 - H_2/2}{2}$, $T_+ = \dfrac{2\pi t/T_1 + 2\pi t/T_2}{2}$, $T_- = \dfrac{2\pi t/T_1 - 2\pi t/T_2}{2}$. (1.1.4)

Moreover, using the additional theorem, Eq.(1.1.3) can be expressed by Eq.(1.1.5).

$$\begin{aligned}\eta &= H_+\left(\cos T_+\cos T_- - \sin T_+\sin T_- + \cos T_+\cos T_- + \sin T_+\sin T_-\right)\\&+ H_-\left(\cos T_+\cos T_- - \sin T_+\sin T_- - \cos T_+\cos T_- - \sin T_+\sin T_-\right)\\&= 2H_+\cos T_+\cos T_- - 2H_-\left[\cos(T_+ - T_-) - \cos T_+\cos T_-\right]\\&= (2H_+ + 2H_-)\cos T_+\cos T_- - 2H_-\cos(T_+ - T_-)\end{aligned} \quad (1.1.5)$$

Then, substituting Eq.(1.1.4) back into Eq.(1.1.5), Eq.(1.1.6) is obtained.

$$\eta = H_1\cos\left[\frac{2\pi}{2T_1T_2/(T_2 - T_1)}t\right]\cos\left[\frac{2\pi}{2T_1T_2/(T_2 + T_1)}t\right] - \frac{H_1 - H_2}{2}\cos\frac{2\pi}{T_2}t \quad (1.1.6)$$

Fig. 1.2 shows the results of synthesising a regular wave with a period of 12.0s and a wave height of 4.0m and a regular wave with a period of 8.0s and a wave height of 2.0m. **Fig. 1.3** shows the results of synthesising a regular wave with a period of 12.0s and a wave height of 4.0m and a regular wave with a period of 10.9s and a wave height of 2.0m. When regular waves are combined, irregularity increases as shown in **Fig. 1.2**, but when regular waves with a small difference in wave period are combined, the period $[2T_1T_2/(T_2 - T_1)]$ of the first cosine function on the right-hand side of Eq. (1.1.6) becomes long-period, resulting in **beat waves (roaring waves)** as shown in **Fig. 1.3**.

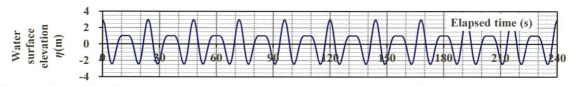

Fig. 1.2 Example of a synthetic wave generated by combining two regular waves with periods of 12.0s and 8.0s.

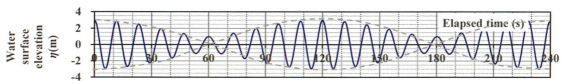

Fig. 1.3 Example of a synthetic wave generated by combining two regular waves with periods of 12.0s and 10.9s.

1. STATISTICAL PROPERTIES AND GENERATION MECHANISMS OF WAVES

2) Irregular Waves and Wave Grouping

Actual ocean waves are irregular, as shown in **Fig. 1.4**, and usually have a series of around 10 regular waves that gradually increase in size from small waves to large waves and back to small waves again. This property is called **wave groupiness.** This is known to create phenomena such as resonance in harbours and floating structures, instability of rubble and blocks in sloping dikes, mean water surface fluctuations near the shoreline, and increases in wave run-up and overtopping.

In the case of regular waves, a wave is counted as the surface profile from one crest to the next crest or one trough to the next trough. But in the case of real seas where waves are irregular as shown in ⑤ and ⑥ of the water surface line of **Fig. 1.4**, this definition can produce waves in which the water surface height η does not cross the still water surface (the horizontal line with a water surface height of 0 m in **Fig. 1.4**) within a single wave. Therefore, the **zero up-crossing method**, which defines a wave as one in which the water surface elevation η rises across the still water surface and then rises again across the still water surface, or **zero down-crossing method**, which defines a wave as one in which the water surface height η descends across the still water surface and then descends across the still water surface again, is used.

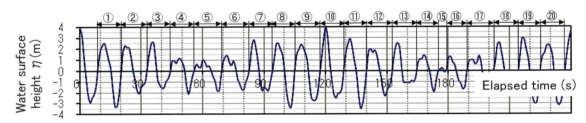

Fig. 1.4 Waveforms of ocean waves and how to count waves using the zero-up crossing method.

Two methods are available to analyse irregular waves: the **wave-by-wave analysis method (= time-domain analysis method)** and the **spectral analysis method (= frequency domain analysis method)**.

3) Wave-by-wave Analysis Method

To ensure wave stationarity, the method of dealing with the distribution of the occurrence probability of wave height and wave period using a continuous time series record of about 20 minutes (about 200 waves) is called the **wave-by-wave analysis method**. This can be adapted to waves with strong non-linearity.

The probability density function of water surface elevation η follows a normal distribution (= Gaussian distribution) when the mean is 0 and the variance is 1 and is expressed by Eq. (1.1.7).

1.1 Statistical Properties of Waves

> The following mathematical expressions, which may be difficult for the reader to understand, are explained here.
> $$y = A \times \exp\{B\} \tag{a.1}$$
> This expression is equivalent to the following expression using Napier numbers ($e = 2.7182\cdots$).
> $$y = A \times e^B \tag{a.2}$$
> Here, if the exponent B is a complex expression, the expression in Eq. (a.2) makes the exponent part difficult to read, so the expression $\exp\{B\}$ is adopted.

$$p(\eta/\sigma_\eta) = \frac{1}{\sqrt{2\pi}} \exp\left\{-\frac{1}{2}\left(\frac{\eta}{\sigma_\eta}\right)^2\right\} \tag{1.1.7}$$

where σ_η is the standard deviation of the water surface elevation η, defined by Eq. (1.1.8) using the number of data n.

$$\sigma_\eta = \sqrt{\frac{\sum_{i=1}^{n} \eta_i^2}{n}} \tag{1.1.8}$$

Moreover, according to Longuet-Higgins [1], the probability density function of wave height H follows the Rayleigh distribution shown in **Fig. 1.5** and is expressed by Eq. (1.1.9).

$$p(H) = \frac{H}{4\sigma_\eta^2} \exp\left\{-\frac{1}{8}\left(\frac{H}{\sigma_\eta}\right)^2\right\}, \quad p\left(\frac{H}{\overline{H}}\right) = \frac{\pi}{2} \cdot \frac{H}{\overline{H}^2} \exp\left\{-\frac{\pi}{4}\left(\frac{H}{\overline{H}}\right)^2\right\} \tag{1.1.9}$$

where, $\overline{H}$ is the mean wave height expressed by using Eq. (1.1.10).

$$\overline{H} = \frac{\int_0^\infty H \cdot p(H) dH}{\int_0^\infty p(H) dH} = \sqrt{2\pi} \sigma_\eta \tag{1.1.10}$$

The probability of occurrence of a wave height greater than a certain wave height H (= the probability of exceedance) can be obtained by integrating Eq. (1.1.9) from H to ∞, which is given by Eq. (1.1.11).

$$P(>H) = \int_H^\infty p(H) dH = \exp\left\{-\frac{1}{8}\left(\frac{H}{\sigma_\eta}\right)^2\right\} = \exp\left\{-\frac{\pi}{4}\left(\frac{H}{\overline{H}}\right)^2\right\} \tag{1.1.11}$$

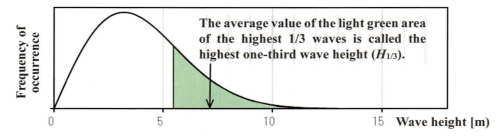

Fig. 1.5 Example of Rayleigh distribution of wave heights and explanation of the significant wave height.

1. STATISTICAL PROPERTIES AND GENERATION MECHANISMS OF WAVES

(1) Mean Wave

The mean wave height ($\overline{H}$) and mean period ($\overline{T}$), determined by averaging over all measured waves, can be obtained using Eq. (1.1.12).

$$\overline{H} = \frac{\sum_{i=1}^{n} H_i}{n}, \quad \overline{T} = \frac{\sum_{i=1}^{n} T_i}{n} \tag{1.1.12}$$

The square root of the arithmetic mean of the squared values of all measured wave heights is called the **root mean square of wave heights (H_{rms})** and is defined by Eq. (1.1.13). It is used for studies that focus on wave energy, as wave energy is proportional to the square of the wave height.

$$H_{rms} = \sqrt{\overline{H^2}} = \left\{ \frac{\int_0^\infty H^2 \cdot p(H) dH}{\int_0^\infty p(H) dH} \right\}^{1/2} = \sqrt{8}\sigma_\eta = \sqrt{\frac{4}{\pi}\overline{H}} \tag{1.1.13}$$

(2) Highest One-third Wave and Significant Wave

As shown in **Fig. 1.5**, the arithmetic mean of the respective heights and periods of the highest one-third waves of a measured wave record is called **highest one-third wave height ($H_{1/3}$)** and **highest one-third wave period ($T_{1/3}$)** respectively. This wave ($H_{1/3}$, $T_{1/3}$) is called the **highest one-third wave**. Since these values are approximately equal to the average of the visual measurements (meaning that more than half of the small waves are missed by an observing human), they are also called **significant wave height (H_s)**, **significant wave period (T_s)** and **significant wave** respectively. For this reason, the wave height and period of a significant wave is usually used to represent the actual wave height and period.

Under the assumption of the Rayleigh distribution, the relationship between significant wave height and mean wave height is as follows.

$$H_{1/3} = H_s = 1.597 \cdots \times \overline{H} \doteqdot 1.60 \times \overline{H} \tag{1.1.14}$$

in which $\overline{H}$ is the mean wave height. $H_{1/3}$ and $T_{1/3}$ are widely used in coastal engineering design.

(3) Highest One-tenth Wave

The arithmetic means of the heights and periods of the top 1/10 waves of a measured wave record are called **highest one-tenth wave height ($H_{1/10}$)** and **highest one-tenth wave period ($T_{1/10}$)** respectively. This wave ($H_{1/10}$, $T_{1/10}$) is called the **highest one-tenth wave**. The relationships between the highest one-tenth wave height, significant wave height and mean wave height, under the assumption of the Rayleigh distribution, are as follows.

$$H_{1/10} \doteqdot 1.271 \times H_s \doteqdot 2.031 \times \overline{H} \tag{1.1.15}$$

(4) Highest Wave

The wave with the highest wave height in a measured wave record is called the **highest wave**.

1.1 Statistical Properties of Waves

The wave height and period of this wave are called **maximum wave height (H_{max})** and **maximum period (T_{max})** respectively and are used in the design of port and marine structures.

The relationship between the maximum wave height and the significant wave height based on a wave record with n number of waves is shown in Eq. (1.1.16) as the highest value of the frequency distribution of these ratios and Eq. (1.1.17) as the mean value of the frequency distribution.

In the case of the highest value:
$$\frac{H_{max}}{H_{1/3}} \fallingdotseq 0.706\sqrt{\ln(n)} \qquad (1.1.16)$$

In the case of the mean value:
$$\frac{H_{max}}{H_{1/3}} \fallingdotseq 0.706\left\{\sqrt{\ln(n)} + \frac{r}{2\sqrt{\ln(n)}}\right\} \qquad (1.1.17)$$

where r is Euler's constant (= 0.5772).

For port structures, number of waves $n = 600$ is adopted in Eq. (1.1.16) or Eq. (1.1.17).

In the case of port structures:
$$H_{max} = 1.8 \times H_{1/3} \qquad (1.1.18)$$

For ocean structures, number of waves $n = 3000$ is adopted.

In the case of ocean structures:
$$H_{max} = 2.0 \times H_{1/3} \qquad (1.1.19)$$

(5) Distribution of Wave Period

According to Bretschneider [2], the probability density function of the period T of wind waves is expressed by Eq. (1.1.20).

$$p(T) = 2.7\frac{T^3}{\overline{T}^4}\exp\left\{-0.675\left(\frac{T}{\overline{T}}\right)^4\right\} \qquad (1.1.20)$$

However, there are cases where wave period distribution with two peaks, such as a composite of wind waves and swell, becomes pronounced and does not have the same general form as in the case of wave heights.

The following relationships between $\overline{T}$, $T_{1/3}$, $T_{1/10}$ and T_{max} have been established:
$$T_{max} \fallingdotseq T_{1/10} \fallingdotseq T_{1/3} \fallingdotseq (1.1 \sim 1.2)\overline{T} \qquad (1.1.21)$$

4) Spectral Analysis Method

As shown in **Fig. 1.6**, which is a superposition of regular waves (component waves 1 to 5) with wave heights (4 m to 0.4 m) and periods (12 s to 3 s), irregular waves with wave grouping properties can be created by synthesising regular waves. In this way, irregular waves are considered synthesises of component waves with various wave heights and periods, similar to the spectrum of light concept.

The method of representing irregular waves by the frequency distribution of the energy possessed by each component wave is called the **spectral analysis method**, which can be applied to linear waves.

1. STATISTICAL PROPERTIES AND GENERATION MECHANISMS OF WAVES

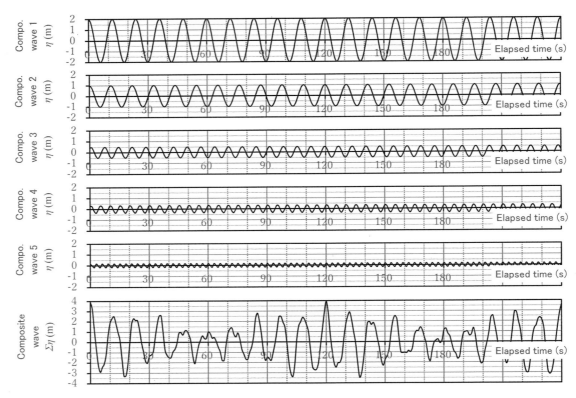

Fig. 1.6 Representation of the generation of irregular waves by synthesising regular waves.

(1) Frequency Spectrum

Sea waves are considered superpositions of component waves (regular waves) with various wave heights and periods, and the energy of each component wave as a function of frequency is called a **frequency spectrum**. The surface elevation of irregular waves η is expressed as a composite of numerous regular wave components shown in Eq. (1.1.22).

$$\eta = \sum_{i=1}^{\infty} a_i \cos(2\pi f_i t + \varepsilon_i) \tag{1.1.22}$$

where a_i, f_i, and ε_i are the amplitude (= $H/2$), frequency, and phase of the i-th component wave respectively, assuming that the amplitude of each component wave is small, there is no mutual interference between the component waves, and the phase is randomly distributed.

Then, from the small amplitude wave theory explained in Section 2.1, the total energy per unit area of the i-th component wave (regular wave), $E(f_i)$ [kg-m/s^2×m/m^2=N×m/m^2], is expressed by Eq. (1.1.23).

$$E(f_i) = \frac{1}{8}\rho g H_i^2 = \frac{1}{2}\rho g a_i^2 \tag{1.1.23}$$

Hence, the degree of total energy of the component waves between frequencies f and $f+df$ is expressed as follows:

$$\sum_{i=f}^{f+df} \frac{1}{2}a_i^2 = S(f)df \qquad (1.1.24)$$

where $S(f)$ is the wave energy per unit frequency corresponding to the frequency f, expressed in unit [$m^2 \times s$], and is called a **frequency energy spectral density function**, or a **frequency spectrum** in short.

(2) Relationship between Frequency Spectrum and Representative Waves

The mean energy $\overline{E}$ of irregular waves can be obtained from Eq. (1.1.25) using time series data of water surface elevation η.

$$\overline{E} = \rho g \frac{\int_0^{T_{term}} \eta^2 dt}{T_{term}} = \rho g \overline{\eta^2} = \rho g \sigma_\eta^2 \qquad (1.1.25)$$

where ρ is the density of seawater, g is the acceleration of gravity, T_{term} is the observed period of water surface elevation, and σ_η is the standard deviation of water surface elevation.

Comparing Eqs. (1.1.23), (1.1.24) and (1.1.25), it can be seen that integrating waves over all frequencies and multiplying it by ρg, Eq. (1.1.25) can be obtained. Hence, we obtain the following equation.

$$\overline{\eta^2} = \sigma_\eta^2 = \int_0^\infty S(f) \, df \qquad (1.1.26)$$

If the frequency spectrum is known, the standard deviation of the water surface elevation can be obtained from Eq. (1.1.26), and the wave heights of various representative waves can be calculated from Eqs. (1.1.10), (1.1.13) - (1.1.19).

In addition, the mean frequency and mean period are defined as follows:

$$\overline{f} = \left[\frac{\int_0^\infty f^2 S(f) df}{\int_0^\infty S(f) df} \right]^{1/2} = \sqrt{\frac{m_2}{m_0}} \qquad (1.1.27)$$

$$\overline{T} = \sqrt{\frac{m_0}{m_2}} \qquad (1.1.28)$$

where m_0 is the zero-th moment $\left(= \int_0^\infty S(f)df \right)$, and m_2 is the second moment $\left(= \int_0^\infty f^2 S(f)df \right)$ with respect to f.

However, the mean period obtained from the wave height data is 1.2 times the value obtained from Eq. (1.1.28).

If wave heights follow the Rayleigh distribution, the following relationship can be obtained from Eqs. (1.1.10), (1.1.14) and (1.1.26). Once the frequency spectrum is established, the significant wave height can be obtained from Eq. (1.1.29).

$$H_{1/3} \doteqdot 4.004 \sqrt{\int_0^\infty S(f)df} = 4.004 \sqrt{\overline{\eta^2}} = 4.004 \eta_{rms} \qquad (1.1.29)$$

where η_{rms} means the root mean square of water surface elevation.

1. STATISTICAL PROPERTIES AND GENERATION MECHANISMS OF WAVES

(3) Typical Frequency Spectra

As wind continues to blow, energy is supplied by the wind to the water masses near the water surface and develops ripples, which then turn into waves with large wave heights and periods. In addition to wind speed, which governs the magnitude of the shear force at the water surface, other important physical quantities involved in the development of gravity waves are **wind duration** (= blowing time) and **fetch** (= blowing distance). The longer the wind duration and fetch, the more the waves develop.

The time required for a wind wave to develop into a steady state at a certain fetch is called the **minimum duration time**. The following spectral models are built on the assumption that the minimum duration time has been reached.

(a) Mitsuyasu Type II Frequency Spectrum

Mitsuyasu [3], using experimental and measured data, expressed the frequency spectrum [m²×s] of wind waves at a finite blowing distance as follows. This is called the **Mitsuyasu type II frequency spectrum**.

$$S(f) = 0.000858 g^2 f^{-5} \left(\frac{gF}{u_*^2}\right)^{-0.312} \times \exp\left\{-1.25 \left(\frac{gF}{u_*^2}\right)^{-1.32} \left(\frac{u_* f}{g}\right)^{-4}\right\} \quad (1.1.30)$$

where g is the acceleration of gravity, f is the frequency, F is the fetch, and u_* is the friction velocity which can be obtained from Eq. (1.1.31).

$$u_* = \sqrt{\frac{\tau}{\rho_a}} = \sqrt{\frac{r_{10}^2 \rho_a U_{10}^2}{\rho_a}} = \sqrt{r_{10}^2} U_{10} \quad (1.1.31)$$

where τ is the shear force at the water surface due to wind, ρ_a is the air density, U_{10} is the mean wind speed 10 m above the sea surface, and r_{10}^2 is the coefficient of friction (if U_{10} is greater than 15 m/s, $r_{10}^2 \fallingdotseq 0.0026$).

An example of the distribution of this spectrum is shown in **Fig. 1.7**. As the fetch increases, the wave energy increases (wave height increases), and the frequency at the energy peak (called the **peak frequency**) decreases. In other words, the wave period of the peak becomes longer.

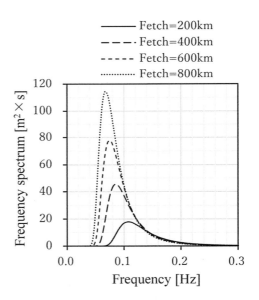

Fig. 1.7 Example of Mitsuyasu type II frequency spectrum.

(b) Bretschneider-Mitsuyasu Frequency Spectrum

Eq. (1.1.30) of Mitsuyasu [3] was rewritten using Eq. (1.1.32) which relates the peak frequency

f_p to the significant period and Eq. (1.1.29) described earlier so that the frequency spectrum [m²×s] can be obtained from the significant wave height and period, which can be easily determined from a given wave record. That is Eq. (1.1.33).

$$T_{1/3} = \frac{1}{1.05 f_p} \quad (1.1.32)$$

$$S(f) = 0.258 H_{1/3}^2 \left(T_{1/3}^{-4} f^{-5}\right) \times \exp\left\{-1.03\left(T_{1/3} f\right)^{-4}\right\} \quad (1.1.33)$$

Eq. (1.1.33) is called the **Bretschneider-Mitsuyasu frequency spectrum** because it agrees with the equation for the frequency spectrum of wind waves at finite fetches proposed by Bretschneider [4]. An example of a significant wave height of 7 m and a significant wave period of 14 s is shown in **Fig. 1.8**.

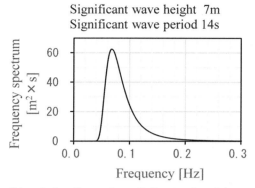

Fig. 1.8 Example of Bretschneider–Mitsuyasu frequency spectrum.

(c) Pierson-Moskovitz Frequency Spectrum

Wave height and period of wind waves increase with increasing fetch at a constant wind speed but eventually reach an equilibrium between the energy loss due to wave breaking and the energy supplied by wind. Pierson and Moskovitz [5], using data from observations in the North Atlantic Ocean, expressed the frequency spectrum [m²×s] of well-developed wind waves in energy equilibrium as follows:

$$S(f) = \frac{0.00810}{(2\pi)^4} g^2 f^{-5} \times \exp\left\{-0.74\left(\frac{g}{2\pi U_{19.5} f}\right)^4\right\} \quad (1.1.34)$$

where $U_{19.5}$ is the mean wind speed 19.5 m above the sea surface, obtained using Eq. (1.1.35).

$$U(z) = U_{10}\left(1 + 5.75\sqrt{r_{10}^2} \log\frac{z}{10}\right) \quad (1.1.35)$$

where z is the height above the sea surface [m], U_{10} is the wind speed at $z = 10$ m, and r_{10}^2 is the friction coefficient (if U_{10} is greater than 15 m/s, $r_{10}^2 \fallingdotseq 0.0026$).

Eq. (1.1.34) is called the **Pierson-Moskovitz frequency spectrum**. An example of Pierson-Moskovitz frequency spectrum for a wind speed of 21 m/s, measured at 19.5 m above sea level is shown in **Fig. 1.9**.

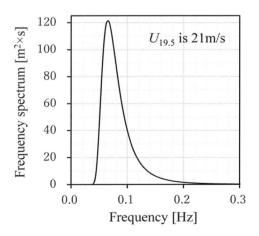

Fig. 1.9 Example of Pierson–Moskovitz frequency spectrum.

1. STATISTICAL PROPERTIES AND GENERATION MECHANISMS OF WAVES

(d) Modified JONSWAP Frequency Spectrum by Goda

Hasselmann et al [6] proposed a spectral density function for wind waves called the JONSWAP frequency spectrum using wave data observed in the North Sea, while Goda [7], based on the results of numerical simulations of irregular waves, proposed Eq. (1.1.36) that can express the frequency spectrum [m²×s] using the significant wave height and significant period. This equation can be used for a wide range of spectral shapes by choosing the index γ [$\fallingdotseq 90 \times (H_{1/3}/L_p)$ as a general rule of thumb, L_p is the wavelength to the peak frequency], which represents the sharpness of the spectrum in the equation.

$$S(f) = \frac{0.06238(1.094 - 0.01915 \times \ln\gamma)\gamma^\alpha}{0.230 + 0.0336\gamma - 0.185(1.9 + \gamma)^{-1}} \quad (1.1.36)$$
$$\times H_{1/3}^2 T_p^{-4} f^{-5} \times \exp\left\{-1.25(T_p f)^{-4}\right\}$$

where α is the exponent obtained from Eq. (1.1.37), T_p is the peak period obtained from Eq. (1.1.38), β is a coefficient that can be set in Eq. (1.1.39), and f_p is the peak frequency (inverse of T_p).

$$\alpha = \exp\left\{\frac{-(T_p f - 1)^2}{2\beta^2}\right\} \quad (1.1.37)$$

$$T_p = \frac{T_{1/3}}{1 - 0.132(\gamma + 0.2)^{-0.559}} \quad (1.1.38)$$

$$\beta = \begin{cases} 0.07 & : f \leq f_p \\ 0.09 & : f \geq f_p \end{cases} \quad (1.1.39)$$

Eq. (1.1.36) is called the **Modified JONSWAP frequency spectrum.** An example of this for a significant wave height of 7 m and a significant wave period of 14 s is shown in **Fig. 1.10**.

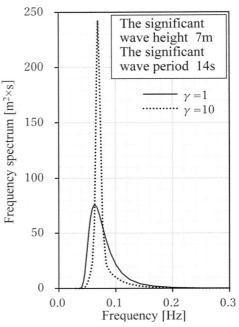

Fig. 1.10 Example of Modified JONSWAP frequency spectrum.

(e) Bi-modal spectrum

The coexistence of wind waves and swell may result in a frequency spectrum with two distinct peaks. The frequency spectra of wind waves and swells can be represented by obtaining the frequency spectrum for each wave component and superimposing them linearly using the modified JONSWAP frequency spectrum, which can be applied to a wide range of spectral shapes with a single peak, or the **Wallops frequency spectrum** proposed by Huang et al [7]. Examples include the work of Tanemoto et al [8] and Ochi and Hubble [9].

(4) Directional Spectrum

Actual waves are composites of component waves that differ not only in wave heights and periods but also in wave direction. Therefore, the energy (proportional to the square of the wave height) of the waves is expressed using Eq. (1.1.40), which is the product of the frequency spectrum $S(f)$ and the **directional distribution function** $G(f, \theta)$.

$$S(f,\theta)=S(f)\times G(f,\theta) \tag{1.1.40}$$

where the function $G(f, \theta)$ represents the distribution in the direction of wave propagation θ.

Eq. (1.1.40) is called **directional spectrum** of waves. The directional distribution function has no dimensions and must satisfy Eq. (1.1.41). There are several studies on the directional distribution functions, such as Arthur [10], but here the directional distribution function proposed by Longuet-Higgins et al [11] and reinforced by Mitsuyasu [12] are introduced.

Longuet-Higgins et al [11] proposed Eq. (1.1.42) based on data observed in the open ocean.

$$\int_0^{2\pi} G(f,\theta)\, d\theta = 1 \tag{1.1.41}$$

$$G(f,\theta)= G_0 \cos^{2S} \frac{\theta}{2} \tag{1.1.42}$$

where S is a parameter representing the concentration of directional energy distribution; G_0 is the coefficient defined in Eq. (1.1.43) to normalise $G(f, \theta)$, expressed by Eq. (1.1.44) when the range of θ_{min} to θ_{max} is $-\pi$ to $+\pi$.

$$G_0 = \left\{ \int_{\theta_{min}}^{\theta_{max}} \cos^{2S} \left(\frac{\theta}{2} \right) d\theta \right\}^{-1} \tag{1.1.43}$$

$$G_0 = \frac{1}{\pi} 2^{2S-1} \frac{\left\{ \Gamma(S+1) \right\}^2}{\Gamma(2S+1)} \tag{1.1.44}$$

where $\Gamma(\ )$ is the gamma function.

Mitsuyasu et al [12] then confirmed the validity of Eq. (1.1.42) using measured data and proposed the following equation to obtain the concentration parameter S of the energy directional distribution.

$$S \doteqdot \begin{cases} 14 f_{p*}^{-6.7} f_*^4 & (f_* \leq f_{p*}) \\ 14 f_*^{-2.7} & (f_* > f_{p*}) \end{cases} \tag{1.1.45}$$

where f_* is the dimensionless frequency defined in Eq. (1.1.46), and f_{p*} is the dimensionless peak frequency defined in Eq. (1.1.47).

$$f_* = \frac{2\pi f\, U_{10}}{g} \tag{1.1.46}$$

$$f_{p*} = \frac{2\pi f_p U_{10}}{g} \doteqdot 18.8 \times \left(\frac{gF}{U_{10}^2} \right)^{-0.330} \tag{1.1.47}$$

where U_{10} is the mean wind speed at 10 m above sea level, and F is the fetch.

Goda and Suzuki [13] proposed Eq. (1.1.48) to express the concentration parameter S of the energy directional distribution in terms of frequency, peak frequency, and the maximum value of the concentration parameter S_{max} defined by Eq. (1.1.49) obtained from the measured data by Mitsuyasu et al.

1. STATISTICAL PROPERTIES AND GENERATION MECHANISMS OF WAVES

$$S = \begin{cases} S_{max} \times (f/f_p)^5 & (f \leq f_p) \\ S_{max} \times (f/f_p)^{-2.5} & (f > f_p) \end{cases} \quad (1.1.48)$$

$$S_{max} = 11.5 \left(\frac{2\pi f_p U_{10}}{g} \right)^{-2.5} \quad (1.1.49)$$

here, the peak frequency f_p is obtained from Eq. (1.1.50) using the significant period.

$$f_p = \frac{1}{1.05 T_{1/3}} \quad (1.1.50)$$

The directional distribution functions obtained from Eqs. (1.1.42), (1.1.44) and Eqs. (1.1.48) to (1.1.50) are called the **Mitsuyasu-type directional distribution function**. Examples of cases with significant wave height of 7 m and S_{max} of 25 are shown in **Fig. 1.11**.

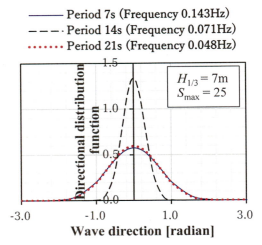

Fig. 1.11 Examples of the relationship between the directional distribution function of Mitsuyasu-type and wave directions.

As the accuracy of Eq. (1.1.49) has not been fully verified and setting the appropriate U_{10} for the target wave is not straightforward, Goda recommends using the following values for the maximum value of the concentration parameter:

For wind waves --------------------------------- S_{max} = 10
For swell with short attenuation distance ---- S_{max} = 25
   (with relatively large wave steepness)
For swell with long attenuation distance ---- S_{max} = 75
   (with relatively small wave steepness)

The relationship between S_{max} obtained by Goda and Suzuki and the wave steepness is shown in **Fig. 1.12**, and the relationship between S_{max} and relative water depth is shown in **Fig. 1.13** (α_{po} in the figure is the predominant incident angle of irregular waves to the offshore contour, and $(S_{max})_o$ is the maximum value of the concentration parameter at the offshore contour).

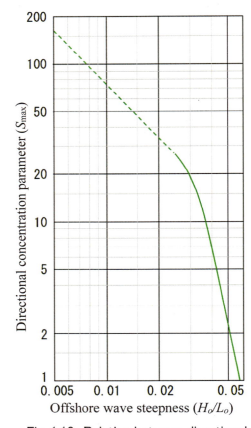

Fig. 1.12 Relation between directional concentration parameters and wave steepness (from Goda and Suzuki [13]).

1.1 Statistical Properties of Waves

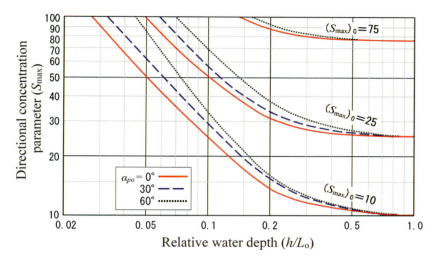

In the figure, a_{po} is the predominant incident angle of irregular waves to the offshore contour and $(S_{max})_o$ is the maximum value of the concentration parameter at the offshore contour.

Fig. 1.13 Relation between directional concentration parameters and relative water depth (from Goda and Suzuki [13]).

1. STATISTICAL PROPERTIES AND GENERATION MECHANISMS OF WAVES

1.2 Generation Mechanism of Sea Waves

1) Basis of Wave Generation Theory

Waves are generated when the propagation velocity of small sea level fluctuations caused by turbulent wind pressure fluctuations matches the velocity of the wind pressure fluctuations. If the energy supply from wind continues, waves develop from **ripples** to large **wind waves**. As they develop further, they begin to break near the wave crest and lose some of that energy. When the energy supplied by wind and the energy lost due to wave breaking is balanced, an equilibrium state is reached. When the energy supply is reduced due to waves leaving the generation area or due to the weakening of the wind field, short-period components of wind waves are attenuated by the viscous effect of the water and are absorbed by larger and longer components. This generates **swells** with relatively long wavelengths. Swell waves lose a little energy but gain significant energy by absorbing short wind waves during their propagation in the sea, and thus can propagate to reach far distances.

Since wave energy is transported by the propagation velocity of the wave group (refer to Section 2.1 7) (d)), an energy conservation equation can be formulated using the wave energy and the propagation velocity of the wave group. Hasselmann [14] and others have proposed an energy equilibrium equation, in which wave energy is replaced by a directional spectrum, for tracing the generation, growth and attenuation of waves. The basic form of this equation is as follows:

$$\frac{\partial S(f,\theta)}{\partial t} + \frac{\partial\left[S(f,\theta)\times C_{gx}\right]}{\partial x} + \frac{\partial\left[S(f,\theta)\times C_{gy}\right]}{\partial y} = S_E \tag{1.2.1}$$

$$S_E = S_{in} + S_{nl} + S_{ds} \tag{1.2.2}$$

where $S(f,\theta)$ is the directional wave spectrum [m²×s], t is the time, C_{gx} and C_{gy} are respectively the x- and y-directional components of the wave group velocity [m/s], x and y are the horizontal distances offshore and alongshore, and S_E is a function of energy supply and dissipation [m²]. In addition, S_{in} is the energy supply from wind to the waves, S_{nl} is the energy transport due to non-linear interaction between wave components, and S_{ds} is the energy dissipation due to wave breaking.

According to the Research Status Review Subcommittee of the Japan Society of Civil Engineers (JSCE) [15] and the WAVEWATCH III Development Group of the National Oceanic and Atmospheric Administration (NOAA) [16], S_E in Eq. (1.2.2) can be expressed by an appropriate evaluation equation to obtain the distribution of the wave energy. Also, the equation relating the wave energy to the significant wave height is used to obtain the distribution of developing and decaying wave height.

(1) Energy Supply

The following theories are well known to describe the amount of energy supply S_{in} from wind to waves:

① **Resonance theory of Phillips** [17] (theory of wind wave generation): Under conditions of negative pressure on the rising water surface and positive pressure on the falling water surface, if the propagation velocity of minute sea level fluctuations caused by turbulent wind pressure fluctuations matches the movement velocity of the wind pressure fluctuations, then the sea level fluctuations grow as a wave. Although this theory is important for explaining the generation of wind waves, the amount of energy transfer due to this is so small that the following theory is needed to explain the development of wind waves.

② **Shear flow theory of Miles** [18, 19] (theory of wind wave development): when waves appear on the sea surface, the air currents above the sea surface are disturbed by the surface waves, creating instability. The pressure is higher upwind of the sea surface and lower downwind, pushing the waves downwind. In this way, wind energy is efficiently supplied to the waves.

There are many studies to formulate S_{in} and a representative formula is introduced here.

$$S_{in} = C_a + C_b \left(u_*/C\right)^2 \omega \times S(f,\theta) \tag{1.2.3}$$

where the first term C_a on the right-hand side of this equation represents linear wave development according to Phillips' resonance theory, the second term is the term proportional to the shear force according to Miles' theory, C_b is the proportional coefficient (around 0.05), u_* is the wind friction velocity, C is the wave celerity, and ω is the wave angular frequency.

(2) Energy Transportation by Non-linear Interaction

As sea waves are irregular with wave grouping characteristics, they can be approximated by superimposing regular waves (component waves) with different wave heights, wave numbers and angular frequencies (linear approximation). However, if the approximation accuracy of the waves is to be improved, the energy exchange due to the non-linear interaction between the component waves cannot be ignored. The first theoretical expression of the energy transfer S_{nl} due to the non-linear interaction between the component waves in a wave group was given by Hasselmann [20] and it is as follows:

$$\begin{aligned} S_{nl}\left(\vec{\kappa}_4\right) = \omega_4 \times \iiint_{-\infty}^{\infty} F\left(\vec{\kappa}_1,\vec{\kappa}_2,\vec{\kappa}_3,\vec{\kappa}_4\right) \times \delta\left(\vec{\kappa}_1+\vec{\kappa}_2-\vec{\kappa}_3-\vec{\kappa}_4\right)\delta\left(\omega_1+\omega_2-\omega_3-\omega_4\right) \\ \times\left[N_1 N_2\left(N_3+N_4\right)-N_3 N_4\left(N_1+N_2\right)\right]d\vec{\kappa}_1 d\vec{\kappa}_2 d\vec{\kappa}_3 \end{aligned} \tag{1.2.4}$$

where ω is the angular frequency; $\vec{\kappa}$ is the wavenumber vector (using the directional angle θ of the component waves, with the x-directional component being $\kappa\cos\theta$ and the y-directional component being $\kappa\sin\theta$); $F(\ )$ is the integral kernel function representing the coupling coefficient of the spectral components; $\delta(\ )$ is the delta function corresponding to the resonance condition; and N [$= S\left(\vec{\kappa}\right)/\omega$, $S\left(\vec{\kappa}\right)$ is the directional spectrum represented by the wavenumber vector] is called **wave action density**. The subscripts denote four sets of component waves.

In deep water (depth/wavelength > 1/2) and in intermediate depths (depth/wavelength between 1/2 and 1/20), the interference between the four sets of component waves affects energy development.

1. STATISTICAL PROPERTIES AND GENERATION MECHANISMS OF WAVES

The energy of the waves shifts to both the low-frequency and the high-frequency sides of the spectrum due to this interference. The peak frequency of the energy shifts to the low-frequency side because the high-frequency energy tends to dissipate and be absorbed by low-frequency energy. As this calculation requires a significant amount of time, an approximate calculation method has been devised. Furthermore, in shallow water (depth/wavelength less than 1/20), the energy transition due to three-wave resonance (where wave energy is transferred from the low-frequency side to the high-frequency side) has a significant effect on the development and attenuation of waves. For example, a single-crest wave spectrum can change to a multi-peak spectrum as waves approach the shore.

(3) Energy Dissipation

The energy dissipation S_{ds} consists of dissipation due to the collapse of the crests of wind waves (called **whitecaps**), friction induced by the seabed, and wave breaking due to the shallowing of the water depth.

An equation for energy dissipation due to whitecaps S_{wc} is given by Eq. (1.2.5), which is improved by Komen et al [21]. The energy dissipation due to seabed friction S_{bf} can be given by Eq. (1.2.6), in which the proportional coefficient of the simplified equation of Hasselmann and Collins [22] is determined from the JONSWAP observations and is presented below. The total energy dissipation in this case is the sum of these two quantities ($S_{ds} = S_{wc} + S_{bf}$).

$$\left. \begin{aligned} S_{wc} &= -C_1 \left[(1-C_2) + C_2 \frac{\kappa}{\tilde{\kappa}} \right] \left(\frac{\tilde{\kappa}\sqrt{S_T}}{0.05496} \right)^{C_3} \tilde{\omega} \frac{\kappa}{\tilde{\kappa}} \times S(f,\theta) \\ \tilde{\omega} &= \left[\frac{\iint S(f,\theta)\omega^{-1}dfd\theta}{S_T} \right]^{-1} \\ \tilde{\kappa} &= \left[\frac{\iint S(f,\theta)\kappa^{-0.5}dfd\theta}{S_T} \right]^{-2} \\ S_T &= \iint S(f,\theta)dfd\theta \end{aligned} \right\} \tag{1.2.5}$$

$$S_{bf} = -f_b \left(\frac{\omega}{g \sinh \kappa d} \right)^2 \times S(f,\theta) \tag{1.2.6}$$

where C_1, C_2 and C_3 are adjustment coefficients given so that the directional spectra are balanced, κ is the wave number, $\tilde{\kappa}$ is the mean wave number, S_T is the total directional spectrum of the waves, $\tilde{\omega}$ is the mean angular frequency, ω is the angular frequency, f_b is the bottom friction coefficient, and d is the water depth.

The estimation of the distribution of wave height, period and direction based on wind data is called **wave forecasting and hindcasting**. The straightforward method, which uses the equations of motion and the continuity equation in two horizontal directions (x, y) to estimate waves over a wide area for some days, is computationally very demanding. Hence, the method of solving the energy equilibrium equation using the directional wave spectrum has become popular, which is widely used

1.2 Generation Mechanism of Sea Waves

in numerical wave models. Also, there are other simplified methods based on empirical equations and calculation diagrams. These wave forecasting methods are described below.

2) Wave Forecasting Models

Wave forecasting is essential for the safe and economical operation of ships, safe operation of fishing vessels, various construction works and leisure activities in the sea.

In wave forecasting, first, wind distribution is predicted. Then, the forecasted wind field is fed to a wave forecasting model to forecast the time variation of wave parameters over the forecast area. Wave forecasting methods can be broadly classified into two types: the significant wave method and the spectral method.

① The **significant wave method** is a method for determining the significant wave height and period by giving the wind speed, fetch and the wind duration to a calculation chart or equations, includes simplified methods such as the **SMB method** (refer to Section 2) (2) (a)) and the **Wilson method** (refer to Section 2) (2) (b)). This is also called **parameter method** because the significant wave parameters are obtained from dimensionless parameters. This method is not suitable for complex wave fields, such as when wind waves and swell are mixed, when the wind field changes rapidly, or when the topography or pressure distribution is complex.

② The **spectral method** is a method for calculating the generation, growth, propagation and attenuation of waves by solving the energy equilibrium equation for waves using the directional spectrum, includes numerical models such as **WAve Model (WAM)** (refer to Section 2) (3) (a)) and the **Simulating WAve Nearshore model (SWAN)** (refer to Section 2) (3) (b)). Although these models demand more computational time than the significant wave method, this method has a clear physical meaning and can be used for complex sea states.

(1) Prediction Method for Wind Distribution

When forecasting waves, the distribution of wind, which is an external force, needs to be predicted in advance. Currently, meteorological data, such as ocean winds predicted with high accuracy using numerical models that solve the turbulent equations of motion in the atmosphere, are readily available from official weather forecasting organisations such as the European Centre for Medium-Range Weather Forecasting (ECMWF), the Japan Meteorological Agency (JMA) and the National Centres for Environmental Prediction (NCEP) in the USA, which enables accurate wave forecasting.

Furthermore, several ways in which readers can easily predict the distribution of offshore winds from weather maps are explained below.

1. STATISTICAL PROPERTIES AND GENERATION MECHANISMS OF WAVES

(a) Relationship between Gradient Wind and Wind at the Sea Level

Since wind blows from high pressure areas to low pressure areas, the higher the pressure gradient, the faster the wind speed. In this case, the wind direction is orthogonal to the isobaric line from the high-pressure to the low-pressure side. Furthermore, the "Coriolis force" (which deflects the wind direction to the right in the Northern Hemisphere) must be considered. The wind generated when the Coriolis force balances the pressure gradient is called **geostrophic wind**.

As isobars are generally curved, centrifugal forces must also be considered. The wind generated when these three forces are balanced is called **gradient wind**. The wind speed of the gradient wind U_{gr} is expressed by Eq. (1.2.7). The direction of the gradient wind is along the isobars because of the Coriolis force and centrifugal force.

$$U_{gr} = \frac{f_{cl} r}{2}\left[-1+\sqrt{1\pm\left(\frac{4}{\rho_a f_{cl}^2 r}\frac{dp}{dr}\right)}\right] \qquad (1.2.7)$$

where f_{cl} is the Coriolis coefficient [= 2 × Earth's angular rate of rotation × sin(latitude), Earth's angular rate of rotation = 7.292×10^{-5} rad/s]; r is the curvature radius of the isobaric line; ρ_a is the density of the atmosphere (1.1×10^{-3} g/cm³); and dp/dr is the pressure gradient; The symbol ($\pm$) will be (+) for low pressure and (-) for high pressure.

The mean wind speed U_{10} at 10 m elevation above the sea surface used for wave forecast is lower than U_{gr} given by Eq. (1.2.7), mainly due to the frictional resistance force at the sea surface. The wind direction α above the sea surface also has a blowing or sucking angle that is a constant crossing angle from the isobar as shown in **Fig. 1.14**, due to the three-dimensional blowing and sucking phenomena.

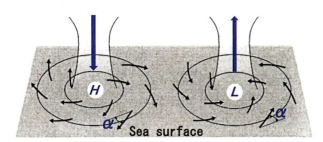

Fig. 1.14 Wind blowing out of an anticyclone (H) and suction into a low pressure system (L) in the Northern Hemisphere.

According to the study of Takahashi [23], the wind speed ratio U_{10}/U_{gr} and wind direction α are shown in **Table 1.1**.

Table 1.1 Relationship between gradient wind and wind over the sea surface.

Latitude (degrees)	10	20	30	40	50
Blow-off and suction angle α (degrees) from a line parallel to the tangent to the isobar	24	20	18	17	15
U_{10}/U_{gr}	0.51	0.60	0.64	0.67	0.70

(b) Wind Distribution during a Typhoon

Since isobars are concentric and the pressure decreases towards the centre of a typhoon, it is

1.2 Generation Mechanism of Sea Waves

relatively easy to formulate the pressure distribution within the typhoon area. Several pressure distribution equations have been proposed. Eq. (1.2.8) of Fujita [24] and Eq. (1.2.9) of Myers [25] are introduced here.

$$p = p_a - \frac{\Delta p}{\sqrt{1+(r/r_0)^2}} \quad (1.2.8)$$

$$p = p_a - \Delta p \left[1 - \exp\left(-\frac{r_0}{r}\right)\right] \quad (1.2.9)$$

where P_a is the atmospheric pressure (=1013hpa), Δp is the pressure drop at the typhoon centre, r_o is the distance from the typhoon centre to the position of maximum wind speed, and r is the linear distance from the typhoon centre to the target position.

Substituting the pressure gradient obtained from Eq. (1.2.8) or Eq. (1.2.9) into Eq. (1.2.7) and using U_{10}/U_{gr} in **Table 1.1**, the wind speed over the sea surface in a typhoon area can be obtained. The wind direction over the sea surface can also be determined using the blowing angles in **Table 1.1**. Additionally, by combining the typhoon wind speeds and directions with the typhoon movement velocity vector (with the consideration of the wind velocity reductions due to frictional resistance at the sea surface, etc.), wind information necessary for the wave forecast can be obtained.

In the semicircle at the right of the typhoon centre, the wind direction coincides with the typhoon movement direction, and thus the wind is stronger as shown in **Fig. 1.15**. Therefore, the semicircle at the right of the typhoon centre is called the **dangerous semicircle**. On the other hand, in the semicircle at the left side of the typhoon centre, the wind direction is opposite to the direction of the typhoon movement, and thus the wind is weaker. Therefore, the semicircle at the left of the typhoon centre is called a **navigable semicircle**.

If the typhoon pressure distribution is elliptical, the following equation of Nonaka et al [26] can be used to calculate the pressure.

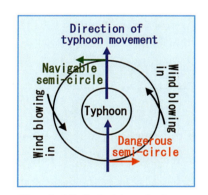

Fig. 1.15 Relation between wind blowing in and direction of typhoon movement.

$$p = p_c + \Delta p \times \exp\left\{\frac{-1}{\sqrt{(x/\alpha)^2+(y/\beta)^2}}\right\} \quad (1.2.10)$$

where P_c is the atmospheric pressure at the typhoon centre (= $P_a - \Delta p$), x and y are the bi-directional distances from the typhoon centre to the target location in rectangular coordinates, and α and β are coefficients for adjusting the major and minor axes of the elliptical distribution (with a length dimension).

1. STATISTICAL PROPERTIES AND GENERATION MECHANISMS OF WAVES

(2) Significant Wave Method

The foundations of the significant wave method were laid during World War II and played an important role in the Allied landing operations at Normandy in June 1944. Even today, although numerical models are widely used, this method is still used for validating numerical results and estimating waves at specific locations.

(a) Wind Wave Forecast Method Part 1 (SMB Method)

A calculation chart that enables the estimation of significant wave parameters from wind data was developed by Sverdrup and Munk [27] during World War II and published after the war in 1947. Bretschneider [28, 29] further improved this method, later known as the SMB method, after their initial attempts. Subsequently, Bretschneider [30] proposed Eqs. (1.2.11) and (1.2.12), which can be used to find the significant wave height $H_{1/3}$ and the significant wave period $T_{1/3}$ of wind waves.

$$\frac{gH_{1/3}}{U_{10}^2} = 0.283 \tanh\left\{0.0125\left(\frac{gF}{U_{10}^2}\right)^{0.42}\right\} \quad (1.2.11)$$

$$\frac{gT_{1/3}}{2\pi U_{10}} = 1.20 \tanh\left\{0.077\left(\frac{gF}{U_{10}^2}\right)^{0.25}\right\} \quad (1.2.12)$$

The minimum duration t_{min} (the minimum time required for waves to develop towards the target waves) is defined by Eq. (1.2.13).

$$t_{min} = \int_0^F \left(\frac{1}{C_g}\right) dx = \int_0^F \frac{4\pi}{gT_{1/3}} dx \quad (1.2.13)$$

The tanh$\{x\}$ in Eqs. (1.2.11) and (1.2.12) is a type of hyperbolic functions. The hyperbolic functions are composite functions with exponential functions e^x and e^{-x} corresponding to the variable x, as shown in the diagram below.

$$\cosh(x) = \frac{e^x + e^{-x}}{2} \quad (b.1)$$

$$\sinh(x) = \frac{e^x - e^{-x}}{2} \quad (b.2)$$

$$\tanh(x) = \frac{\sinh(x)}{\cosh(x)} = \frac{e^x - e^{-x}}{e^x + e^{-x}} \quad (b.3)$$

Also, differentiating hyperbolic functions yield the following equations:

$$\frac{d\cosh(x)}{dx} = \sinh(x) \quad (b.4)$$

$$\frac{d\sinh(x)}{dx} = \cosh(x) \quad (b.5)$$

$$\frac{d\tanh(x)}{dx} = \frac{1}{\cosh^2(x)} \quad (b.6)$$

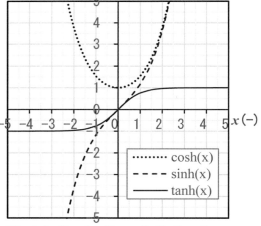

Fig. b.1 Graph of hyperbolic functions.

1.2 Generation Mechanism of Sea Waves

where g is the acceleration of gravity, U_{10} is the mean wind speed at 10 m above the sea surface, F is the fetch, and C_g is the propagation velocity of the wave group.

The wind-wave calculation diagram produced using Eqs. (1.2.11) to (1.2.13) is shown in **Fig. 1.16**. The method of using this diagram is as follows:

(i) Plot the wind speed U_{10} and fetch F on the calculation diagram and read off the significant wave height and period.

(ii) Plot the wind speed U_{10} and duration t on the calculation diagram and read off the significant wave height and period.

(iii) Compare the two and select the smaller of the two.

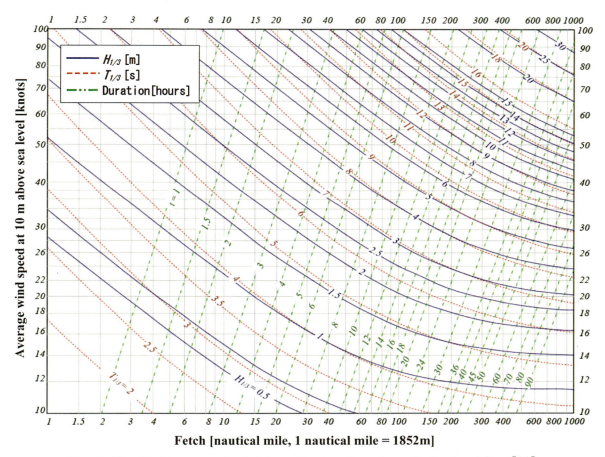

Fig. 1.16 Wind wave calculation diagram based on Bretschneider [30].

(b) Wind Wave Forecast Method Part2 (Wilson Method)

Wilson [31] proposed the following equations, which can be used as a simple calculation method to obtain the significant wave parameters in a wind wave domain from the average wind speed at 10 m above the sea surface and the fetch. The minimum duration t_{min} is obtained from Eq. (1.2.13).

1. STATISTICAL PROPERTIES AND GENERATION MECHANISMS OF WAVES

$$\frac{gH_{1/3}}{U_{10}^2} = 0.30\left[1-\left\{1+0.004\left(\frac{gF}{U_{10}^2}\right)^{1/2}\right\}^{-2}\right] \quad (1.2.14)$$

$$\frac{gT_{1/3}}{2\pi U_{10}} = 1.37\left[1-\left\{1+0.008\left(\frac{gF}{U_{10}^2}\right)^{1/3}\right\}^{-5}\right] \quad (1.2.15)$$

The wind wave calculation diagram produced using Eqs. (1.2.13) - (1.2.15) is shown in **Fig. 1.17**. The estimated wave heights in this figure are almost the same as in **Fig. 1.16**, but the estimated periods tend to be smaller. The way to use this diagram is the same as **Fig. 1.16**: the smaller of the two is chosen by comparing the significant wave height and the significant wave period calculated from the wind speed U_{10} and fetch F with those calculated from the wind speed U_{10} and duration t.

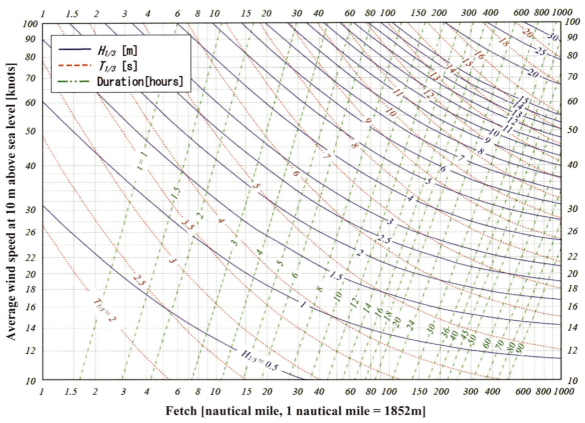

Fig. 1.17 Wind wave calculation diagram based on Wilson [31].

Goda [32, 33] obtained the following equations based on Wilson's formula and proposed a simplified estimation method that does not use calculation diagrams.

$$t_{min} = F^{0.73}\, U_{10}^{-0.46} \quad \text{or} \quad F_{min} = t^{1.37}\, U_{10}^{0.63} \quad (1.2.16)$$

$$T_{1/3} \fallingdotseq 3.3 \times H_{1/3}^{0.63} \quad (1.2.17)$$

Note that in Eq. (1.2.16), t and t_{min} are the duration [hours] and the minimum duration [hours]

1.2 Generation Mechanism of Sea Waves

respectively; F and F_{min} are the fetch [km] and the minimum fetch [km] respectively; U_{10} [m/s] is the average wind speed at 10 m above sea level. In Eq. (1.2.17), $T_{1/3}$ [s] is the significant wave period; and $H_{1/3}$ [m] is the significant wave height. It is necessary to pay attention to the units.

The calculation procedure is as follows:

(ⅰ) Determine the wind range and set U_{10} [m/s], F [km] and t [hours].

(ⅱ) Find the minimum fetch F_{min} from the second equation of Eq. (1.2.16) and, if $F > F_{min}$, since wave development is defined by the duration, use F_{min} for the fetch to obtain the significant wave height from Eq. (1.2.14) and the significant wave period from Eq. (1.2.15).

(ⅲ) Conversely, if $F < F_{min}$, since wave development is defined by the fetch, use F to obtain the significant wave height from Eq. (1.2.14) and the significant wave period from Eq. (1.2.15).

The significant wave period can be obtained from Eq. (1.2.17) using the previously determined significant wave height.

(c) Wind Wave Forecast Method in Shallow Water (Bretschneider Method)

Bretschneider [34] provides a diagram to calculate wind waves in shallow water where the frictional resistance due to the seabed and the suppression effect of wave development due to seabed penetration of wave pressure fluctuations cannot be ignored. **Fig. 1.18** shows the case of constant water depth and **Fig. 1.19** shows the case of seabed gradients of 1/200 to 1/500.

Significant wave heights are obtained by fitting the average wind speed U_{10}, the water depth h in the target area and the fetch F to the target point to each diagram. Here, the sea-bottom friction coefficient, $f_b = 0.01$ is assumed to be appropriate based on comparison with measured values.

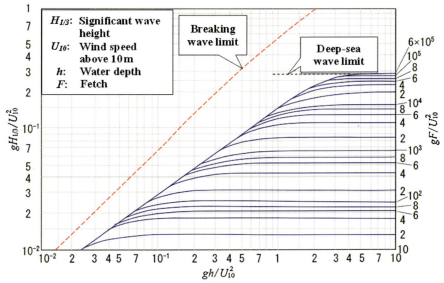

Fig. 1.18 Wind wave calculation diagram in shallow water (for constant water depth) based on Bretschneider [34].

1. STATISTICAL PROPERTIES AND GENERATION MECHANISMS OF WAVES

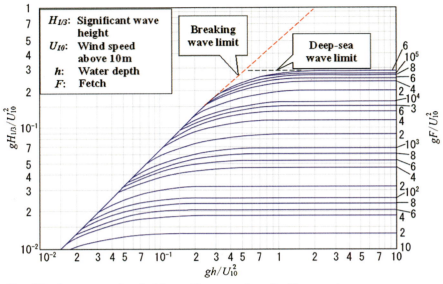

Fig. 1.19 Wind wave calculation diagram in shallow water (for seabed gradients of 1/200 to 1/500) based on Bretschneider [34].

Shallow water deformation and wave breaking due to changes in water depth, as explained in Chapter 2, are likely to occur in shallow water areas. Therefore, corrections for those must be added to the calculated significant wave height. As only the significant wave heights can be obtained from these diagrams, the period of the significant waves can be obtained from the following equation.

$$T_{1/3} = 3.86 \times \sqrt{H_{1/3}} \qquad (1.2.18)$$

(d) Method for Swell Forecasting

When waves leave the wind field, they are no longer supplied with energy by wind and they propagate as swell.

Sverdrup and Munk [27] expressed the tendency of swell waves losing energy due to atmospheric resistance and attenuating the significant wave height $H_{1/3}$, while increasing the significant wave period $T_{1/3}$ based on Eq. (1.2.19). The swell arrival time t_r to a travel distance D from the wind field boundary is also expressed by Eq. (1.2.20).

$$\frac{T_{1/3}}{T_o} = \left[1 + 0.000595 \frac{D}{gT_o^2}\right]^{1/2}, \quad \frac{H_{1/3}}{H_o} = \left[\frac{T_{1/3}}{T_o}\right]^{-2.65} \qquad (1.2.19)$$

$$t_r = \int_F^{F+D} \frac{1}{C_g} dx = 42200 \times \left(\frac{T_{1/3} - T_o}{T_o}\right) T_o \qquad (1.2.20)$$

where T_o and H_o are the significant wave height and period at the boundary where waves depart from the wind field respectively, and F is the fetch within the wind field.

Bretschneider [35] also proposed Eq. (1.2.21) to calculate the significant wave period $T_{1/3}$ and the significant wave height $H_{1/3}$ at a distance D from the wind field boundary.

The arrival time t_r to travel distance D is obtained by Eq. (1.2.22).

$$\frac{H_{1/3}}{H_o} = \left[\frac{04F}{0.4F+D}\right]^{1/2}, \quad \frac{T_{1/3}}{T_o} = \left[2.0 + \frac{H_{1/3}}{H_o}\right]^{1/2} \tag{1.2.21}$$

$$t_r = \frac{D}{C_g} = \frac{4\pi D}{gT_{1/3}} \tag{1.2.22}$$

A comparison of Sverdrup and Munk [27] and Bretschneider [35] formulae is shown in the **Fig. 1.20** for the case of a significant wave height and period of 10 m and 12 s at the wind field boundary respectively and a fetch of 500 km.

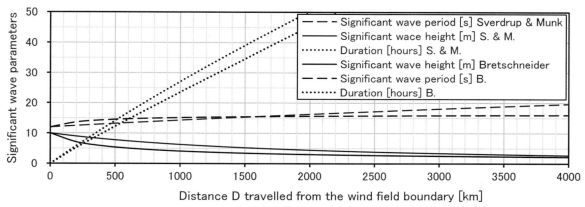

Fig. 1.20 Examples of swell calculations using Sverdrup and Munk's formula and Bretschneider's formula.

(3) Spectral methods

Although the significant wave calculation method is easy to apply, the evaluation of swell is not clear, and the treatment of multi-directional waves is ambiguous. So, Pierson, Neumann, and James [36] developed the **PNJ method**, which can predict waves using a directional spectrum consisting of a frequency spectrum with wind speed as a parameter proposed by Neumann and a square-type cosine directional distribution function. However, the spectral method was not commonly used in coastal projects because of its unreliability at the time.

Nevertheless, the spectral method was superior in its physical assembly. As computer performance improved, full-scale numerical models incorporating wave generation, growth, and attenuation mechanisms into the wave energy equilibrium equation were developed which can be applied to complex seabed bathymetries and handle multi-component directional spectra.

Since the Significant Wave method and the PNJ method estimate waves that develop and decay in a constant wind field, they are equivalent to solving a relational equation that integrates the energy equilibrium equation under a constant wind speed. Hence, there are inherent limitations in applying these methods when the wind field is changing over time. Under such a situation, it is better to use numerical models that solve the energy equilibrium equation, which have been improved through the

1. STATISTICAL PROPERTIES AND GENERATION MECHANISMS OF WAVES

following process.

(i) First generation models (1960s to early 1970s)

A numerical model in which only the energy supply from wind to waves S_{in} and the energy dissipation due to wave breaking S_{ds} are considered as the energy supply and dissipation functions S_E in the energy equilibrium equation. It is also called the **Decoupled Propagation Method** because the component waves that make up the directional spectrum are considered to propagate independently. However, since the development coefficients obtained from measured data are adopted as the supply quantity S_{in}, the amount of energy transport S_{nl} due to the non-linear interactions between the component waves within a wave field has been incompletely included. A representative example is the **Meteorological Research Institute model (MRI)** of Isozaki and Uji [37] of the Japan Meteorological Agency (JMA).

(ii) Second generation models (1970s to early 1980s)

It became clear that the amount of energy transport S_{nl} due to the non-linear interaction between the component waves of in a wave field has a significant influence on the wave development process. Therefore, in addition to the supply and dissipation quantities S_{in} and S_{ds}, the spectral similarity of wind waves was used to describe the spectral variability of wind waves using several parameters, so that the energy transport due to the non-linear interaction could be considered indirectly. However, since the swell component could not be represented using the parameters, the **Coupled Hybrid Model** which uses the decoupled propagation model for the swell component has emerged. The JMA's **MRI-II** is a representative example of this model.

The **Coupled Discrete Model**, which directly considers energy transport S_{nl} due to non-linear interactions without separating the wind and swell components, has also emerged. However, since it is complex and difficult to calculate the transport S_{nl}, the second-generation coupled discrete model was devised so that the transport S_{nl} could be calculated using a simple scheme (e.g. by rigorously calculating the transport S_{nl}, expressing it in terms of a few parameters and applying it to a variety of wave spectra).

(iii) Third generation models (late 1980s to present)

The coupled discrete model has been improved to calculate the transport quantity S_{nl}, the supply quantity S_{in}, the dissipation quantity S_{ds}, and the energy supply and dissipation functions S_E as accurately as possible. These third-generation models include **WAM** model developed by the WAve Model Development and Implementation (WAMDI) Group [38] and the **MRI-III** model by Ueno and Ishizaka [39] of the JMA. Moreover, Booij et al. of the Delft University of Technology [40] developed the **SWAN** model based on WAM, which can be applied to very shallow waters. The WAVEWATCH III Development Group [16] also developed the **WAVEWATCH-III** model with improved estimation accuracy based on WAM. WAM and SWAN models, for which calculation programmes are publicly available, are introduced below.

(a) WAM

WAM is a numerical model for deepwater wave estimation, published in 1988 by the WAMDI Group [38], organised by Hasselmann et al., and summarised in detail by the JSCE Research Status Review Subcommittee [15]. Manuals and other information on the new version are available at https://github.com/mywave/WAM or http://mywave.github.io/WAM/.

Two types (orthogonal and spherical) of coordinate system are provided in WAM, and the amount of energy S_{in} supplied by the wind to the waves, the amount of energy transport S_{nl} due to the non-linear interaction of four-wave resonances (the transport of wave energy from the spectral peak to the low and high frequency sides), and the energy dissipation S_{ds} due to whitecap breaking and seabed friction are taken into account.

The energy equilibrium equation for the spherical coordinate system is expressed as follows:

$$\frac{\partial S(f,\theta)}{\partial t} + \frac{1}{\cos\theta_x} \frac{\partial\left[S(f,\theta)\dfrac{C_g\cos\theta\cos\theta_x}{R}\right]}{\partial\theta_x} + \frac{\partial\left[S(f,\theta)\dfrac{C_g\sin\theta}{R\cos\theta_x}\right]}{\partial\theta_y}$$

$$+ \frac{\partial\left[S(f,\theta)\dfrac{C_g\sin\theta\tan\theta_x}{R}\right]}{\partial\theta} = S_{in} + S_{nl} + S_{ds} \tag{1.2.23}$$

where $S(f,\theta)$ is the wave direction spectrum, θ_x is latitude, θ_y is longitude, θ is wave direction, and R is the radius of the earth.

The energy supply from wind to waves S_{in} is neglected in WAM, as it is easier to obtain reasonable results without Phillips' generation term. Hence, if the initial wave spectrum is zero, no waves will be generated, so the directional spectrum calculated from the fetch and the initial wind speed must be given in advance.

The following equation of the Snyder model was adopted for the developmental term of Miles.

$$S_{in} = B \times S(f,\theta) = \max\left[0,\ 0.25\frac{\rho_a}{\rho_w}\left\{28\frac{u_*}{C}\cos(\theta-\theta_w)-1\right\}\right]\omega \times S(f,\theta) \tag{1.2.24}$$

The following mathematical expressions, which may be difficult for readers to understand, are explained here.

$$y = \max\left(A,B\right) \tag{c.1}$$

$$y = \min\left(A,B,C\right) \tag{c.2}$$

$y = \max\left(A,B\right)$ means that the larger of the values of A and B is adopted as y.

$y = \min\left(A,B,C\right)$ means that the smallest value among A, B and C is adopted as y.

1. STATISTICAL PROPERTIES AND GENERATION MECHANISMS OF WAVES

where B is the coefficient of the Miles development term, ρ_a is the density of the atmosphere, ρ_w is the density of seawater, u_* is the friction velocity, C is the phase velocity, θ_w is the wind direction, and ω is the angular frequency.

The latest version also introduces the following Janssen model.

$$S_{in} = B \times S(f,\theta) = \frac{\rho_a}{\rho_w} \beta \chi^2 \omega \times S(f,\theta) \tag{1.2.25}$$

where β and χ are the parameters expressed using the following equations.

$$\left.\begin{aligned}\beta &= \frac{1.2}{k^2} \mu \times (\ln \mu)^4 \quad (\mu \leq 1) \\ \beta &= 0 \quad\quad\quad\quad\quad (\mu > 1)\end{aligned}\right\} \tag{1.2.26}$$

$$\chi = \left(\frac{u_*}{C}\right)\cos(\theta - \alpha) \tag{1.2.27}$$

where k is the Karman constant (0.41) and μ is the parameter expressed as follows:

$$\mu = \frac{gz_o}{C^2} \times \exp\left(\frac{k}{\chi}\right) \tag{1.2.28}$$

$$z_o = \frac{0.01u_*^2}{g\sqrt{1 - \dfrac{\tau_w}{\rho_a u_*^2}}}, \quad \tau_w = \rho_w \iint \left[BS(f,\theta) \times \omega \cos(\theta - \theta_w)\right] dfd\theta \tag{1.2.29}$$

where z_o is the roughness length obtained by Eq. (1.2.29); and τ_w is the wave-induced stress, its ratio to the total stress $\rho_a u_*^2$, i.e. $\tau_w / (\rho_a u_*^2)$ approaches 1 for small waves of small wave age and 0.5 for well-developed waves of large wave age.

These systems of equations are closed and can be solved by successive approximation methods.

Eq. (1.2.4) cannot be easily used to calculate S_{nl} because there are innumerable combinations of the four component waves. Therefore, instead of numerical integration over countless resonant combinations of four waves, Eq. (1.2.4) is simplified to Eq. (1.2.31) by using the resonant four-wave combination expressed in Eq. (1.2.30) with the parameter λ_{nl} in the WAM.

$$\left.\begin{aligned}\omega_1 &= \omega_2 = \omega, \quad \omega_3 = \omega(1 + \lambda_{nl}) = \omega^+, \quad \omega_4 = \omega(1 - \lambda_{nl}) = \omega^- \\ \theta_1 &= \theta_2 = \theta = 0°, \quad \theta_3 - \theta = \pm 11.5°, \quad \theta_4 - \theta = \mp 33.6° \\ \lambda_{nl} &= 0.25 (= \text{const.})\end{aligned}\right\} \tag{1.2.30}$$

$$\begin{Bmatrix}\delta S_{nl} \\ \delta S_{nl}^+ \\ \delta S_{nl}^-\end{Bmatrix} = \begin{Bmatrix}-2(\Delta\omega\Delta\theta)/(\Delta\omega\Delta\theta) \\ (1 + \lambda_{nl})(\Delta\omega\Delta\theta)/(\Delta\omega^+\Delta\theta) \\ (1 - \lambda_{nl})(\Delta\omega\Delta\theta)/(\Delta\omega^-\Delta\theta)\end{Bmatrix} \tag{1.2.31}$$

$$\times C_{nl}\omega^{11}g^{-4}\left[S(f,\theta)^2\left\{\frac{S(f,\theta)^+}{(1 + \lambda_{nl})^4} + \frac{S(f,\theta)^-}{(1 - \lambda_{nl})^4}\right\} - 2\frac{S(f,\theta)S(f,\theta)^+ S(f,\theta)^-}{(1 - \lambda_{nl}^2)^4}\right]$$

where δS_{nl}, δS_{nl}^+, and δS_{nl}^- are the changes per unit time of the nonlinear energy transport in the directional spectra $S(f,\theta)$, $S(f,\theta)^+$ and $S(f,\theta)^-$ respectively; $\Delta\omega$, $\Delta\omega^+$, and $\Delta\omega^-$ are the angular frequency grid widths for the angular frequencies ω, ω^+ and ω^- respectively; $\Delta\theta$ is the grid width of the wave direction; and C_{nl} is a dimensionless constant adjusted so that the results from this

1.2 Generation Mechanism of Sea Waves

approximation method match the exact calculated values.

Adding up the terms in the left-hand side of Eq. (1.2.31) over all frequencies and wave directions yields the transport S_{nl} in the deepsea region. This approximation is called the **Discrete Interaction Approximation (DIA)**. As the effect of this non-linear interaction is greater in shallow water, the transport S_{nl} in shallow water is determined by the following equation using the amplification factor R.

$$S_{nl} \text{ in the shallow area } = R \times S_{nl} \text{ in the deep sea region}$$

$$R = \begin{cases} \left[1 + \dfrac{5.5}{\kappa_p d} \left(1 - \dfrac{5}{6} \kappa_p d \right) \exp\left(-1.25\kappa_p d \right) \right] & (\kappa_p d > 0.5) \\ 4.43 & (\kappa_p d \leq 0.5) \end{cases} \tag{1.2.32}$$

where d is the water depth; κ_p is the peak wavenumber of the JONSWAP spectrum, but to be able to be used for any spectral shape, the mean wave number $\tilde{\kappa}$ in Eq. (1.2.5) is used, and $\kappa_p = 0.75\tilde{\kappa}$.

Finally, energy dissipation due to whitecap breaking S_{wc} is considered in Eq. (1.2.5) and the energy dissipation due to seabed friction S_{bf} is considered in Eq. (1.2.6).

WAM is the most popular numerical wave prediction model in the world. However, as WAVEWATCH-III, which was developed based on WAM, is considered better in terms of prediction accuracy, the reader is recommended to use WAVEWATCH-III, for which the manual and programme are publicly (https:// github.com/NOAA-EMC/WW3).

(b) SWAN Model

SWAN is a wave generation model based on WAM developed for deep waters, and was modified by Booij et al [40, 41] in 1996 for shallow water applications, upto wave breaking point. Manuals and other information on the new version of SWAN are available from the Delft University of Technology at http://www.swan.tudelft.nl.

In this model, the amount of energy supplied by the wind to the waves S_{in}, the amount of energy transported by the non-linear interaction of four-wave resonances S_{nl}, and the amount of energy dissipation due to whitecap breaking and seabed friction S_{ds}, are considered. The amount of energy transported by the non-linear interaction of three-wave resonances in very shallow waters (the phenomenon where wave energy is transported from low frequency to high frequency in very shallow waters) S_{nl3} is also considered. Moreover, the amount of energy dissipation due to wave breaking in very shallow waters S_{ds}, and transmission and reflection by structures are also considered.

SWAN provides both Cartesian and spherical coordinate systems. To be able to take the flow into account, the wave action equilibrium equation, which is a conservation law for the **wave action density $N(\sigma, \theta)$** [$= S(\sigma, \theta) / \sigma$: the directional spectrum divided by the **relative frequency (σ)** (= angular frequency in a coordinate system moving with the flow)], is used. They are expressed as follows.

1. STATISTICAL PROPERTIES AND GENERATION MECHANISMS OF WAVES

In the case of Cartesian coordinate system (x,y):

$$\frac{\partial N(\sigma,\theta)}{\partial t} + \frac{\partial\left[N(\sigma,\theta)C_x\right]}{\partial x} + \frac{\partial\left[N(\sigma,\theta)C_y\right]}{\partial y}$$

$$+ \frac{\partial\left[N(\sigma,\theta)C_\sigma\right]}{\partial\sigma} + \frac{\partial\left[N(\sigma,\theta)C_\theta\right]}{\partial\theta} = \frac{S_{in}+S_{nl}+S_{ds}}{\sigma} \tag{1.2.33}$$

where θ is the wave direction, C_x and C_y are the propagation velocities in real space (x,y), and C_σ and C_θ are the propagation velocities in spectral space (σ,θ), and they are expressed as follows:

$$\left.\begin{array}{l} C_x = C_g\cos\theta+U, \quad C_y = C_g\sin\theta+V, \quad C_\sigma = \dfrac{\partial}{\partial t}\left(\sqrt{g\kappa\tanh\kappa d}-\vec{\kappa}\vec{U}\right) \\[3mm] C_\theta = \dfrac{C_g}{C}\left(\sin\theta\dfrac{\partial C}{\partial x}-\cos\theta\dfrac{\partial C}{\partial y}\right)-\left(\sin\theta\dfrac{\partial}{\partial x}-\cos\theta\dfrac{\partial}{\partial y}\right)\left(\dfrac{\kappa}{\kappa}\vec{U}\right) \end{array}\right\} \tag{1.2.34}$$

where C is the phase velocity of each component wave, C_g is the group velocity, $\vec{U}=(U,V)$ is the steady flow velocity vector, $\vec{\kappa}=(\kappa_x,\kappa_y)$ is the wavenumber vector, $\kappa=|\vec{\kappa}|$ is the wavenumber, and d is the water depth.

In Eq. (1.2.33), the first term on the left-hand side represents the time variation of the action density spectrum; the second and third terms represent the spatial variation due to propagation in the x and y directions; the fourth term represents the relative frequency variation due to changes in water depth and flow over time; and the fifth term represents the wave refraction variation.

In the case of spherical coordinate system (θ_x,θ_y):

$$\frac{\partial N(\sigma,\theta)}{\partial t} + \frac{1}{\cos\theta_x}\frac{\partial\left[N(\sigma,\theta)\dfrac{C_g\cos\theta+U}{R}\cos\theta_x\right]}{\partial\theta_x} + \frac{\partial\left[N(\sigma,\theta)\dfrac{C_g\sin\theta+V}{R\cos\theta_x}\right]}{\partial\theta_y}$$

$$+ \frac{\partial\left[N(\sigma,\theta)\dfrac{\partial\left(\sqrt{g\kappa\tanh\kappa d}-\vec{\kappa}\vec{U}\right)}{\partial t}\right]}{\partial\sigma} \tag{1.2.35}$$

$$+ \frac{\partial\left[N(\sigma,\theta)\left\{\dfrac{C_g\sin\theta\tan\theta_x}{R}+\dfrac{1}{\kappa R}\left(\sin\theta\dfrac{\partial}{\partial\theta_x}-\dfrac{\cos\theta}{\cos\theta_x}\dfrac{\partial}{\partial\theta_y}\right)\left(\sqrt{g\kappa\tanh\kappa d}-\vec{\kappa}\vec{U}\right)\right\}\right]}{\partial\theta}$$

$$= \frac{S_{in}+S_{nl}+S_{ds}}{\sigma}$$

where θ_x is the latitude, θ_y is the longitude, C_g is the velocity of the wave group, $\vec{U}=(U,V)$ is the steady flow velocity vector, R is the radius of the earth, d is the water depth, $\kappa=|\vec{\kappa}|$ is the wavenumber, and $\vec{\kappa}=(\kappa_x,\kappa_y)$ is the wavenumber vector.

Phillips' generation term C_a in Eq. (1.2.3), Eq. (1.2.36), which is the equation of Cavaleri and Malanotte-Rizzoli [42] with a filter to remove wave growth in the frequency band below the peak frequency of the Pierson-Moskowitz spectrum, is adopted to determine the energy supply from wind

1.2 Generation Mechanism of Sea Waves

to waves S_{in}. It also allows the user to choose between Eq. (1.2.24) and Eq. (1.2.25) as the Miles development term.

$$C_a = \frac{0.0015}{2\pi g^2}\left[u_* \max\{0,\ \cos(\theta-\theta_w)\}\right]^4 \times \exp\left\{-\left(\frac{\sigma}{0.009286\pi g/u_*}\right)^{-4}\right\} \quad (1.2.36)$$

where u_* is the friction velocity and is determined by the following equation.

$$u_*^2 = C_D \times u_{10}^2 \quad (1.2.37)$$

$$C_D = \begin{cases} 1.2875\times10^{-3} & (u_{10} < 7.5\text{m/s}) \\ (0.8+0.065\times u_{10})\times10^{-3} & (u_{10} \geq 7.5\text{m/s}) \end{cases} \quad (1.2.38)$$

where C_D is the drag coefficient of Wu [43], and u_{10} is the mean velocity at 10 m above the sea surface.

The energy transport due to the non-linear interaction of the four-wave resonance in the shallow-water region S_{nl} is then obtained by multiplying the amount of transport in the deep-water region by the amplification factor R given in the Eq. (1.2.39), which consists of the peak wavenumber κ_p and water depth d. Here, the amount of transport in the deep-water region is obtained by using the Discrete Interaction approximation, which adds up the left-hand side of Eq. (1.2.31) over all frequencies and wave directions.

$$R = \begin{cases} 1+\dfrac{5.5}{\kappa_p d}\left(1-\dfrac{6}{7}\kappa_p d\right)\exp\left(-1.25\kappa_p d\right) & (\kappa_p d > 0.5) \\ 4.43 & (\kappa_p d \leq 0.5) \end{cases} \quad (1.2.39)$$

Eq. (1.2.39) is an expression with different coefficients in Eq. (1.2.32) of the WAM.

In very shallow waters, the energy transport due to the non-linear interaction of the three-wave resonance S_{nl3} cannot be neglected. When the three component waves satisfy the relationships $\vec{\kappa}_1 \pm \vec{\kappa}_2 = \vec{\kappa}_3$, and $\sigma_1 \pm \sigma_2 = \sigma_3$, the component wave with wavenumber κ_3 and relative frequency σ_3 grows by receiving energy from the other component waves. In SWAN, this transport quantity is obtained by the approximate solution method **Lumped Triad Approximation (LTA)** of Eldeberky [44] as follows:

$$\left.\begin{aligned} S_{nl3}(\sigma,\theta) &= S_{nl3}^+(\sigma,\theta) + S_{nl3}^-(\sigma,\theta) \\ S_{nl3}^+(\sigma,\theta) &= \max\left[0,\ C_5 2\pi CC_g J^2|\sin\beta|\times\left\{S^2\left(\tfrac{\sigma}{2},\theta\right)-2S\left(\tfrac{\sigma}{2},\theta\right)S(\sigma,\theta)\right\}\right] \\ S_{nl3}^-(\sigma,\theta) &= -2S_{nl3}^+(2\sigma,\theta) \end{aligned}\right\} \quad (1.2.40)$$

where C_5 is an adjustment coefficient, usually 0.1, and J is a value defined by the following equation and called the interaction coefficient.

$$J = \frac{\kappa_{\sigma/2}^2\left(gd+2C_{\sigma/2}^2\right)}{\kappa_\sigma d\left(gd+\dfrac{2}{15}gd^3\kappa_\sigma^2-\dfrac{2}{5}\sigma^2 d^2\right)} \quad (1.2.41)$$

β is the phase angle of the bispectrum of the third-order correlation function between each component wave, expressed using the Ursell number U_r as follows:

1. STATISTICAL PROPERTIES AND GENERATION MECHANISMS OF WAVES

$$\beta = -\frac{\pi}{2} + \frac{\pi}{2}\tanh\left(\frac{0.2}{U_r}\right), \qquad U_r = \frac{g}{8\sqrt{2}\pi^2}\frac{H_{1/3}\overline{T}^2}{d^2} \tag{1.2.42}$$

where $\overline{T} = 2\pi/\overline{\sigma}$ is the mean period.

The energy transitions of the three-wave resonance are calculated when the Ursell number is $0.1 < U_r < 10$.

Finally, for the energy dissipation S_{ds}, the energy dissipation S_{wc} due to whitecap breaking is considered in Eq. (1.2.5). The energy dissipation due to seabed friction S_{bf} can be taken into account by selecting from three types: Eq. (1.2.6), an equation based on the resistance law by Collins [45] and the eddy viscosity model by Madsen et al. [46]. In SWAN, the energy dissipation due to shallow water wave breaking S_{wb} is taken into account using the bore model of Battjes and Janssen [47], given by Eq. (1.2.43). The total energy dissipation S_{ds} in this case is $S_{wc}+S_{bf}+S_{wb}$.

$$\left. \begin{aligned} S_{wb} &= -\frac{1}{4}C_6 Q_b \left(\frac{\overline{\sigma}}{2\pi}\right)\frac{H_b^2}{S_T}\times S(f,\theta) \\ Q_b &= 8\frac{S_T}{H_b^2}\ln Q_b + 1 \\ \overline{\sigma} &= \frac{\iint S(\sigma,\theta)\sigma d\sigma d\theta}{S_T} \\ H_b &= 0.73d, \quad S_T = \iint S(\sigma,\theta)d\sigma d\theta \end{aligned} \right\} \tag{1.2.43}$$

where C_6 is the adjustment factor, Q_b is the probability of wave breaking calculated assuming a Rayleigh wave height distribution, $\overline{\sigma}$ is the mean relative frequency, H_b is the wave breaking height, d is the water depth, and S_T is the omnidirectional spectrum of waves.

It should be noted that numerical calculations with the SWAN model use an approximate implicit method of iterative convergence calculations and therefore take longer than WAM.

List of References in Chapter 1

1) Longuet-Higgins, M. S.: On the Statistical Distributions of the Heights of Sea Waves, *Jour. Marine Res.*, Vol. 9, No. 3, 1952, pp.245-266.

2) Bretschneider, C. L.: Wave Variability and Wave Spectra for Wind-generated Gravity Waves, *Tech. Memo.*, No. 118, Beach Erosin Board, 1959.

3) Mitsuyasu, H.: Development of the Spectrum of Wind Waves (2) - on the shape of the spectrum of wind waves within a finite fetch -, *Proc. of the 17th Coastal Engineering Conference*, JSCE, 1970, pp.1-7 (in Japanese).

4) Bretschneider, C. L.: Significant Waves and Wave Spectrum, *Ocean Industry*, Feb. 1968, pp.40-46.

5) Pierson, W. J., Jr and Moskowitz, L.: A Proposed Spectral Form for Fully Developed Wind Seas Based on the Similarity Theory of S. A. Kitaigorodskii, *Journal Geophys. Res.*, Vol. 69, No.24, 1964, pp.5181-5190.

6) Hasselmann, K. et al.: Measurements of Wind Wave Growth and Swell Decay during the Joint North Sea Wave Project (JONSWAP), *Deutsche Hydrogr. Zeit.*, Vol.A8, No.12, 1973.

7) Huang, N., Hwang, P., Wang, H., Long, S., and Bliven, L.: A Study on the Spectral Models for Waves in the Finite Water Depth, *Journal Geophys. Res.*, Vol.C14, 1983, pp.9579-9587.

8) Tanemoto, J., Ishihara, T. and Yamaguchi, A.: *A Study on a Spectral Model for the Combined Waves of Wind Waves and Swell*, https://www.jstage.jst.go.jp/article/jweasympo/38/0/38_81/_pdf/-char /ja (in Japanese).

9) Ochi, M. K. and Hubble, E. N.: On Six-parameter Wave Spectra, *Proc. 15th Int. Conf. Coastal Eng.*, ASCE, 1976, pp.301-328.

10) Arthur, R. S.: Variability in Direction of Wave Travel in Ocean Surface Waves, *Ann. New York Acad. Sci.*, Vol.51, No.3, 1949, pp.511-522.

11) Longuet-Higgins, M. S., Cartwright, D. E. and Smith, N. D.: Observations of the Directional Spectrum of Sea Waves Using the Motions of a Floating Buoy, *In Ocean Wave Spectra, Proc. Conf.*, Easton, Prentice-Hall, 1961, pp.111-132.

12) Mitsuyasu, H., Mizuno, S., Honda, T. and Rikiishi, K.: On the Directional Spectrum of Ocean Waves (Continued), *Proc. of the 21st Coastal Engineering Conference*, 1974, pp.261-265 (in Japanese).

13) Goda, Y. and Suzuki, Y.: *Computation of Refraction and Diffraction of Sea Waves with Mitsuyasu's Directional Spectrum*, Technical Note of the Port and Habour Research Institute, Ministry of Transport, Japan, No.230, 1975, 45 p. (in Japanese).

14) Hasselmann, K.: Grundgleichungen der Seegangsvoraussage, *Schiffstechnik*, 7, 1960, pp.191-195.

1. STATISTICAL PROPERTIES AND GENERATION MECHANISMS OF WAVES

15) Research status review subcommittee: *New Wave Calculation Methods and Future Design Methods for Marine Facilities* (*towards the establishment of performance design methods*), Coastal Engineering Committee, JSCE, 2001, 256 p. (in Japanese).

16) The WAVEWATCH III Development Group: *User Manual and System Documentation of WAVEWATCH III Version 607*, Technical Note, 333, National Centers for Environmental Prediction, National Weather Service, NOAA, 2019, 320 p.

17) Phillips, O. M.: On the Generation of Waves by Turbulent Wind, *Jour. Fluid Mech.*, Vol.2, 1957, pp.417-445.

18) Miles, J. M.: On the Generation of Surface Waves by Shear Flows, *Jour. Fluid Mech.*, Vol.3, No.2, 1957, pp.185-204.

19) Miles, J. M.: On the Generation of Surface Waves by Turbulent Shear Flows, *Jour. Fluid Mech.*, Vol.7, No.3, 1960, pp.469-478.

20) Hasselmann, K.: On the Non-linear Energy Transfer in a Gravity-wave Spectrum. Part1. General theory, *Jour. Fluid Mech.*, 12, 1962, pp.481-500.

21) Komen, G. J., Hasselmann, S. and Hasselmann, K.: On the Existence of a Fully Developed Wind Sea Spectrum, *Jour. Phys. Oceanogr.*, 3, 14, 1984, pp.1271-1285.

22) Hasselmann, K. and Collins, J. I.: Spectral Dissipation of Finite-depth Gravity Waves due to Turbulent Bottom Friction, *Jour. Maritime Research*, 26, 1968, pp.1-12.

23) Takahashi, K.: *Study on Quantitative Weather Forecasting Based on Extrapolation* (*Part 1*), Research Bulletin, No. 13, JMA, 1947, 20p. (in Japanese).

24) Fujita, T.: Pressure Distribution within Typhoon, *Geophys. Mag.*, 23, 1952, pp.437-451.

25) Myers, V. A.: *Characteristics of U.S. Hurricanes Pertinent to Levee Design for Lake Okechobee*, Florida Hydromet Rep., U.S. Weather Bureau, No.32, 1954, 106 p.

26) Nonaka, K., Yamaguchi, M., Hatada, Y. and Ito, Y.: Extreme Value Estimation System of Wave Height Using Extended Probabilistic Typhoon Model, *Journal of Coastal Engineering*, Vol. 47, 2000, pp. 271-275 (in Japanese).

27) Sverdrup, H. and Munk, W. H.: *Wind, Sea, and Swell: Theory of relations for forecasting*, U. S. Navy Hydrographic Office, Washington, No.601, 1947.

28) Bretschneider, C. L.: The Generation and Decay of Wind Waves in Deep Water, *Trans. A.G.U.*, 33(3), 1952, pp.381-389.

29) Bretschneider, C. L.: Revision in Wave Forecasting; Deep Water and Shallow Water, *Proc. 6th Conf. Coastal Eng.*, 1958, pp.30-67.

30) Bretschneider, C. L.: Forecasting Relations for Wave Generation, *Look Lab. Hawaii*, 1 (3), 1970, pp.31-41.

31) Wilson, B. W.: Numerical Prediction of Ocean Waves in The North Atlantic for December 1959, *Deutche Hydrographisch Zeit*, Vol.18, No.3, 1965, pp.114-130.

32）Goda, Y.: On the Simple Calculation of Waves by the Wilson Estimation Formula, *ECOH/YG Technical Bulletin*, No.1, 2002, 3 p. https://www.ecoh.co.jp/tech/techlist/pdf/wilsonsprediction.pdf

33) Goda, Y.: Revisiting Wilson's Formulas for Simplified Wind-wave Prediction. *Journal of Waterway, Port, Coastal and Ocean Engineering*, Vol. 129, No. 2, 2003, pp. 93-95.

34）Bretschneider, C. L.: Generation of Wind Waves over a Shallow Bottom, *Tech. Memo. Beach Erosion Board*, No. 51, 1954.

35）Bretschneider, C. L.: Decay of Ocean Waves. Fundamentals of Ocean Engineering-Part8b, *Ocean Industry*, April, 1968, pp.45-50.

36）Pierson, W. J., Neumann, G. and James, R. W.: *Observing and Forecasting Ocean Waves by means of Wave Spectra and Statistics*, U. S. Naval Oceanogr. Office, Pub. No.603, 1955, 284 p.

37）Isozaki, I. and Uji, T.: Numerical Prediction of Ocean Wind Wave, *Paper Meteorol. Geophys.*, 24, 1973, pp.207-231.

38）WAMDI Group: The WAM Model － a third generation ocean wave prediction model, *Jour. Phys. Oceanogr.*, 18, 1988, pp.1775-1810.

39）Ueno, K. and Ishizaka, M.: Efficient Calculation Method of Nonlinear Energy Transfer of Wind and Waves, *Weather Service Bulletin*, Vol. 64, 1997, pp.137-140.

40）Booij, N., Holthuijsen, L. H. and Ris, R. C.: The "SWAN" Wave Model for Shallow Water, *Proc. 25th Int. Conf. Coastal Eng.*, 1996, pp.668-676.

41）Booij, N., Ris, R. C. and Holthuijsen, L. H.: A Third-generation Wave Model for Coastal Regions, 1. Model description and validation, *Jour. Geophys. Res.*, 104(C4), 1999, pp.7649-7666.

42）Cavaleri, L. and Malanotte-Rizzoli, P.: Wind Wave Prediction in Shallow Water: Theory and application, *Jour. Geophys. Res.*, 86(C11), 1981, pp.10961-10973.

43) Wu, J.: Wind-stress Coefficients over Sea Surface from Breeze to Hurricane, *Jour. Geophys Res.*, 87(C12), 1982, pp.9704-9706.

44) Eldeberky, Y.: *Nonlinear Transformation of Wave Spectra in the Nearshore Zone*, Ph.D. thesis, Delft University of Technology, Department of Civil Engineering, The Netherland, 1996.

45) Collins, J. I.: Prediction of Shallow Water Spectra, *Jour. Geophys. Res.*, 77(15), 1972, pp.2693-2707.

46) Madsen, O. S., Poon, Y.-K. and Graber, H. C.: Spectral Wave Attenuation by Bottom Friction (Theory), *Proc. 21st Int. Conf. Coastal Eng.*, ASCE, 1988, pp.492-504.

47) Battjes, J. A. and Janssen, J. P. F. M.: Energy Loss and Set-up due to Breaking of Random Waves, *Proc. 16th Int. Conf. Coastal Eng.*, ASCE, 1978, pp.569-587.

Chapter:2

WAVE THEORIES AND, PROPAGATION AND DEFORMATION

2.WAVE THEORIES AND, PROPAGATION AND DEFORMATION

This chapter describes the deformation of waves that originate offshore and propagate towards the shore.

Ocean waves are irregular and usually propagate in groups. However, to simplify the theoretical governing equations for wave propagation in the sea, the theoretical equations for regular waves are used in this chapter. Therefore, the wave characteristics obtained theoretically by solving the linear equations of motion for regular waves are first clarified. Next, the main analytical solutions to the non-linear equations of motion are presented, as the influence of the non-linear terms in the equations of motion cannot be ignored when the water depth becomes shallow. Furthermore, for phenomena such as wave refraction, diffraction, shoaling and wave breaking, which are introduced in Section 2.3, evaluation methods that consider the irregularity of waves are also introduced, as non-negligible differences between the measured values and those obtained by fitting the regular wave theory appear. Then, main numerical simulation models of wave fields applicable to arbitrary topography are presented in Section 2.4.

2.1 Small Amplitude Wave Theory

According to Lamb [1], the first to elucidate the motion of water surface waves with mathematical equations was Gerstner, who in 1802 published the trochoidal wave theory derived from Lagrange's Equation of motion. However, these waves have vorticity and cannot be regarded as naturally occurring gravity water waves. Airy [2] was the first to derive a theory of small amplitude waves based on velocity potentials under conditions of zero vorticity.

1) Assumptions for Derivation
(1) Eulerian Treatment

When analysing fluid motion, there are two methods: the **Lagrangian method**, which focuses on a single particle and tracks this particle movement in time, and the **Eulerian method**, which considers a small rectangular region (called a control volume) in Cartesian coordinates in the fluid and considers the behaviour of the fluid in this region. In this chapter, the Eulerian method, which is common in the field of fluid mechanics, is adopted.

(2) Assumption of an Ideal Fluid (Perfect Fluid)

In the case of liquids such as seawater, unlike gases such as air, contraction and expansion can be ignored for pressure changes of around 10 atmospheres. Also, the viscosity of seawater cannot be ignored when the generation and development of waves due to wind is concerned, but the viscosity can be ignored when dealing with the normal motion of waves. Therefore, in this section, the seawater is treated as an **ideal fluid** or **perfect fluid**, i.e., the compressibility and viscosity of seawater are ignored.

2.1 Small Amplitude Wave Theory

Neglecting viscosity and assuming no vorticity, we can introduce a physical quantity ϕ, called the **velocity potential**, which is defined by the following equations.

$$u = -\frac{\partial \phi}{\partial x} \quad \text{and} \quad w = -\frac{\partial \phi}{\partial z} \qquad (2.1.1)$$

where u and w are the horizontal and vertical velocities, and x and z are the horizontal and vertical coordinates.

The velocity potential ϕ is a physical quantity that, when differentiated in spatial coordinates, results in a velocity in the direction of those coordinates, making it convenient to handle. In analogy to the reduction of potential energy into kinetic energy, the reduction of the velocity potential in the direction of the spatial coordinate changes into the velocity in that coordinate direction.

(3) Assumptions underlying Small Amplitude Waves

When the wave height is less than about 1/5 of the water depth or 1/50 of the offshore wavelength, the non-linear terms in the equation of motion can be ignored. Airy derived his theoretical equation for wave motion on the assumption that the wave's amplitude (one-half of the wave height) is sufficiently small compared to the water depth and wavelength. This is called the **small amplitude wave theory**. If the wave height is greater than about 1/5 of the water depth or 1/50 of the offshore wavelength, the **finite amplitude wave theory** should be applied, which assumes that the wave amplitude is finite.

(4) Other Assumptions

The seawater does not peel away from the seawater surface. There is no seepage flow from the seabed surface into the ground and the velocity in the subsurface direction is zero at the seabed surface. Furthermore, **surface tension** and the **Coriolis force** can be ignored.

2) Derivation of the Water Particle Velocity Equations

To clarify the motion of waves, it is necessary to determine the horizontal velocity u and vertical velocity w of water particles in addition to the water surface height above the still water surface (water surface elevation) η. The relevant equations for these three unknowns, the Euler's equations of motion in the horizontal and vertical directions, (2.1.2) and (2.1.3), and the Continuity equation (2.1.4) representing the conservation of mass in seawater, can be solved (see **Fig. 2.1** for coordinates and parameters). However, Eqs (2.1.2) and (2.1.3) are not easy to solve because they are non-linear differential equations with advection terms on the left-hand side of each equation.

2. WAVE THEORIES AND, PROPAGATION AND DEFORMATION

$$\frac{\partial u}{\partial t} + u\frac{\partial u}{\partial x} + w\frac{\partial u}{\partial z} = -\frac{1}{\rho}\frac{\partial p}{\partial x} \quad (2.1.2)$$

$$\frac{\partial w}{\partial t} + u\frac{\partial w}{\partial x} + w\frac{\partial w}{\partial z} = -\frac{1}{\rho}\frac{\partial p}{\partial z} - g \quad (2.1.3)$$

$$\frac{\partial u}{\partial x} + \frac{\partial w}{\partial z} = 0 \quad (2.1.4)$$

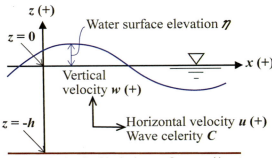

Fig. 2.1 Definition of coordinates and parameters.

where t is time, x is the horizontal coordinate, z is the vertical coordinate, ρ is the density of seawater, P is the water pressure and g is the acceleration of gravity.

Eqs. (2.1.2), (2.1.3) and (2.1.4) are difficult to solve in conjunction, but assuming small amplitude waves, the absolute value of the vertical velocity w can be regarded as small compared to the absolute value of the horizontal velocity u, and the second and third terms on the left-hand side of Eq. (2.1.3) can be ignored. Eq. (2.1.3) can therefore be simplified as follows:

$$\frac{\partial w}{\partial t} = -g - \frac{1}{\rho}\frac{\partial p}{\partial z} \quad (2.1.5)$$

Integrating Eq. (2.1.5) in the vertical direction yields Eq. (2.1.6).

$$\int_0^z \frac{\partial w}{\partial t} dz = -\int_0^z g\, dz - \frac{1}{\rho}\int_0^z \frac{\partial p}{\partial z} dz$$

$$= -[gz]_0^z - \frac{1}{\rho}[p]_0^z = -gz - \frac{1}{\rho}[p(z) - p(0)] \quad (2.1.6)$$

For small amplitude waves, the water surface $z = \eta \approx 0$ can be approximated, so Eq. (2.1.7) is given.

$$\left[\int_0^z \frac{\partial w}{\partial t} dz\right]_{z=\eta} = [-gz]_{z=\eta} \quad (2.1.7)$$

If the vertical velocity w in Eq. (2.1.7) is integrated over z, it is equal to $-\phi$ from Eq. (2.1.1). So, Eq. (2.1.7) can be rewritten as follows:

$$\left[\frac{\partial \phi}{\partial t}\right]_{z=\eta} = g\eta \quad (2.1.8)$$

Substituting Eq. (2.1.1) for the velocity potential ϕ into Eq. (2.1.4), the following two dimentional Laplace's equation is obtained.

$$\frac{\partial^2 \phi}{\partial x^2} + \frac{\partial^2 \phi}{\partial z^2} = 0 \quad (2.1.9)$$

Once the velocity potential is determined from Eqs. (2.1.8) and (2.1.9), the velocity w and u can then be obtained using Eq. (2.1.1).

The velocity potential is proportional to the result of integrating the water surface height η with time t from Eq. (2.1.8) and can therefore be expressed as follows with reference to Eq. (1.1.2).

$$\phi = Z(z)\sin(\kappa x - \omega t) \quad (2.1.10)$$

Substituting Eq. (2.1.10) into Eq. (2.1.9) yields the following second-order differential equation, since $\dfrac{\partial^2 \phi}{\partial x^2} = -\kappa^2 Z(z)\sin(\kappa x - \omega t)$.

$$\frac{d^2 Z(z)}{dz^2} - \kappa^2 Z(z) = 0 \tag{2.1.11}$$

The general solution to Eq. (2.1.11) is expressed by using the constants A and B as follows:

$$Z = Ae^{\kappa z} + Be^{-\kappa z} \tag{2.1.12}$$

Substituting Eq. (2.1.12) into Eq. (2.1.10), the velocity potential is expressed as follows:

$$\phi = (Ae^{\kappa z} + Be^{-\kappa z})\sin(\kappa x - \omega t) \tag{2.1.13}$$

Next, the constants A and B are determined using a physical boundary condition. If the seawater is assumed not to percolate at the seabed surface ($z = -h$, h is the water depth) and the vertical velocity w is zero, the following boundary conditions are obtained.

$$[w]_{z=-h} = \left[-\frac{\partial \phi}{\partial z}\right]_{z=-h} = 0 \tag{2.1.14}$$

Substituting Eq. (2.1.13) into Eq. (2.1.14) yields the following relationship.

$$A = Be^{2\kappa h} \tag{2.1.15}$$

Moreover, if the hyperbolic function $\cosh(x) = (e^x + e^{-x})/2$ is also used, the velocity potential is expressed as follows:

$$\phi = 2Be^{\kappa h}\cosh\kappa(h + z)\sin(\kappa x - \omega t) \tag{2.1.16}$$

Using Eq. (2.1.8) and connecting Eq. (2.1.16) with Eq. (1.1.2), the following relationship is obtained. Here, from the assumption of small amplitude waves ($\eta < h$), $h + \eta$ can be approximated as equal to h.

$$2Be^{\kappa h} = -\frac{Hg}{2\omega \cosh\kappa(h + \eta)} \approx -\frac{Hg}{2\omega \cosh\kappa h} \tag{2.1.17}$$

Substituting Eq. (2.1.17) into Eq. (2.1.16), the velocity potential is expressed as follows:

$$\phi = -\frac{Hg}{2\omega}\frac{\cosh\kappa(h + z)}{\cosh\kappa h}\sin(\kappa x - \omega t) \tag{2.1.18}$$

Furthermore, substituting Eq. (2.1.18) into Eq. (2.1.1) yields the following equations for velocity u and w.

$$u = -\frac{\partial \phi}{\partial x} = \frac{Hg\kappa}{2\omega}\frac{\cosh\kappa(h + z)}{\cosh\kappa h}\cos(\kappa x - \omega t) \tag{2.1.19}$$

$$w = -\frac{\partial \phi}{\partial z} = \frac{Hg\kappa}{2\omega}\frac{\sinh\kappa(h + z)}{\cosh\kappa h}\sin(\kappa x - \omega t) \tag{2.1.20}$$

3) Equations for Wavelength and Wave Celerity

Here, equations for determining wavelength L and wave celerity C are introduced.

The vertical velocity w of a water particle and the vertical velocity of the water surface elevation are assumed to be equal near the water surface under the assumption of small amplitude waves. That is to say:

2.WAVE THEORIES AND, PROPAGATION AND DEFORMATION

$$\left[w\right]_{z=\eta} = \left[-\frac{\partial\phi}{\partial z}\right]_{z=\eta} = \left[\frac{\partial\eta}{\partial t}\right]_{z=\eta} \tag{2.1.21}$$

Using Eq. (2.1.21) and connecting Eqs. (2.1.20) and (1.1.2), the following relationship is obtained.

$$\frac{Hg\kappa}{2\omega}\frac{\sinh\kappa h}{\cosh\kappa h}\sin(\kappa x - \omega t) = \frac{H}{2}\omega\sin(\kappa x - \omega t) \tag{2.1.22}$$

where $h + \eta$ was approximated as h.

If Eq. (2.1.22) is rearranged, the following equation, called the **dispersion relation formula**, is obtained:

$$\omega^2 = g\kappa\tanh\kappa h \tag{2.1.23}$$

Substituting $\omega = 2\pi/T$ and $\kappa = 2\pi/L$ into the dispersion relation formula, and rewriting it as an expression for the relationship between wavelength and wave period, the following relationship is obtained.

$$L = \frac{gT^2}{2\pi}\tanh\left(\frac{2\pi h}{L}\right) \tag{2.1.24}$$

To find the wavelength from this equation, the approximate value of the wavelength must be substituted into the right-hand side and the calculation repeated until the value agrees with the value on the left-hand side within a permissible error.

Furthermore, if Eq. (2.1.23) or Eq. (2.1.24) is rewritten using the wave celerity $C = L/T = \omega/\kappa$, the following relationship can be obtained.

$$C = \frac{gT}{2\pi}\tanh\left(\frac{2\pi h}{L}\right) \tag{2.1.25}$$

From this equation, it can be seen that in the case of irregular waves, dispersion of the waves occurs because the waves with longer periods move faster. Hence, these equations are called **dispersion relation equations**.

Furthermore, multiplying both sides of Eq. (2.1.25) by the wave celerity C gives the following equation.

$$C = \sqrt{\frac{gL}{2\pi}\tanh\left(\frac{2\pi h}{L}\right)} = \sqrt{\frac{g}{\kappa}\tanh\kappa h} \tag{2.1.26}$$

(1) Case of Deep Sea Waves (Offshore Waves)

Since $\tanh\kappa h$ approaches 1 as close as possible for the deep sea where the variable κh is greater than 3, the following important relations are obtained.

$$\left(\frac{2\pi h}{L} \geq 3 \Rightarrow \frac{h}{L} \geq \frac{1}{2}\right) : \left.\begin{array}{l} L_o = \dfrac{gT^2}{2\pi} = 1.56T^2\,[m] \\[2mm] C_o = \dfrac{gT}{2\pi} = 1.56T\,[m/s] \end{array}\right\} \tag{2.1.27}$$

Here, the condition $h \geq L/2$ can be regarded as the condition for sufficiently deepwater, considering that the wavelength L of ordinary waves is about 80 m to 400 m. The waves in this area are referred to as **deep sea waves (offshore waves)**. The subscript 'o' means the offshore wave parameters.

(2) Case of Shallow Sea Waves (Long Waves)

Since tanhκh approaches kh as close as possible for shallow water where the variable κh is less than 0.3, the following important relations are obtained.

$$\left(\frac{2\pi h}{L} < 0.3 \Rightarrow \frac{h}{L} < \frac{1}{20} \right) : \quad \left. \begin{array}{l} L = \sqrt{gh}T \\ C = \sqrt{gh} \end{array} \right\} \qquad (2.1.28)$$

Here, the condition $h < L/20$ can be regarded as a condition of sufficiently shallow water depth, since the wavelength L of ordinary waves is about 80 m to 400 m. The waves in this area are called **shallow sea waves**. In the case of **long waves** such as **tsunamis**, which are generated by earthquakes and have wavelengths more than several thousand metres, so this condition is also satisfied offshore.

(3) Case of Intermediate Sea Waves

The condition $L/2 > h \geq L/20$ is positioned as intermediate depth between deep water and shallow water, and the waves in this area are called **intermediate sea waves**, the wavelength for this condition can be obtained from Eq. (2.1.24) and the wave celerity for this condition from Eq. (2.1.25) or Eq. (2.1.26).

4) Acceleration and Orbit of Water Particles

Substituting Eq. (2.1.23) into Eqs. (2.1.19) and (2.1.20) gives the following equations.

$$\left. \begin{array}{l} u = \dfrac{H\omega}{2} \cdot \dfrac{\cosh \kappa (h+z)}{\sinh \kappa h} \cos(\kappa x - \omega t) \\[3mm] w = \dfrac{H\omega}{2} \cdot \dfrac{\sinh \kappa (h+z)}{\sinh \kappa h} \sin(\kappa x - \omega t) \end{array} \right\} \qquad (2.1.29)$$

If the advection term is regarded as negligible because of the small amplitude wave, the acceleration of the water particles (horizontal direction a_x, vertical direction a_z) can be obtained by differentiating Eq. (2.1.29) with time.

$$\left. \begin{array}{l} a_x = \dfrac{H\omega^2}{2} \cdot \dfrac{\cosh \kappa (h+z)}{\sinh \kappa h} \sin(\kappa x - \omega t) \\[3mm] a_z = -\dfrac{H\omega^2}{2} \cdot \dfrac{\sinh \kappa (h+z)}{\sinh \kappa h} \cos(\kappa x - \omega t) \end{array} \right\} \qquad (2.1.30)$$

Furthermore, because of the small amplitude wave, if the movement of the water particles is regarded as small and the average position (x_o, z_o) of the water particles is used for (x,z) in Eq. (2.1.29), the travel distance ($\xi = x - x_o$, $\zeta = z - z_o$) can be obtained by integrating Eq. (2.1.29) over time as follows:

2. WAVE THEORIES AND, PROPAGATION AND DEFORMATION

$$\left. \begin{array}{l} \xi = -\dfrac{H}{2} \cdot \dfrac{\cosh \kappa (h+z_0)}{\sinh \kappa h} \sin(\kappa x_0 - \omega t) \\ \zeta = \dfrac{H}{2} \cdot \dfrac{\sinh \kappa (h+z_0)}{\sinh \kappa h} \cos(\kappa x_0 - \omega t) \end{array} \right\} \quad (2.1.31)$$

Substituting relational equations for trigonometric functions obtained from Eq. (2.1.31) into trigonometric formula $\sin^2(\kappa x_0 - \omega t) + \cos^2(\kappa x_0 - \omega t) = 1$ yields the following equation.

$$\left(\dfrac{\xi}{H \cosh \kappa (h+z_0)/(2 \sinh \kappa h)} \right)^2 + \left(\dfrac{\zeta}{H \sinh \kappa (h+z_0)/(2 \sinh \kappa h)} \right)^2 = 1 \quad (2.1.32)$$

Eq. (2.1.32) shows an elliptical orbit with horizontal length $H \cosh \kappa (h+z_0)/\sinh \kappa h$ as the major axis and vertical length $H \sinh \kappa (h+z_0)/\sinh \kappa h$ as the minor axis. The orbits of the water particles are shown in **Fig. 2.2** for the deepsea waves and for the intermediate or shallow sea waves.

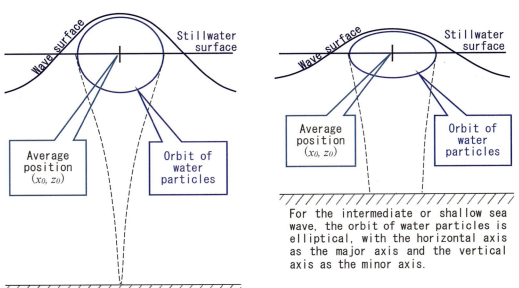

For the deepsea wave, the orbit of water particles is close to circular and the lengths of the horizontal and vertical axes are comparatively the same.

For the intermediate or shallow sea wave, the orbit of water particles is elliptical, with the horizontal axis as the major axis and the vertical axis as the minor axis.

Fig. 2.2 Illustration of the orbit of water particles based on the small amplitude wave theory.

5) Method for Measuring Hydrodynamic Pressure and Wave Height

For the vertical equation of motion (2.1.6), the vertical flow velocity w integrated with z is equal to the velocity potential ϕ with a minus sign, and $P(0)$ is the atmospheric pressure, so if the water pressure is expressed as gauge pressure, the total water pressure $P(z)$ is obtained as follows:

$$P(z) = \rho \dfrac{\partial \phi}{\partial t} - \rho g z = \dfrac{H}{2} \rho g \dfrac{\cosh \kappa (h+z)}{\cosh \kappa h} \cos(\kappa x - \omega t) - \rho g z \quad (2.1.33)$$

In this equation, the first term on the right-hand side is the hydrodynamic pressure due to waves expressed in Eq. (2.1.34), and the second term on the right-hand side is the hydrostatic pressure.

$$P_d(z) = \frac{H}{2}\rho g \frac{\cosh \kappa(h+z)}{\cosh \kappa h} \cos(\kappa x - \omega t) \qquad (2.1.34)$$

The relationship between the hydrodynamic pressure $P_d(z)$ and the wave height H is expressed using Eq. (2.1.34), so if the maximum hydrodynamic pressure $P_{d\,max}$ is obtained from a water-pressure gauge installed on the seabed, the wave height can be obtained from Eq. (2.1.35).

$$H = \frac{2 \times P_{d\,max}}{\left[\rho g \dfrac{\cosh \kappa(h+z)}{\cosh \kappa h}\right]} \qquad (2.1.35)$$

This is the principle of an underwater pressure-type wave gauge. However, to apply this method of determining wave heights from measured water pressures using a theoretical formula, it is naturally necessary to install the water-pressure gauge at a water depth deeper than the breaking water depths for the waves of interest.

[How to determine a wave height using data from an underwater pressure-type wave gauge]

Now, if the maximum hydrodynamic pressure of 37.9 kN/m² and the period of 15.0 sec are obtained from the time series data of a wave gauge located on the seabed at a depth of 15.0 m, then, Eq. (d.1) of the wave height is got by using Eq. (2.1.35).

$$H = \frac{2 \times 37.9\,\text{kN/m}^2}{\left[1030\,\text{kg/m}^3 \times 9.8\,\text{m/s}^2 \times \dfrac{\cosh \kappa(15.0\,\text{m} - 15.0\,\text{m})}{\cosh(\kappa \times 15.0\,\text{m})}\right]} \qquad (\text{d}.1)$$

Where, to obtain the wave number κ, the wavelength of the deep-sea waves is obtained using Eq. (2.1.27).

$$L_o = 1.56T^2 = 1.56 \times (15.0)^2 \fallingdotseq 351\,\text{m} \qquad (\text{d}.2)$$

Since $h/L_o = 1/23.4 < 1/20$, the waves in this condition can be regarded as shallow sea waves. Therefore, from Eq. (2.1.28), the wavelength at the depth of 15.0 m is expressed as follows:

$$L = \sqrt{gh} \times T = \sqrt{9.8 \times 15.0} \times 15.0 \fallingdotseq 182\,\text{m} \qquad (\text{d}.3)$$

Consequently, $\kappa = 2 \times 3.14 / L \fallingdotseq 0.0345\,\text{m}^{-1}$ is got, the wave height is become as follows:

$$H = \frac{2 \times 37.9\,\text{kN/m}^2}{\left[1030\,\text{kg/m}^3 \times 9.8\,\text{m/s}^2 \times \dfrac{\cosh(0)}{\cosh(0.0345 \times 15)}\right]} \fallingdotseq 8.5\,\text{m} \qquad (\text{d}.4)$$

2. WAVE THEORIES AND, PROPAGATION AND DEFORMATION

6) Wave Group Velocity

Considering the sum of the two waves of wavenumber κ_1 and angular frequency ω_1, and wave number $\kappa_1+\Delta\kappa$ and angular frequency $\omega_1+\Delta\omega$, where $\Delta\kappa$ and $\Delta\omega$ are small deviations from κ_1 and ω_1 respectively, the water surface elevation η of the composite wave above the still water surface can be expressed as follows:

$$\eta = \frac{H}{2}\cos\left[(\kappa_1+\Delta\kappa)x-(\omega_1+\Delta\omega)t\right]+\frac{H}{2}\cos\left[\kappa_1 x-\omega_1 t\right] \qquad (2.1.36)$$
$$= H\cos\left(\frac{\Delta\kappa}{2}x-\frac{\Delta\omega}{2}t\right)\cos\left(\frac{2\kappa_1+\Delta\kappa}{2}x-\frac{2\omega_1+\Delta\omega}{2}t\right)$$

The waveform of this composite wave is shown in **Fig. 2.3**. In the last equation of Eq. (2.1.36), the cosine function in the first half represents the envelope wave of wavenumber $\Delta\kappa/2$ and angular frequency $\Delta\omega/2$ (dashed line in **Fig. 2.3**) and the cosine function in the second half represents the carrier wave of wavenumber $(2\kappa_1+\Delta\kappa)/2$ and angular frequency $(2\omega_1+\Delta\omega)/2$ (solid line in **Fig. 2.3**).

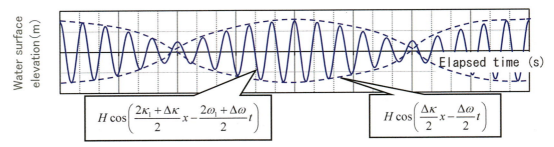

Fig. 2.3 Examples of waves with typical wave grouping properties.

where the propagation velocity C (= angular frequency/wavenumber) of the carrier wave, represented by the solid line, is expressed by the following equation

$$C = \frac{2\omega_1+\Delta\omega}{2\kappa_1+\Delta\kappa} \doteq \frac{\omega_1}{\kappa_1} \qquad (2.1.37)$$

The propagation velocity C_g of this wave group is also expressed by the following equation using the wavenumber and angular frequency of the envelope wave represented by the dashed line.

$$C_g = \frac{\Delta\omega}{\Delta\kappa} \qquad (2.1.38)$$

Substituting $\omega = C\kappa$ into Eq. (2.1.38), the following approximate expression is obtained.

$$C_g = \frac{\Delta C\kappa}{\Delta\kappa} \doteq \frac{dC\kappa}{d\kappa} = C+\kappa\frac{dC}{d\kappa} \qquad (2.1.39)$$

On the other hand, squaring Eq. (2.1.26) and differentiating it by the wavenumber yields Eq. (2.1.40).

$$2C\frac{dC}{d\kappa} = -\frac{g}{\kappa^2}\tanh\kappa h+\frac{gh}{\kappa(\cosh\kappa h)^2} \qquad (2.1.40)$$

Simplifying Eq. (2.1.40) by dividing it by C^2 yields the following relation.

$$\frac{2}{C}\frac{dC}{d\kappa} = -\frac{1}{\kappa} + \frac{h}{\left(\cosh \kappa h\right)^2 \tanh \kappa h} = -\frac{1}{\kappa} + \frac{2h}{\sinh 2\kappa h} \qquad (2.1.41)$$

Substituting Eq. (2.1. 41) into Eq. (2.1. 39) yields the following equation.

$$\left.\begin{array}{l} C_g = \dfrac{C}{2}\left(1 + \dfrac{2\kappa h}{\sinh 2\kappa h}\right) = C \times n \\[3mm] n = \dfrac{C_g}{C} = \dfrac{1}{2}\left(1 + \dfrac{2\kappa h}{\sinh 2\kappa h}\right) \end{array}\right\} \qquad (2.1.42)$$

The velocity C_g of the wave group is called the **group velocity**.

In deep water, since the ratio n becomes *1/2* due to $\sinh 2\kappa h \rightarrow \infty$, the group velocity is half the value of the carrier wave (component wave).

In shallow water, since the ratio n becomes *1* due to $\sinh 2\kappa h \rightarrow 2\kappa h$, the group velocity is equal to that of the carrier wave (component wave).

7) Wave Energy and Power

Waves propagate towards the shore, but on the offshore side deeper than the breaking water depth, the motion of the water particles is closed in elliptical orbits. This means that the energy of the waves is transmitted towards the shore but the water particles move in elliptical orbits within one wavelength and do not move outwards (no mass transport of water occurs on the offshore side deeper than the depth of the breaking wave). In this section, the energy content of the wave is estimated.

(a) Potential Energy

If the water depth from the seabed to the still water surface is represented by h and the water surface elevation from the still water surface to the wave surface by η, the potential energy dE_{p1} of the control volume consisting of the total water depth $h + \eta$ and the base with horizontal length $\Delta x \times$ unit width 1 is expressed by the following equation.

$$dE_{p1} = \text{mass below the wave surface} \times g \times \text{mean depth below the wave surface}$$

$$= \rho(h+\eta)\Delta x \times g \times \frac{(h+\eta)}{2} = \frac{1}{2}\rho g (h+\eta)^2 \Delta x \qquad (2.1.43)$$

The potential energy dE_{p2} of the control volume consisting of the water depth h and the base with horizontal length $\Delta x \times$ unit width 1 is expressed by the following equation.

$$dE_{p2} = \text{mass below the still water surface} \times g \times \text{mean depth below the still water surface}$$

$$= \rho h \Delta x \times g \times \frac{h}{2} = \frac{1}{2}\rho g h^2 \Delta x \qquad (2.1.44)$$

The average potential energy over a wavelength using the potential energy dE_p $(= dE_{p1} - dE_{p2})$ with reference to the still water surface is then given by the following equation.

2.WAVE THEORIES AND, PROPAGATION AND DEFORMATION

$$E_p = \frac{1}{L}\int_x^{x+L}\left(dE_{p1} - dE_{p2}\right)dx = \frac{1}{L}\int_x^{x+L}\left\{\rho g h \eta + \frac{1}{2}\rho g \eta^2\right\}dx$$

$$= \frac{\rho g}{2}\cdot\frac{1}{L}\int_x^{x+L}\left\{Hh\cos\left(\kappa x - \omega t\right) + \frac{H^2}{4}\cos^2\left(\kappa x - \omega t\right)\right\}dx = \frac{1}{16}\rho g H^2 \tag{2.1.45}$$

(b) Kinetic Energy

The kinetic energy dE_k of the control volume with horizontal length Δx × unit width 1 × vertical height Δz is expressed by the following equation.

$$dE_k = \frac{1}{2}\times\text{mass}\times\left(u^2 + w^2\right) = \frac{1}{2}\times\rho\Delta x\Delta z\times\left(u^2 + w^2\right)$$

$$= \frac{\rho\Delta x\Delta z}{2}\cdot\frac{g\kappa H^2}{4\sinh 2\kappa h}\left[\cosh 2\kappa\left(h + z\right) - \cos 2\left(\kappa x - \omega t\right)\right] \tag{2.1.46}$$

Integrating dE_k vertically from the seabed to the still water surface, the average kinetic energy over one wavelength is obtained as follows:

$$E_k = \frac{1}{L}\int_x^{x+L}\left(\int_{-h}^0 dE_k dz\right)dx = \frac{1}{16}\rho g H^2 \tag{2.1.47}$$

(c) Total Average Energy

The total average energy (per unit area) is given as follows:

$$E = E_p + E_k = \frac{1}{8}\rho g H^2 \qquad [\text{Nm/m}^2] \tag{2.1.48}$$

Here, the important finding "the total average energy of the wave is proportional to the square of the wave height" is obtained.

(d) Wave Power

Next, the power dW per unit area due to the wave can be expressed as follows:

$$dW = \text{force per unit area due to the wave}\times\frac{\text{travel distance}}{\text{action time}} = p_d \times u \tag{2.1.49}$$

where the force per unit area is the hydrodynamic pressure P_d and the travel distance / action time is the velocity u.

Therefore, the average value W of the power (= generating power) per unit width due to the wave in one cycle is as follows:

$$W = \frac{1}{T}\int_t^{t+T}\int_{-h}^0 p_d u\,dz\,dt = \frac{\rho g H^2}{8}\cdot\frac{\omega}{\kappa}\cdot\frac{1}{2}\left(1 + \frac{2\kappa h}{\sinh 2\kappa h}\right) \tag{2.1.50}$$

Substituting Eq. (2.1.48) of the total average energy per unit area of the wave, the wave celerity $C = \omega/\kappa$ and Eq. (2.1.42) to Eq. (2.1.50), the next important relationship is obtained.

$$W = E\times C\times n = E\times C_g \tag{2.1.51}$$

The important finding "the mean energy of the wave is carried by the group velocity" is obtained from Eq. (2.1.51).

2.1 Small Amplitude Wave Theory

8) Standing Waves

When incident waves with the wave height H, the wave number κ, the angular frequency ω, and the water surface elevation η_I strike a vertical wall such as a breakwater at right angles and are reflected without energy loss, reflected waves with water surface elevation η_R are in the opposite direction and have the same wave height, wave number, and angular frequency. The composite water surface elevation η from the incident and reflected waves is expressed by the following equation.

$$\eta = \eta_I + \eta_R = \frac{H}{2}\cos(\kappa x - \omega t) + \frac{H}{2}\cos(\kappa x + \omega t) \quad (2.1.52)$$

$$= H\cos\kappa x \cos\omega t = H\cos\frac{2\pi x}{L}\cos\frac{2\pi t}{T}$$

The variation of the water surface elevation η according to Eq. (2.1.52) is shown in **Fig. 2.4**. Every half wavelength ($L/2$), there is a loop (= anti-node) where the water surface oscillates up and down with both amplitudes H to $-H$, and every middle of this half wavelength there is a node where the water surface does not vary. These composite waves are stationary waves whose waveforms do not progress and are called **standing waves**.

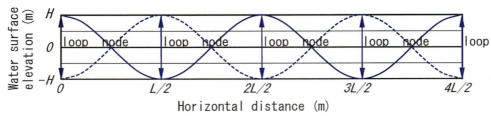

Fig. 2.4 Illustration of water surface variation between $t = 0$ and $T/2$ for standing waves.

Using Eq. (2.1.52), the velocity (u,w) of the water particle, the travel distance ($\xi = x-x_o$, $\zeta = z-z_o$) from its mean position (x_o,z_o), the hydrodynamic pressure P_d and the energy E per unit area of the wave can be found as follows

$$\left. \begin{array}{l} u = H\omega \dfrac{\cosh\kappa(h+z)}{\sinh\kappa h}\sin\kappa x \sin\omega t \\[2mm] w = -H\omega \dfrac{\sinh\kappa(h+z)}{\sinh\kappa h}\cos\kappa x \sin\omega t \end{array} \right\} \quad (2.1.53)$$

$$\left. \begin{array}{l} \xi = -H\dfrac{\cosh\kappa(h+z_0)}{\sinh\kappa h}\sin\kappa x_0 \cos\omega t \\[2mm] \zeta = H\dfrac{\sinh\kappa(h+z_0)}{\sinh\kappa h}\cos\kappa x_0 \cos\omega t \end{array} \right\} \quad (2.1.54)$$

$$P_d = \rho g H \frac{\cosh\kappa(h+z)}{\cosh\kappa h}\sin\kappa x \sin\omega t \quad (2.1.55)$$

$$E = \frac{1}{4}\rho g H^2 \quad (2.1.56)$$

2. WAVE THEORIES AND, PROPAGATION AND DEFORMATION

9) Partial Standing Waves

When incident waves with the wave height H_1, the wavenumber κ, the angular frequency ω, and the water surface elevation η_I hit an obstacle and are reflected with energy loss (partial reflection), reflected waves with water surface elevation η_R become waves with a different wave height H_2 in the opposite direction. The composite water surface elevation η from the incident and reflected waves is expressed by the following equation.

$$\begin{aligned}\eta = \eta_I + \eta_R &= \frac{H_1}{2}\cos(\kappa x - \omega t) + \frac{H_2}{2}\cos(\kappa x + \omega t) \\ &= \frac{1}{2}(H_1 + H_2)\cos\kappa x \cos\omega t + \frac{1}{2}(H_1 - H_2)\sin\kappa x \sin\omega t \\ &= \frac{1}{2}(H_1 - H_2)\cos(\kappa x - \omega t) + H_2 \cos\kappa x \cos\omega t\end{aligned} \quad (2.1.57)$$

The variation of the water surface elevation η according to Eq. (2.1.57) is a composite waveform of the standing waves consisting of a cosine function of the wave height (H_1+H_2) and the standing waves consisting of a sine function of the wave height (H_1-H_2), as shown in **Fig. 2.5**. It is also a composite waveform of travelling waves of the wave height (H_1-H_2) and standing waves of the cosine function of the wave height H_2, and is called **partial standing waves**.

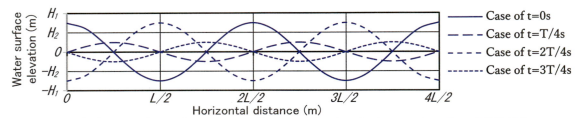

Fig. 2.5 Illustration of water surface variation between $t = 0$ and $3T/4$ for partial standing waves.

In partial standing waves, the maximum wave height H_{max} and minimum wave height H_{min} are as follows:

$$\left.\begin{aligned}\text{Maximum wave height:} \ H_{max} &= H_1 + H_2 \\ \text{Minimum wave height:} \ H_{min} &= H_1 - H_2\end{aligned}\right\} \quad (2.1.58)$$

Therefore, if the maximum wave height H_{max} and minimum wave height H_{min} can be measured, the incident wave height H_1 and the reflected wave height H_2 can be obtained from the following equation.

$$H_1 = \frac{H_{max} + H_{min}}{2}, \quad H_2 = \frac{H_{max} - H_{min}}{2} \quad (2.1.59)$$

2.1 Small Amplitude Wave Theory

Here, the basic dimensionless quantities related to waves are introduced.

ⅰ) Wave steepness:

The wave steepness **H/L**, is the wave height divided by the wavelength and represents the steepness of the wave. Wind waves that develop in a wind area due to strong winds are waves with a larger wave steepness than swells that decay from wind waves as they exit the wind area or as the strong winds cease. Waves with smaller wave steepness (longer waves, typically tsunamis) are more likely to run up higher on shore and have larger reflected waves.

ⅱ) Relative water depth:

The relative water depth **h/L**, is the water depth divided by the wavelength, and indicates the strength of wave dispersion. *h/L > 1/2* is called deep sea waves (= deep water waves), *1/2 > h/L > 1/20* intermediate sea waves (= intermediate water waves) and *1/20 > h/L* shallow sea waves (= shallow water waves) or long waves.

ⅲ) Relative wave height or wave height-to-depth ratio:

The relative wave height (wave height-to-depth ratio) **H/h**, is the wave height divided by the water depth and indicates the strength of the wave nonlinearity; when *H/h > 1/5*, the nonlinearity cannot be ignored.

ⅳ) Ursell number:

The Ursell number is a dimensionless quantity expressed in Eq. (e.1); the larger this value, the smaller the water depth relative to the wave size, the stronger the influence of the seabed, or the steeper the waves become, meaning they are outside the scope of application of small amplitude wave theory. When the Ursell number exceeds 2, the wave non-linearity becomes stronger, and it is better to switch from small amplitude wave theory to finite amplitude wave theory.

$$Ur = \frac{HL^2}{h^3} = \frac{H/h}{\left(h/L\right)^2} = \frac{H/L}{\left(h/L\right)^3} \tag{e.1}$$

If $L = \sqrt{gh} \times T$ is substituted into Eq. (e.1) for application in shallow waters, the Ursell number becomes as follows:

$$Ur = \frac{gHT^2}{h^2} \tag{e.2}$$

2.WAVE THEORIES AND, PROPAGATION AND DEFORMATION

2.2 Finite Amplitude Wave Theory
1) Characteristics and Applicable Range of Finite Amplitude Wave Theory

Finite amplitude wave theory is a theory that does not assume small wave heights and does not neglect the nonlinear terms in the equations of motion. In addition to the **Stokes wave theory** and the **Cnoidal wave theory**, which are based on the perturbation method to solve the nonlinear equations of motion, there is the **Stream Function Method (SFM)**, which estimates the series solution of the stream function from the Stokes wave solution and determines the constants of the series solution from the boundary conditions at the water surface.

Representative examples of the waveforms of Stokes waves, cnoidal waves, waves using the stream function method (SFM), and Airy waves (small amplitude waves) by Horikawa [3] are shown in **Fig. 2.6**, where the waveforms of the three types of finite amplitude waves have sharp peaks and flat valleys compared to those of small amplitude waves.

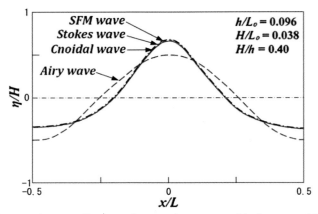

Fig. 2.6 Comparison of waveforms between finite-amplitude and small-amplitude waves (from Horikawa [3]).

According to Horikawa [3], the applicable range of each theory for finite amplitude waves classified by the relative wave height, the relative water depth, and the Ursell number is shown in **Fig. 2.7**.

In this figure, the Stokes wave is the fifth-order approximate solution, the cnoidal wave is the third-order approximate solution and SFM19 is the 19th-order approximate solution using SFM. The wave breaking limit (the relative wave height at which waves can exist without breaking) according to Goda [4] is shown in this figure as a single dotted line for seabed slopes $\tan\beta = 1/10, 1/20, 1/30$, and $1/50$ to 0. The wave breaking limits for $\tan\beta = 1/50$ to 0 coincide with the wave height limits calculated numerically by Yamada and Shiotani [5].

According to this figure, the Stokes waves (5th order approximate solution) is highly applicable on the lower right side and the cnoidal waves (3rd order approximate solution) on the upper left side

2.2 Finite Amplitude Wave Theory

of the figure, bounded by Ursell number Ur =25. The SFM (19th-order approximate solution) is applicable for the whole area below the breaking limit line with the seabed slope $\tan\beta$ = 1/50 to 0, especially for $H/h > 0.4$, only the SFM (19th-order approximate solution) is applicable.

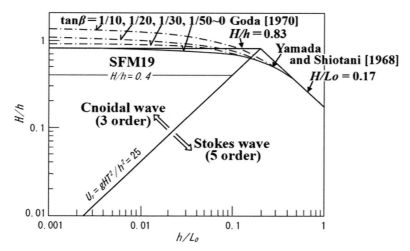

Fig. 2.7 Applicable scope of each theory of finite amplitude waves (from Horikawa [3]).

As numerical models for obtaining accurate numerical solutions from the governing equations for arbitrary seabed topography have become more widespread, the use of analytical solutions based on finite amplitude wave theory has become less common, so only an overview of Stokes and cnoidal waves is presented below.

2) Stokes Wave Theory

Stokes waves, first analysed by Stokes [6] in 1847, are finite-amplitude waves obtained under the conditions of vortex-free motion and sufficiently small wave steepness and can be adapted to small relative wave height and large relative water depth. According to Isobe et al [7], the fifth-order approximate solution for the water surface elevation η above the still water surface and the horizontal velocity u and vertical velocity w of water particles can be expressed as follows:

$$\eta = \sum_{n=1}^{5} \frac{A_n}{\kappa} \cos n(\kappa x - \omega t) \tag{2.2.1}$$

$$u = -B_0 C_0 + \sum_{n=1}^{5} B_n C_0 \cosh n\kappa (h+z) \cos n(\kappa x - \omega t) \tag{2.2.2}$$

$$w = \sum_{n=1}^{5} B_n C_0 \sinh n\kappa (h+z) \sin n(\kappa x - \omega t) \tag{2.2.3}$$

where A_n and B_n are the coefficients obtained by Isobe et al (refer to Horikawa [3]), κ is the wave number, ω is the angular frequency, C_0 is the wave celerity in small amplitude wave theory, h is the water depth, x is the horizontal coordinate, z is the vertical coordinate, and t is time.

2. WAVE THEORIES AND, PROPAGATION AND DEFORMATION

Although the solution by Isobe et al. is highly accurate, the computational method is complicated, so a third-order approximate solution in shallow waters by Skjelbreia [8], which can be calculated relatively easily on a personal computer, is presented below.

$$\eta = a\cos(\kappa x - \omega t) + \frac{a^2 \kappa}{2} \frac{\cosh \kappa h (\cosh 2\kappa h + 2)}{2(\sinh \kappa h)^3} \cos 2(\kappa x - \omega t)$$

$$+ \frac{a^3 \kappa^2}{4} \frac{24(\cosh \kappa h)^6 + 3}{16(\sinh \kappa h)^6} \cos 3(\kappa x - \omega t) \qquad (2.2.4)$$

$$C = \sqrt{\frac{g}{\kappa} \tanh \kappa h \left\{ 1 + (a\kappa)^2 \frac{\cosh 4\kappa h + 8}{8(\sinh \kappa h)^4} \right\}} \qquad (2.2.5)$$

$$L = \frac{gT^2}{2\pi} \tanh \kappa h \left\{ 1 + (a\kappa)^2 \frac{\cosh 4\kappa h + 8}{8(\sinh \kappa h)^4} \right\} \qquad (2.2.6)$$

where a is the amplitude of the small amplitude wave and its relation to the wave height H of the finite amplitude wave is expressed by Eq. (2.2.7). C is the wave celerity, L is the wavelength, g is the acceleration of gravity, and T is the wave period.

$$H = 2a + \frac{3a^3 \kappa^2 \left\{ 8(\cosh \kappa h)^6 + 1 \right\}}{32(\sinh \kappa h)^6} \qquad (2.2.7)$$

Although the first approximate solution of Stokes waves equals the solution of small amplitude waves, Eqs. (2.2.5) and (2.2.6) show that the wave celerity and wavelength of third-order approximate solution of Stokes waves are larger than those of the small amplitude waves. In addition, as shown in **Fig. 2.8**, the waveform of third-order approximate solution of Stokes waves is sharper at the peaks and flatter at the troughs than that of the small amplitude wave. Furthermore, the motion orbit of the water particle in third-order approximate solution of Stokes waves does not follow an elliptical path as in the case of the small amplitude wave, but a spiral path, moving slightly in the direction of wave travel at each cycle. In other words, in areas where Stokes waves exist, the seawater is moving in the direction of wave travel, and suspended matter in the water is also transported little by little. This phenomenon is called the **drift phenomenon** or **mass transport**, and the **mass transport velocity** of a water particle is expressed by Eq. (2.2.8).

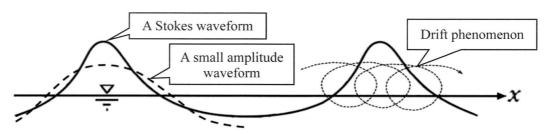

Fig. 2.8 Illustration of Stokes waveforms and the drift phenomenon.

$$U_m = \frac{1}{2} C a^2 \kappa^2 \frac{\cosh 2\kappa (h+z_0)}{(\sinh kh)^2} \qquad (2.2.8)$$

where U_m is the mass transport velocity [m/s], and z_0 is the height of the mean position of the water particles [m].

3) Cnoidal Wave Theory

Cnoidal waves, first analysed by Korteweg de Vries [9] in 1895, are finite-amplitude waves that are represented by Jacobi's elliptic function and can be adapted to small relative water depths and large relative wave heights. According to Isobe et al [7], the third-order approximate solution for the water surface elevation η above the still water surface and the horizontal velocity u and vertical velocity w of water particles can be expressed as follows:

$$\eta = \sum_{n=0}^{3} C_n h \times cn^{2n} \left(\frac{2K_1}{L} x - \frac{2K_1}{T} t, \lambda \right) \qquad (2.2.9)$$

$$u = \sum_{n=0}^{3} \sum_{m=0}^{2} \sqrt{gh} D_{nm} \left(\frac{h+z}{h} \right)^{2m} cn^{2n} \left(\frac{2K_1}{L} x - \frac{2K_1}{T} t, \lambda \right) \qquad (2.2.10)$$

$$\begin{aligned}
w = {} & cn \left(\frac{2K_1}{L} x - \frac{2K_1}{T} t, \lambda \right) \times sn \left(\frac{2K_1}{L} x - \frac{2K_1}{T} t, \lambda \right) \times dn \left(\frac{2K_1}{L} x - \frac{2K_1}{T} t, \lambda \right) \\
& \times \sum_{n=1}^{3} \sum_{m=0}^{2} \sqrt{gh} \frac{4nK_1 h/L}{2m+1} D_{nm} \left(\frac{h+z}{h} \right)^{2m+1} cn^{2(n-1)} \left(\frac{2K_1}{L} x - \frac{2K_1}{T} t, \lambda \right)
\end{aligned} \qquad (2.2.11)$$

where C_n and D_{nm} are the coefficients obtained by Isobe et al (refer to Horikawa [3]); h is the water depth; cn, sn, and dn are Jacobi's elliptic functions; λ is the parameter of the elliptic function; K_1 is the complete elliptic integral of the first kind; and z is the vertical coordinate.

Although the solution by Isobe et al. is highly accurate, the calculation method is complicated, so a second-order approximate solution by Laitone [10] (modified by Iwagaki and Sawaragi [11]), which can be calculated relatively easily on a personal computer, is presented below.

$$\eta = H cn^2 \left(\frac{2K_1}{L} x - \frac{2K_1}{T} t, \lambda \right) \left[1 - \frac{3}{4} \frac{H/h}{1-(\delta/H)(H/h)} \left\{ 1 - cn^2 \left(\frac{2K_1}{L} x - \frac{2K_1}{T} t, \lambda \right) \right\} \right] \qquad (2.2.12)$$

$$\begin{aligned}
C = {} & \sqrt{gh \left(1 - \frac{\delta}{H} \frac{H}{h} \right)} \left[1 + \frac{H/h}{1-(\delta/H)(H/h)} \frac{1}{\lambda^2} \left(\frac{1}{2} - \frac{K_2}{K_1} \right) \right. \\
& \left. + \frac{(H/h)^2}{\{1-(\delta/H)(H/h)\}^2} \frac{1}{\lambda^4} \left\{ \frac{K_2}{K_1} \left(\frac{K_2}{K_1} + \frac{3}{4} \lambda^2 - 1 \right) - \left(\frac{\lambda^4 + 14\lambda^2 - 9}{40} \right) \right\} \right]
\end{aligned} \qquad (2.2.13)$$

$$L = \frac{4\lambda K_1 h}{\sqrt{3}} \left(1 - \frac{\delta}{H} \frac{H}{h} \right)^{3/2} \left(\frac{h}{H} \right)^{1/2} \left\{ 1 - \frac{H/h}{1-(\delta/H)(H/h)} \frac{7\lambda^2 - 2}{8\lambda^2} \right\}^{-1} \qquad (2.2.14)$$

$$\frac{\delta}{H} = \frac{1}{\lambda^2} \left(\frac{K_2}{K_1} + \lambda^2 - 1 \right) + \frac{H/h}{1-(\delta/H)(H/h)} \frac{1}{4\lambda^4} \left\{ 2(1-\lambda^2) - (2-\lambda^2) \frac{K_2}{K_1} \right\} \qquad (2.2.15)$$

2.WAVE THEORIES AND, PROPAGATION AND DEFORMATION

where, H is the wave height, cn is Jacobi's elliptic function, K_1 is the complete elliptic integral of the first kind, K_2 is the complete elliptic integral of the second kind, λ is the parameter number, and h is the water depth.

Moreover, If the wavelength $L \to \infty$, i.e. the parameter $\lambda \to 1$, the complete elliptic integral $K_1 \to \infty$ and the complete elliptic integral $K_2 \to 1$ are assumed, a non-periodic finite amplitude wave with only one peak is obtained. This is called a **solitary wave**, and the water surface elevation η and wave celerity C are given by the equations below.

$$\eta = \frac{H}{\cosh^2 \zeta} \left\{ 1 - \frac{3}{4} \frac{H}{h} \left(1 - \frac{1}{\cosh^2 \zeta} \right) \right\} \tag{2.2.16}$$

$$\zeta = \sqrt{\frac{3H}{4h^3}} \left(1 - \frac{5}{8} \frac{H}{h} \right)(x - Ct) \tag{2.2.17}$$

$$C = \sqrt{gh} \left\{ 1 + \frac{1}{2} \frac{H}{h} - \frac{3}{20} \left(\frac{H}{h} \right)^2 \right\} \tag{2.2.18}$$

2.3 Wave Propagation and Deformation
1) Shallow Water Deformation and Refraction Deformation of Waves

Sea waves undergo refraction deformation in the same way as light waves when propagating towards the shore. As shown in **Fig.2.9**, the wave propagation direction lines (blue arrow lines in **Fig. 2.9**) align to be perpendicular to the depth contours. As a result, sea waves gather at the headland where the depth contour is convex offshore and spread at the back of the bay where the depth contour is concave, resulting in higher wave height at the headland than at the back of the bay. This phenomenon is called **wave refraction**. Moreover, the shallower the water depth, the slower the wave velocity becomes under conditions where wave energy is conserved, and thus the higher the wave height becomes. This phenomenon is called **wave shoaling**.

In **Fig. 2.9**, based on the conservation law of energy, because the total power of the wave at the spacing b_o of the offshore wave direction line coincides with the total power of the wave at the spacing b of the shore side wave direction line, the following equation is obtained.

$$\frac{1}{8}\rho g H^2 C_g \times b = \frac{1}{8}\rho g H_o^2 C_{go} \times b_o \quad (2.2.19)$$

where H is the wave height on the shore side, H_o is the wave height offshore (referred to as **offshore wave height**), C_g is the group velocity on the shore side, C_{go} is the group velocity offshore, b is the wave line spacing on the shore side, and b_o is the wave line spacing offshore. Eq. (2.2.19) can be transformed as follows:

$$\frac{H}{H_o} = \sqrt{\frac{C_{go}}{C_g}}\sqrt{\frac{b_o}{b}} = K_s \times K_r \quad (2.2.20)$$

As the group velocity slows down as the water depth gets shallower, $\sqrt{C_{go}/C_g}$ indicates that the closer the waves are to the shore, the higher the wave height, which is denoted by K_s and referred to as the **shoaling coefficient**. Substituting the formula defining the group velocity into Eq. (2.2.20), the shoaling coefficient can be obtained from the following equation.

$$K_s = \sqrt{C_{go}/C_g}$$
$$= \frac{1}{\sqrt{\left[1+\dfrac{4\pi h/L}{\sinh(4\pi h/L)}\right]\tanh\left(\dfrac{2\pi}{L}h\right)}} \quad (2.2.21)$$

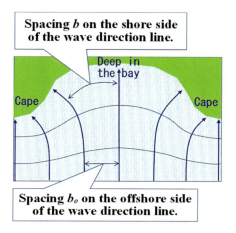

Fig. 2.9 Illustration of wave refraction and shoaling.

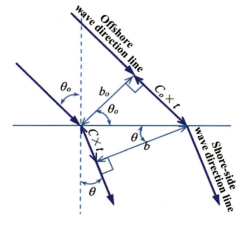

Fig. 2.10 Illustration of refraction deformation.

2. WAVE THEORIES AND, PROPAGATION AND DEFORMATION

$\sqrt{b_o/b}$ indicates that the wave height becomes lower as the wave direction line spacing increases, and this value is denoted by K_r and referred to as the **refraction coefficient**.

In **Fig. 2.10**, the following geometrical relationship is obtained between the wave direction angle θ_o, the wave velocity C_o and the wave direction line spacing b_o on the offshore side of the depth contour, and the wave direction angle θ, the wave velocity C, and the wave direction line spacing b on the shore side.

$$\frac{\sin\theta_o}{C_o} = \frac{\sin\theta}{C} \text{ (Snell's law)}, \qquad \frac{b_o}{\cos\theta_o} = \frac{b}{\cos\theta} \qquad (2.2.22)$$

Substituting the offshore wave direction angle θ_o, the wave velocity C_o, the wave direction spacing b_o and the shore side wave velocity C for the target depth contour into Eq. (2.2.22), the shore side wave direction angle θ and the wave direction spacing b can be obtained. Eqs. (2.2.21) and (2.2.20) can then be used to obtain the shoaling coefficient K_s and the refraction coefficient K_r. By proceeding to the next depth contour on the shore side, the wave direction angle, the shoaling coefficient, and the refraction coefficient can be obtained one after another.

However, the accuracy of the small amplitude wave theory is poor in extremely shallow waters due to the strong nonlinearity, so the relationship between the shoaling coefficient and relative water depth obtained by Goda [13] based on Suto's nonlinear shallow water theory [12] is shown in **Fig. 2.11**. In this figure, the wave steepness $H_o/L_o = 0$ for offshore waves corresponds to the value of the small amplitude wave theory.

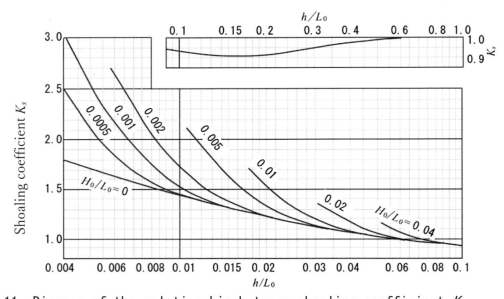

Fig. 2.11 Diagram of the relationship between shoaling coefficient K_s and relative water depth based on non-linear shallow water theory [12].

Furthermore, actual sea waves are not regular, but irregular, and the superposition of the refraction deformation of component waves shows non-negligible differences from the values

2.3 Wave Propagation and Deformation

obtained by the theoretical equation of regular waves for the significant wave parameters. It is therefore advisable to use the SWAN model introduced in Section 1.2 or the numerical models for irregular waves introduced in Section 2.4.

For engineers who cannot use such numerical models, Goda and Suzuki [14] obtained the refraction coefficient K_r and the predominant refracted wave direction α_p, which is a superposition of component waves with a myriad of frequencies and wave directions, using a combination of Bretschneider-Mitsuyasu frequency spectrum and Mitsuyasu-type directional distribution function for the direction spectrum, based on Eq. (2.2.23). It should be noted that their formulations are applicable to coasts with parallel seabed contours only. The relationship of the refraction coefficient and the predominant refracted wave direction to the relative water depth is then shown in **Figs. 2.12** and **2.13** respectively.

$$\left. \begin{aligned} K_r &= \sqrt{\frac{1}{m_{so}} \int_0^\infty \int_{\theta_{min}}^{\theta_{max}} S(f,\theta) \times K_s^2(f) \times K_r^2(f,\theta) \, d\theta df} \\ m_{so} &= \int_0^\infty \int_{\theta_{min}}^{\theta_{max}} S(f,\theta) \times K_s^2(f) \, d\theta df \end{aligned} \right\} \qquad (2.2.23)$$

where $S(f,\theta)$ is the directional spectrum, $K_s(f)$ is the shoaling coefficient of component (regular) waves, and $K_r(f,\theta)$ is the refraction coefficient of component (regular) waves.

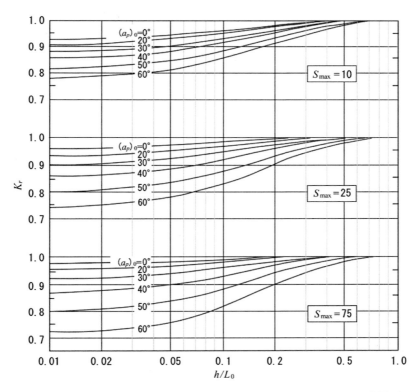

Fig. 2.12 Diagram of the relationship between the refraction coefficient K_r and relative water depth for irregular waves according to Goda and Suzuki [14].

2.WAVE THEORIES AND, PROPAGATION AND DEFORMATION

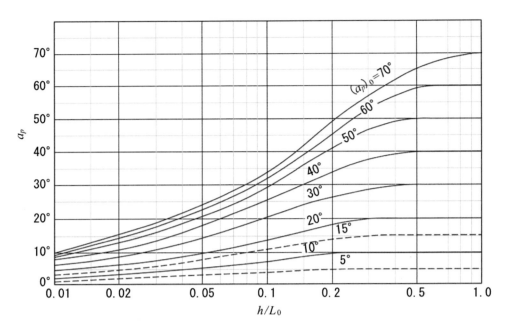

Fig. 2.13 Diagram of the relationship between the predominant refraction wave direction α_p and relative water depth for irregular waves by Goda and Suzuki [14].

2) Diffraction Deformation of Waves

If there is a barrier in the sea, such as an island or a breakwater, which prevents waves from travelling towards the shore, as in the case of light, wave wrapping-around occurs in the shielded area. However, because the supply of wave energy into the shielded area is restricted, the wave height decreases as the waves move further into the shielded area. This phenomenon is called **wave diffraction**. The ratio of the wave height on the shore side of the barrier, excluding the effects of shoaling and refraction deformation, to the offshore wave height is denoted by K_d and called the **diffraction coefficient**.

Substituting the wave solution into the Laplace equation (equation of continuity), which assumes the existence of a velocity potential, gives the Helmholtz equation, and the deformation of regular waves due to diffraction phenomena can be found as a solution to the Helmholtz equation (Sommerfeld's solution). However, the diffraction deformation of irregular waves is also considerably different from that of regular waves. Therefore, the irregularity should be considered and is better calculated using the numerical models for irregular waves presented in Section 2.4.

For engineers who cannot use such numerical models, Goda and Suzuki [14] obtained the diffraction coefficient K_d for a superposition of component waves with a myriad of frequencies and wave directions using a combination of the Bretschneider-Mitsuyasu frequency spectrum and Mitsauyasu-type directional distribution function for the direction spectrum, based on Eq. (2.2.24). The distribution of the diffraction coefficients obtained for a semi-infinite length breakwater is shown

2.3 Wave Propagation and Deformation

by solid lines in **Fig.2.14**, and the case of waves of wavelength L entering a breakwater with opening width B (=$4 \times L$) is shown as solid lines on the right side of **Figs.2.15** and **2.16**.

$$\left. \begin{array}{c} K_d = \sqrt{\dfrac{1}{m_o} \int_0^\infty \int_{\theta_{\min}}^{\theta_{\max}} S(f,\theta) \times K_d^{\,2}(f,\theta) \, d\theta df} \\ m_o = \int_0^\infty \int_{\theta_{\min}}^{\theta_{\max}} S(f,\theta) \, d\theta df \end{array} \right\} \quad (2.2.24)$$

where $S(f,\theta)$ is the directional spectrum, and $K_d(f,\theta)$ is the diffraction coefficient of the component waves (regular waves).

Furthermore, when the directional function varies with wave frequency, the wave period T in the shielded area also varies regardless of the wave period T_o on the offshore side of the breakwater. Goda and Suzuki [14] obtained the period ratio T/T_o using the same directional spectrum based on Eq. (2.2.25). The distribution of the period ratios for the breakwater of semi-infinite length are shown as dashed lines in **Fig.2.14** and for the breakwater with opening width B (=$4 \times L$) as solid lines on the left side of **Figs.2.15** and **2.16**.

$$\frac{T}{T_o} = \sqrt{\frac{\sqrt{1.029\pi} \int_0^\infty \int_{\theta_{\min}}^{\theta_{\max}} S(f,\theta) \times K_d^{\,2}(f,\theta) \, d\theta df}{\int_0^\infty \int_{\theta_{\min}}^{\theta_{\max}} f^2 S(f,\theta) \times K_d^{\,2}(f,\theta) \, d\theta df}} \quad (2.2.25)$$

Note that improved diffraction distribution diagrams for a breakwater of semi-infinite length and a breakwater with opening width B (four types of waves with wavelengths $L=B$, $B/2$, $B/4$, and $B/8$) are given in Goda [15].

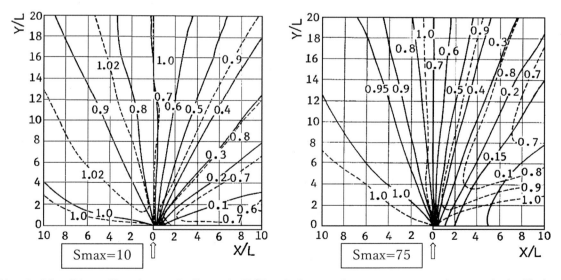

Fig. 2.14 Distribution of K_d and T/T_o of irregular waves at the semi-infinite length breakwater by Goda and Suzuki [14].
(Solid lines in the figures are K_d, dashed lines are T/T_o, X and Y are the distances from the breakwater diffraction start point, L is the wavelength)

2. WAVE THEORIES AND, PROPAGATION AND DEFORMATION

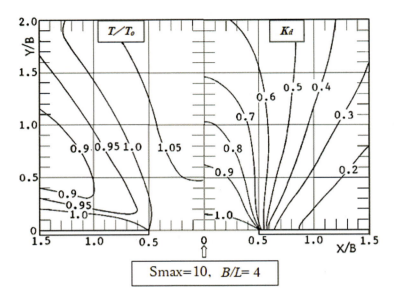

Fig. 2.15 Distribution of K_d and T/T_o for irregular waves of Smax=10 with opening width B according to Goda and Suzuki [14].

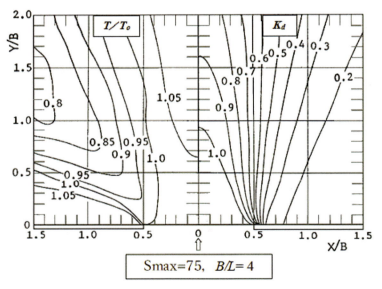

Fig. 2.16 Distribution of K_d and T/T_o for irregular waves of Smax=75 with opening width B according to Goda and Suzuki [14].

From the above, the shore-side significant wave height $H_{1/3}$, after considering shoaling, refraction and diffraction deformation of waves, is expressed by the following equations as a function of the offshore significant wave height H_o.

$$H_{1/3} = K_s \times K_r \times K_d \times H_o \tag{2.2.26}$$

$$H_o' = K_r \times K_d \times H_o \tag{2.2.27}$$

$$H_{1/3} = K_s \times H_o' \tag{2.2.28}$$

where H_o' is the offshore wave height, which considers the refraction and diffraction deformations in two horizontal dimensions, expressed in Eq. (2.2.27), and is called **equivalent offshore wave height**. Therefore, $H_{1/3}$ can also be expressed in Eq. (2.2.28), after considering the shoaling deformation in the vertical direction.

3) Transmission and Reflection of Waves

If there is a transmissive obstacle on the way of wave propagation in the sea, the transmitted wave (refraction angle θ_T) and the reflected wave (reflection angle $\theta_R = \theta_I$) are generated from an incident wave with an incident angle θ_I, as shown in **Fig.2.17** and the loss of wave energy occurs in the wave propagation route. The energy balance can be given as follows:

$$\left(EC_g\right)_I = \left(EC_g\right)_T + \left(EC_g\right)_R + \Delta E_C \qquad (2.2.29)$$

where E is the average energy per unit area of the wave ($= \rho_w g H^2 / 8$), C_g is the group wave velocity, ΔE_C is the amount of energy loss of the wave, and the subscripts I, T, and R stand for the quantities related to the incident wave, transmitted wave, and reflected wave respectively.

In Eq. (2.2.29), since all group velocities can be regarded as the same when there is no difference in water depth, dividing both sides by the energy transport of the incident wave gives the following equation.

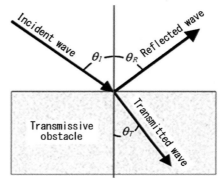

Fig. 2.17 Illustration of transmitted and reflected waves.

$$1 = \left(H_T/H_I\right)^2 + \left(H_R/H_I\right)^2 + \left(\Delta E_C \Big/ \frac{1}{8}\rho_w g H_I^2 C_g\right) \qquad (2.2.30)$$

where, H_T/H_I is referred to as the **transmission coefficient** and is denoted by K_T. H_R/H_I is called the **reflection coefficient** and is denoted by K_R. Furthermore, the third term on the right-hand side is called the **energy loss rate** and is expressed as follows:

$$\left(\Delta E_C \Big/ \frac{1}{8}\rho_w g H_I^2 C_g\right) = K_{loss} \qquad (2.2.31)$$

Eq. (2.2.30) is therefore expressed as follows:

$$1 = K_T^2 + K_R^2 + K_{loss} \qquad (2.2.32)$$

(1) Wave Transmission Coefficient

When installing wave-dissipative structures, the transmission coefficient needs to be estimated in advance. The evaluation methods for the main wave-dissipative structures are described below.

2. WAVE THEORIES AND, PROPAGATION AND DEFORMATION

(a) Case of a Sloping Breakwater of Deformed Concrete Blocks

The structure shown in **Fig. 2.18**, consisting of deformed concrete blocks stacked on a rubble foundation, is called the **wave absorbing breakwater** or the **block mound breakwater** when it is built for wave breaking, and the **detached breakwater** when made to store sand in the backshore area or to restore the sandy beach.

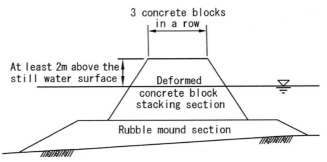

Fig. 2.18　Illustration of the sloping breakwater.

The top of a breakwater is called the **crown** or the **crest**. In the case of wave absorbing breakwaters and detached breakwaters, the height above the water surface to the crest should be at least 2 m to ensure a sufficient wave-absorbing effect, and the width at the crest should be generally three concrete blocks in a row so that they mesh tightly together. The wave transmission coefficient (K_T) in this case includes the component of waves overtopping the crest of the breakwater. Transmission coefficients by Hattori [16] are shown in **Fig.2.19** containing data observed on local coasts (for transmission coefficients, a comparison with field observation data is important, as the dimensional effect of an experimental model cannot be ignored).

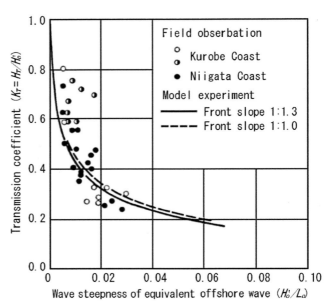

Fig. 2.19　Relationship between transmission coefficient and wave steepness for wave absorbing breakwaters and detached breakwaters. (from Hattori [16])

Moreover, Numata [17] proposed the following equation to obtain the transmission coefficient of regular waves for the case where the crest of wave-absorbing breakwaters and detached breakwaters are so high that overtopping does not occur and the transmitted waves are only the component passing through the breakwater.

$$K_T = \frac{1}{\left[1 + 1.135\left(B_{SWL}/D_{eff}\right)^{0.66}\sqrt{H_I/L}\right]^2} \quad (2.2.33)$$

2.3 Wave Propagation and Deformation

where B_{SWL} is the width of the breakwater at the still water surface, D_{eff} is the effective height of the deformed concrete block, H_I is the incident wave height, and L is the wavelength.

Note that the applicable range of this equation is $3 < B_{SWL}/D_{eff} < 10$.

Furthermore, Takayama et al [18] proposed the following equation to obtain the transmission coefficient of irregular waves for wave-absorbing breakwaters and detached breakwaters with wide crown widths and low crown heights consisting of deformed concrete blocks.

$$\frac{H_T}{H_{in}} = -0.92\frac{B}{L_o} - 0.42\frac{h_c}{H_o{'}} + 3.8\frac{H_o{'}}{L_o} + 0.51 \qquad (2.2.34)$$

where B is the crown width, h_c is the crown height from the still water surface, $H_o{'}$ is the equivalent offshore wave height, and L_o is the offshore wavelength.

Note that the transmission coefficient in this equation is the non-dimensionalised value of the transmitted wave height H_T in terms of the incident wave height H_{in} at the position of the breakwater.

(b) Case of Two-tiered Rubble Breakwater

Van der Meer et al [19] analysed experimental irregular wave data from the EU DELOS project and proposed the following two formulae for the transmission coefficient $K_T (=H_T/H_I)$ of a two-tiered rubble breakwater (large and small stones) with a low crown height shown in **Fig.2.20**, for the relative crown width $B/H_I = 10$.

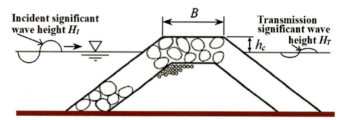

Fig. 2.20   Illustration of the two-tiered rubble breakwater (large stones and small stones).

$$\frac{B}{H_I} \leq 10: \quad K_T = \min\left\{\begin{array}{l}\left[-0.006(B/H_I)+0.93\right], \\ \max\left[0.05, -0.4\frac{h_c}{H_I}+0.64\left(\frac{B}{H_I}\right)^{-0.31} \times\left(1-e^{-0.5\xi}\right)\right]\end{array}\right. \qquad (2.2.35)$$

$$\frac{B}{H_I} \geq 10: \quad K_T = \min\left\{\begin{array}{l}\left[-0.006(B/H_I)+0.93\right], \\ \max\left[0.05, -0.35\frac{h_c}{H_I}+0.51\left(\frac{B}{H_I}\right)^{-0.65} \times\left(1-e^{-0.41\xi}\right)\right]\end{array}\right. \qquad (2.2.36)$$

where, B is the crown width, h_c is the crown height above the still water surface, and H_I is the incident wave height. ξ is called the **Iribarren number** and is defined using the slope gradient β of a breakwater section and the front wavelength L as follows:

$$\xi = \frac{\tan\beta}{\sqrt{H_I/L}} \qquad (2.2.37)$$

2. WAVE THEORIES AND, PROPAGATION AND DEFORMATION

(c) Composite Breakwaters

The breakwater shown in the upper right-hand corner of **Fig.2.21** consists of upright impermeable bodies (concrete caissons or concrete rectangular blocks) with the depth d below the still water surface and the crown height h_c above the still water surface, on the rubble foundation built at the front water depth h and is called the **composite breakwater**. In this case, the transmission coefficient is dominated by overtopping components of waves, rather than the components passing through the breakwater.

Goda [20] considered the relative crown height h_c/H_I and the depth ratio d/h as the main parameters of the transmission coefficient K_T (H_T/H_I) and obtained the relationship shown in **Fig.2.21**. Although this graph was obtained from regular wave experiments, Goda et al [21] confirmed that it was also in close agreement with experimental values from irregular waves. According to Goda et al [21], the ratio T_T/T_I between the period T_T of transmitted waves and the period T_I of incident waves decreases to a value between 1.0 and 0.5 because of the generation of short period waves when the overtopping water mass hits the water surface inside the breakwater.

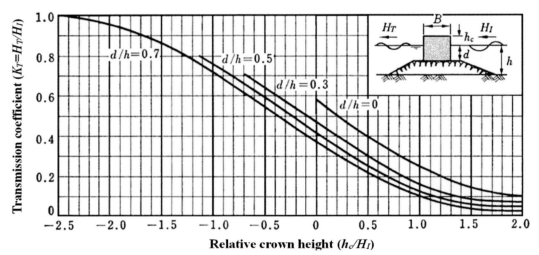

Fig. 2.21 Diagram of the relationship between the wave transmission coefficient K_T and relative crown height h_c/H_I according to Goda [20].

Using regular wave experiments, Kondo and Sato [22] obtained Eq. (2.2.38) to express the coefficient K_T for composite breakwaters and Eqs. (2.2.38) and Eq. (2.2.39) to express that for composite breakwaters with wave absorbers. These expressions are also applicable to irregular waves.

Composite breakwaters without wave absorbers:
$$K_T = 0.3 \times (1.5 - h_c/H_I) \qquad [0.0 < h_c/H_I \leq 1.25] \qquad (2.2.38)$$

Composite breakwaters with wave absorbers:
$$K_T = 0.3 \times (1.1 - h_c/H_I) \qquad [0.0 < h_c/H_I \leq 0.75] \qquad (2.2.39)$$

2.3 Wave Propagation and Deformation

(2) Wave Reflection Coefficient

Wave data measured closer the shore or in the vicinity of coastal structures always include reflected waves, and the method for determining the wave reflection coefficient is presented below.

The reflection coefficient for an upright impermeable breakwater is proportional to the crown height above the water surface and ranges from $K_R = 0.6$ to 1.0. The reflection coefficient for a deformed concrete block breakwater and a rubble breakwater is $K_R = 0.3 - 0.6$, and that for sandy beaches is $K_R = 0.05 - 0.2$.

(a) Healy's Calculation Method

The reflection coefficient K_R of regular waves is obtained from the relationship in Eqs. (2.1.57) to (2.1.59), using the maximum and minimum wave heights H_{max} and H_{min} of measured partial standing waves, as follows:

$$K_R = \frac{H_R}{H_I} = \frac{H_{max} - H_{min}}{H_{max} + H_{min}} \tag{2.2.40}$$

This method of determining the reflection coefficient is called **Healy's method**.

(b) Calculation Method of Goda et al.

The method of Goda et al [21] for obtaining the reflection coefficient K_R of regular and irregular waves is introduced below.

Goda et al. assumed that under multiple reflections of regular waves, the measured waves can be represented by the incident waveform η_I and the reflected waveform η_R shown in Eq. (2.2.41), and that the waveform η_1 at measurement position No. 1 and η_2 at measurement position No. 2, which is about l metres away, can be represented by Eq. (2.2.42).

$$\left. \begin{aligned} \eta_I &= \frac{H_I}{2} \cos\left(\kappa x - \omega t + \varepsilon_I\right) \\ \eta_R &= \frac{H_R}{2} \cos\left(\kappa x + \omega t + \varepsilon_R\right) \end{aligned} \right\} \tag{2.2.41}$$

$$\left. \begin{aligned} \eta_1 &= \left(\eta_I + \eta_R\right)_1 = A_1 \cos\left(\omega t\right) + B_1 \sin\left(\omega t\right) \\ \eta_2 &= \left(\eta_I + \eta_R\right)_2 = A_2 \cos\left(\omega t\right) + B_2 \sin\left(\omega t\right) \end{aligned} \right\} \tag{2.2.42}$$

where H_I is the incident wave height, H_R is the reflected wave height, κ is the wavenumber, ω is the angular frequency, ε_I is the phase angle of the incident wave, ε_R is the phase angle of the reflected wave, A_i and B_i are the amplitudes of the cosine and sine waves of the waveform η_i respectively.

From the relationship between Eqs. (2.2.41) and (2.2.42), the incident and reflected wave heights can be expressed by Eq. (2.2.43). Here, the wavenumber κ can be determined from the small amplitude wave theory, and the distance l between measurement positions No. 1 and No. 2 must be set so that $\sin(\kappa l)$ is not close to zero.

2.WAVE THEORIES AND, PROPAGATION AND DEFORMATION

$$\left.\begin{aligned}
\frac{H_I}{2} &= \frac{\left[\left(A_2 - A_1 \cos \kappa l - B_1 \sin \kappa l\right)^2 + \left(B_2 + A_1 \sin \kappa l - B_1 \cos \kappa l\right)^2\right]^{1/2}}{2|\sin \kappa l|} \\
\frac{H_R}{2} &= \frac{\left[\left(A_2 - A_1 \cos \kappa l + B_1 \sin \kappa l\right)^2 + \left(B_2 - A_1 \sin \kappa l - B_1 \cos \kappa l\right)^2\right]^{1/2}}{2|\sin \kappa l|}
\end{aligned}\right\} \tag{2.2.43}$$

The actual measured waveforms η_1 and η_2 contain frequency components that are twice, three times, ..., n times higher than the fundamental periodic component expressed in Eq. (2.2.42), but if the amplitudes A_i and B_i are obtained from Fourier Analysis, the measured waves can be separated into incident waves and reflected waves using Eqs. (2.2.41) and (2.2.43).

In the case of irregular waves, they can be considered as a superposition of numerous component waves (regular waves), and the above calculation method can be applied to each component wave and the results synthesised. The FFT method is a suitable method for breaking up irregular waves into their component waves, and the calculated cosine and sine wave amplitudes can be used to obtain the respective incident and reflected wave heights from Eq. (2.2.43). If the frequency spectra of the incident and reflected waves are then obtained, the average reflection coefficient can be obtained from the following Eqs. (2.2.44) and (2.2.45).

$$\left.\begin{aligned}
E_I &= \int_{f_{\min}}^{f_{\max}} S_I(f)\,df = \int_{f_{\min}}^{f_{\max}} \left\{ \frac{1}{2}\left[\frac{H_I(m)}{2}\right]^2 n\Delta t \right\} df \\
E_R &= \int_{f_{\min}}^{f_{\max}} S_R(f)\,df = \int_{f_{\min}}^{f_{\max}} \left\{ \frac{1}{2}\left[\frac{H_R(m)}{2}\right]^2 n\Delta t \right\} df
\end{aligned}\right\} \tag{2.2.44}$$

$$K_R = \sqrt{E_R / E_I} \tag{2.2.45}$$

where $f_{\max}$ is the maximum frequency of the frequency band present, $f_{\min}$ is the minimum frequency of the frequency band present, H_I (m) is the incident wave height of the mth component wave expressed as a Fourier series, H_R (m) is the reflected wave height of the same component wave, n is the number of wave measurement data, and Δt is the measurement time interval.

For the distance l between measurement positions No. 1 and No. 2, the following conditions must be satisfied in order not to reduce the calculation accuracy.

$$\left. \text{The lower limit: } \frac{l}{L_{\max}} = 0.05, \quad \text{The upper limit: } \frac{l}{L_{\min}} = 0.45 \right\} \tag{2.2.46}$$

where $L_{\max}$ is the wavelength corresponding to the minimum frequency in the frequency band present, and $L_{\min}$ is the wavelength corresponding to the maximum frequency in the same frequency band.

The incident wave height H_I and reflected wave height H_R of the measured (irregular) waves can be obtained from the following equations using the representative wave height H_c and the average reflection coefficient K_R of the measured waves. For example, if the significant wave height of the measured waves is used for H_c, the significant wave heights of the incident and reflected waves can be obtained.

2.3 Wave Propagation and Deformation

$$H_I = \frac{1}{\sqrt{1+K_R^2}} \times H_c, \quad H_R = \frac{K_R}{\sqrt{1+K_R^2}} \times H_c \quad (2.2.47)$$

(c) Reflection Coefficient on a Uniform Slope

Miche [23] proposed a method for determining the reflection coefficient K_R on the uniform slope using the following equation consisting of a correction factor χ_1 due to slope roughness and permeability and a coefficient χ_2 relating the slope gradient to the reflection coefficient.

$$K_R = \chi_1 \chi_2 \quad (2.2.48)$$

where the coefficient χ_1 is 1.0 for impermeable smooth surfaces, $\chi_1 = 0.7 - 0.9$ for impermeable rough surfaces, $\chi_1 \approx 0.8$ for sandy beaches, and $\chi_1 = 0.3 - 0.6$ for rubble slopes based on regular wave model experiments.

The coefficient χ_2 relates to the energy loss due to wave breaking and is given as follows:

$$\chi_2 = \begin{cases} \dfrac{(H_0/L_0)_c}{H_0/L_0} & : H_0/L_0 > (H_0/L_0)_c \\ 1 & : H_0/L_0 \leq (H_0/L_0)_c \end{cases} \quad (2.2.49)$$

where, $(H_0/L_0)_c$ is the limiting wave steepness of deep-sea waves that are perfectly reflected from the slope and is expressed by the following equation according to small amplitude wave theory. In the equation, β is the angle (in radians) between the horizontal plane and the uniform slope.

$$(H_0/L_0)_c = \sqrt{\frac{2\beta}{\pi} \frac{\sin^2 \beta}{\pi}} \quad (2.2.50)$$

Greslou and Mahe [24] also obtained the reflection coefficient on a uniform slope due to regular waves by Healy's method and obtained **Fig.2.22**. This figure shows that as the slope gradient decreases, the reflection coefficient decreases rapidly due to wave breaking on the slope.

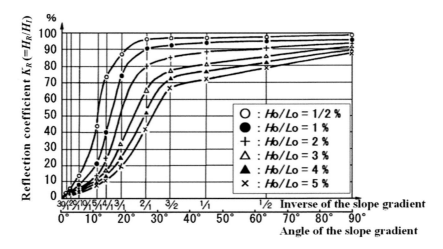

Fig. 2.22 Reflection coefficient on uniform slopes based on regular wave experiments by Greslou and Mahe [24].

2. WAVE THEORIES AND, PROPAGATION AND DEFORMATION

Note that in cases with large wave steepnesses, the effect of finite amplitude cannot be neglected and the actual reflection coefficient is considered to be larger.

(d) Reflection Coefficient for a Circular Perforated Caisson Breakwater

The general reflection coefficients of upright breakwaters are $K_R = 0.3 - 0.8$, and the results of irregular wave experiments conducted by Tanimoto et al [25] on a circular perforated caisson breakwater are shown in **Fig.2.23**.

4) Wave Breaking

The closer the waves are to the shore, the shallower the water depth becomes, so the wave height will be greater, and the wavelength will be smaller. Hence, the wave steepness increases, and waves break when the wave form exceeds a limiting wave steepness. It can be said that the shallower the water depth, the slower the wave propagates, but the faster the horizontal velocity of the water particles.

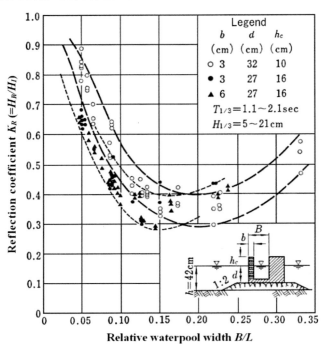

Fig. 2.23 Reflection coefficient against a circular perforated caisson breakwater based on irregular wave experiments by Tanimoto et al [25].

If the horizontal velocity of the water particles is faster than the wave propagation velocity (wave celerity), the wave will break.

(1) Wave Breaking Types and Existence Range

Wave breaking can be classified into the following three types (see **Fig.2.24**)

(a) Spilling Breaker

The asymmetry of the waveform is small and appears when waves with a large wave steepness enter a beach with a loose seabed gradient.

(b) Plunging Breaker

The asymmetry of the waveform is relatively large, the energy loss occurs rapidly and appears when waves of medium wave steepness enter a relatively steep beach.

2.3 Wave Propagation and Deformation

(c) Surging Breaker

The asymmetry of the waveform is very large and appears when waves with a small wave steepness enter a beach with a steep gradient.

Moreover, for the existence range of each wave breaking type, Battjes [26] proposed Eq. (2.2.51) based on the Iribarren number shown in Eq. (2.2.52) using the seabed slope tan β and the wave steepness of the offshore waves.

$$\left. \begin{array}{l} \text{Spilling breaker:} \quad 0.46 \geq \xi_o \\ \text{Plunging breaker:} \; 3.3 \geq \xi_o \geq 0.46 \\ \text{Surging breaker:} \quad \xi_o \geq 3.3 \end{array} \right\} \quad (2.2.51)$$

$$\xi_o = \tan\beta / \sqrt{H_o / L_o} \quad (2.2.52)$$

The Iribarren number is an important indicator for wave transmission and reflection, as can be seen from Eqs. (2.2.35), (2.2.36) and (2.2.50), and it is also called the **surf similarity parameter** because it can estimate the type of wave breaking, as shown in Eq. (2.2.51).

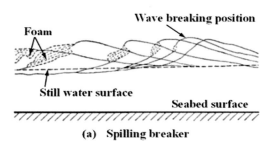
(a) Spilling breaker

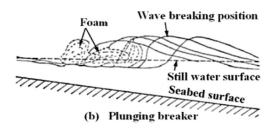
(b) Plunging breaker

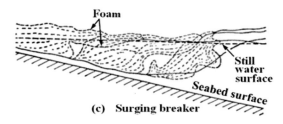
(c) Surging breaker

Fig. 2.24 Types of wave breaking. (Wiegel, R. L., 1953)

(2) Conditions for Wave Breaking Limits

Michell [27] derived Eq. (2.2.53) which can be used to determine the wave breaking limits of deepwater waves. For shallow water waves, Street and Camfield [28] derived Eq. (2.2.54) and Yamada and Shiotani [29] derived Eq. (2.2.55). For intermediate water waves that lie between these equations, Miche [30] derived Eq. (2.2.56), and Hamada [31] also derived Eq. (2.2.56) and confirmed its validity through experiments. These conditions of wave breaking limits can be applied regardless of the type of wave breaking if the seabed gradient is loose.

$$\text{Deep water waves:} \quad \left(\frac{H_0}{L_0}\right)_b = 0.142 \fallingdotseq \frac{1}{7} \quad (2.2.53)$$

$$\text{Shallow water waves:} \quad \gamma = \frac{H_b}{h_b} = 0.75 \quad (2.2.54)$$

$$\gamma = \frac{H_b}{h_b} = 0.83 \quad (2.2.55)$$

$$\text{Intermediate water waves:} \quad \frac{H_b}{L_b} = 0.142 \tanh\frac{2\pi h_b}{L_b} \quad (2.2.56)$$

2.WAVE THEORIES AND, PROPAGATION AND DEFORMATION

where the subscript b means the wave breaking parameter.

Substituting for the deep water wave condition $\tanh(2\pi h_b/L_b) \to 1$ into Eq. (2.2.56), the result agrees with Eq. (2.2.53), but substituting for the shallow water wave condition $\tanh(2\pi h_b/L_b) \to 2\pi h_b/L_b$, the result becomes $H_b/h_b = 0.89$, which does not agree with Eqs. (2.2.54) and (2.2.55).

(a) Effect of Seabed Gradient on Wave Breaking Limit Conditions

The wave breaking limit conditions in Eqs. (2.2.53) - (2.2.56) are not applicable when the seabed gradient is steeper than about 1/20, as they underestimate breaking limit. Therefore, Goda [4] proposed **Fig.2.25** for wave breaking limit conditions that considers the effect of seabed gradient, as well as **Fig.2.26**, which is related to offshore wave heights.

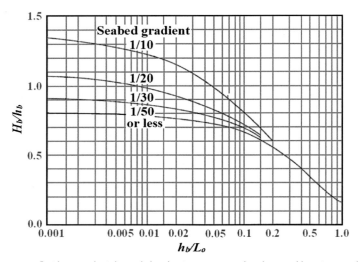

Fig. 2.25 Diagram of the relationship between seabed gradient, relative water depth and wave breaking height based on the experimental results by Goda [4].

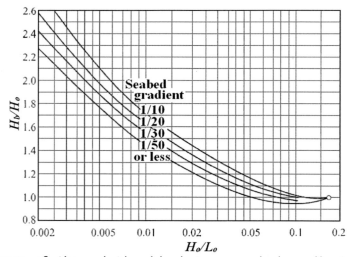

Fig. 2.26 Diagram of the relationship between seabed gradient, wave steepness and wave breaking height based on the experimental results by Goda [4].

Battjes [32] also proposed Eq. (2.2.57), which takes into account the effect of seabed gradient, and Ostendrof and Madsen [33] proposed Eq. (2.2.58) for the coefficient γ_c, which is generally consistent with Eq. (2.2.53) for the condition of deep water waves and Eqs. (2.2.54) and (2.2.55) for the condition of shallow water waves.

$$\frac{H_b}{L_b} = 0.140 \tanh\left(\frac{\gamma_c}{0.88}\frac{2\pi h_b}{L_b}\right) \tag{2.2.57}$$

$$\left.\begin{aligned}\text{In the case of } \tan\beta < 0.1 : \quad \gamma_c = 0.8 + 5\tan\beta\\ \text{In the case of } \tan\beta \geq 0.1 : \quad \gamma_c = 1.3\end{aligned}\right\} \tag{2.2.58}$$

where the subscript b means the wave breaking parameters, and $\tan\beta$ is the seabed slope.

Goda [13] proposed the following equation, which is related to the wavelength of offshore waves, based on a study using finite amplitude wave theory and experimental data with steep seabed gradients.

$$\frac{H_b}{L_0} = 0.17 \times \left[1 - \exp\left\{-1.5\frac{\pi h_b}{L_0}\left(1 + 15\tan^{4/3}\beta\right)\right\}\right] \tag{2.2.59}$$

Moreover, the wave breaking limit condition is also affected by currents and reflected waves.

(b) Effect of currents on Wave Breaking Limit Conditions

For the effect of currents, from Eq. (2.2.60), which relates the wave velocity C_{MC} from the coordinate moving at the same speed as the flow to the wave velocity C from the fixed coordinate, the relationship between the angular frequency σ from the coordinate moving at the same speed as the flow and the angular frequency ω from the fixed coordinate can be expressed as Eq. (2.2.61).

$$C_{MC} = C - U\cos\alpha \tag{2.2.60}$$
$$\sigma = \omega - U\kappa\cos\alpha \tag{2.2.61}$$

where U is the flow velocity, κ is the wavenumber, and α is the angle of intersection between the waves and the current.

In Eq. (2.2.61), for example, if the wave is incident head-on to the flow (the intersection angle $\alpha = 180°$), σ is larger than ω. This means that the wavelength is shorter and therefore wave breaking is more likely to occur. By connecting the wave breaking limit conditions such as Eq. (2.2.56) with Eq. (2.2.61), an equation for the wave breaking condition that takes the flow into account can be developed.

Sakai et al [34] proposed Eqs. (2.2.62) - (2.2.64) to obtain the wave breaking depth h_{bc} and the wave breaking height H_{bc} when the flow rate per unit width of the backflow is considered, using the wave breaking depth h_b and the wave breaking height H_b without flow obtained from Goda's wave breaking limit condition.

$$\frac{h_{bc}}{h_b} = \begin{cases} 0.93 + 170C_b & : \quad C_b \geq 0.0004 \\ 1.0 & : \quad C_b < 0.0004 \end{cases} \tag{2.2.62}$$

2.WAVE THEORIES AND, PROPAGATION AND DEFORMATION

$$\frac{H_{bc}}{H_b}=\begin{cases}0.96+30C_b & : & C_b \geq 0.0013 \\ 1.0 & : & C_b < 0.0013\end{cases} \tag{2.2.63}$$

$$C_b=\begin{cases}\dfrac{q}{g^2T^3}\dfrac{\sqrt[4]{\tan\beta}}{H_o/L_o} & : & \dfrac{H_o}{L_o}\leq 0.05 \\[4mm] \dfrac{q}{g^2T^3}\dfrac{\sqrt[4]{\tan\beta}}{0.05} & : & \dfrac{H_o}{L_o}> 0.05\end{cases} \tag{2.2.64}$$

where q is the flow rate per unit width and $\tan\beta$ is the seabed gradient.

(c) Effect of Reflected Waves on Wave Breaking Limit Conditions

Iwata et al [35] proposed Eq. (2.2.66), which takes into account the effect of reflected waves (the reflection coefficient K_R), using Eq. (2.2.65) of the wave breaking limit condition for perfect standing waves by Wiegel [36]. When the reflection coefficient K_R is zero, Eq. (2.2.66) agrees with Eq. (2.2.56) for incident waves only.

$$\frac{H_b}{L_b}=0.218\tanh\left(\frac{2\pi h_b}{L_b}\right) \tag{2.2.65}$$

$$\frac{H_b}{L_b}=\left(0.218-0.076\frac{1-K_R}{1+K_R}\right)\tanh\left(\frac{2\pi h_b}{L_b}\right) \tag{2.2.66}$$

(d) Wave Breaking Limit Conditions for Irregular Waves

Since 1975, several studies have been carried out on the wave breaking limit conditions for irregular waves and the following characteristics have been identified

(i) The breaking positions of irregular waves are distributed over a wider area than those of regular waves, but as the seabed gradient becomes gentle, the waves tend to break more in combination than individually, and the breaking waves cannot be represented by a simple superposition of regular waves.

(ii) The wave breaking index H_b/h_b of irregular waves is smaller than that of regular waves due to the influence of waves before and after them and unsteady return currents.

Goda [13], using a theoretical model whose validity was confirmed through comparison between experimental irregular wave data and field observation data, organised the wave height at the definition of significant waves and the change in wave height after wave breaking for each seabed slope in relation to water depth, as shown in **Fig.2.27**. The height of the highest wave and the wave height change after wave breaking are given in **Fig.2.28** in relation to the water depth.

2.3 Wave Propagation and Deformation

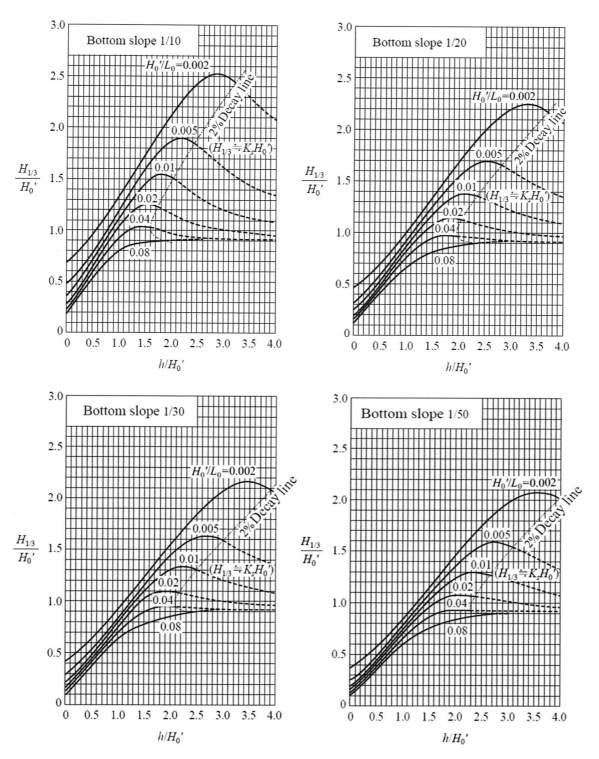

Fig. 2.27 Diagram of the relationship between wave height in significant waves and wave height change after wave breaking and water depth according to Goda [13].

2.WAVE THEORIES AND, PROPAGATION AND DEFORMATION

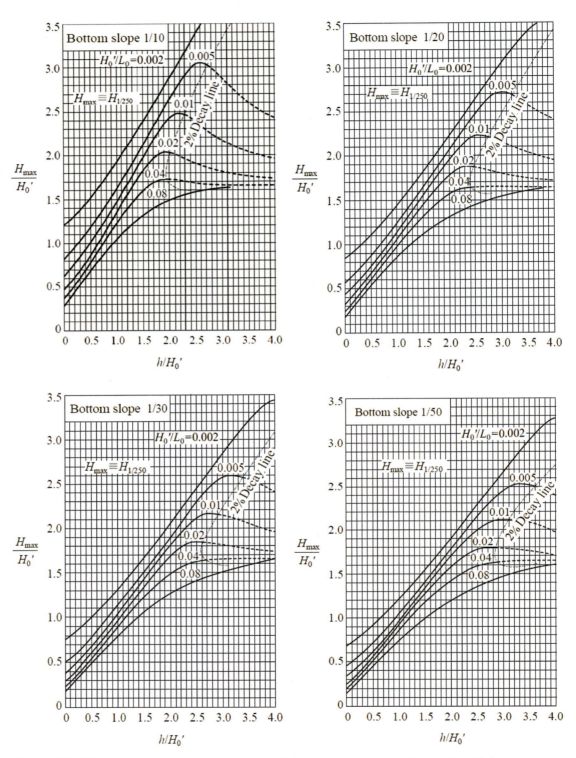

Fig. 2.28 Diagram of the relationship between wave height at the highest wave and wave height change after wave breaking and water depth according to Goda [13].

(e) Wave Breaking Limit Conditions for Superimposed Waves

When multiple waves with different incident directions overlap, wave heights fluctuate locally due to their mutual phase relationship, and the wave breaking limit conditions introduced so far are not suitable to determine wave breaking. Therefore, Watanabe et al [37] applied small amplitude wave theory to the wave breaking limit conditions of Goda [4] and produced a wave breaking limit condition diagram based on the water particle horizontal velocity u_b and wave velocity C_b at the still water surface below the wave crest at the wave breaking position. Isobe [38] approximated this wave breaking limit condition diagram by the following equation.

$$\frac{u_b}{C_b} = 0.53 - 0.3\exp\left(-3\sqrt{h_b/L_o}\right) + 5\tan^{3/2}\beta\exp\left\{-45\left(\sqrt{h_b/L_o} - 0.1\right)^2\right\} \qquad (2.2.67)$$

where h_b is the wave breaking depth, $\tan\beta$ is the seabed gradient.

(3) Dynamic Models of Wave Breaking

Goda [13] proposed an early wave breaking model for irregular waves, which fits to f Rayleigh distribution, and propagate on a uniformly sloping seabed topography, the wave heights exceeding the breaking limit conditions for each depth are decreased within the wave breaking zone and added to the low wave height distribution. Currently, it is common to consider wave breaking phenomena in numerical models based on the wave energy equilibrium equation and the equations of motion by adding a term that estimates the energy loss associated with the vortex motion generated during wave breaking, so that the model can be applied to arbitrary seabed topography. The main energy loss terms are introduced below.

(a) Energy Equilibrium Equation and Wave Breaking Models

The simplest energy equilibrium equation for the one-dimensional, stationary case is expressed in Eq. (2.2.68).

$$\frac{\partial\left[E_s \times C_g\right]}{\partial x} = \triangle E_b \qquad (2.2.68)$$

where E_s is the wave energy per unit area or directional spectrum, C_g is the group velocity, and ΔE_b is the energy loss term due to breaking waves.

There have been many studies on the energy loss term due to breaking waves.
For example, if the loss term ΔE_b in Eq. (2.2.68) is assumed to be proportional to E_s with respect to wave energy, and if the relative wave height (wave height - depth ratio) in the breaking zone is constant due to a uniform seabed gradient, or if the long wave approximation for small amplitude waves can be applied, Eq. (2.2.68) can be solved and the energy loss term can be determined relatively easily. Moreover, Mizuguchi et al [39], Dibainia et al [40] and others have devised ways to apply this loss term to arbitrary seabed topography.

2.WAVE THEORIES AND, PROPAGATION AND DEFORMATION

Modelling to mimic a **bore** (= stepped wave) was also attempted by LeMéhauté [41] and others. A typical example of this is Eq. (1.2.43) based on the bore model of Battjes and Janssen [42], which is used in the numerical model SWAN, which has been shown to be highly accurate for wave forecasting in shallow to very shallow water areas.

$$
\left.
\begin{aligned}
\varDelta E_b &= -\frac{1}{4} C_6 Q_b \left(\frac{\bar{\sigma}}{2\pi} \right) \frac{H_b^{\,2}}{S_T} \times S(f,\theta) \\
Q_b &= 8 \frac{S_T}{H_b^{\,2}} \ln Q_b + 1 \\
\bar{\sigma} &= \frac{\iint S(\sigma,\theta)\sigma d\sigma d\theta}{S_T} \\
H_b &= 0.73d, \quad S_T = \iint S(\sigma,\theta)d\sigma d\theta
\end{aligned}
\right\}
\qquad (1.2.43)
$$

where C_6 is the adjustment factor, Q_b is the probability of wave breaking calculated assuming the Rayleigh distribution for the wave height distribution, $\bar{\sigma}$ is the mean relative frequency, H_b is the wave breaking height, S_T is the all-round wave spectrum, and d is the water depth.

(b) Motion Equation and Wave Breaking Models

Many numerical models have been proposed that incorporate a loss term similar to the energy loss term due to breaking waves in the energy equilibrium equation into the equation of motion to calculate the wave field including in the breaking wave zone. Representative examples are Watanabe and Maruyama [43], Watanabe et al [44] and Sato and Suzuki [45].

In numerical models using the equations of motion, it is necessary to solve a total of four equations in three-dimensional space (a continuity equation and equations of motion in three directions) and three equations in a two-dimensional plane (the continuity equation and equations of motion in two horizontal directions) in a simultaneous system, and what is directly obtained is the change over time of the water surface elevation and flow velocity (or the flow rate per unit width = the flow velocity × the water depth). Therefore, if the objective is only to obtain the wave height distribution, the method based on the energy equilibrium equation is more advantageous. However, if the **nearshore current**, which is the flow generated by wave deformation and wave breaking mainly in shallow waters, and the run-up and -down motions of waves are also to be determined simultaneously, highly accurate equations of motion incorporating wave breaking models should be used.

As an example, the proposal by Vu et al [46], where the energy loss due to breaking waves is considered with a turbulence model in the two-dimensional Boussinesq equations on the horizontal plane by taking into account nonlinearities and dispersion, is presented as follows: Eqs. (2.2.69) and (2.2.70) are the equations of motion in the x- and y-directions respectively, and Eq. (2.2.71) is the continuity equation.

2.3 Wave Propagation and Deformation

$$\frac{\partial q_x}{\partial t} + \frac{\partial}{\partial x}\left(\frac{q_x^2}{d}\right) + \frac{\partial}{\partial y}\left(\frac{q_x q_y}{d}\right) + gd\frac{\partial \eta}{\partial x} + \frac{h^3}{6}\left[\frac{\partial^3}{\partial x^2 \partial t}\left(\frac{q_x}{h}\right) + \frac{\partial^3}{\partial x \partial y \partial t}\left(\frac{q_y}{h}\right)\right]$$

(2.2.69)

$$-\frac{h^2}{2}\left(\frac{\partial^3 q_x}{\partial x^2 \partial t} + \frac{\partial^3 q_y}{\partial x \partial y \partial t}\right) - M_{bx} + \frac{f_b}{d^2}|q_x|q_x = 0$$

$$\frac{\partial q_y}{\partial t} + \frac{\partial}{\partial x}\left(\frac{q_x q_y}{d}\right) + \frac{\partial}{\partial y}\left(\frac{q_y^2}{d}\right) + gd\frac{\partial \eta}{\partial y} + \frac{h^3}{6}\left[\frac{\partial^3}{\partial y^2 \partial t}\left(\frac{q_y}{h}\right) + \frac{\partial^3}{\partial x \partial y \partial t}\left(\frac{q_x}{h}\right)\right]$$

(2.2.70)

$$-\frac{h^2}{2}\left(\frac{\partial^3 q_y}{\partial y^2 \partial t} + \frac{\partial^3 q_x}{\partial x \partial y \partial t}\right) - M_{by} + \frac{f_b}{d^2}|q_y|q_y = 0$$

$$\frac{\partial q_x}{\partial x} + \frac{\partial q_y}{\partial y} + \frac{\partial \eta}{\partial t} = 0$$

(2.2.71)

where, q_x and q_y are the flow rates per unit width which are defined using $q_x = \int u dz$ and $q_y = \int v dz$ (u is the velocity in x direction, v is the velocity in y direction, z is the coordinate in the direction of water depth); η is the water surface elevation; h is the mean water depth; d is the total water depth (= $h + \eta$); g is the gravitational acceleration; f_b is the frictional coefficient on the seabed; t is the time; x is the coordinate in the offshore direction; y is the coordinate in the alongshore direction; and M_{bx} and M_{by} are the terms of wave energy loss expressed using Eq. (2.2.72).

$$\left.\begin{aligned} M_{bx} &= \frac{\partial}{\partial x}\left(f_D v_t \frac{\partial q_x}{\partial x}\right) + \frac{\partial}{\partial y}\left(f_D v_t \frac{\partial q_x}{\partial y}\right) \\ M_{by} &= \frac{\partial}{\partial x}\left(f_D v_t \frac{\partial q_y}{\partial x}\right) + \frac{\partial}{\partial y}\left(f_D v_t \frac{\partial q_y}{\partial y}\right) \end{aligned}\right\}$$

(2.2.72)

where v_t is the eddy viscosity coefficient; and f_D is the empirical coefficient (usually 1.5), which takes into account that part of the wave energy lost during wave breaking is converted to turbulent energy.

When calculating the development of a turbulent boundary layer and turbulent energy transport on the seabed due to wave breaking, the distribution of turbulent energy within the wave breaking zone should be considered not only in the offshore direction but also across the water column. However, according to Ting and Kirby [47], [48] and Nadaoka et al [49], the offshore transport timescale of turbulent energy within the wave breaking zone is longer than the vertical transport timescale, and because the wave equation is integrated in the depth direction, only the transport in the offshore and alongshore directions is considered. The depth-integrated, planar, two-dimensional equations for the generation and dissipation of turbulent energy within the breaking zone and its transport in the offshore direction is expressed using Eqs. (2.2.73) – (2.2.75).

$$\frac{\partial k}{\partial t} + \frac{\partial \tilde{u} k}{\partial x} + \frac{\partial \tilde{v} k}{\partial y} = P_r - \varepsilon + \frac{\partial}{\partial x}\left(\frac{v_t}{\sigma_t}\frac{\partial k}{\partial x}\right) + \frac{\partial}{\partial y}\left(\frac{v_t}{\sigma_t}\frac{\partial k}{\partial y}\right)$$

(2.2.73)

$$\frac{\partial \varepsilon}{\partial t} + \frac{\partial \tilde{u} \varepsilon}{\partial x} + \frac{\partial \tilde{v} \varepsilon}{\partial y} = \frac{\partial}{\partial x}\left(\frac{v_t}{\sigma_\varepsilon}\frac{\partial \varepsilon}{\partial x}\right) + \frac{\partial}{\partial y}\left(\frac{v_t}{\sigma_\varepsilon}\frac{\partial \varepsilon}{\partial y}\right) + \frac{\varepsilon}{k}(C_{1\varepsilon}P_r - C_{2\varepsilon}\varepsilon)$$

(2.2.74)

$$v_t = C_\varepsilon k^2 / \varepsilon$$

(2.2.75)

where, k and ε are the turbulence energy integrated in the depth direction and its dissipation rate,

2.WAVE THEORIES AND, PROPAGATION AND DEFORMATION

respectively; and $\tilde{u}$ and $\tilde{v}$ are the ensemble mean velocity in the x and y directions averaged across the water column respectively; σ_t (=1), σ_ε (=1.3) and C_ε (=0.09) are the closure coefficients from Jones and Launder [50]; $C_{1\varepsilon}$ and $C_{2\varepsilon}$ denote the generation and decay rates of the turbulence energy dissipation rate ε. In the standard k-ε model, using the assumption that the turbulence energy is dissipated by the cascade process, the equation for the dissipation rate of the turbulence energy is derived, where $C_{1\varepsilon}$ is 1.44 and $C_{2\varepsilon}$ is 1.92. However, in breaking waves, the turbulence scale is smaller than that of normal shear turbulence, so the dissipation rate of turbulence energy is considered to be larger than that of the standard k-ε model.

The turbulent energy generation P_r is described by the following equation, consisting of generation due to bottom resistance, generation due to shear stress, and generation due to breaking waves.

$$P_r = P_{rb} + P_{rs} + P_{rw} \tag{2.2.76}$$

where P_{rb}, P_{rs} and P_{rw} are the generation of turbulence energy due to bottom resistance, shear stress and breaking waves respectively. The generation of turbulence energy due to bottom resistance is considered to be small and is ignored in this model. The generation of turbulence energy due to shear stress is expressed by the following equation.

$$P_{rs} = \nu_t d \left[\left(\frac{\partial u}{\partial y} \right)^2 + \left(\frac{\partial v}{\partial x} \right)^2 \right] \tag{2.2.77}$$

The generation of turbulence energy due to wave breaking P_{rw} is assumed to be proportional to the kinetic energy in the surface roller and is expressed by the following equation.

$$P_{rw} = \alpha_p g C \delta \beta_d / d \tag{2.2.78}$$

where C is the wave celerity; α_p (= 0.33) and β_d (= 0.08) are constants; δ is the thickness of the surface roller, estimated by the method of Schaffer et al [51]. The wave breaking position is assumed to be where the water surface gradient exceeds the limit value. The surface roller then moves towards the shore with damping. When the surface gradient in front of the surface roller falls below a limit value, the surface roller disappears and the wave recovers. This method reproduces the formation, deformation, and movement of the surface roller from moment to moment, as well as the loss of wave energy due to wave breaking, and the generation, transport, and dissipation of turbulent energy.

5) Wave Height Reduction Due to Seabed Friction and Other Factors

Wave height decreases due to energy loss as waves propagate. Other causes of wave height reduction besides wave breaking and headwinds include: 1) loss due to friction on the seabed, 2) loss due to seepage at the bottom of sand and gravel beds, and 3) loss due to the internal viscosity of the seawater itself.

2.3 Wave Propagation and Deformation

(1) Friction Loss on a Seabed Surface

Although the frictional boundary layer on the seabed surface in a wave field is less developed than in a field with currents in only one direction, the horizontal reciprocal motion of water particles on the seabed surface is more pronounced in very shallow waters, so the thickness of this frictional boundary layer δ is a significant quantity and is expressed by Eq. (2.2.79).

$$\delta = 4.1\sqrt{\frac{\nu T}{\pi}} \tag{2.2.79}$$

where ν is the kinematic viscosity and T is the period of the incident wave.

Wave energy loss due to bottom friction occurs within the boundary layer, and this loss increases as the boundary layer develops. From this equation, the energy loss due to bottom friction cannot be neglected when long period waves enter a shallow beach. The equation for wave height reduction due to bottom friction by Bretschneider and Reid [52] is presented as follows:

The amount of energy loss due to friction per unit area and unit time at the seabed surface is equal to the power (W_f) of the frictional stress.

$$W_f = \tau_b u_b = f_b \rho u_b^{3} \tag{2.2.80}$$

where τ_b is the frictional stress (shear stress) at the seabed, f_b is the friction coefficient at the seabed, ρ is the density of seawater, and u_b is the horizontal velocity of water particles at the seabed ($z = -h$) according to the small amplitude wave theory.

Hence, the average energy loss (per unit area and per unit time) during one wave cycle is expressed by Eq. (2.2.81).

$$\Delta E_f = \frac{2}{T} f_b \rho \int_{-T/4}^{T/4} u_b^{3} dt = \frac{4}{3} \pi^2 \frac{f_b \rho}{\sinh^3 \kappa h}\left(\frac{H}{T}\right)^{3} = \frac{4}{3\pi} f_b \rho \hat{u}_b^{3} \tag{2.2.81}$$

where κ is the wave number, H is the wave height, and $\hat{u}_b$ is the amplitude of the horizontal velocity of water particles at the seabed surface ($z = -h$) according to the small amplitude wave theory.

The relationship between the difference in wave energy transport and the amount of energy loss between offshore position 1 and onshore position 2 on a shallow beach is expressed by Eq. (2.2.82).

$$E_2 C_{g2} - E_1 C_{g1} = \Delta\left(E C_g\right) = -\Delta E_f \Delta x \tag{2.2.82}$$

where E is the wave energy per unit area, C_g is the group velocity, and the subscripts 1 and 2 denote the offshore and shore side positions. Δx is the distance between positions 1 and 2.

Substituting the relation based on the small amplitude wave theory, Eq. (2.2.81) and the shallow water coefficient K_s into Eq. (2.2.82), and integrating with constant water depth between positions 1 and 2, Eq. (2.2.83) is obtained as the wave height ratio between these positions.

$$\frac{H_2}{H_1} = \left[1 + \frac{64}{3} \cdot \frac{\pi^3}{g^2} \cdot \frac{f_b H_1 \Delta x}{h^2}\left(\frac{h}{T^2}\right)^2 \frac{K_s^{2}}{\sinh^3 \kappa h}\right]^{-1} = K_f \tag{2.2.83}$$

where H_2 is the wave height at the shore side position, H_1 is the wave height at the offshore position, and K_f is the wave height reduction factor (reduction coefficient) due to bottom friction.

2.WAVE THEORIES AND, PROPAGATION AND DEFORMATION

The following formula (2.2.84) from Jonsson [53] can be used to estimate the friction coefficient f_b. However, the relationship between the friction coefficient f_w according to Jonsson and the friction coefficient f_b in Eq. (2.2.83) is shown in Eq. (2.2.85)

$$\left.\begin{array}{rl} \dfrac{1}{4\sqrt{f_w}} + \log_{10}\dfrac{1}{4\sqrt{f_w}} = -0.08 + \log_{10}\dfrac{a_m}{k_e}: & \dfrac{a_m}{k_e} \geq 1.57 \\[2mm] f_w = 0.30 \qquad\qquad : & \dfrac{a_m}{k_e} \leq 1.57 \\[2mm] a_m = \dfrac{H}{2}\left(\sinh\dfrac{2\pi h}{L}\right)^{-1} & \end{array}\right\} \tag{2.2.84}$$

$$\hat{\tau}_b = \frac{f_w}{2}\rho\hat{u}_b{}^2 \cong f_b\rho\hat{u}_b{}^2 \tag{2.2.85}$$

where a_m is the horizontal amplitude of the water particle orbit at the seabed surface; k_e is the equivalent roughness, which is taken as three times the height of the sand ripple formed on the seabed; $\hat{\tau}_b$ is the amplitude of the frictional stress; and $\hat{u}_b$ is the amplitude of the horizontal velocity of the water particles at the seabed surface.

(2) Loss Due to Seepage at a Sand and Gravel Bed Bottom

When the thickness of sand and gravel layer on the seabed is more than 0.3 times the wavelength, seepage flow into the sand and gravel layer also causes wave energy loss, although this loss is smaller than that due to friction on the seabed. Reid and Kajiura [54] proposed the following equation as the amount of energy loss per unit area and unit time due to seepage.

$$\Delta E_p = \frac{\pi}{4}\rho g \frac{k_c H^2}{L\cosh^2 2\kappa h} \tag{2.2.86}$$

where k_c is the hydraulic conductivity within the sand and gravel layer, L is the wavelength, and κ is the wave number.

Substituting Eq. (2.2.86) into the energy equilibrium equation expressed by Eq. (2.2.87) and solving it gives the wave height after reduction.

$$\frac{\partial\left[E_s \times C_g\right]}{\partial x} = -\Delta E_p \tag{2.2.87}$$

where E_s is the wave energy per unit area and unit time, and C_g is the group velocity.

(3) Loss Due to Viscosity of Seawater

The internal viscosity of the seawater itself also causes wave energy loss. This is negligible for waves of general period but becomes non-negligible when the period is considerably shorter.

Lamb [1] presented the following equation for the amount of energy loss per unit area and unit time due to internal viscosity.

2.3 Wave Propagation and Deformation

$$\Delta E_i = \frac{\mu\beta}{2}\left(\frac{\pi H}{T}\right)^2 \frac{1}{\sinh^2 \kappa h} \times \frac{2\kappa}{\beta}\sinh 2\kappa h$$

$$\left.\beta = \sqrt{\frac{\rho\omega}{2\mu}} \right\}$$

(2.2.88)

where μ is the viscosity coefficient of seawater, κ is the wave number, ρ is the density of the seawater, and ω is the angular frequency of the wave.

Substituting Eq. (2.2.88) into the energy equilibrium equation expressed by Eq. (2.2.89) and solving it, the wave height after reduction can be obtained.

$$\frac{\partial\left[E_s \times C_g\right]}{\partial x} = -\Delta E_i$$

(2.2.89)

2.WAVE THEORIES AND, PROPAGATION AND DEFORMATION

2.4 Numerical Models for Calculating Wave Fields

To numerically simulate wave fields on arbitrary seabed topography, the following numerical models have been presented since around 1970, in line with the improvement of computer computing power.

1) Models Based on Energy Conservation Law

The most basic method for calculating wave shoaling and refraction is the **wave ray method**, which is based on Eq. (2.2.90).

$$\frac{\partial\left[E_s \times \left(C_g \cos\theta\right)\right]}{\partial x} + \frac{\partial\left[E_s \times \left(C_g \sin\theta\right)\right]}{\partial y} = 0 \tag{2.2.90}$$

However, since the irregularity of waves cannot be neglected in refraction calculations, the SWAN model is introduced as a numerical spectral wave model for solving the energy equilibrium equation for irregular waves in the section 1.2, 2), (3), (b). The SWAN model incorporates a wave breaking term and, although not considered in the theory, can reproduce quasi-diffraction phenomena.

2) Models Based on the Equation of Motion or the Mild Slope Equation

(1) Numerical Wave Analysis Method

The method of solving the energy equilibrium equation allows a wide range of calculations due to the relatively low computational load, but there is no guarantee that the diffraction is correctly reproduced. Ito and Tanimoto [55] have therefore developed the **numerical wave analysis method** that can calculate the refraction, diffraction, and reflection of regular waves by solving the plane two-dimensional linear equation of motion (2.2.91) integrated across water depth and the Continuity equation (2.2.92) simultaneously, to reduce the computational load on a computer.

$$\frac{\partial q_x}{\partial t} = -C^2\frac{\partial\eta}{\partial x}, \quad \frac{\partial q_y}{\partial t} = -C^2\frac{\partial\eta}{\partial y} \tag{2.2.91}$$

$$\frac{\partial\eta}{\partial t} = \frac{\partial q_x}{\partial x} + \frac{\partial q_y}{\partial y} \tag{2.2.92}$$

where q_x and q_y are the flow rates per unit width defined by $q_x = \int u\,dz$ and $q_y = \int v\,dz$ (u is the x-directional velocity, v is the y-directional velocity, and z is the coordinate in the direction of water depth); C is the wave celerity; η is the water surface elevation; and x and y are the horizontal coordinates.

(2) Models based on the Mild Slope Equation

Berkhoff [56] introduced an elliptic equation (2.2.93) for the water surface amplitude, called the **mild slope equation**, using the velocity potential ϕ under the assumption of a mild seabed slope and small amplitude waves, and proposed a numerical model that can calculate wave shoaling as well as refraction, diffraction and reflection of regular waves by solving Eq. (2.2.93).

2.4 Numerical Models for Calculating Wave Fields

$$\frac{\partial\left(CC_g \frac{\partial f_{xy}}{\partial x}\right)}{\partial x} + \frac{\partial\left(CC_g \frac{\partial f_{xy}}{\partial y}\right)}{\partial y} + \kappa^2 CC_g f_{xy} = 0, \qquad \eta = \mathrm{Re}\left\{f_{xy} \times e^{-i\omega t}\right\} \qquad (2.2.93)$$

where C_g is the group wave velocity; f_{xy} is the spatial distribution function of the amplitude of the water surface elevation η (Re means the real part of the complex number, i is the imaginary unit, ω is the angular frequency); and κ is the wave number.

Eq. (2.2.93) becomes the Helmholz equation, the governing equation for diffraction phenomena on constant water depth.

Furthermore, Kubo et al [57] extended this mild slope equation to an unsteady mild slope equation that can also handle irregular waves.

Watanabe and Maruyama [43] rewrote the unsteady mild slope equation into a system of linear unsteady motion equations and the continuity equation to facilitate the incorporation of wave breaking terms and boundary condition treatment. This model is regarded as the one that extended the numerical wave analysis method to include wave shoaling and wave breaking.

(3) Models Based on the Parabolic Wave Equation

The computational load of numerical models based on the equation of motion or the mild slope equation is heavier than that of numerical models based on the energy conservation law. Therefore, Radder [58] reduced the computational load by using Eq. (2.2.94) called the **parabolic wave equation**, which was derived from the mild slope equation by considering only travelling waves.

$$\frac{\partial}{\partial y}\left(CC_g \frac{\partial f_{xy}}{\partial y}\right) + 2i\kappa CC_g \frac{\partial f_{xy}}{\partial x} + \left\{i \frac{\partial \kappa CC_g}{\partial x} + 2\kappa^2 CC_g\right\} f_{xy} = 0 \qquad (2.2.94)$$

(4) Models Based on the Nonlinear Unsteady Motion Equation

The numerical models described in (1) to (3) enable the calculation of wave fields in arbitrary seabed topography, but if nearshore currents and wave run-up motion caused by wave deformation and breaking waves are to be calculated simultaneously and with high accuracy, the nonlinear unsteady equations of motion must be solved. The Boussinesq equation in two planar dimensions and the Navier-Stokes equation in three dimensions are representative of the equations of motion that satisfy this condition.

(a) The Boussinesq Equation

The Boussinesq equation is an equation of motion that takes dispersion into account in the plane two-dimensional nonlinear long-wave equation obtained by integrating the Navier-Stokes equation in the vertical direction using the long-wave approximation to reduce the computational load. Numerical models based on the Boussinesq equation only became popular in the 1990s, when the computational

2.WAVE THEORIES AND, PROPAGATION AND DEFORMATION

power of computers began to catch up, because their computational load is heavier than that of the numerical models in (1) to (3).

As an example of a numerical model based on the Boussinesq equation, the model proposed by Vu et al [46] is presented in the section 2.3, 4), (3), (b), and the detailed use of this model is explained in the section 2 of the Appendix.

(b) The Navier-Sokes Equation

Numerical models based on the three-dimensional Navier-Stokes equations only became popular in coastal engineering in the 2000s. As an example of a numerical model based on the Navier-Stokes equations, the governing equations of the **CADMAS SURF** developed by Isobe et al [59] are presented below. The detailed use of this model is described by Isobe et al [59] and the source program can be downloaded from the website (https://github.com/CADMAS-SURF/, https://www.cdit.or.jp /program/cadmas-download.html) of Coastal Development Institute of Technology, Japan.

$$
\begin{aligned}
&\lambda_v \frac{\partial u}{\partial t} + \frac{\partial \lambda_x uu}{\partial x} + \frac{\partial \lambda_y vu}{\partial y} + \frac{\partial \lambda_z wu}{\partial z} = -\frac{\gamma_v}{\rho}\frac{\partial p}{\partial x} \\
&+\frac{\partial}{\partial x}\left\{\gamma_x v_e\left(2\frac{\partial u}{\partial x}\right)\right\} + \frac{\partial}{\partial y}\left\{\gamma_y v_e\left(\frac{\partial u}{\partial y}+\frac{\partial v}{\partial x}\right)\right\} + \frac{\partial}{\partial z}\left\{\gamma_z v_e\left(\frac{\partial u}{\partial z}+\frac{\partial w}{\partial x}\right)\right\} - \gamma_v D_x u - R_x + \gamma_v S_u \\
&\lambda_v \frac{\partial v}{\partial t} + \frac{\partial \lambda_x uv}{\partial x} + \frac{\partial \lambda_y vv}{\partial y} + \frac{\partial \lambda_z wv}{\partial z} = -\frac{\gamma_v}{\rho}\frac{\partial p}{\partial y} \\
&+\frac{\partial}{\partial x}\left\{\gamma_x v_e\left(\frac{\partial v}{\partial x}+\frac{\partial u}{\partial y}\right)\right\} + \frac{\partial}{\partial y}\left\{\gamma_y v_e\left(2\frac{\partial v}{\partial y}\right)\right\} + \frac{\partial}{\partial z}\left\{\gamma_z v_e\left(\frac{\partial v}{\partial z}+\frac{\partial w}{\partial y}\right)\right\} - \gamma_v D_y v - R_y + \gamma_v S_v \\
&\lambda_v \frac{\partial w}{\partial t} + \frac{\partial \lambda_x uw}{\partial x} + \frac{\partial \lambda_y vw}{\partial y} + \frac{\partial \lambda_z ww}{\partial z} = -\frac{\gamma_v}{\rho}\frac{\partial p}{\partial z} \\
&+\frac{\partial}{\partial x}\left\{\gamma_x v_e\left(\frac{\partial w}{\partial x}+\frac{\partial u}{\partial z}\right)\right\} + \frac{\partial}{\partial y}\left\{\gamma_y v_e\left(\frac{\partial w}{\partial y}+\frac{\partial v}{\partial z}\right)\right\} + \frac{\partial}{\partial z}\left\{\gamma_z v_e\left(2\frac{\partial w}{\partial z}\right)\right\} - \gamma_v D_z w - R_z + \gamma_v S_w - \gamma_v \frac{\rho^*}{\rho} g
\end{aligned}
\tag{2.2.95}
$$

$$
\frac{\partial \gamma_x u}{\partial x} + \frac{\partial \gamma_y v}{\partial y} + \frac{\partial \gamma_z w}{\partial z} = \gamma_v S_p
\tag{2.2.96}
$$

where x and y are horizontal and z are vertical coordinates; u, v and w are velocities in the x, y and z directions; λ_v, λ_x, λ_y and λ_z are the coefficients expressed in Eq. (2.2.97) using the inertia force coefficient C_M (the second term on the righthand side is the effect of the inertia force received from the structure). γ_v is the porosity; p is the pressure; ρ is the reference density of seawater; γ_x, γ_y and γ_z are area permeabilities in the x, y and z directions; v_e is the sum of the molecular and eddy kinematic viscosity coefficients; D_x, D_y and D_z are the coefficients for the energy attenuation zone; R_x, R_y and R_z are the resistance terms expressed in Eq. (2.2.98) using the drag coefficient C_D. Moreover, S_u, S_v, S_w and S_p are the terms for generating the input waves; ρ^* is the density considering the buoyancy force of seawater.

2.4 Numerical Models for Calculating Wave Fields

$$\left.\begin{aligned}
\lambda_v &= \gamma_v + (1-\gamma_v)C_M, \\
\lambda_x &= \gamma_x + (1-\gamma_x)C_M, \\
\lambda_y &= \gamma_y + (1-\gamma_y)C_M, \\
\lambda_z &= \gamma_z + (1-\gamma_z)C_M
\end{aligned}\right\} \tag{2.2.97}$$

$$\left.\begin{aligned}
R_x &= \frac{1}{2}\frac{C_D}{\Delta x}(1-\gamma_x)u\sqrt{u^2+v^2+w^2} \\
R_y &= \frac{1}{2}\frac{C_D}{\Delta y}(1-\gamma_y)v\sqrt{u^2+v^2+w^2} \\
R_z &= \frac{1}{2}\frac{C_D}{\Delta z}(1-\gamma_z)w\sqrt{u^2+v^2+w^2}
\end{aligned}\right\} \tag{2.2.98}$$

The free surface is modelled using the Volume Of Fluids (VOF) method, which uses a VOF function F to represent the free surface sharply, and the advection equation for the VOF function F is expressed in Eq. (2.2.99).

$$\gamma_v\frac{\partial F}{\partial t}+\frac{\partial \gamma_x uF}{\partial x}+\frac{\partial \gamma_y vF}{\partial y}+\frac{\partial \gamma_z wF}{\partial z}=\gamma_v S_F \tag{2.2.99}$$

Here S_F is the term (= $F{\times}S_P$) for generating the input waves.

2.WAVE THEORIES AND, PROPAGATION AND DEFORMATION

List of References in Chapter 2

1) Lamb, H.: *Hydrodinamics*, 6th ed., Cambridge University Press, 1932, 738p.

2) Airy, G. B.: Tides and Waves, *Encyclopaedia Metropolitana*, London, 1845.

3) Horikawa, K. (Editor): *Nearshore Dynamics and Coastal Processes*, University of Tokyo Press, 1988, 522p.

4) Goda, Y.: A Synthesis of Breaker Indices, *Proceedings of Japan Society of Civil Engineers*, No. 180, 1970, pp.39-49. (in Japanese)

5) Yamada, H. and Shiotani, T.: On the Highest Water Waves of Permanent Type, *Bull. Disaster Prevention Res. Inst.*, Kyoto University, 18, 1968, pp.1-22.

6) Stokes, G. G.: On the Theory of Oscillatory Waves, *Trans. Cambridge Phil. Soc.*, 8, 1847, pp.441-455.

7) Isobe, M., Nishimura, H. and Horikawa, K.: Indication of Perturbation Solutions of Conserved Waves by Wave Heights, *Proceedings of the 33rd Annual Meeting*, II-394, JSCE, 1978, pp. 760-761. (in Japanese)

8) Skjelbreia, L.: *Gravity Waves, Stokes' Third Order Approximations, Tables of Function*, Council on Wave Research, Engineering Foundation, University of California, Berkeley, 1959.

9) Korteweg, D. J. and de Vries, G.: On the Change of Form of Long waves Advancing in a Rectangular Canal and on a New Type of Long Stationary waves, *Phil. Mag.*, Ser. 5, Vol. 39, 1895, pp.422-443.

10) Laitone, E. V.: The Second Approximation to Cnoidal and Solitary Waves, *J. Fluid Mech.*, 9, 1960, pp.430-444.

11) Iwagaki, Y. and Sawaragi, T.: *Coastal Engineering*, Civil Engineering Vol.25 in University Lectures, Kyoritsu Shuppan Co., Ltd., 1979, 473p. (in Japanese)

12) Suto, N.: Deformation of Non-linear Long Waves - the Case of Varying Channel Width and Depth -, *Proc. of the 21st Coastal Engineering Conference*, JSCE, 1974, pp.57-63. (in Japanese)

13) Goda, Y.: Deformation of Irregular Waves Due to Depth-controlled Wave Breaking, *Report of the Port and Harbour Research Institute*, Vol.14, No.3, 1975, pp.59-106. (in Japanese)

14) Goda, Y. and Suzuki, Y.: *Computation of Refraction and Diffraction of Sea Waves with Mitsuyasu's Direction Spectrum*, Technical Note of the Port and Harbour Research Institute, No.230, 1975, 45p. (in Japanese)

15) Goda, Y.: *Random Seas and design of Maritime Structures*, University of Tokyo Press, 1985, 323p.

16) Hattori, M.: Coastal Development and Wave Control, *JSCE Hydraulic Committee Hydraulic Engineering Series*, 75-B-2, 1975, pp.B-2-1-B-2-24. (in Japanese)

17) Numata, A.: Experimental Study on the Wave Dissipating Effect of a Block Breakwater, *Proc. of the 22nd Coastal Engineering Conference*, JSCE, 1975, pp.501-505. (in Japanese)

18) Takayama, T., Nagai, T. and Sekiguchi, T.: Experiment on Irregular Wave Concerning Wave Breaking Effects by Submerged Breakwater with Wide Levee Crown, *Proc. of the 32nd Coastal Engineering Conference*, JSCE, 1985, pp.545-549. (in Japanese)

19) Van der Meer, J. W., Briganti, R., Zanuttingh, B., and Wang, B. X.: Wave Transmission and Reflection at Low-crested Structures: Design Formulae, Oblique Wave Attack and Spectral Change, *Coastal Engineering*, Vol.52, 2005, pp.915-929.

20) Goda, Y.: Re-analysis of Laboratory Data on Wave Transmission over Breakwaters, *Rept. Port and Harbour Res. Inst.*, Vol.8, No.3, 1969, pp.3-18.

21) Goda, Y., Suzuki, Y., Kishira, Y. and Kikuchi, O.: Estimation of Incident and Reflected Waves in Random Wave Experiments, *Technical Note of the Port and Harbour Research Institute*, No.248, 1976, 24p. (in Japanese)

22) Kondo, H. and Sato, I.: Study on the Crown Height of a Breakwater, *Monthly Report of the Civil Engineering Laboratory*, Hokkaido Development Bureau, No. 117, 1964, pp.1-15. (in Japanese)

23) Miche, M.: Le Pouvoir Réfléchissant des Ouvrages Maritime Exposés à l'Action de la Houle, *Annales des Ponts et Chaussees*, 121e Annee, 1951, pp.285-319.

24) Gleslou, L. and Mahe, Y.: Etude du Coefficient de Reflexion d'une Houlle sur un Obstacle Constitute par un Plane Incline, *Proc. 5 th Conf. on Coastal Eng.*, 1954.

25) Tanimoto, K, Haranaka, S, Takahashi, S., Komatsu, K., Todoroki, M. and Osato, M.: An Experimental Investigation of Wave Reflection, Over Topping and Wave Force for Several Types of Breakwaters and Sea Walls, *Technical Note of the Port and Harbour Research Institute*, No.246, 1976, 38p. (in Japanese)

26) Battjes, J.A.: Surf Similarity, *Proc. 14th Inter. Conf. Coastal Eng.*, 1974, pp.466-480.

27) Michell, J.H.: On the Highest Waves in Water, *Phil. Mag.* Ser.5, Vol.36, 1893, pp.430-437.

28) Street, R.L. and Camfield, F.E.: Observations and Experiments on Solitary Wave Deformation, *Proc. 10th Inter. Conf. Coastal Eng.*, 1966, pp.284-299.

29) Yamada, H. and Shiotani, T.: On the Highest Water Waves of Permanent Type, *Bull. Disaster Prevention Res. Inst.*, Kyoto Univerity, Vol.18, part 2, No.135, 1968, pp.1-22.

30) Miche, M. : Movements Ondulatoires de la Mer en Profondeur Constante ou Décroissante, *Annales des Ponts et Chaussées*, Vol. 114, 1944, pp.25-406.

31) Hamada,T.: *Breakers and Beach Erosion*, Rept. Transportation Tech. Res. Inst., Ministry of Transportation, No.1, 1951, 165p.

32) Battjes, J.A.: Set-up due to Irregular Waves, *Proc. 13th Inter. Conf. Coastal Eng.*, 1972, pp.1993-2004.

33) Ostendrof, D.W. and Madsen, O.S.: *An Analysis of Longshore Currents and Associated Sediment Transport in the Surf Zone*, Rept. No. MITSG 79-13, MIT, 1978, 169p.

2.WAVE THEORIES AND, PROPAGATION AND DEFORMATION

34) Sakai, S., Hirayama, K. and Saeki, H.: An Expression for Influences of Opposing Current on Wave Breaking of Shoaling Wave on Uniform Slope, *Journal of Japan Society of Civil Engineers*, Vol.393, II-9, 1988, pp.43-48. (in Japanese)

35) Iwata, K. and Seino, H.: Experimental Study on the Breaking Limit of Partially Standing Waves, *Proc. 30th Coastal Engineering Conference*, JSCE, 1983, pp.1-4. (in Japanese)

36) Wiegel, R.L.: *Oceanographical Engineering*, Englewood Cliffs, N.J., Prentice-Hall, 1964, 532p.

37) Watanabe, A., Hara, A. and Horikawa, K.: On Wave Breaking in a Polymerised Wave Field, *Proc. 30th Coastal Engineering Conference*, JSCE, 1983, pp.5-9. (in Japanese)

38) Isobe, M.: Method for Calculating Refraction, Diffraction and Wave Breaking Deformation of Irregular Waves Using the Parabolic Equation, *Proc. 33rd Coastal Engineering Conference*, JSCE, 1986, pp.134-138. (in Japanese)

39) Mizuguchi, M., Tsujioka, K. and Horikawa, K.: A Study of Wave Height Changes after Wave Breaking, *Proc. 25th Coastal Engineering Conference*, JSCE, 1978, pp.155-159. (in Japanese)

40) Dibajnia, M. and Watanabe, A.: Numerical Model for simulating wave fields and beach profile changes, *Proc. 34th Coastal Engineering Conference*, JSCE, 1987, pp.291-295. (in Japanese)

41) LeMéhauté, B.: On Non-saturated Breakers and the Wave Run-up, *Proc. 8th Int. Conf. Coastal Eng.*, ASCE, 1962, pp.77-92.

42) Battjes, J.A. and Janssen. J.P.F.M.: Energy Loss and Set-up Due to Breaking of Random Waves, *Proc. 16th Int. Conf. Coastal Eng.*, ASCE, 1978, pp.569-587.

43) Watanabe, A. and Maruyama, K.: Numerical Analysis Method for Wave Fields Including Refraction, Diffraction and Dissipation Due to Wave Breaking, *Proc. 31st Coastal Engineering Conference*, JSCE, 1984, pp.103-107. (in Japanese)

44) Watanabe, A., Isobe, M., Izumiya, T. and Nakano, H.: Analysis of Shallow Water Wave Breaking Deformation of Irregular Waves by Unsteady Mild Slope Equations, *Proc. 35th Coastal Engineering Conference*, JSCE, 1988, pp.173-177. (in Japanese)

45) Sato, S. and Suzuki, H.: Method for Evaluating Bottom Flow Velocity Fluctuation Waveforms in Wave Breaking Zones, *Journal of Coastal Engineering*, JSCE, Vol. 37, 1990, pp.51-55. (in Japanese)

46) Vu, T.C., Yamamoto, Y., Tanimoto, K., and Arimura, J.: Simulation on Wave Dynamics and Scouring near Coastal Structures by a Numerical Model, *Proc.28th Inter.Conf. on Coastal Engineering,* ASCE, 2002, pp.1817-1829.

47) Ting, F.C.K. and Kirby, J.T.: Dynamics of surf zone turbulence in a strong plunging breaker, *Coastal Engineering*, Vol. 24, 1995, pp.177-204.

48) Ting F.C.K. and Kirby, J.T.: Dynamics of surf zone turbulence in a spilling breaker, *Coastal Engineering*, Vol.27, 1996, pp.131-160.

49) Nadaoka, K., Hino, M. and Koyano, Y.: Structure of the Turbulent Flow Field under Breaking

Waves in the Surf Zone, *J. Fluid Mechanics*, Vol. 204, 1989, pp.359-387.

50) Jones, W.P. and Launder, B.E.: The Prediction of Laminarization with a Two-equation Model of Turbulence, *Int. J. Heat Mass Transfer.*, Vol. 15, 1972, pp.301-314.

51) Schaffer, H.A., Madsen, P.A. and Deigaard, R.: A Boussinesq Model for Waves Breaking in Shallow Water, *Coastal Engineering*, Vol. 20, 1993, pp.185-202.

52) Bretschneider, C.L. and Reid, R.O.: Modification of Wave Height Due to Bottom Friction, and Reflection, *B.E.B. Tech. Memo.*, No. 45, 1954.

53) Jonsson, I.G.: Wave Boundary Layer and Friction Factors, *Proc. 10th Int. Conf. on Coastal Eng.*, 1966, pp.127-148.

54) Reid, R.O. and Kajiura, K.: On the Damping of Gravity Waves over a Permeable Sea Bed, *Trans. AGU*, Vol. 38, No.5, 1957, pp.662-666.

55) Ito, Y. and Tanimoto, K.: Numerical Wave Analysis Method and Its Applications - Waves around Structures -, *Proc. 18th Coastal Engineering Conference*, JSCE, 1971, pp.67-70. (in Japanese)

56) Barkhoff, J.C.W.: Computation of Combined Refraction - Diffraction, *Proc. 13th Int. Conf. on Coastal Eng.*, 1972, pp.471-490.

57) Kubo, Y., Kotake, Y., Isobe, M. and Watanabe, A.: On the Unsteady Mild Slope Irregular Wave Equation, *Journal of Coastal Engineering*, JSCE, Vol.38, 1991, pp.46-50. (in Japanese)

58) Radder, A.C.: On the Parabolic Equation Method for Water-wave Propagation, *J. Fluid Mechanics*, Vol. 95, 1979, pp.159-176.

59) Isobe, M., et al. 37: *Coastal Technology Library No.39 CADMAS-SURF/3D Research and Development of a Numerical Wave Tank*, Coastal Development Institute of Technology, 2010, 235p. (in Japanese)

Chapter:3

WAVE RUN-UP,
WAVE OVERTOPPING AND WAVE FORCES

3.WAVE RUN-UP, WAVE OVERTOPPING AND WAVE FORCES

3.1 Wave Run-up and Wave Overtopping

1) Influence of Wave Irregularity on Wave Run-up

The irregularity of sea waves should be considered when determining wave run-up on a beach. An initial consideration of the influence of wave irregularity on wave run-up was to apply a wave height-period coupling function to the run-up height of regular waves to obtain the run-up height of irregular waves, assuming that the individual waves of an irregular wave train have the same run-up motion as regular waves with the same wave height and period, as in Battjes [1] and Sawaragi et al [2].

Later, Mase [3], Sawaragi and Iwata [4], and some others have shown that the simple combination of the run-up heights of regular waves cannot correctly determine the run-up height of irregular waves because the following effects become more prominent as the seabed gradient becomes gentler.

(a) Influence of Return Flows on Wave Run-up

The return flow from run-up of the first wave hinders the run-up from the next. This phenomenon can occur even in the case of regular waves. In the case of irregular waves, if the first wave is smaller than the second wave, since the return flow from the first wave will be weak, the run-up height of the second wave will not be small, whereas the opposite occurs if the first wave is larger than the second wave [3].

(b) Influence of Wave Absorption and Catch-up Phenomena on Run-up Waves

When irregular waves propagate in the sea, faster propagating longer waves catch up and absorb the energy of slower propagating shorter waves. Due to this process, not all waves in an incoming wave train run-up on a beach. So, the wave train of run-up waves and that of offshore waves may no longer correspond wave by wave (like that of **Fig. 3.2**) [3].

(c) Influence of Long-period Waves on Run-up Waves

Long-period gravity waves with period of several ten seconds to several minutes called **infragravity waves** or **surf beat** are observed in the wave breaking zone. Because of the presence of the run-up component of the surf beat, it is often impossible to evaluate the actual run-up waves based only on the run-up of short-period waves alone.

According to studies by Mase [3], Kubota [5], Yamamoto and Tanimoto [6], and others, these phenomena are more pronounced when the mean seabed slope in the wave breaking zone is less than about 1/10.

When the mean seabed slope of the wave breaking zone is between about 1/10 and 1/20, the width of the wave breaking zone is larger than that when the seabed slope is steeper than 1/10. Therefore, the return flows of short-period waves and the effect of wave absorption and catch-up phenomena on run-up are more likely to appear on gentle slopes. However, the width of the wave-

3.1 Wave Run-up and Wave Overtopping

breaking zone on a steep beach may not be large enough for the short-period waves to decay sufficiently after breaking and to generate long-period gravity waves. Mase and Kobayashi [7] and Kubota et al [8] studied run-up waveform, while Sawaragi and Iwata [4] and Mase et al [9] studied run-up height in such situations.

When the mean seabed slope of the wave-breaking zone is less than about 1/20, the width of the breaking zone is sufficiently large and the short-period waves are considerably dissipated. In such situations, wave run-up is dominated by long-period waves. Studies of wave run-up in this case have been carried out by Sawaragi and Iwata [4], Mase et al [9], and Kato [10]. However, in hydraulic model experiments, the effect of multiple reflections of long-period waves by wave-making plates cannot be ignored. Therefore, the run-up height distribution formula of Sawaragi and Iwata [4] and the run-up height formula of Mase et al [9], which are based on hydraulic model experiments, are considered to include the effect of multiple reflections for gentle slope beaches where long-period wave components are dominant. On the other hand, the run-up formula of Kato [10] is based on field observation data and is free from multiple reflections although it is limited to an average seabed slope of about 1/60.

As actual sea waves have wave grouping characteristics as shown in **Fig.1.4**, irregular waves can be approximated by a bichromatic wave train (see **Fig.1.3**), which is a composite of two regular wave components. Hydraulic model experiments were conducted to check the effect of different seabed slopes on the run-up characteristics. The experimental results are shown in **Figs. 3.1** and **3.2**.
Fig.3.1 shows wave run-up on a 1/10th gradient slope, where the return flows of short-period waves and wave absorption/catch-up phenomena are observed.
Fig.3.2 shows the run-up waves on a 1/30th gradient slope, where the long-period gravity waves are dominant.

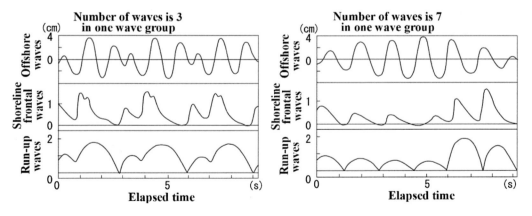

Fig. 3.1 Deformation during run-up of incident waves on a 1/10th gradient slope.
(from Yamamoto and Tanimoto [6])

3. WAVE RUN-UP, WAVE OVERTOPPING AND WAVE FORCES

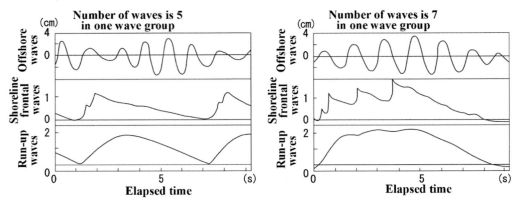

Fig. 3.2 Deformation during run-up of incident waves on a 1/30th gradient slope. (from Yamamoto and Tanimoto [6])

For further understanding of run-up phenomena caused by irregular waves, it is necessary to study long-period gravity waves within the wave breaking zone, which can have a significant influence on run-up waves.

2) Long-period Gravity Waves Due to Wave Grouping

If we are concerned with two-dimensional conditions, there are two theories for the generation of long-period gravity waves.

(a) Theory of Bound Long Waves (BLW)

The pioneering research on wave group induced bound long waves has been done by Longuet-Higgins and Stewart [11]. According to their research and the research done by some others [12, 13, 14], the mean water level outside the wave breaking zone varies periodically due to wave grouping, resulting in long period waves whose propagation velocity is bounded by the wave group velocity. These long-period waves propagate to the shore as free long waves after breaking because the wave grouping propagation velocity becomes equal to the propagation velocity of long waves near the breaking zone.

(b) Theory of Time-varying Breakpoint-forced Long Waves (BFLW)

This theory was put forward by Symonds et al [15]. They found that wave breaking position varies periodically in the cross-shore direction due to wave grouping which causes periodic changes in the mean water level within the wave breaking zone resulting in the generation of long period (infragravity) waves. The theory of Symonds et al [15] and its modifications are discussed in the following sections.

(1) Theory of Symonds et al.

Fig.3.3 shows the relationship between the movement of the wave breaking position and the elapsed time in a surf beat wave, where the vertical t-axis represents time, the horizontal x-axis

3.1 Wave Run-up and Wave Overtopping

represents the horizontal distance measured from the intersection of the still water surface, and the seabed slope, x_{b1} is the inner-most wave breaking position, and x_{b2} is the outer-most wave breaking position. The *H*-axis on the right side represents the wave height of the incident wave. The change in wave height over time is depicted along the *t*-axis.

Due to the time delay t_b in which wave breaking position of a small wave shifts from x_{b2} to x_{b1} in shallow water with changing water depth, the actual relationship between the movement of wave breaking position and the elapsed time is shown by the thick solid line in **Fig. 3.3(a)**. Symonds et al [15] used the relationship given by the dashed in **Fig. 3.3**, ignoring the time delay t_b, which could only give a qualitative explanation. Kato et al [16] and Nakamura and Katoh [17] showed that by taking t_b into account in the theory of Symonds et al [15], the surf beat in the wave breaking zone can be reproduced well. The shaded area in the figure indicates a situation where propagation of the small wave is superseded by the breaking of a large wave so that the contribution of the breaking of the small wave to the generation of the BFLW is disregarded. In this case, the linear variation of wave breaking position at *t=0* in **Fig. 3.3 (b)** is adopted.

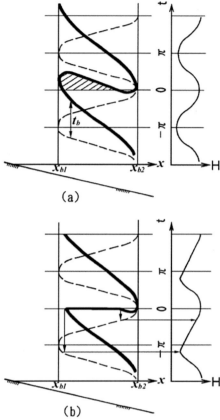

Thick solid line: the actual curve of wave breaking position.
Dashed line: the curve of wave breaking position ignoring the time delay.

Fig. 3.3 Relationship between the movement of wave breaking position and the elapsed time in a beat wave.

The offshore distribution of wave heights of long period waves using this modified model is obtained as shown by the dashed line in **Fig.3.4**. The horizontal axis in the figure is the dimensionless value of the horizontal distance *x* on the left side of **Fig.3.3**, non-dimensionalised using x_m (the mean value of x_{b1} and x_{b2}) and the vertical axis $H_{L1/3}$ and H_{Lm} are the significant wave height and mean wave height of long- period gravity waves, respectively.

The top three graphs in **Fig.3.4** compare calculated values with observed data by Kato et al [17] at Hasaki Coast, Ibaraki Prefecture, while the lower two graphs compare the experimental data by Yamamoto and Tanimoto [6].

In **Fig.3.4**, the calculated long-period wave height becomes abnormally large near the shoreline. Yamamoto and Tanimoto [6] obtained an analytical solution from Eq. (3.1), which takes into account the mean water surface elevation η_0 in the basic equation. The result from their analytical solution is given by the solid lines in **Fig. 3.4**.

3. WAVE RUN-UP, WAVE OVERTOPPING AND WAVE FORCES

$$\frac{\partial u}{\partial t}+g\frac{\partial \eta}{\partial x}=-\frac{1}{\rho(h+\eta_0)}\frac{\partial S_{xx}}{\partial x}, \quad \frac{\partial(h+\eta_0)u}{\partial x}+\frac{\partial \eta}{\partial t}=0 \tag{3.1}$$

where u and η are the velocity and the water surface elevation of the long-period gravity wave, t is the time, g is the acceleration of gravity, x is the offshore coordinate, ρ is the density of seawater, h is the depth from the still water surface, and S_{xx} is the depth-integrated and phase-integrated excess momentum flux caused by the surface gravity waves, called **radiation stress**. The variation of radiation stress is the external force that generates long-period gravity waves.

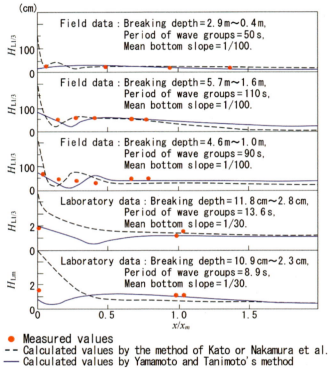

Fig. 3.4 Comparison of measured and calculated offshore distribution of long-period gravity waves.

The calculation of long-period gravity waves can be performed by solving a non-linear equation of motion with a wave breaking term; but to obtain an analytical solution, a linear equation of motion is adopted here, where the external force is the radiation stress obtained from the wave height after taking into account wave breaking.

The radiation stress can be obtained by integrating the momentum transport equation of finite amplitude waves over one wave period, and considering the second-order terms. This can therefore be called the excess momentum flux that is not cancelled out over one wave period. This excess momentum flux generates a shoreward flow from the wave breaking point, causing the mean water surface to rise towards the shoreline within the wave breaking zone. Eq. (3.1) takes this mean water surface elevation η_0 into account.

3.1 Wave Run-up and Wave Overtopping

(2) Solution to Eq. (3.1) with the Mean Water Surface Elevation

Assuming that the amplitude of the input wave within the breaking zone is expressed as the amplitude-depth ratio γ (which approaches 0.50 from 0.39 as the seabed gradient steepens) multiplied by the water depth, and using results of the Airy wave theory, the amplitude A_{mp} and the radiation stress S_{xx} of the input wave on a uniform slope with seabed gradient $\tan\beta$ are expressed by Eq. (3.2).

$$A_{mp} = \gamma \times \left(x \times \tan\beta + \eta_0\right), \qquad S_{xx} = \frac{3}{2} \times \frac{1}{8} \rho g H^2 = \frac{3}{4} \rho g \left(A_{mp}\right)^2 \tag{3.2}$$

Next, Eq. (3.1) is nondimensionalized by Eqs. (3.3).

$$\left.\begin{array}{l} A = \dfrac{a}{x_{bm}}, \quad X = \dfrac{x}{x_{bm}}, \quad X_{b1} = \dfrac{x_{b1}}{x_{bm}}, \quad X_{b2} = \dfrac{x_{b2}}{x_{bm}}, \quad \tau = \dfrac{2\pi}{T_L} t \\[4mm] U(X,\tau) = \dfrac{u(x,t)}{1.5\gamma^2 \dfrac{2\pi}{T_L} X}, \quad Z(X,\tau) = \dfrac{\eta(x,t)}{1.5\gamma^2 x_{bm} \tan\beta} \end{array}\right\} \tag{3.3}$$

where a is $\left(x_{b1} - x_{b2}\right)\big/2$; and x_{bm} is $\left(x_{b1} + x_{b2}\right)\big/2$.

From non-dimensionalized basic equations, the differential equation for $Z(X,\tau)$ is found by eliminating $U(X,\tau)$ and solved to obtain the following equation for the dimensionless water surface elevation $Z(X,\tau)$.

$$Z(X,\tau) = Z_0(X) + \sum_{n=1}^{\infty} Z_n(X,\tau) \tag{3.4}$$

(a) From Shore to X_{b1}

$$\left.\begin{array}{l} Z_0(X) = \dfrac{1 - X + 1.5\gamma^2 A}{1 + 1.5\gamma^2} \\[4mm] Z_n(X,\tau) = -\left(I_{bJ} + I_{aN}\right) J_0(z)\cos(n\tau) - I_{bN} J_0(z)\sin(n\tau) \\[4mm] z = 2n \dfrac{2\pi}{T_L} \left(\dfrac{x_{bm}}{g\tan\beta}\right)^{\!\!1/2} \left(1 + 1.5\gamma^2\right)\left(X + 1.5\gamma^2 Z_0\right)^{\!1/2} \end{array}\right\} \tag{3.5}$$

where $J_0(z)$ is a zero-order Bessel function.

(b) From X_{b1} to X_{b2}

$$\left.\begin{array}{l} Z_0(X) \doteqdot \left\{(1-X)\cos^{-1}\left[(X-1)\big/A\right] - \left[A^2 - (X-1)^2\right]^{1/2}\right\}\Big/\pi \\[3mm] Z_n(X,\tau) = \left[-\left(I_{bJ} + I_{aN}\right)J_0(z) + C_n N_0(z) + \eta_{pa}\right]\cos(n\tau) + \left[-I_{bN}J_0(z) + \eta_{pb}\right]\sin(n\tau) \\[3mm] C_n = 0 \quad \text{(shore side)} \;, \qquad C_n = I_{aJ} \quad \text{(offshore side)} \\[3mm] z = 2n\dfrac{2\pi}{T_L}\left(\dfrac{x_{bm}}{g\tan\beta}\right)^{\!\!1/2}\left\{1 + 1.5\gamma^2\cos^{-1}\left[(X-1)\big/A\right]\big/\pi\right\}\left(X + 1.5\gamma^2 Z_0\right)^{\!1/2} \\[3mm] \eta_{pa} \doteqdot 2\pi\left[\int_{xb1}^{x} X_{an}N_0(z)\,dX \times J_0(z) - \int_{xb1}^{x} X_{an}J_0(z)\,dX \times N_0(z)\right] \\[3mm] \eta_{pb} \doteqdot 2\pi\left[\int_{xb1}^{x} X_{bn}N_0(z)\,dX \times J_0(z) - \int_{xb1}^{x} X_{bn}J_0(z)\,dX \times N_0(z)\right] \\[3mm] X_{an} = d\left[\left(X + 1.5\gamma^2 Z_0\right)a_n\right]\big/dX \\[3mm] X_{bn} = d\left[\left(X + 1.5\gamma^2 Z_0\right)b_n\right]\big/dX \end{array}\right\} \tag{3.6}$$

where $N_0(z)$ is a zero-order Neumann function.

101

3.WAVE RUN-UP, WAVE OVERTOPPING AND WAVE FORCES

(c) From X_{b2} to Offshore

$$\left.\begin{array}{l} Z_0(X)=0 \\ Z_n(X,\tau)=-I_{bJ}J_0(z)\cos(n\tau)-I_{bJ}N_0(z)\sin(n\tau) \\ z=2n\dfrac{2\pi}{T_L}\left(\dfrac{x_{bm}}{g\tan\beta}\right)^{1/2}X^{1/2} \end{array}\right\} \tag{3.7}$$

I_{aJ}, I_{aN}, I_{bJ} and I_{bN} of Eqs. (3.5) – (3.7) are defined by Eq. (3.8) using X_{an}, X_{bn} and z of Eq. (3.6).

$$\left.\begin{array}{ll} I_{aJ}\doteqdot2\pi\int_{xb1}^{xb2}X_{an}J_0(z)dX, & I_{aN}\doteqdot2\pi\int_{xb1}^{xb2}X_{an}N_0(z)dX \\ I_{bJ}\doteqdot2\pi\int_{xb1}^{xb2}X_{bn}J_0(z)dX, & I_{bN}\doteqdot2\pi\int_{xb1}^{xb2}X_{bn}N_0(z)dX \end{array}\right\} \tag{3.8}$$

The method of integration by parts is used to perform the integrals in the above equations. The results of the calculations using Eq. (3.2) and Eqs. (3.4) - (3.8) are the solid lines shown in Fig.3.4, and the calculated values near the shores are in very good agreement with the measured values (including components due to BLW).

(3) Numerical Model and Generation Characteristics of Long-period Gravity Waves

Goda et al. [18], List [19, 20], and Yamamoto and Tanimoto [21] proposed a numerical model in which the external force is the radiation stress described in Section 4.1 1).

In particular, Yamamoto and Tanimoto [21] calculated long-period gravity waves on an arbitrary sea bottom topography using the numerical model based on equation (3.9) (the distribution of the radiation stress is given only within the wave breaking zone for the BFLW calculation, and the distribution of the radiation stress is given only outside the wave breaking zone for the BLW calculation. Furthermore, the BLW based on Longuet-Higgins and Stewart's theory is given at the offshore boundary). The results show that BLW is dominant when the width of the wave-breaking zone is narrow and a shallow water area where the group velocity is close to the phase velocity of the long wave extends outside of the wave-breaking zone, while BFLW is dominant when the width of the wave breaking zone is wide on a shelf-like shallow beach.

$$\frac{\partial q}{\partial t}+g(h+\eta)\frac{\partial\eta}{\partial x}=-\frac{1}{\rho}\frac{\partial S_{xx}}{\partial x}, \qquad \frac{\partial q}{\partial x}+\frac{\partial\eta}{\partial t}=0 \tag{3.9}$$

where q is the flow rate per unit width [$= u(h+\eta)$] of long-period gravity waves, u is the horizontal velocity, h is the water depth from the still water surface, η is the water surface elevation, t is the time, g is the acceleration of gravity, x is the offshore coordinate, ρ is the density of seawater, and S_{xx} is the radiation stress.

(4) Empirical Expression for Long-period Gravity Wave Parameters

As a simple way to determine the characteristics of long-period gravity waves in the wave breaking zone without using the analytical solutions or numerical models introduced earlier, Yamamoto and Tanimoto [6] proposed empirical equations by combining their experimental data with

3.1 Wave Run-up and Wave Overtopping

measured data by Goda [22], Yamaguchi and Hatada [23], Hirose and Hashimoto [24], Kato [10, 25], Kato et al. [26], Iwata et al. [27], Mase and Kobayashi [7], and Mase et al. [9]. These equations can be used as a guide, as the correlation with actual measured values is 70-90%, which is hardly good enough.

$$\frac{H_{L1/3}}{H_{o1/3}} \doteqdot \frac{0.066 \times (\tan \beta)^{1/6}}{\sqrt{(H_{o1/3}/L_{o1/3})(1+h/H_{o1/3})}}, \quad H_{Lm} \doteqdot \frac{H_{L1/3}}{1.5} \quad (1/10 \geq i \geq 1/70) \tag{3.10}$$

$$\left.\begin{array}{c} T_{L1/3} \doteqdot 305.7\sqrt{\dfrac{H_{L1/3}}{g}} \\[3mm] T_{Lm} \doteqdot \dfrac{T_{L1/3}}{1.3} \doteqdot 288.0\sqrt{\dfrac{H_{Lm}}{g}} \end{array}\right\} \quad (1/20 \geq i \geq 1/60 \text{ and } H_{om}/L_{om} \geq 0.015) \tag{3.11}$$

$$\left.\begin{array}{cc} \dfrac{R_{Lm}}{H_{om}} \doteqdot \dfrac{1.52 \times \tan \beta}{\sqrt{H_{om}/L_{om}}}, & \dfrac{R_{Lm}}{H_{o1/3}} \doteqdot \dfrac{\tan \beta}{\sqrt{H_{o1/3}/L_{o1/3}}} \\[3mm] R_{L\max} \doteqdot 2.4 R_{Lm}, & R_{L1/3} \doteqdot 1.5 R_{Lm} \end{array}\right\} \quad (1/20 \geq i \geq 1/60) \tag{3.12}$$

where $H_{L1/3}$, H_{Lm}, $T_{L1/3}$, T_{Lm}, R_{Lm}, R_{Lmax}, and $R_{L1/3}$ are the significant wave height, the mean wave height, the significant period, the mean period, the mean run-up height, the maximum run-up height and the significant run-up height of long-period gravity waves, respectively; $H_{o1/3}$, H_{om}, $L_{o1/3}$, and L_{om} are the significant offshore wave height, the mean offshore wave height, the significant offshore wavelength, and the mean offshore wavelength, respectively. Moreover, $\tan\beta$ is the average seabed slope within the wave breaking zone, h is the water depth, and g is the acceleration of gravity.

3) Calculation of Wave Run-up Height

Calculation of run-up height from large waves is important to determine the crown height of coastal dikes or seawalls that can prevent damage from wave overtopping on land. However, they can be uneconomical if the crown height is determined to ensure absolutely no wave overtopping. Therefore, in practice, the crown height is determined so that unacceptable damage does not occur. Incidentally, a **coastal dike** is an embankment built to prevent seawater intrusion on a low-lying coastal area, and to prevent beach erosion by waves. The seaward surface of a coastal dike is usually covered with concrete or other materials (see **Fig. 7.1**). On a high-lying coastal area, a wall-like structure called a **seawall** is built of concrete or other materials to protect the seaward edge of the coast from the erosion by waves (see **Fig. 7.2**).

(1) Calculation Methods for Regular Waves

Among many methods for calculating run-up heights of regular waves, experimental formulas of Takada [28] and Hunt [29] were developed for the case that the water depth at the toes of coastal dikes and seawalls is deeper than the wave breaking water depth. Moreover, the virtual slope method of Saville [30] and the improved virtual slope method of Nakamura et al. [31] were developed for the

3.WAVE RUN-UP, WAVE OVERTOPPING AND WAVE FORCES

case that the water depth at the toes of coastal dikes and seawalls is shallower than the wave breaking water depth.

(a) Takada's Formula (for the case that the frontal water depth is deeper than the wave breaking depth)

When the water depth at the toe of a coastal dike or a seawall is deeper than the wave breaking water depth, in addition to the incident wave parameters, the surface slope of the target structure has a significant effect on the wave run-up height. Takada [28] proposed Eq. (3.13) to obtain wave run-up height R in this situation. Eq. (3.13) is applicable when $\cot\alpha < 8$.

$$
\left.
\begin{aligned}
&\left[\text{For the case that waves do not break due to a steep slope of the surface}\right]\\
&\alpha_c \le \alpha: \quad \frac{R}{H_o{}'} = \left(\sqrt{\frac{\pi}{2\alpha}} + \frac{h_o}{H}\right) K_s\\
&\left[\text{For the case that waves break due to a gentle slope of the surface}\right]\\
&\alpha_c > \alpha: \quad \frac{R}{H_o{}'} = \left(\sqrt{\frac{\pi}{2\alpha_c}} + \frac{h_o}{H}\right) K_s \left(\frac{\cot\alpha_c}{\cot\alpha}\right)^{2/3}
\end{aligned}
\right\}
\tag{3.13}
$$

$$
\frac{H_o{}'}{L_o} = \sqrt{\frac{2\alpha_c}{\pi}} \frac{\sin^2\alpha_c}{\pi}
\tag{3.14}
$$

$$
\frac{h_o}{H} = \pi \frac{H}{L} \coth \frac{2\pi h}{L} \left[1 + \frac{3}{4\sinh^2\left(2\pi h/L\right)} - \frac{1}{4\cosh^2\left(2\pi h/L\right)}\right]
\tag{3.15}
$$

where $H_o{}'$ is the equivalent offshore wave height; L_o is the offshore wavelength; α is the surface slope angle of the coastal dike or the seawall; α_c is the critical surface slope angle of the coastal dike or the seawall when wave breaking occurs, which can be obtained from Miche's perfect reflection condition of Eq. (3.14); and h_o/H is the ratio between the crest height above the still water level of standing waves and the incident wave height, which can be obtained from Eq. (3.15). H, L, and h are respectively the incident wave height, the incident wavelength, and the water depth at the toe of the coastal dike or the seawall; and K_s is the shoaling coefficient.

(b) Hunt's formula (for the case that the frontal water depth is deeper than the wave breaking depth)

When the water depth at the toe of the coastal dike or the seawall is deeper than the wave breaking depth, and if waves break on a tilted surface of the target structure, the mean surface slope from the wave breaking point to the wave run-up height and incident wave parameters have a significant effect on the wave run-up height. Hunt [29] proposed Eq. (3.16) to obtain wave run-up height R in this situation.

$$
\frac{R}{H} = 1.01 \frac{\tan\alpha}{\sqrt{H/L_o}}
\tag{3.16}
$$

where α is the mean slope angle of the surface from the wave breaking depth to the wave run-up height, H is the incident wave height at the toe of the target structure, and L_o is the offshore wavelength.

3.1 Wave Run-up and Wave Overtopping

(c) Saville's Virtual Slope Method (for the case that the frontal water depth is shallower than the wave breaking depth)

If the water depth at the toe of the coastal dike or the seawall is shallower than the wave breaking water depth, it is necessary to consider two factors: the surface slope of the target structure and the cross-shore seabed topography between wave breaking point and the toe of the seawall. Therefore, for target structures with composite slopes shown in **Fig.3.5**, Saville [30] proposed a diagram to obtain wave run-up height (**Fig.3.6**) using the average surface slope angle α (called the **virtual slope angle**). The run-up height should be obtained iteratively by taking wave breaking depth as the first approximation of wave run-up height. The iterations should be repeated until the result converges within a given allowable error. **Fig. 3.6** shows that when the wave steepness H_o'/L_o is greater than 0.004, the wave run-up height for $1.4 < \cot\alpha < 4$ tends to be larger than that for $\cot\alpha < 1.0$. Therefore, the sloping walls tend to have larger run-up heights than near-vertical walls.

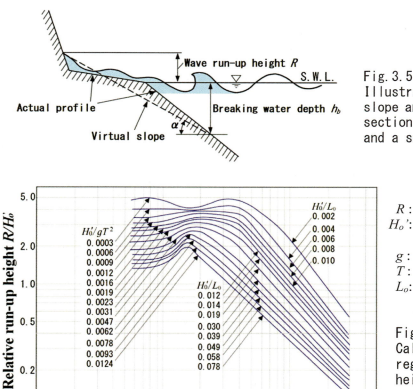

Fig. 3.5 Illustration of the virtual slope angle α on a composite section consisting of a dike and a sandy beach.

R : Wave run-up height.
H_o' : Incident offshore wave height.
g : Acceleration of gravity.
T : Wave period.
L_o : Incident offshore wavelength.

Fig. 3.6 Calculation diagram of regular wave run-up heights by Saville [30].

If beach erosion progresses with time, the sandy beach in front of a seawall can disappear and the slope can become steeper up to the wave breaking depth. This process may lead to the formation of a gentle slope offshore of the immediate front of the structure as shown in **Fig. 3.7**.

3. WAVE RUN-UP, WAVE OVERTOPPING AND WAVE FORCES

In this case, as understood from the definition of the virtual slope shown in **Fig. 3.5**, the maximum run-up height tends to occur when the wave breaks at the location of the sudden change in the seabed slope shown in **Fig. 3.7**.

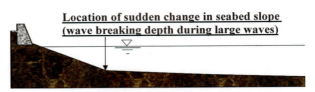

Fig. 3.7 Typical beach cross-sectional profile of an eroding beach.

(d) Improved Virtual Slope Method of Nakamura et al. (for the case that the frontal water depth is shallower than the wave breaking depth)

Saville's virtual slope method ignores the effect of the cross-shore topography between the wave run-up height and the wave breaking depth. Therefore, Nakamura et al. [31] defined the **improved virtual slope angle** α obtained by Eq. (3.17). In **Fig. 3.8**, the area of the triangle **abc** formed using the improved virtual slope line (dashed line **ab**) is equal to the shaded area A between the wave run-up height R and the wave breaking depth h_b.

The wave run-up height R can be obtained from **Fig. 3.9**, created based on extensive experimental data and one field observation, using the improved virtual slope angle α and the wave steepness H_o'/L_o. Similarly to Saville's calculation method, the calculation must be repeated until convergence.

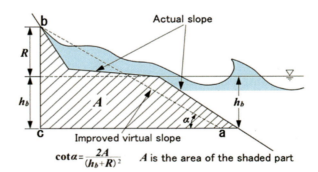

$$\cot\alpha = \frac{2A}{(h_b + R)^2} \quad (3.17)$$

Fig. 3.8
Illustration of the improved virtual slope angle α on a composite section consisting of a dike and a sandy beach.

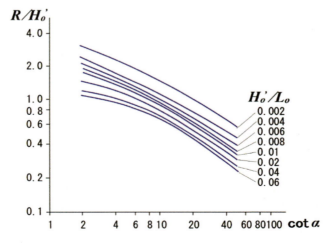

R : Wave run-up height.
H_o': Incident offshore wave height.
L_o: Incident offshore wavelength.

Fig. 3.9
Calculation diagram of wave run-up heights by Nakamura et al [31].

3.1 Wave Run-up and Wave Overtopping

Calculating the improved virtual slope angle α using Eq. (3.17) means that the average slope angle α is calculated by fixing the point b at the wave run-up height R in **Fig. 3.8**. Therefore, in **Fig. 3.7**, the improved virtual slope angle tends to be larger than the Saville's virtual slope angle when the beach section erodes to the lower left, and conversely, the improved virtual slope angle tends to be smaller than the virtual slope angle when the beach section is deposited to the upper right.

(e) Yamamoto's Formula (for the case that the frontal water depth is shallower than the wave breaking depth)

Yamamoto [32] proposed a calculation method to generate **Fig. 3.9**. The calculations were done using a computer thereby greatly reducing the iterative calculation effort required by the improved virtual method of Nakamura et al. First, based on the energy conservation law, Eq. (3.18) relating the run-up height R to the maximum run-up velocity u_s on the shoreline, and Eq. (3.19) relating the velocity u_s and the maximum water surface elevation η_s on the shoreline to the breaking wave height H_b were obtained. Next, after Eq. (3.20) relating the wave breaking height to offshore wave parameters by Sunamura and Horikawa [33] was substituted to Eq. (3.19), Eq. (3.21) was obtained by determining an unknown coefficient using **Fig. 3.9**. With this equation, the run-up height of regular waves after breaking on the composite beach cross-section can be obtained, and the correlation coefficient between Eq. (3.21) and **Fig. 3.9** is 0.992, which is sufficiently accurate.

$$R = \left(1 - C_f\right)\frac{u_s^2}{2g} \tag{3.18}$$

$$u_s = C_s\sqrt{g\eta_s}\cos\alpha, \qquad \eta_s = 0.8\left(\tan\alpha\right)^{0.6} H_b \tag{3.19}$$

$$H_b = \left(\tan\alpha\right)^{0.2}\left(H_o/L_o\right)^{-1/4} H_o \tag{3.20}$$

$$\left.\begin{array}{c} R = 1.25\left(\cos\alpha\right)^2\left(\tan\alpha\right)^{0.6}\left(\dfrac{H_o}{L_o}\right)^{-\frac{1}{4}} H_o \qquad \left[1/3 \geq \tan\alpha \geq 1/50\right] \\[2ex] \alpha = \tan^{-1}(R + h_b)^2\big/(2A) \end{array}\right\} \tag{3.21}$$

where R is the wave run-up height, C_f is the energy loss coefficient, g is the acceleration of gravity, C_s is the correction factor, α is the improved virtual slope angle of Nakamura et al., H_o is the incident offshore wave height and L_o is the incident offshore wavelength, h_b is the wave breaking depth, and A is the area of the shaded part in **Fig. 3.8**.

(f) Comparison of Saville's Method with Nakamura et al.'s Method (Yamamoto's Formula)

Comparing **Fig. 3.6** and **Fig. 3.9**, the wave run-up heights calculated by Saville are larger for $\cot\alpha$ values of up to about 10, while the wave run-up heights calculated by Nakamura et al. tend to be larger for $\cot\alpha$ values of 10 or greater. As a result, the tendency shown in **Fig. 3.10** is recognized. The area shaded in yellow in **Fig. 3.10** indicates the approximate range of the ratio of the value calculated using Saville's method to the measured value, and the area shaded in black indicates the

3. WAVE RUN-UP, WAVE OVERTOPPING AND WAVE FORCES

approximate range of the ratio of the value calculated using Nakamura et al.'s method (calculated using Yamamoto's formula) to the measured value.

The figure shows that Saville's method is more accurate when the mean seabed slope from wave breaking depth to wave run-up height is steep than 1/14, and Nakamura et al.'s method (Yamamoto's formula) is more accurate when the mean seabed slope is between 1/15 and 1/24. When the mean seabed slope is less than 1/25, Saville's method underestimates the wave run-up height too much.

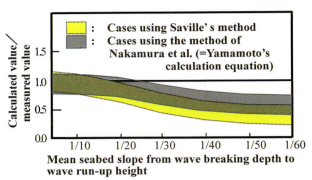

Fig. 3.10 Relationship between the ratio of calculated to measured wave run-up heights and the mean seabed slope.

Generally, when the seabed slope is smaller than 1/20, the calculation method based on regular waves underestimates the wave run-up height because the effect of long-period gravity waves is larger. Yamamoto et al [34] clarified the following:

(ⅰ) When the improved virtual slope is smaller than 1/20, by adding half of the wave height at the shoreline of the long-period gravity wave obtained from Eq. (3.10) to the water depth, the wave run-up height calculated using the method of Nakamura et al. becomes approximately equal to the measured value.

(ⅱ) When the improved virtual slope is smaller than 1/40, the measured wave run-up height is almost equal to the run-up height of the long-period gravity wave.

(g) Reduction Effect of Gravel and Stone Layers on Wave Run-up Heights

In wave run-up experiments on sand and gravel slopes by Savage [35] and others, the reduction effect of the permeability and the surface roughness of a sand layer on wave run-up height is not significant, although the reduction effect of a gravel layer cannot be neglected.

It is common to ignore the effect of the permeability and surface roughness of the beach on reducing wave run-up height when the front of the coastal dike or the seawall is sandy. However, when there is a gravel beach or a stone layer in front of the structure, the reduction effect cannot be ignored. Since the reduction effect depends on the ratio between the total area and the area covered by gravel or stone in the wave run-up zone, the layer thickness and median grain size of gravel or stone, etc., this effect should be confirmed by model experiments using irregular waves.

3.1 Wave Run-up and Wave Overtopping

(2) Calculation Methods for Irregular Waves

If the mean seabed slope is smaller than 1/20, the wave run-up height should be calculated using a method based on irregular waves with wave grouping characteristics, and the following are typical calculation methods for irregular waves.

(a) Mase's formula

Mase [36] proposed Eq. (3.22), which can calculate wave run-up height on an impermeable smooth slope, using experimental data based on irregular waves.

$$
\left.
\begin{aligned}
&\frac{R}{H_o{}'} = a\left(\frac{\tan\alpha}{\sqrt{H_o{}'/L_o}}\right)^b \qquad \left[1/5 \geq \tan\alpha \geq 1/30,\ H_o{}'/L_o \geq 0.007\right] \\[4pt]
&\text{For the maximum run-up height}\,(R = R_{\max}): \quad a = 2.32,\ b = 0.77 \\
&\text{For the run-up height in excess of 2\%}\,(R = R_{2\%}): \quad a = 1.86,\ b = 0.71 \\
&\text{For the highest one-tenth run-up height}\,(R = R_{1/10}): \quad a = 1.70,\ b = 0.71 \\
&\text{For the highest one-third run-up height}\,(R = R_{1/3}): \quad a = 1.38,\ b = 0.70 \\
&\text{For the average run-up height}\,(R = R_{mean}): \quad a = 0.88,\ b = 0.69
\end{aligned}
\right\} \tag{3.22}
$$

where $H_o{}'$ and L_o are respectively the significant wave height and wavelength in equivalent offshore waves, and $\tan\alpha$ is the slope of the beach cross-section.

(b) Van der Meer and Stam's Formula

Van der Meer and Stam [37] proposed formulae for determining wave run-up height on a uniform rubble slope using experimental data based on irregular waves. The upper 2% excess run-up height $R_{2\%}$ on the rubble slope can be obtained using Eq. (3.23). The highest one-third run-up height $R_{1/3}$ on the rubble slope can be obtained using Eq. (3.24).

$$
\left.
\begin{aligned}
&\frac{R_{2\%}}{H_{1/3}} = 0.96\times\left(\frac{\tan\alpha}{\sqrt{H_{1/3}/L_{om}}}\right) \qquad \left[\frac{\tan\alpha}{\sqrt{H_{1/3}/L_{om}}} \leq 1.5\right] \\[6pt]
&\frac{R_{2\%}}{H_{1/3}} = 1.17\times\left(\frac{\tan\alpha}{\sqrt{H_{1/3}/L_{om}}}\right)^{0.46} \qquad \left[\frac{\tan\alpha}{\sqrt{H_{1/3}/L_{om}}} \geq 1.5\right]
\end{aligned}
\right\} \tag{3.23}
$$

$$
\left.
\begin{aligned}
&\frac{R_{1/3}}{H_{1/3}} = 0.72\times\left(\frac{\tan\alpha}{\sqrt{H_{1/3}/L_{om}}}\right) \qquad \left[\frac{\tan\alpha}{\sqrt{H_{1/3}/L_{om}}} \leq 1.5\right] \\[6pt]
&\frac{R_{1/3}}{H_{1/3}} = 0.88\times\left(\frac{\tan\alpha}{\sqrt{H_{1/3}/L_{om}}}\right)^{0.41} \qquad \left[\frac{\tan\alpha}{\sqrt{H_{1/3}/L_{om}}} \geq 1.5\right]
\end{aligned}
\right\} \tag{3.24}
$$

where $H_{1/3}$ is the incident significant wave height at the toe of the coastal dike or the seawall, L_{om} is the mean wavelength of offshore waves ($=1.56T_m^2$: T_m is the mean period), and $\tan\alpha$ is the mean slope of the coastal cross-section.

3. WAVE RUN-UP, WAVE OVERTOPPING AND WAVE FORCES

(c) Calculation Method of De Waal and Van der Meer

De Waal and Van der Meer [38] proposed Eq. (3.25) for calculating the upper 2% excess run-up height $R_{2\%}$ on a composite coastal section using experimental data based on irregular waves.

$$\frac{R_{2\%}}{H_{1/3}} = \begin{cases} 1.5\gamma_h\gamma_\theta\gamma_b\gamma_f \dfrac{\tan\alpha}{\sqrt{H_{1/3}/L_{om}}} & \left[0.5 < \dfrac{\tan\alpha}{\sqrt{H_{1/3}/L_{om}}} \leq 2\right] \\ 3.0\gamma_h\gamma_\theta\gamma_f & \left[2 < \dfrac{\tan\alpha}{\sqrt{H_{1/3}/L_{om}}}\right] \end{cases} \quad (3.25)$$

where $H_{1/3}$ is the incident significant wave height at the toe of the coastal dike or the seawall, L_{om} is the mean wavelength of offshore waves, and $\tan\alpha$ is the mean surface slope between the surface slope of the structure and the slope of the beach cross-section from the wave breaking depth to the front of the structure (refer to **Fig. 3.11**).

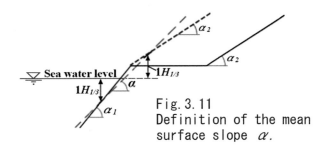

Fig. 3.11 Definition of the mean surface slope α.

γ_h is the reduction factor due to shoaling deformation as the wave height distribution deviates from the Rayleigh distribution and is obtained from Eq. (3.26).

$$\gamma_h = \frac{H_{2\%}}{1.4 H_{1/3}} \quad (3.26)$$

where $H_{2\%}$ is the wave height in excess of the upper 2%.

γ_θ is the reduction factor due to obliquely incident waves shown in **Fig. 3.12** and is obtained using Eq. (3.27) in the case of unidirectional waves and Eq. (3.28) in the case of multi-directional waves.

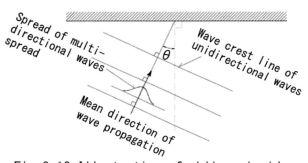

Fig. 3.12 Illustration of oblique incidence of waves.

For unidirectional waves:
$$\gamma_\theta = \begin{cases} 1.0 & \left[0° \leq \theta \leq 10°\right] \\ \cos(\theta - 10°) & \left[10° < \theta \leq 63°\right] \\ 0.6 & \left[63° < \theta\right] \end{cases} \quad (3.27)$$

For multi-directional waves:
$$\gamma_\theta = 1 - 0.0022\theta \quad (3.28)$$

γ_b is the reduction factor due to the convex shape (root protection work, etc.) shown in **Fig. 3.13**, and is obtained from Eq. (3.29).

$$\gamma_b = 1 - r_B(1 - r_{dB}), \quad r_B = 1 - \frac{\tan\alpha_{eq}}{\tan\alpha}, \quad r_{dB} = 0.5(d_B/H_{1/3})^2 \quad (3.29)$$

where $\gamma_b = 1.0 - 0.6$, $r_{dB} = 1.0 - 0$.

3.1 Wave Run-up and Wave Overtopping

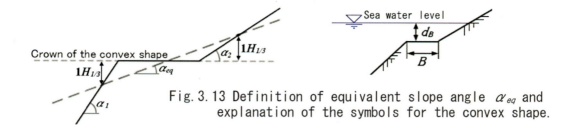

Fig. 3.13 Definition of equivalent slope angle α_{eq} and explanation of the symbols for the convex shape.

γ_f is the reduction factor due to the roughness of the sloping surface and ranges from 1.0 to 0.5 as shown in **Table 3.1**.

Table 3.1 Relationship between conditions of the slope and reduction factor γ_f

Conditions of the slope	γ_f
Smooth concrete or asphalt	1.0
Smooth block seawall	1.0
Grass (3cm long)	0.90 – 1.0
1) One layer of stones of diameter D ($H_{1/3}/D$ = 1.5 – 3.0)	0.55 – 0.6
2) Multi-layers of stones of diameter D ($H_{1/3}/D$ = 1.5 – 6.0)	0.50 – 0.55
Square block (length = width = b, height = h) h/b　　$b/H_{1/3}$　　　　Target Area 0.88　　0.12 – 0.19　　　　1/9	0.70 – 0.75
0.88　　0.12 – 0.24　　　　1/25	0.75 – 0.85
0.44　　0.12 – 0.24　　　　1/25	0.85 – 0.95
0.88　　0.12 – 0.18　　　　1/25 (on the sea level)	0.85 – 0.95
0.18　　0.55 – 1.10　　　　1/4	0.75 – 0.85
Rib-shape (length = width = b, height = h) h/b　　$b/H_{1/3}$　　　　Target Area 1.00　　0.12 – 0.19　　　　1/7.5	0.60 – 0.70

(3) Effect of Wave Direction on Wave Run-up height

The more oblique the incident wave angle to the cross-shore direction of the coastal dike or the seawall, the lower the wave run-up height becomes. Hosoi and Suto [39] obtained the results shown in **Fig. 3.14** from wave run-up experiments with different incident angles. From this figure, within the range shown in Eq. (3.30), the wave run-up height can be reduced using the reduction ratio K_θ determined from the wave incident angle θ (the intersection angle between the line perpendicular to the wall surface and the wave direction line) and the wave steepness H_o'/L_o.

$$\frac{1+\cos\theta}{2} \geq \text{reduction ratio } K_\theta \geq \cos\theta \qquad (3.30)$$

3. WAVE RUN-UP, WAVE OVERTOPPING AND WAVE FORCES

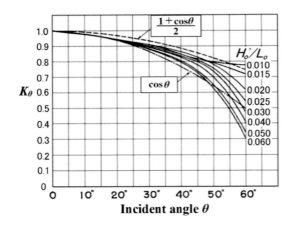

Fig. 3.14
Relationship between the incident angle of waves and the reduction ratio of wave run-up heights according to Hosoi and Suto [39].

θ is the intersection angle between the line perpendicular to the wall surface and the wave direction line.
K_θ is the ratio of the wave run-up height at oblique angle to the wave run-up height at right incident angle.

(4) Effect of Wind on Wave Run-up Height

Tailwinds increase wave run-up heights. Sibul and Tickner [40] conducted wind tunnel experiments and obtained **Fig. 3.15**, which shows that the wave run-up height increases with increasing wind speed. Furthermore, Yamamoto et al [41] showed that, when there is a tailwind of 5 m/s or more, a calculated run-up height close to the measured value can be obtained if the calculated wave run-up height at no wind is multiplied by the increased ratio of wave run-up height due to wind obtained from **Fig. 3.15**.

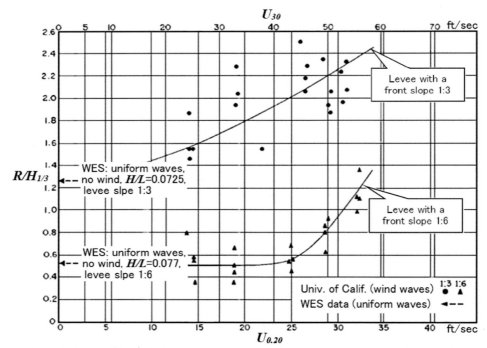

U_{30}: wind speed about 30 ft above the mean water surface in a logarithmic law distribution,
$U_{0.20}$: wind speed about 0.20 ft above the mean water surface.

(For example, for a front slope 1:3, $R/H_{1/3}$=1.26 of no wind increases to $R/H_{1/3}$=1.68 at U_{30}=30ft/sec)

Fig. 3.15 Relationship between wind speed and wave run-up height in the wind tunnel experiment by Sibul and Tickner [40].

3.1 Wave Run-up and Wave Overtopping

4) Calculation of Wave Overtopping Rate

The design crown height of a coastal dike or a seawall should be high enough to keep wave overtopping below the allowable wave overtopping rate for the target coast. On the other hand, the actual crown height is also required to be as low as possible to maintain low construction costs, easy access to the coast, and prevention of landscape deterioration.

Actual waves are irregular, and the maximum wave height is approximately twice the significant wave height. Therefore, when calculating the wave overtopping rate using regular waves, it is easy to underestimate the damage caused by wave overtopping if the significant wave height is used, and overestimate the damage caused by wave overtopping if maximum wave height is used for calculations. In addition, when the mean seabed slope is less than 1/20, the calculated wave overtopping rate in regular waves is likely to be underestimated. Therefore, the calculation of the wave overtopping rate should be based on irregular waves with wave grouping characteristics.

When a vertical wall or a wall with a surface slope steeper than 1:1 is constructed on a uniformly sloping seabed surface, the calculation diagrams of Goda et al [42], and Goda and Kishira [43] are used. When a sloping wall with a surface slope smaller than 1:1 or if the seabed topography between the wave breaking depth and the structure has two distinct slopes like in **Fig. 3.8**, the calculation method of Yamamoto and Horikawa [44] can be used. In the case of composite structures or beaches with composite slopes with two distinct slopes where surface roughness cannot be neglected, the calculation method of Van der Meer and Jansen [45] can be used. Furthermore, for an arbitrary coastal cross-section, numerical models using higher-order equations of motion (e.g., Yamamoto et al [41]) can be applied, while CADMAS-SURF of Isobe et al. [46] can be applied for an arbitrary coastal cross-section with structures whose permeability cannot be ignored.

(a) Goda's Calculation Diagrams in Irregular Waves

Goda et al [42], using the results of model experiments with irregular waves, developed diagrams for calculating the average wave overtopping rate per unit width q [m³/sec/m= m²/sec] of an upright seawall, constructed on a seabed with a uniform slope of 1/10 or 1/30, shown in **Fig. 3.16** and **Fig. 3.17**. Goda et al [42] also conducted experiments on wave overtopping from irregular waves of an upright wall with a wave breaker, where the crown width is two blocks and the crown height is about $0.1H_o'$ lower than the crown height of the wall, and then developed diagrams for calculating the average wave overtopping rate q per unit width of the upright wall with the wave breaker, constructed on the seabed with the uniform slope 1/10 or 1/30, shown in **Fig. 3.18** and **Fig. 3.19**.

In these diagrams, H_o' is the equivalent offshore wave height, L_o is the offshore wavelength, h_c is the crown height, h is the depth of the wall toe, and g is the acceleration of gravity. Since the wave overtopping rate when the slope of the wall is steeper than 1:1 is not significantly different from that of an upright wall, these calculation diagrams can be applied to any wall with slopes steeper than 1:1.

113

3. WAVE RUN-UP, WAVE OVERTOPPING AND WAVE FORCES

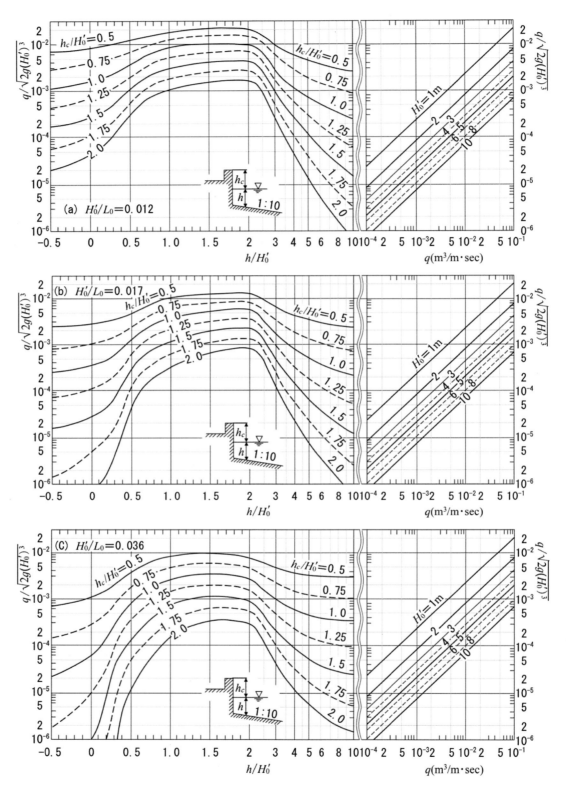

Fig. 3.16 Calculation diagram of the average overtopping rate of irregular waves for the upright wall on the seabed slope of 1/10 by Goda et al [42].

3.1 Wave Run-up and Wave Overtopping

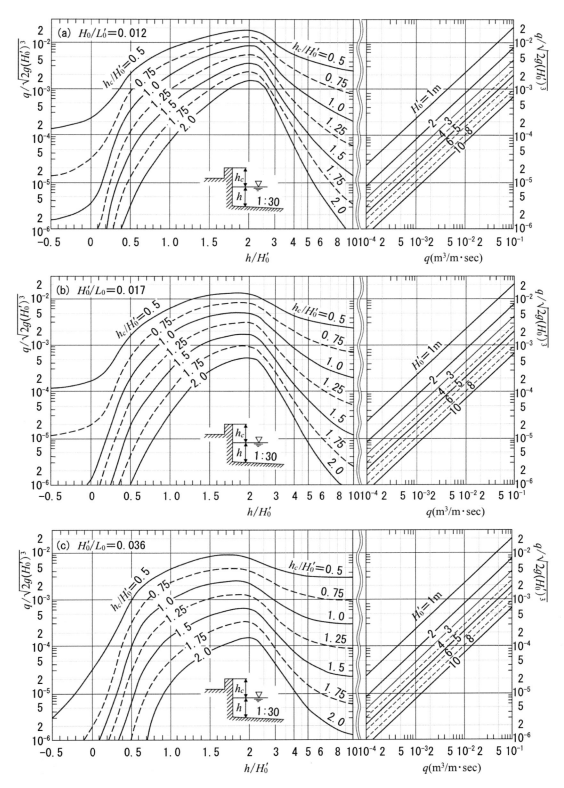

Fig. 3.17 Calculation diagram of the average overtopping rate of irregular waves for the upright wall on the seabed slope of 1/30 by Goda et al [42].

3. WAVE RUN-UP, WAVE OVERTOPPING AND WAVE FORCES

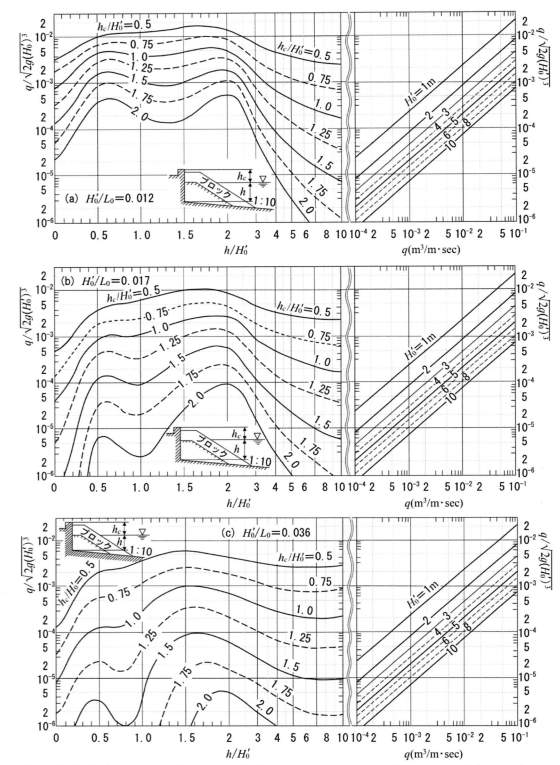

Fig. 3.18 Calculation diagram of the average overtopping rate of irregular waves for the upright wall according to Goda et al [42]. (When the wave breaker, where the crown height is $0.1H_0'$ lower than the crown height of the upright wall and the crown width is 2 blocks, is attached to the upright wall on the 1/10 seabed slope)

3.1 Wave Run-up and Wave Overtopping

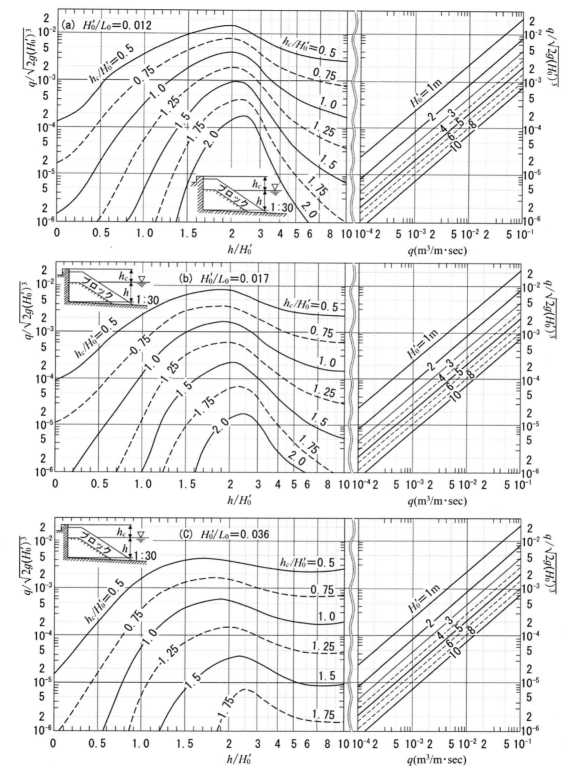

Fig. 3.19 Calculation diagram of the average overtopping rate of irregular waves for the upright wall according to Goda et al [42]. (When the wave breaker, where the crown height is $0.1H_0'$ lower than the crown height of the upright wall and the crown width is 2 blocks, is attached to the upright wall on the 1/30 seabed slope)

3. WAVE RUN-UP, WAVE OVERTOPPING AND WAVE FORCES

Goda and Kishira [43] obtained the ratio of crown heights (h_c/h_{co}) between a vertical seawall with wave breakers (h_c) and a vertical seawall (h_{co}) required for the same wave overtopping rates for different offshore wave steepness by using various experiment data, and then proposed **Figs. 3.20** and **3.21** showing the relationship between (h_c/h_{co}) and depth-to-wave height ratio (h/H_o').

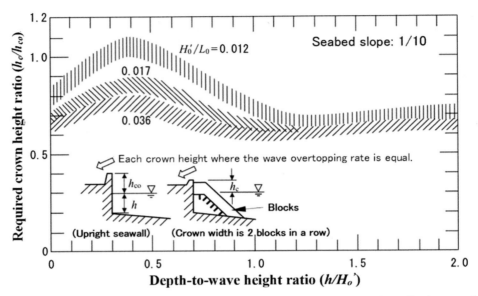

Fig. 3.20 Relationship between the required crown height ratio of the upright wall with the wave breaker having the same wave overtopping rate to the upright wall and the depth-to-wave height ratio, for the seabed slope of 1/10. (According to Goda and Kishira [43])

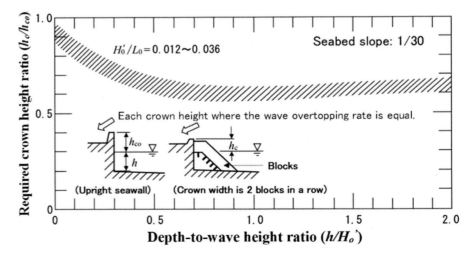

Fig. 3.21 Relationship between the required crown height ratio of the upright wall with the wave breaker having the same wave overtopping rate to the upright wall and the depth-to-wave height ratio, for the seabed slope of 1/30. (According to Goda and Kishira [43])

3.1 Wave Run-up and Wave Overtopping

In the case where the seabed slope is 1/10, **Fig. 3.20** shows that when the water depth-to-wave height ratio h/H_o' is larger than 0.7, the wave overtopping reduction effect of the wave breaker can be substantial; but when h/H_o' is less than 0.7, the wave overtopping reduction effect becomes small, especially in the case of $H_o'/L_o = 0.012$. For $0.3 < h/H_o' < 0.6$, little or no wave overtopping reduction effect is expected.

In the case where the seabed slope is 1/30, **Fig. 3.21** shows that the wave overtopping reduction effect of the wave breaker can be seen when h/H_o' is larger than 0.1, and the effect of wave steepness on wave overtopping reduction is negligible.

Goda and Kishira [43] also studied the effects of the horizontal number of blocks at the crown of the wave breaker on reducing the crown height of the upright wall. The results of the study are summarized in **Fig. 3.22**. In the figure, h_{c2} is the crown height of the wall when the horizontal number of blocks at the crown of the wave breaker is 2, and h_{cN} is the crown height of the wall when the horizontal number of blocks at the crown of the wave breaker is N.

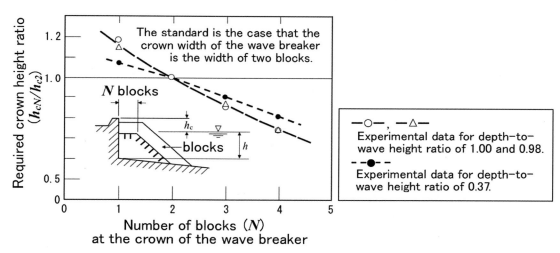

Fig. 3.22 Relationship between the required crown height ratio of the upright wall with the wave breaker of *N* crown blocks having the same wave overtopping rate to the upright wall with the wave breaker of two crown blocks and the number of blocks at the crown of the wave breaker, according to Goda and Kishira [43].

(b) Yamamoto and Horikawa's Calculation Method in Irregular Waves

Yamamoto and Horikawa [44] proposed the following method for calculating wave overtopping rates in irregular waves.

(i) When regular waves do not break before reaching a seawall, wave overtopping rates per unit width of the seawall and per wave q_r [m³/m/T] and wave run-up heights R are calculated using Eq. (3.31) proposed by Takada [28].

3.WAVE RUN-UP, WAVE OVERTOPPING AND WAVE FORCES

(ii) When regular waves break before reaching the seawall, wave overtopping rates per unit width and per wave q_r [m³/m/T] and wave run-up heights R are calculated using Eqs. (3.32) and (3.21).

(iii) Using the regular wave overtopping rate q_r obtained above, the irregular wave overtopping rate per unit width of the wall and per wave q_T [m³/m/T] can be obtained from Eq. (3.33). p is a distribution function of wave heights and wave periods, which can be obtained from Eq. (3.34) proposed by Watanabe and Kawahara [47].

Furthermore, when the improved virtual slope is less than 1/20, the accuracy of this calculation method can be improved by adding half of the wave height at the shoreline of the long-period gravity wave by Eq. (3.10) to the water depth to calculate the run-up height by using Eq. (3.21).

$$\left. \begin{aligned} q_r &= 0.65\left(R - h_c\right)^2 \\ R &= \left[1.0 + \pi\frac{H}{L}\coth\left(\frac{2\pi}{L}\right)h\right]H_o \end{aligned} \right\} \tag{3.31}$$

where h_c is the crown height of the sloping wall, H and L are the incident wave height and the wavelength at the water depth h of the sloping wall toe, and H_o is the offshore wave height.

$$\left. \begin{aligned} q_r &= 0.1\left(\frac{L_o}{H_b}\right)^{1/3}\frac{\cos\alpha + \cos\beta}{2}\left\{\cot\left[\alpha - \tan^{-1}\left(\frac{h_m}{R\sin\alpha}\right)\right] - \cot\alpha\right\}\frac{\left(R - h_c\right)^2}{2} \\ \frac{h_m}{H_b} &= 0.7\left[\frac{0.1589\left(\tan\alpha\right)^{0.5}}{\left(0.8H_b/L_o\right)^{0.25}} + 0.8\left(\tan\alpha\right)^{0.6}\right] \end{aligned} \right\} \tag{3.32}$$

where L_o is the offshore wavelength; H_b is the regular wave breaking height in non-overtopping condition, and is obtained from Eq. (3.20); α is the improved virtual slope angle by Nakamura et al [31], and is obtained using Eq. (3.17); β is the mean surface slope angle of the sloping wall; and h_m is the thickness from the intersection of the sloping wall surface and the still water level to the surface of the run-up wave.

$$q_T = \int_0^\infty \int_0^\infty q_r \times p\,dHdT \tag{3.33}$$

$$\left. \begin{aligned} p &= p(\tau)p(\chi_f|\tau)/\chi_m(\tau) \\ p(\tau) &= \frac{\sqrt{1+\lambda^2}}{1+\sqrt{1+\lambda^2}}\frac{\lambda^2}{\left[\lambda^2 + (\tau - 1)^2\right]^{3/2}}, \quad p(\chi_f|\tau) = \left(\frac{32}{\pi^2}\right)\chi_f^2\exp\left(\frac{-4\chi_f^2}{\pi}\right) \\ \tau &= \frac{T}{T_m}, \quad \chi = \frac{H}{H_m}, \quad \chi_f = \frac{\chi}{\chi_m(\tau)}, \quad \chi_m(\tau) = \frac{\sqrt{S(f)f}}{\int_0^\infty \sqrt{S(f)f}\,p(\tau)d\tau} \\ \lambda &= \sqrt{\left(m_0 m_2/m_1^2\right) - 1} \end{aligned} \right\} \tag{3.34}$$

where H is the wave height component; T is the periodic component; H_m is the mean wave height; T_m is the mean period; f is the frequency (inverse of the period); $S(f)$ is the Bretschneider-Mitsuyasu frequency spectrum; and m_0, m_1, and m_2 are respectively the zeroth, first, and second order moments using the same frequency spectrum.

3.1 Wave Run-up and Wave Overtopping

(c) Van der Meer and Janssen's Calculation Method

Van der Meer and Janssen [45] proposed Eq. (3.35) to calculate the average wave overtopping rate per unit width q [m³/m/s] from experimental data of irregular waves, by considering the reduction effect of oblique wave incidence, convex shape (root protection work, etc.), and surface roughness of the sloping wall surface.

$$\left.\begin{aligned}\frac{q}{\sqrt{gH_{1/3}^3}} &= 0.06\sqrt{\frac{\tan\alpha}{\sqrt{H_{1/3}/L_{op}}}} \times \exp\left(\frac{-5.2}{\gamma_h\gamma_\theta\gamma_b\gamma_f}\frac{h_c}{H_{1/3}}\frac{\sqrt{H_{1/3}/L_{op}}}{\tan\alpha}\right) &\left[\frac{\tan\alpha}{\sqrt{H_{1/3}/L_{op}}}\leq 2\right] \\ \frac{q}{\sqrt{gH_{1/3}^3}} &= 0.2\times\exp\left(\frac{-2.6}{\gamma_h\gamma_\theta\gamma_b\gamma_f}\frac{h_c}{H_{1/3}}\right) &\left[\frac{\tan\alpha}{\sqrt{H_{1/3}/L_{op}}}\geq 2\right]\end{aligned}\right\} \quad (3.35)$$

Where $H_{1/3}$ is the significant wave height at the toe of the sloping wall ; g is the gravity acceleration; $\tan\alpha$ is the average gradient obtained by linearizing the cross-section; L_{op} is the peak offshore wavelength; h_c is the crown height of the sloping wall; γ_h is the reduction factor due to shoaling deformation, which is obtained from Eq. (3.26); γ_θ is the reduction factor due to oblique wave incidence, which has slightly higher reduction effect than that due to the wave run-up height, which can be obtained from Eqs. (3.36) and (3.37); γ_b is the reduction factor due to convexity (root protection work, etc.), which is obtained from Eq. (3.29); γ_f is the reduction factor due to the surface roughness of the sloping wall, which is obtained from **Table 3.1**.

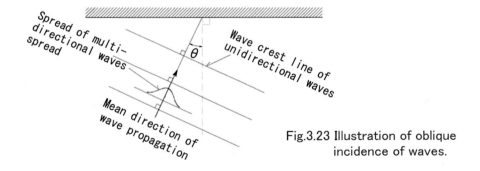

Fig.3.23 Illustration of oblique incidence of waves.

For unidirectional waves:

$$\gamma_\theta = \begin{cases} 1.0 & \left[0°\leq\theta\leq 10°\right] \\ \left(\cos(\theta-10°)\right)^2 & \left[10°<\theta\leq 50°\right] \\ 0.6 & \left[50°<\theta\right] \end{cases} \quad (3.36)$$

For multi-directional waves: $\quad \gamma_\theta = 1 - 0.0033\theta \quad (3.37)$

Since 2000, the reliability of various experimental formulae for determining wave run-up heights and wave overtopping rates has been tested in Western Europe, and several new empirical

3.WAVE RUN-UP, WAVE OVERTOPPING AND WAVE FORCES

formulae have been proposed in the EurOtop manual [48] to calculate wave overtopping with sufficient accuracy for a variety of structures with different surface geometries and materials. In this manual, an assessment of the overtopping calculation error is considered, which provides useful information for reliable design.

(d) Numerical Calculation Models

Numerical models based on nonlinear equations of motion are used for calculating wave overtopping rates of arbitrarily shaped beaches or seawall cross-sections. They can accurately calculate the effects of long-period waves, generated by the groupness of irregular waves. The following is an example of numerical models that use the Boussinesq equation and the continuity equation by Yamamoto et al. [41] or Vu et al. [49].

$$
\left.
\begin{aligned}
\frac{\partial q}{\partial t}+\frac{\partial}{\partial x}\left(\frac{q^2}{d}\right)+gd\frac{\partial \eta}{\partial x}+\frac{d^3}{3}\frac{\partial^3}{\partial x^2 \partial t}\left(\frac{q}{d}\right)-\frac{h^2}{2}\frac{\partial^3 q}{\partial t \partial x^2}-W_b+\frac{f_b}{d^2}|q|q=0 \\
\frac{\partial q}{\partial x}+\frac{\partial \eta}{\partial t}=0
\end{aligned}
\right\}
\tag{3.38}
$$

where q is the horizontal flow rate per unit width (horizontal velocity integrated from the sea bottom to the water level), η is the water surface elevation, d is the time-averaged water depth, x is the horizontal distance off the shore, and t is the elapsed time. W_b is the dissipation term due to wave breaking, the model based on the eddy viscosity coefficient proposed by Sato and Suzuki [50] is employed, and the empirical equation of Goda [22] is used to determine the wave breaking position. f_b is the friction coefficient on the seabed, and empirical formulas such as Yamamoto et al. [34] can be used.

In this numerical model, the time-varying water level fluctuation of irregular waves with wave grouping is an input condition provided at the offshore boundary. The spatial derivatives are calculated on a staggered grid, using an upwind difference scheme for non-linear terms. The time derivative is discretized using the Crank-Nicolson method. On the landward boundary, the same process as in Okayasu et al. [51] is used, and the overtopping flow rate is removed as it is already in the calculation of wave overtopping rate.

(e) Numerical Calculation Method Using CADMAS-SURF

The numerical model CADMAS-SURF (2D, 3D) developed by Isobe et al. [46] consists of the Navier Stokes equations of motion and the continuity equation for incompressible viscous fluids extended by a turbulence model (based on the high Re-type k-ε 2 equations) and a porous model [see Eqs. (2.2.95) to (2.2.98)], and solved with the addition of a free surface treatment model using the volume of the fluid method [see Eq. (2.2.99)]. Therefore, CADMAS-SURF can be used to accurately calculate wave overtopping rates of coastal dikes and seawalls constructed on arbitrary coastal topography and permeable add-ons such as wave breakers. However, this 3D numerical model is very

3.1 Wave Run-up and Wave Overtopping

computationally expensive and requires complicated input data preparation. Therefore, 2D numerical models are preferred.

The calculation method of CADMAS-SURF is explained in the Japanese manual "Research and Development of Numerical Wave Tank CADMAS-SURF/3D" by Isobe et al [46], and the source program can be downloaded from the following websites.

https://github.com/CADMAS-SURF/, or

https://www.cdit.or.jp/program/cadmas-download.html.

3. WAVE RUN-UP, WAVE OVERTOPPING AND WAVE FORCES

3.2 Wave Forces

This section describes methods for evaluating wave forces acting on upright walls of upright and composite type structures, the stability of covered stones and concrete blocks that make up slope type structures, and wave forces acting on legged-type structures.

1) Wave Forces Acting on Upright Walls

Wave forces can be obtained by integrating the wave pressure obtained using the calculation equations presented in this section over the area where it acts. According to previous research results, the magnitude of the wave pressure and its variation with time depends largely on the relative wave height (H/h) of incident waves. The variation of wave pressure over time, as summarised by Horikawa [52] based on many experiments at different relative wave heights, is shown in **Fig. 3.24** (horizontal axis: time t, vertical axis: wave pressure p).

Coastal structures can be roughly classified, according to the methods of resistance to wave forces and the methods used to calculate the wave forces acting on them, as follows:

(i) a **gravity type** which withstands the wave forces by its own weight,

(ii) a **legged type** that withstands the wave forces by driving columns into the ground.

Gravity-type structures can be divided into two types. a **sloping type** in which rubble stones or deformed concrete blocks are piled up in a trapezoidal cross-section, and an **upright type** in which concrete cube blocks or caissons are used. There is also a **composite type** in which concrete cube blocks or caissons are placed on top of a foundation made of rubble stones or flat concrete blocks.

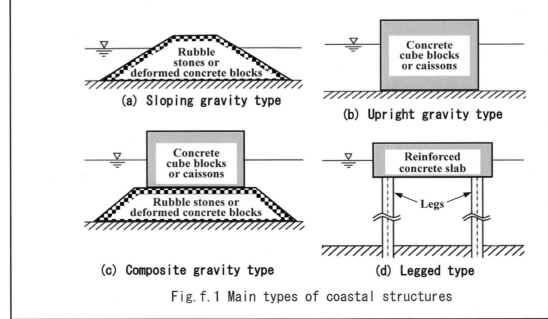

Fig.f.1 Main types of coastal structures

3.2 Wave Forces

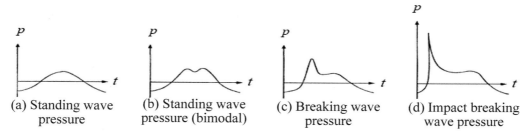

Fig. 3.24 Time-varying wave pressure acting on an upright wall over one wave cycle. (According to Horikawa [52])

There were many studies in the 1900s on the relationship between the wave pressure and the incident wave parameters, and the commentary to **Fig. 3.24** summarised by Horikawa [52] is presented as follows:

(a) When the relative wave height is small ($H/h < 1/2$), the wave in front of an upright wall becomes a standing wave and the wave pressure gradually increases with the rise of the water surface, as shown in **Fig. 3.24 (a)**.

(b) As the wave steepness increases, the wave pressure component of the doubled frequency becomes more pronounced, and the relative wave height becomes higher, the wave pressure distribution changes to bimodal as shown in **Fig. 3.24 (b)**.

(c) When the relative wave height exceeds the breaking limit of the standing wave (this limit is smaller than the limit of the travelling wave), the first peak of the bimodal wave pressure becomes greater than the second peak, which is called the **breaking wave pressure**, as shown in **Fig 3.24 (c)**.

(d) When the relative wave height exceeds the breaking limit of the travelling wave, the first peak of the bimodal wave pressure is even higher than the second peak, which is called the **impact breaking wave pressure** (when a breaking wave strikes an upright wall with an air layer entrapped, an impact pressure similar to a hammer strike is generated, followed by an impact due to the compressed air) pointed out by Bagnold [53], as shown **Fig. 3.24 (d)**.

(e) In addition, in a situation where a wave after broken strikes the upright wall, a pressure distribution similar to what is shown in **Fig. 3.24 (c)** can be observed, and it is called the **broken wave pressure**.

(1) Sainflou's Formula for Standing Wave Pressure (Relative Wave Height $H/h \leq 1/2$)

Among several standing wave pressure formulae found in literature, Sainflou's [54] formula is introduced here because it is easy to use and reliable.

Sainflou [54] obtained a wave pressure calculation formula for standing waves based on Trochoidal wave theory, a type of finite amplitude wave when $H/h \leq 1/2$. As shown in **Fig. 3.25**, the wave pressure from $p = 0$ at the wave peak to $p = p_1$ ($p = p_1'$) and the wave pressure from $p = p_1$ ($p =$

3. WAVE RUN-UP, WAVE OVERTOPPING AND WAVE FORCES

p_1') to $p = p_2$ ($p = p_2'$) are linearized for simplicity. Eq. (3.39) is used when a wave crest reaches the wall (the force pushing on the wall) and Eq. (3.40) is used when a wave trough reaches the wall (the force pulling the wall occurs).

Furthermore, uplift pressure acts upwards at the bottom of the wall section, given by a triangular distribution as shown in **Fig. 3.26**. Therefore, it is necessary to consider this uplift pressure and the buoyancy force, which is equals to the weight of water equivalent to the volume of the upright section under the water surface of the breakwater (area shaded in light blue), acting upward. The maximum value of this uplift pressure p_u is equal to p_2.

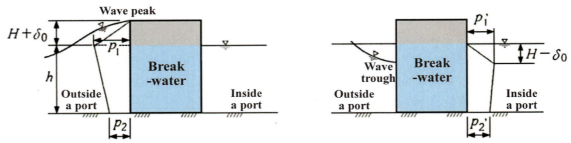

Fig. 3.25 Distribution of standing wave pressure according to Sainflou's formula [54].

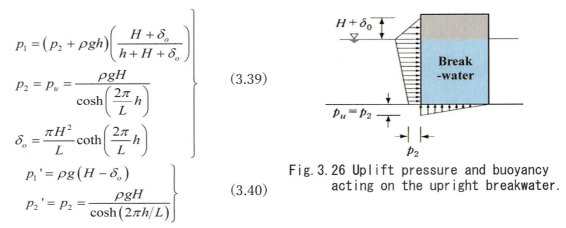

$$\left. \begin{array}{l} p_1 = (p_2 + \rho g h)\left(\dfrac{H + \delta_o}{h + H + \delta_o}\right) \\[6pt] p_2 = p_u = \dfrac{\rho g H}{\cosh\left(\dfrac{2\pi}{L}h\right)} \\[10pt] \delta_o = \dfrac{\pi H^2}{L}\coth\left(\dfrac{2\pi}{L}h\right) \end{array} \right\} \quad (3.39)$$

$$\left. \begin{array}{l} p_1' = \rho g (H - \delta_o) \\[6pt] p_2' = p_2 = \dfrac{\rho g H}{\cosh(2\pi h/L)} \end{array} \right\} \quad (3.40)$$

Fig. 3.26 Uplift pressure and buoyancy acting on the upright breakwater.

where p_1, p_2, p_1' and p_2' are the wave pressures per unit length [N/m³/m]; ρ is the density of seawater; g is the acceleration of gravity; h is the water depth in front of the breakwater; and H and L are respectively the incident wave height (the type of wave height used is not specified) and wavelength at the water depth h.

Since offshore structures are often destroyed by very occasional large wave forces, the use of significant wave heights in Sainflou's equation is likely to underestimate the wave pressure. Therefore, according to Goda [55], besides the significant wave height, the highest one-tenth wave height, the upper 1% excess wave height, etc. have also been used. In Japan, although the significant wave height is used, as an improvement measure, the range from $H/2$ below the design water surface to $H/2$ above the same surface in the wave pressure distribution obtained by Sainflou's formula had been replaced using the wave pressure by Hiroi's [56] formula introduced next (this is wave breaking pressure, which is greater than the wave pressure by Sainflou's formula) till around 1975.

(2) Hiroi's Formula for Breaking Wave Pressure (Relative Wave Height $H/h > 1/2$)

When $H/h > 1/2$, the wave pressure changes from standing wave pressure to breaking wave pressure as shown in **Fig. 3.24**.

There have been several wave pressure formulas for breaking waves since Gaillard's hydrodynamic pressure formula (1928). Hiroi's [56] formula is introduced here because of its simplicity and direct relationship to incident wave height, which is a representative parameter of the external force.

$$p = 1.5\rho g H_{1/3} \tag{3.41}$$

where p is the wave pressure per unit length [N/m³/m], ρ is the density of seawater, g is the acceleration of gravity, and $H_{1/3}$ is the significant wave height of the incident wave at the water depth in front of the breakwater.

When the crown height h_c of the breakwater is higher than $1.25H_{1/3}$ above the still water surface, the wave pressure obtained from Eq. (3.41) acts from the sea bottom up to $1.25H_{1/3}$ above the still water surface as shown in the top figure of **Fig. 3.27**.

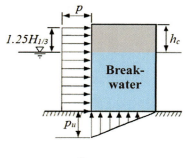

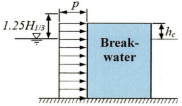

Fig. 3.27 Distribution of breaking wave pressure by Hiroi [56].

Furthermore, buoyancy is calculated by considering the area of the upright wall below the still water level (area shown in light blue). A triangular distribution of uplift pressure acts upward at the bottom of the upright section. The maximum value of uplift pressure p_u is obtained from Eq. (3.42).

$$p_u = 1.25\rho g H_{1/3} \tag{3.42}$$

When the crown height h_c of the breakwater is lower than $1.25H_{1/3}$ above the still water surface, the wave pressure obtained from Eq. (3.41) is considered to act from the seabed to the crown height of the upright wall as shown in the lower figure of **Fig. 3.27**. Furthermore, as wave overtopping can occur in this situation, the uplift pressure is considered negligible, although the buoyancy force should be calculated for the entire area of the cross-section shown in light blue.

Although the distribution of breaking wave pressure by Hiroi's formula is considerably different from the actual wave pressure distribution, his formula can satisfactorily calculate the average value of breaking wave pressure (refer to Horikawa [52] or Goda [55]). As Hiroi's formula is easy to handle, it had been widely used in East Asia for about 50 years until the appearance of Goda's formula which can be used continuously from standing waves to breaking waves.

3. WAVE RUN-UP, WAVE OVERTOPPING AND WAVE FORCES

(3) Goda's Formula for Wave Pressure in Two Ranges

Breaking wave pressures (including broken wave pressures) induced by breaking waves differ considerably from pressures induced by standing waves in front of an upright breakwater. Therefore, breakwater cross-section sizes required to withstand different types of wave pressure are different. However, it is preferable to change the cross-section sizes of breakwaters continuously. Therefore, a formula for calculating the wave pressure per unit length [N/m³/m] that can be applied to both regions and that provides a reasonable value was proposed by Goda [57] based on experimental data.

Goda [57] set up the wave pressure distribution as shown in **Fig. 3.28** and proposed a new set of formulas for calculating the wave pressure p_1 above the seawater surface, the wave pressure p_2 above the seabed surface, the wave pressure p_3 above the bottom of the breakwater, and the maximum uplift pressure p_u. Tanimoto et al [58] improved the evaluation method for the wave pressure reduction coefficient due to oblique incident waves (refer to **Fig. 3.29**), and Takahashi et al [59] proposed an evaluation method for the wave pressure reduction coefficient due to wave breakers. Eqs. (3.43) and (3.44) give the wave pressure and maximum lift pressure by combining Goda [57], Tanimoto et al [58], and Takahashi et al [59].

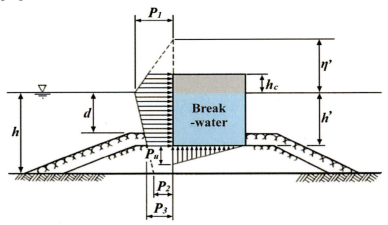

Fig. 3.28 Distribution of the wave pressure by Goda [57].

$$\left. \begin{array}{l} p_1 = \dfrac{1}{2}(1+\cos\theta')\left(\alpha_1 \lambda_1 + \alpha_2 \lambda_2 \cos^2 \theta'\right)\rho g H_{max} \\[6pt] p_2 = \dfrac{p_1}{\cosh(2\pi h/L)} \\[6pt] p_3 = \alpha_3 p_1 \\[6pt] \alpha_1 = 0.6 + \dfrac{1}{2}\left[\dfrac{4\pi h/L}{\sinh(4\pi h/L)}\right]^2 \\[6pt] \alpha_2 = \min\left[\dfrac{h_b - d}{3h_b}\left(\dfrac{H_{max}}{d}\right)^2, \dfrac{2d}{H_{max}}\right] \\[6pt] \alpha_3 = 1 - \dfrac{h'}{h}\left[1 - \dfrac{1}{\cosh(2\pi h/L)}\right] \end{array} \right\} \quad (3.43)$$

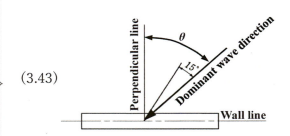

Fig. 3.29 Definition of wave direction angle θ.

$$p_u = \frac{1}{2}(1+\cos\theta')\alpha_1\alpha_3\lambda_3\rho g H_{max} \qquad (3.44)$$

$$\eta' = 0.75(1+\cos\theta')\lambda_1 H_{max} \qquad (3.45)$$

$$h_b = h + 5\times H_{1/3}\tan\beta \qquad (3.46)$$

where θ' is the angle at which the wave pressure is greatest in the incident angle range $\theta\pm15°$; λ_1 is a correction factor for breakwaters covered with wave dissipating blocks, and λ_2 and λ_3 are correction factors related to the structure type, which for a normal upright breakwater, $\lambda_1=\lambda_2=\lambda_3=1$ respectively; ρ is the density of seawater; g is the gravitational acceleration; H_{max} and L are respectively the maximum incident wave height and the significant wavelength at the depth h in front of the breakwater; d is the depth from the still water surface to the surface of the foundation mound; h' is the depth from the still water surface to the bottom of the concrete cube blocks or caissons; η' is the height from the still water surface to the top of the wave pressure as determined by Eq. (3.45); and h_b is the water depth at a location about 5 times the significant wave height $H_{1/3}$ offshore from the front of the breakwater, calculated from Eq. (3.46) using the seabed slope $\tan\beta$.

It would be economical if wave forces could be reduced by turning an upright wall into a sloping wall, as shown in **Fig. 3.30**. Therefore, Tanimoto and Kimura [60] confirmed through model experiments that wave pressure acting on an impermeable sloping wall can be obtained by projecting the wave pressure distribution obtained from Goda's wave pressure calculation formula onto an inclined plane. Here, since the uplift pressure is smaller than that of an upright wall, the value of λ_3 is set from Eq. (3.47) when determining the maximum uplift pressure p_u from Eq. (3.44).

$$\lambda_3 = \exp\left[-2.26\left(\frac{7.2\delta}{L}\right)^3\right] \qquad (\alpha\geq 70°,\ \delta<0.1L) \qquad (3.47)$$

where $\delta = h'\times\cot\alpha$, L is the wavelength at the front of the breakwater.

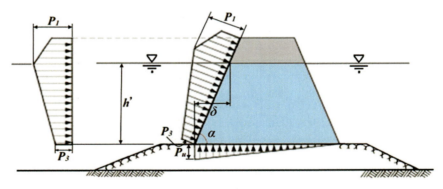

Fig. 3.30 Distribution of the wave pressure acting on impermeable sloping breakwater according to Tanimoto and Kimura [60].

3.WAVE RUN-UP, WAVE OVERTOPPING AND WAVE FORCES

(4) Treatment Methods for Impact Breaking Wave Pressure

When the breaking wave pressure satisfies certain conditions, the first peak of the impact breaking wave pressure becomes unusually large, as shown in (d) of **Fig. 3.24**. According to PRObabilistic design tools for VERtical BreakwaterS (PROVERBS [61], the research project in Western Europe for the impact breaking wave pressure assessment), when high waves break in front of the upright wall and the breaking waves take in a sufficient volume of air, as shown in **Fig. 3.31**, the first peak of the impact breaking wave force becomes abnormally large due to the adiabatic compression of the air mass and the oscillations of the air mass (pocket) may also cause damping oscillations in the wave force.

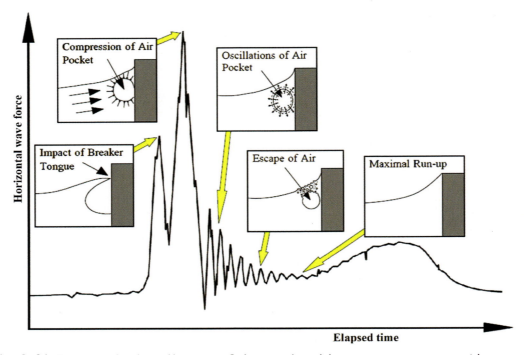

Fig. 3.31 Time variation diagram of impact breaking wave pressure acting on an upright wall over one wave cycle. (According to PROVERBS[61])

The occurrence of impact breaking wave pressure, where wave forces become very large, means that upright breakwaters become very dangerous, and many researchers in Europe, the USA, and Japan have studied the mechanisms of generation and calculation methods for the impact breaking wave pressure. The following sections introduce the initial formula for calculating the impact breaking wave pressure by Minikin [62], a method to determine the occurrence of the impact breaking wave pressure, and a formula for calculating this pressure based on comprehensive research by PROVERBS [61], and a method for coping with the impact breaking wave pressure by Tanimoto [63].

3.2 Wave Forces

(a) Minikin's Formula for Impact Breaking Wave Pressure

Minikin [62] proposed formulas for the distribution of an impact breaking wave pressure shown in **Fig. 3.32** based on measured data.

The distribution given in Eq. (3.48) expresses the impact breaking wave pressure p (with a maximum value of p_{max}), that exists in the range of $\pm H/2$ above the still water surface.

The hydrostatic pressure p_s (incremental hydrostatic pressure due to wave-induced water surface elevation) is given by Eq. (3.49). p_s increases uniformly from the highest water surface ($z = H/2$, $p_s = 0$) to the still water surface ($z = 0$, p_s = the maximum value p_{sm}), and becomes constant below the still water surface.

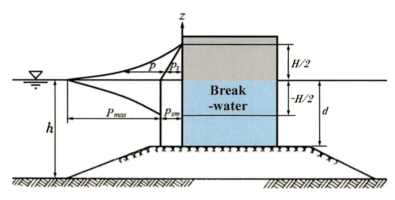

Fig. 3.32 Distribution of Impact breaking wave pressure according to Minikin [62].

(i) Impact Breaking Wave Pressure:

$$\left. \begin{array}{l} p = p_{\max} \left(\dfrac{H_b - 2|z|}{H_b} \right)^2 \quad \left(\dfrac{H_b}{2} \geq z \geq -\dfrac{H_b}{2} \right) \\[2mm] p_{\max} = 102.4 \rho g d \left(1 + \dfrac{d}{h} \right) \dfrac{H_b}{L} \end{array} \right\} \quad (3.48)$$

(ii) Hydrostatic Water Pressure:

$$\left. \begin{array}{l} p_s = \rho g \left(\dfrac{H_b}{2} - z \right) \quad \left(\dfrac{H_b}{2} \geq z \geq 0 \right) \\[2mm] p_{sm} = \rho g \dfrac{H_b}{2} \quad (0 \geq z) \end{array} \right\} \quad (3.49)$$

where p, p_s, and p_{sm} are wave pressure components per unit length [N/m³/m]; H_b and L are the breaking wave height and the wavelength of incident waves at the depth h in front of the upright breakwater; z is the vertical coordinate with the origin at the still water surface and positive upward, and d is the water depth from the still water surface to the top of the foundation mound.

Although Minikin's formula has been widely used in Europe and the USA for nearly half a century since its publication, the facts that this formula easily underestimates the impact breaking wave pressure and that the decrease of horizontal wave forces calculated by this formula with increasing wavelength is unnatural has led to its disuse in recent years.

3.WAVE RUN-UP, WAVE OVERTOPPING AND WAVE FORCES

(b) PROVERBS' Formula for Impact Breaking Wave Pressure

The research project PROVERBS [61] in Western Europe proposed **Fig. 3.33**, which can judge the occurrence of the impact breaking wave pressure, based on a large number of data and Eq. (3.50) to obtain the maximum horizontal force F_{hmax} due to the impact breaking wave pressure on the front face of an upright breakwater.

$$F_{h\max} = F^*_{h\max} \times \rho g H_{bc}^2, \quad F^*_{h\max} = \frac{\alpha}{\gamma}\left\{1 - \left[-\ln P\left(F^*_{h\max}\right)\right]\gamma\right\} + \beta \qquad (3.50)$$

$$H_{bc} = \left(0.1025 + 0.0217\frac{1-C_r}{1+C_r}\right)L_p \tanh\left(k_b \frac{2\pi}{L_p}h\right)$$

$$k_b = 0.0076\left(\frac{B_{eq}}{d}\right)^2 - 0.1402\left(\frac{B_{eq}}{d}\right) + 1 \qquad (3.51)$$

$$B_{eq} = B_b + \left(\frac{h_b}{2\tan\alpha}\right) \qquad (3.52)$$

where $F^*_{h\max}$ is the non-dimensional impact breaking wave force, which can be expressed using the Generalized Extreme Value (GEV) distribution [the second equation of Eq. (3.50)]; ρ is the density of seawater; g is the acceleration of gravity; H_{bc} is the incident breaking wave height at the water depth h in front of the upright breakwater, which can be obtained from Eq. (3.51); $P(F^*_{h\max})$ is the non-exceedance probability of the non-dimensional impact breaking wave force, which generally may be taken as 90%; and α, β, and γ are the statistical parameters obtained from **Table 3.2**. Moreover, C_r can be obtained from **Table 3.3**, L_p is the wavelength corresponding to the peak period at the water depth h, B_{eq} is the effective berm width obtained from Eq. (3.52), and d is the water depth from the still water surface to the crown of the berm. Furthermore, B_b and h_b are the width and the height of the berm respectively, and $\tan\alpha$ is the frontal slope of the berm.

Table 3.2 Values of α, β and γ for GEV distribution of non-dimensional horizontal force.

Bed slope	Num. waves	α	β	γ
1/7	116	2.896	6.976	-0.526
1/10	159	10.209	12.761	-0.063
1/20	538	3.745	7.604	-0.295
1/50	3321	1.910	3.268	-0.232

Table 3.3 Values of the parameter C_r.

Conditions on the structure combination	C_r
Simple vertical walls and small mounds, high crest.	0.95
Low-crest walls $(0.5 < h_c /H_{1/3} <1.0)$ h_c is the crown height above the still water surface.	$0.8+0.1 \times h_c /H_{1/3}$
Composite walls, large mounds, and heavy breaking	0.5 to 0.7

3.2 Wave Forces

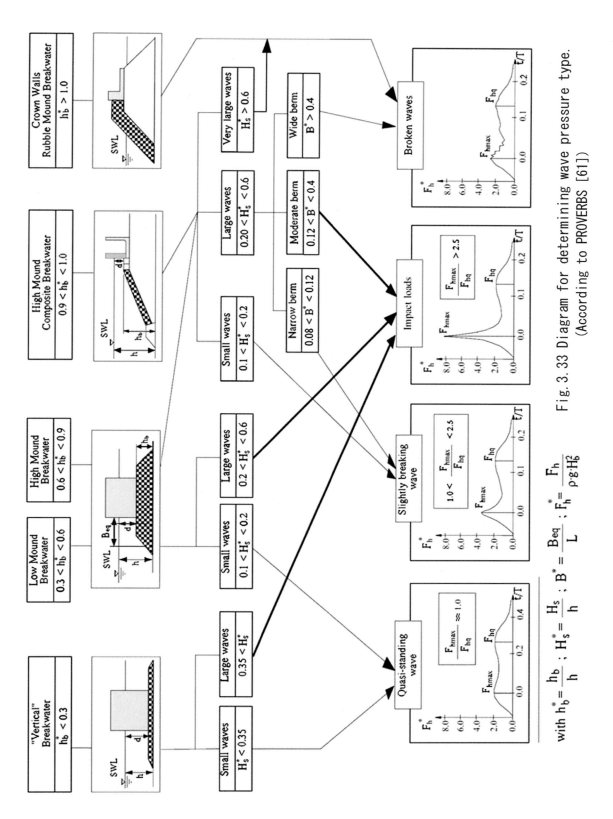

Fig. 3.33 Diagram for determining wave pressure type. (According to PROVERBS [61])

with $h_b^* = \dfrac{h_b}{h}$; $H_s^* = \dfrac{H_s}{h}$; $B^* = \dfrac{B_{eq}}{L}$; $F_h^* = \dfrac{F_h}{\rho \cdot g \cdot H_b^2}$

As the horizontal maximum wave force F_{hmax} obtained from Eq. (3.50) tends to overestimate the impact breaking wave pressure, Allosp and Vicinanza, members of this project, have proposed Eq.

3.WAVE RUN-UP, WAVE OVERTOPPING AND WAVE FORCES

(3.53) for the horizontal maximum wave force due to the impact breaking wave pressure, so that the calculated value is not too large.

$$F_{h\max} = 15\rho gh^2 \left(\frac{H_{sb}}{h}\right)^{3.134} \tag{3.53}$$

where H_{sb} is the significant breaking wave height of incident waves at water depth h in front of the upright breakwater

    If the wave pressure distribution corresponding to the horizontal maximum wave force $F_{h\max}$ is desired, it can be obtained by adopting **Fig. 3.28** (Goda's wave pressure distribution) as follows.

(i) Because the value obtained by integrating the horizontal wave pressure component of **Fig. 3.28** is equal to the horizontal maximum wave force obtained from Eq. (3.50) or Eq. (3,53), the relational equation shown in Eq. (3.54) is established.

(ii) When Eq. (3.55), which expresses the relationships of the height η' from the still water surface to the top of the horizontal wave pressure distribution and the value p_3 at the bottom of the horizontal wave pressure distribution, is substituted into Eq. (3.54), an equation to find the maximum horizontal wave pressure p_1 can be obtained.

(iii) When the maximum horizontal wave pressure p_1 obtained from this equation and the breaking wave height H_{bc} obtained from (3.51) are substituted into Eq. (3.55), p_3 and η' are obtained, and the distribution diagram of the horizontal wave pressure is completed.

$$F_{h\max} = \frac{1}{2}\left(p_1 + \frac{p_1}{\eta'}(\eta' - h_c)\right) \times h_c + \frac{1}{2}(p_1 + p_3) \times h' \tag{3.54}$$

$$\eta' = 0.8 \times H_{bc}, \quad p_3 = 0.45 \times p_1 \tag{3.55}$$

(c) Tanimoto's Method for Coping with the Impact Breaking Wave Pressure

    Tanimoto [63] studied the conditions that need to be satisfied for the generation of impact breaking wave pressure, which were later summarized by Goda [55] in **Fig. 3.34**. If the answer "Yes" continues to the end of both routes A and B in the flowchart shown in **Fig. 3.34**, there is a high risk of the impact breaking wave force generation.

    Next, an effective countermeasure to avoid the impact breaking wave pressure is to install a wave breaker using deformed blocks with a minimum number of two blocks at the crown of the breakwater and the surface slope of 1:1.5, which completely covers the surface of the breakwater. Tanimoto [63] proposed a formula to obtain the broken wave pressure for this case. It is to substitute the correction factors λ_1, λ_2, and λ_3 obtained from Eq. (3.56) when calculating using Eqs. (3.43) - (3.46).

$$\lambda_1 = \lambda_3 = \begin{cases} 1.0 & (H_{\max}/h < 0.3) \\ 1.2 - \dfrac{2}{3}\dfrac{H_{\max}}{h} & (0.3 \leq H_{\max}/h \leq 0.6), \\ 0.8 & (0.6 < H_{\max}/h) \end{cases} \qquad \lambda_2 = 0.0 \tag{3.56}$$

3.2 Wave Forces

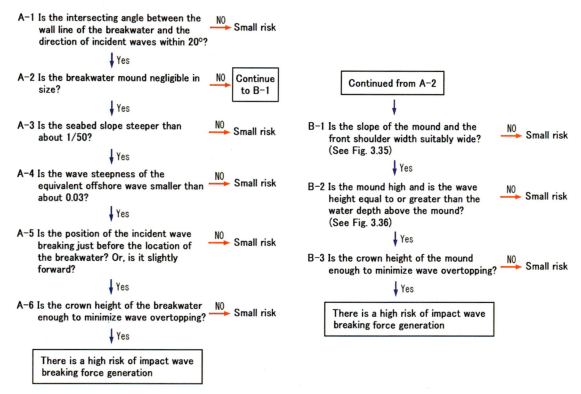

Fig. 3.34 Check flow for danger of impact breaking wave force generation based on Tanimoto [63].

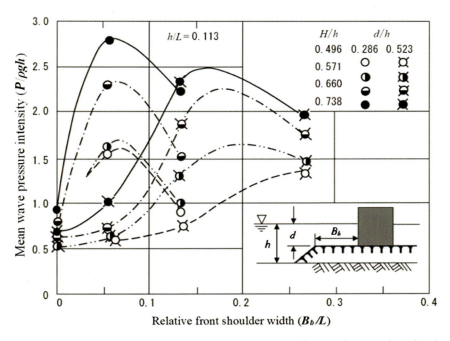

Fig. 3.35 Relationship between mean wave pressure intensity and relative front shoulder width according to Tanimoto [63].

3. WAVE RUN-UP, WAVE OVERTOPPING AND WAVE FORCES

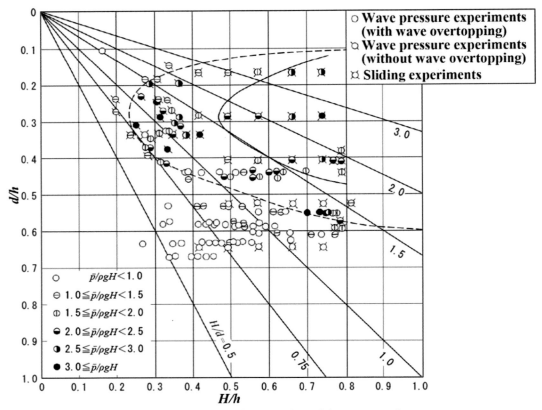

Fig. 3.36 Dangerous range of impact breaking wave force according to Tanimoto [63].

(5) Calculation of Wave Pressure Using CADMAS-SURF

CADMAS-SURF [see Eqs. (2.2.95) - (2.2.99)], introduced in Chapter 2, is the numerical model that analyses the motion of incompressible viscous fluids with the three-dimensional Navier-Stokes equations of motion, which can predict the distributions of velocity, pressure, and water surface around coastal structures, including porous bodies such as wave breakers. Therefore, CADMAS-SURF can be used to calculate the standing, breaking, and broken wave forces acting on coastal structures with complex geometries, such as the case of wave breakers attached to the front of upright walls. However, it is not possible to calculate the impact breaking wave pressures correctly.

For the calculation method using CADMAS-SURF, the readers can refer the Japanese manual "Research and Development of Numerical Wave Tank CADMAS-SURF/3D" by Isobe et al [46], and the source program can be downloaded from the following website of the Coastal Development Institute of Technology.

https://github.com/CADMAS-SURF/, or

https://www.cdit.or.jp/program/cadmas-download.html.

It should be noted that setting up appropriate input data that can be expected to provide high prediction accuracy when using CADMAS-SURF is not simple. Therefore, it is recommended to conduct hydraulic model experiments on representative cases to generate model validation data, confirm the appropriate data setting method through validation calculations using the obtained data, and then conduct comparative simulations for multiple cases.

2) Stability of Stones or Blocks Covering a Sloping Breakwater

When the surface of a sloping breakwater is covered with stones as shown in **Fig. 3.37**, it is necessary to consider the balance between the dead weight of each stone and wave-based hydrodynamic forces (drag and lift) that try to move the stone back and forth or upwards. In this case, the reduction of hydrodynamic forces due to the shape of the stone and the interlocking effect between stones can be expected. In particular, when deformed blocks such as tetrapods are used as covering materials, a considerable interlocking effect between the blocks can be expected. Henceforth, these covering materials are referred to as **armour units**.

The ratio of the drag force F_D and the lift force F_L, which are exerted by the waves to move individual armour units, to the dead weight W, which is the force to resist them, is considered. By adopting the following assumptions:

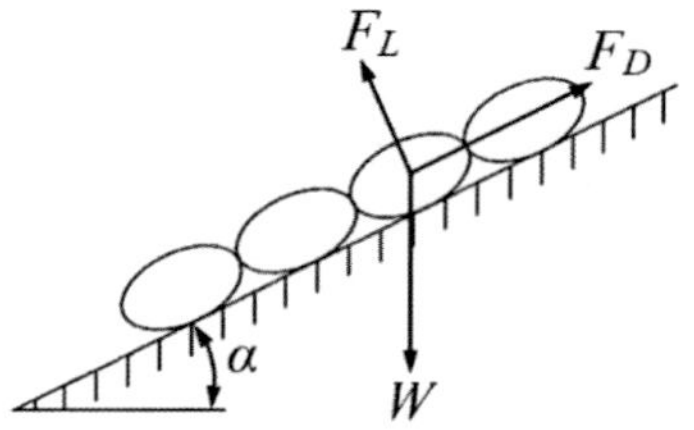

(i) the water satisfies shallow water conditions, i.e. the wave propagation velocity is $v = \sqrt{gh}$ (h is the water depth below the still water surface),

(ii) the incident wave height can be expressed using $H = \gamma h$,

(γ is the wave height-to-depth ratio at the wave breaking limit)

(iii) the drag coefficient C_D, the lift coefficient C_L, and γ can be considered constants,

Fig. 3. 37 Forces acting on armour units of the sloping breakwater.

The dimensionless quantities relating to the stability of armour units shown on the right-hand side of Eq. (3.57) can be obtained.

$$\frac{F_D + F_L}{W} = \frac{(C_D + C_L)\rho v^2 D^2}{(\rho_s - \rho)gD^3} = \frac{(C_D + C_L)H}{\gamma\left(\dfrac{\rho_s}{\rho} - 1\right)D} \propto \frac{H}{\left(\dfrac{\rho_s}{\rho} - 1\right)D} \tag{3.57}$$

where ρ is the density of seawater, ρ_s and D are the density and the representative dimension (the cubic root of the volume) of the covering stones or deformed blocks, v is the wave propagation velocity, and H is the incident wave height (such as significant wave height).

If hydraulic model experiments are conducted to examine the stability of the covering material by changing various representative values of the external forces and the covering material, and the data of the external force and the covering material at the threshold where the covering material is just stable are used to obtain the dimensionless quantity at the right end of Eq. (3.57), the representative dimension of the covering material stable for the design wave height can be obtained.

3.WAVE RUN-UP, WAVE OVERTOPPING AND WAVE FORCES

(1) Hudson's Formula for Calculating the Required Mass of Covering Material

Although Iribarren [64] was the first to propose a formula for determining the weight required for the stability of covering stones in a sloping breakwater, using extensive experimental data, Hudson [65] found that the dimensionless quantity related to stability at the right hand side of Eq. (3.57) is a function of K_D (a physical quantity specific to the covering material represented by the interlocking effect) and the surface slope angle α of the sloping breakwater, as expressed in Eq. (3.58). The required mass M of the covering material for stabilization is obtained from Eq. (3.59) based on Eq. (3.58).

$$\frac{H}{\left(\frac{\rho_s}{\rho}-1\right)D} = \left(K_D \cot\alpha\right)^{1/3} \qquad \text{(during non-overtopping)} \qquad (3.58)$$

$$M = \rho_s D^3 = \frac{\rho_s H^3}{K_D\left(\frac{\rho_s}{\rho}-1\right)^3 \cot\alpha} \qquad \text{(during non-overtopping)} \qquad (3.59)$$

The K_D values of the covering stones are shown in **Table 3.4**.

The K_D values for deformed blocks are usually around 10, but the values published by the respective developers can be used.

When incident waves overtop the crown of a breakwater, the covering materials placed at the crown (especially at the four corners and the top edge of the back slope) can be easily dislodged if they are not sufficiently interlocked. This can lead to the destruction of the breakwater. Therefore, the crown height should be set at a height that will not allow wave overtopping or the covering materials at the crown should be designed to weigh 50% heavier than everywhere else.

Table 3.4 K_D Value of Covering Stones.

K_D value for surface slope angle $1.5 \leq \cot\alpha \leq 3.0$ at $H=H_{1/3}$. [from Shore Protection Manual (1977)]

Shape of stone	Placement	Degree of damage			
		0-5%		5-10%	10-15%
		Breaking wave[1]	Non-breaking wave[2]	Non-breaking wave	Non-breaking wave
Smooth and round	Randomly piled	2.1	2.4	3.0	3.6
Rough and angular	Randomly piled	3.5	4.0	4.9	6.6
Rough and angular	Special[3]	4.8	5.5		

K_D value for $H=H_{1/10}$. [from Shore Protection Manual (1984)]

Shape of stone	Placement	Degree of damage = 0-5%	
		Breaking wave[1]	Non-breaking wave[2]
Smooth and round	Randomly piled	1.2	2.4
Rough and angular	Randomly piled	2.0	4.0
Rough and angular	Special[3]	5.8	7.0

1) Breaking waves refer to waves limited by water depth.
   That is, breaking waves occur in front of the covered slope.

2) Depth-limited breaking waves do not occur in front of the covered slope.

3) A special installation in which the long axis of the stone is perpendicular to the slope.

3.2 Wave Forces

(2) Van der Meer's Formulae for Calculating the Required Mass of Covering Material

Van der Meer [66] proposed Eqs. (3.60), (3.61) for determining the representative dimension D of a stable covering stone for non-overtopping waves. This equation considers the effect of the number of incident storm waves and their period. The required mass M can be obtained from Eq. (3.62) by using D obtained from Eq. (3.60).

$$\frac{H}{\left(\frac{\rho_s}{\rho}-1\right)D} = \begin{cases} \dfrac{6.2 A_s^{0.2} P^{0.18}}{N_w^{0.1} \xi_m^{0.5}} & (\xi_m < \xi_{mc}: \text{plunging waves}, \cot\alpha \geq 4.0) \\ \dfrac{1.0 A_s^{0.2} \sqrt{\cot\alpha}\, \xi_m^P}{P^{0.13} N_w^{0.1}} & (\xi_m > \xi_{mc}: \text{surging waves}) \end{cases} \quad (3.60)$$

$$\xi_m = \frac{\tan\alpha}{\sqrt{H/L_o}} = \frac{\tan\alpha}{\sqrt{H/(gT_m^2/2\pi)}}, \quad \xi_{mc} = \left(6.2 P^{0.31} \sqrt{\tan\alpha}\right)^{1/(P+0.5)} \quad (3.61)$$

$$M = \rho_s D^3 \quad (3.62)$$

where H is the incident wave height [$H_{1/3}$ is used for non-breaking waves and $H_{2\%}/1.4$ for breaking waves ($H_{2\%}$ is reduced by breaking waves)], ρ is the density of seawater, ρ_s and D are the density and the representative dimension of covering stones (cube root of the volume of deformed blocks), A_s can be set from **Table 3.5**. P is the permeability coefficient that can be set from **Fig. 3.38**, N_w is the number of incident storm waves [$N_w \leq 7,500$]; ξ_m and ξ_{mc} are respectively the wave breaking similarity parameter defined in Eq. (3.61) and the threshold of wave breaking type, α is the surface slope angle of the sloping breakwater, and T_m is the mean period.

Table 3.5 Correspondence between the degree of damage and the damage coefficient A_s in the case of two-layer covered stones.

Surface slope	Initial damage	Intermediate damage	Destruction
1 : 1.5	2	3–5	8
1 : 2	2	4–6	8
1 : 3	2	6–9	12
1 : 4 – 1 : 6	3	8–12	17

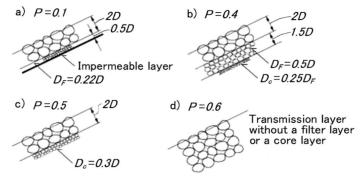

Fig. 3.38 Correspondence between the structure of a covered layer and the transmission coefficient P. (from Van der Meer [66])

3.WAVE RUN-UP, WAVE OVERTOPPING AND WAVE FORCES

The surface area directly exposed to storm waves should be covered with stones of weight W (= $g \times M$) obtained from the calculation formulae introduced earlier. In the case of wave overtopping, the weight of the stone should be increased by 50% or more of the calculated value.

However, the landward side and the base of the breakwater, which are not directly exposed to storm waves, are likely to be stable even if the stones have a smaller weight than the calculated value. In addition, if the stones are installed on normal sandy ground that has not been consolidated sufficiently, liquefaction of the ground during storm waves is likely to cause the covering stones to sink easily. Therefore, a foundation should be constructed on the ground by placing a few centimeters thick layer of crushed stones, followed by a layer of cobble or rubble stone the size of a fist, and larger stones on top of this layer as shown in **Fig. 3.39**, so that the foundation is in good contact with the ground.

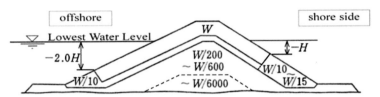

Fig. 3.39 Example of recommended combination of required weight of covering material and filler material.
(modified the illustration of Engineering Research Development Center [67])

Then, it is desirable to use stones or concrete blocks of $W/10$ - $W/15$ closer to the base of the wall, and stones or concrete blocks of weight W in the remaining areas of the seaside of the breakwater as shown in **Fig. 3.39**.

Van der Meer [68] and d'Angremond et al. [69] proposed Eq. (3.63) to determine the representative dimension D of a stable tetrapod (a type of deformed concrete armour block) for a sloping breakwater with a surface slope of 1:1.5, under non-overtopping conditions. The required mass M can be obtained from Eq. (3.62) by using D determined from Eq. (3.63).

$$\frac{H}{\left(\frac{\rho_s}{\rho}-1\right)D} = \left(3.75\frac{N_{od}^{0.5}}{N_w^{0.25}}+0.85\right)\left(\frac{2\pi H}{gT_m^2}\right)^{-0.2} \quad (3.5<\xi_m<6) \quad (3.63)$$

where H is the incident wave height [$H_{1/3}$ is used for non-breaking waves and $H_{2\%}/1.4$ for breaking waves], ρ is the density of seawater, ρ_s and D are respectively the density and the representative dimension (cubic root of the volume) of the tetrapod, N_{od} is the number of blocks that have moved from within the vertical stripe of width D to outside the covering layer, N_w is the incident number of storm waves, and T_m is the mean period.

3.2 Wave Forces

(3) Formulae for Calculating the Required Mass of Various Covering Materials

There are various types of deformed blocks and the equations for calculating the stability of the representative dimensions of major blocks are compiled in the "Coastal Engineering Manual" [67]. Moreover, Kashima et al. [70] and Hanzawa et al. [71] proposed equations for calculating the representative dimension of covered blocks for composite gravity type wave breakers. Brebner and Donnelly [72] proposed Eq. (3.64) for calculating the required mass M of covering material for the foundation mound of composite breakwaters.

$$M = \frac{\rho_s H_{1/3}^3}{N_s^3 \left(\frac{\rho_s}{\rho} - 1\right)^3} \tag{3.64}$$

where N_s is a dimensionless quantity that expresses the degree of stability, called the **Stability Number**, that can be evaluated by using Eqs. (3.65), proposed by Tanimoto et al. [73] and Takahashi et al. [74].

$$\left. \begin{array}{l} N_s = \max\left[1.8, 1.3\dfrac{(1-\kappa_a)}{\kappa_a^{1/3}}\dfrac{h'}{H_{1/3}} + 1.8\exp\left\{-1.5\dfrac{(1-\kappa_a)^2}{\kappa_a^{1/3}}\dfrac{h'}{H_{1/3}}\right\}\right] \\[2mm] \kappa_a = \dfrac{4\pi h'/L'}{\sinh(4\pi h'/L')}\sin^2(2\pi B_b/L') \end{array} \right\} \tag{3.65}$$

Where h' is the water depth from the still water surface to the top of the foundation mound, L' is the wavelength at h', B_b is the front shoulder width (see **Fig. 3.40**).

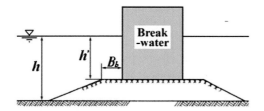
Fig. 3.40 Cross-sectional illustration of a composite breakwater.

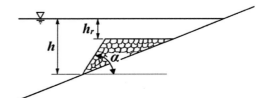
Fig. 3.41 Cross-sectional illustration of an artificial reef.

In the case of a fully submerged breakwater (**artificial reef**) shown in **Fig. 3.41**, the covering material mass M can be determined using Brebner and Donnelly's Eq. (3.64) if the stability number N_s can be properly set by paying attention to the relationship between the water depth on the crown of the breakwater and the incident wave height and the wave period (the longer the wave period, the bigger the damage rate).

Uda et al. [75] stated that even if the frontal water depth, the crown water depth, and the incident wave height are the same, different wavelengths result in different velocities on the artificial reef. Therefore, it is unreasonable to use the incident wave height to determine the maximum velocity which governs the stable mass of the covering material. They proposed Eq. (3.66) for calculating the required mass of the covering material using the water depth at the shore end of the artificial reef ($= h_r + \bar{\eta}$, the

3. WAVE RUN-UP, WAVE OVERTOPPING AND WAVE FORCES

water depth from the still water surface to the crown of the artificial reef + the average water surface elevation from the still water surface at the reef shore end) and the dimensionless quantity K_L representing the stability degree that took the maximum flow velocity u_{max} into account.

$$M = \frac{K_L}{(\cos\phi)^3} \frac{\rho_s (h_r + \bar{\eta})^3}{[(\rho_s/\rho)-1]^3}, \quad K_L = S_n^3 f_u^6 K_v, \quad f_u = \frac{u_{max}}{\sqrt{g(h_r + \bar{\eta})}} \qquad (3.66)$$

Where ϕ is the angle between the reef surface at the point of maximum flow velocity on the artificial reef and the horizontal plane, when the crown depth h_r is greater than 0.7 times the incident wave height H_o', the maximum flow velocity occurs above the crown, so ϕ can be expressed by Eq. (3.67).

$$\left.\begin{array}{l}\cos\phi = \cos\alpha \quad (h_r/H_o' \leq 0.7) \\ \cos\phi = 1 \quad (h_r/H_o' > 0.7)\end{array}\right\} \qquad (3.67)$$

S_n is the stability number (= 0.9 for natural stones); f_u is the dimensionless coefficient of the maximum flow velocity u_{max} on the reef surface, and can be obtained from **Fig. 3.42**; K_v is the coefficient on covering material shape (= 0.5 for natural stones with ρ_s of 2.65ton/m^3).

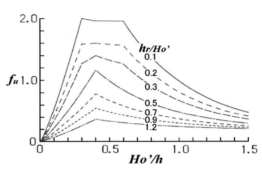

Fig. 3.42 Diagram for f_u according to "Guide to the Design of Artificial Reefs" [76].

In the case of concrete blocks, K_L should be determined by hydraulic model experiments, and as the longer the wave period, the bigger the damage rate, cases with long-period waves are desirable in hydraulic model experiments.

The average water surface elevation on a submerged artificial reef $\bar{\eta}$ is a function of the water depth on the crown h_r, the incident wave height H_o', and the wavelength L_o, and can be obtained from **Fig. 3.43**.

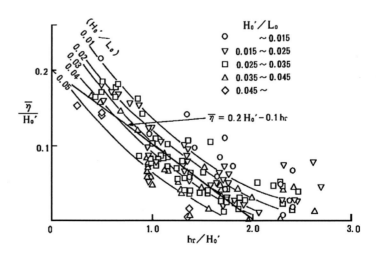

Fig. 3.43 Diagram for $\bar{\eta}$ according to "Guide to the Design of Artificial Reefs" [76].

3.2 Wave Forces

The calculated mass of covering materials M is used on the front surface and the offshore side of the crown of the artificial reef. The covering material mass to be used on the crown can be reduced to M_x according to the shoreward distance x from the offshore edge of the crown as follows:

$$\left. \begin{array}{l} 5 \leq x/H_o' < 10 \;\text{-----}\; M > M_x \geq 0.4M \\ 10 \leq x/H_o' \;\text{--------}\; 0.4M > M_x \geq 0.1M \end{array} \right\} \quad (3.68)$$

3) Wave Forces Acting on a Pillar Structure

The wave force dF acting on a small height dz of the pillar structure shown in **Fig. 3.44** is expressed as the sum of the drag force component dF_D and inertia force component dF_I.

$$dF = dF_D + dF_I = \frac{1}{2}\rho C_D u|u|dA + \rho C_M a dV \quad (3.69)$$

where ρ is the density of seawater, C_D is the drag coefficient, C_M is the mass coefficient, u and a are the horizontal velocity and horizontal acceleration of water particles due to wave motion, dA is the projected area in the incident direction of a wave within dz of the pillar structure ($= dz \times$ column diameter D), and dV is the volume within dz of the pillar structure.

Since wave force oscillates with a passing wave, the following equation is used to find the maximum value of dF_{max}.

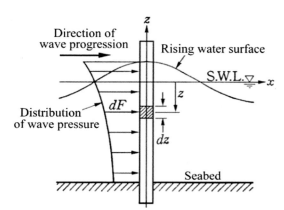

Fig. 3.44 Wave force acting on a pillar structure.

$$dF_{max} = \begin{cases} dF_D + \dfrac{dF_I^2}{4dF_D} & \text{(in the case of } 2dF_D > dF_I\text{)} \\ dF_I & \text{(in the case of } 2dF_D < dF_I\text{)} \end{cases} \quad (3.70)$$

where dF_D and dF_I are the maximum drag and maximum inertia force components of Eq. (3.71) expressed by using the amplitude u_{max} of the horizontal velocity of the water particle and the amplitude a_{max} of the horizontal acceleration, respectively.

$$dF_D = \frac{1}{2}\rho C_D u_{max}^2 D dz, \quad dF_I = \rho C_M a_{max} \frac{\pi D^2}{4} dz \quad (3.71)$$

The drag coefficient C_D and the mass coefficient C_M are functions of the Reynolds number for viscosity and the Keulegan-Carpenter number ($u_{max} \times T/D$) for vibration, and values of 1 to 2 are usually used.

For a legged structure, if the predominant wave period of the incident waves becomes equal to the natural period of the structure, there is a risk of resonance phenomena occurring, so dynamic analysis is necessary during the design.

143

3.WAVE RUN-UP, WAVE OVERTOPPING AND WAVE FORCES

(1) Wave Forces Acting on Small-diameter Cylindrical Objects

For wave forces acting on small-diameter cylindrical objects, Morison [77] proposed the following equation.

$$dF = \frac{1}{2}\rho C_D u |u| D dz + \rho C_M \frac{du}{dt}\frac{\pi D^2}{4} dz \tag{3.72}$$

where D is the diameter of the cross section of the cylindrical member.

The drag coefficient C_D can be determined from Eq. (3.73), and the mass coefficient C_M is generally set to 2.0 based on the potential theory of perfect fluids.

$$C_D = \begin{cases} 1.2 & \left(R_e \leq 2.0\times 10^5\right) \\ 1.2 - 0.5\times \dfrac{R_e - 2.0\times 10^5}{3.0\times 10^5} & \left(2.0\times 10^5 < R_e < 5.0\times 10^5\right) \\ 0.7 & \left(5.0\times 10^5 \leq R_e\right) \end{cases} \tag{3.73}$$

Where R_e is the Reynolds number $\left(= u_{max} \times D/v\right)$ calculated using the maximum flow velocity u_{max}, the diameter of the cross section of the cylindrical member D, and the kinematic viscosity coefficient v.

The ratio of the total drag force to the total inertia force obtained by integrating the drag force and inertia force components of Eq. (3.72) from the seabed to the water surface, respectively, indicates that the larger H/D, the ratio of incident wave height H to member diameter D, and smaller the relative water depth h/L, the more dominant the drag force becomes.

(2) Wave Forces Acting on Large-diameter Cylindrical Objects

As the diameter of a cylindrical object increases, the drag force due to vortex motion becomes relatively negligible (in general, the drag force can be neglected when the ratio of the diameter D of the object to the wavelength L of the incident wave is greater than 0.1), and the wave force for non-breaking waves can be calculated using velocity potential theory. This has been studied by MacCamy and Fuchs [78] and others. From the mass coefficient equation by MacCamy and Fuchs [78], the larger D/L is, the smaller the mass coefficient C_M becomes.

List of References in Chapter 3

1) Battjes, J. A.: Run-up Distributions of Waves Breaking on Slopes, *Proc. ASCE*, WW1, 1971, pp. 91-113.

2) Sawaragi, T., Iwata, K. and Morino, A.: On the Characteristics of Wave Run-up on a Gentle Slope, *Proc. 23rd Coastal Engineering Conference*, JSCE, 1976, pp.164-169. (in Japanese)

3) Mase, H.: Run-up Spectra of Random Waves on Sloping Beaches, *Journal of JSCE*, No.357/II-3, 1985, pp.197-205. (in Japanese)

4) Sawaragi, T. and Iwata, K.: A Nonlinear Model of Irregular Wave Run-up Height and Period Distributions on Gentle Slopes, *Proc. 19 th Coastal Eng. Conf.*, ASCE, 1984, pp. 415-434.

5) Kubota, S.: *Study on Mechanism Elucidation and Prediction of Local Run-up Waves*, Doctoral Dissertation, Chuo University, 1991, 232p. (in Japanese)

6) Yamamoto, Y. and Tanimoto, K.: A Study on Long Period Runup Waves Caused by Irregular Incident Waves, *Journal of JSCE*, No.503/II-29, 1994, pp.109-118. (in Japanese) https://www.jstage.jst.go.jp/article/jscej1984/1994/503/1994_503_109/_pdf/-char/ja

7) Mase, H. and Kobayashi, N.: Low-Frequency Swash Oscillation, *Journal of JSCE*, No.461/II-22, 1993, pp. 49-57. (in Japanese)

8) Kubota, S., Mizuguchi, M. and Takezawa, M.: Prediction Model of Run-up and Reflected Wave Distribution, *Journal of Coastal Engineering*, JSCE, Vol. 39, 1992, pp. 21-25. (in Japanese)

9) H. Mase, H. Doi and Y. Iwagaki: Experimental Study on the Effect of Wave Grouping on the Run-up Characteristics of Irregular Waves, *Proc. 30th Coastal Engineering Conference*, JSCE, 1983, pp. 114-118. (in Japanese)

10) Katoh, K.: *Study on Influence of Infragravity Waves on Sand Transport and Beach Nearshore Process*, Doctoral Dissertation, Tokyo Institute of Technology, 1990, pp.85-110. (in Japanese)

11) Longuet-Higgins, M. S. and Stewart, R. W.: Radiation Stress and Mass Transport in Gravity Waves, with Application to "Surf Beat", *Jour. Fluid Mech.*, Vol. 13, 1962, pp. 481-504.

12) Goda, Y.: Irregular Wave Deformation in the Surf zone, *Coast. Eng. Japan*, JSCE, Vol.18, 1975, pp. 13–26.

13) Guza, R.T. and Thornton, E.B.: Observations of Surf beat. *Jour. Geophys. Res.*, Vol. 90, C2, 1985, pp. 3162–3172.

14) Battjes, J.A., Bakkenes, H.J., Janssen, T.T. and van Dongeren, A.R.: Shoaling of Subharmonic Gravity Waves. *Jour. Geophys. Res.*, Vol.109, C2, 2004. https://doi.org/10.1029/2003JC001863

15) Symonds, G., Huntley, D. A. and Bowen, A. J.: Two-dimensional Surf beat: Long Wave Generation by a Time-varying Breakpoint, *Jour, Geophys. Res.*, Vol. 87, C1, 1982, pp.492-498.

16) Katoh, K., Nakamura, S. and Ikeda, N.: Estimation of Infragravity Waves in Consideration of Wave Groups, *Reort of the Port and Harbour Research Institute*, Vol.30, No.1, 1991, pp.137-

3.WAVE RUN-UP, WAVE OVERTOPPING AND WAVE FORCES

163. (in Japanese)

17) Nakamura, S. and Katoh, K.: Generation of Infragravity Waves in Breaking Process of Wave Groups, *Proc. 23 rd Coastal Eng. Conf.*, ASCE, 1992, pp.990-1003.

18) Goda, Y., Isayama, T. and Sato, S.: Experiments and Field Observations on the Development Mechanism of Infragravity Waves in the Wave Breaking Zone, *Journal of Coastal Engineering*, JSCE, Vol.37, 1990, pp.96-100. (in Japanese)

19) List, J.H.: A Model for the Generation of Two-dimensional Surf Beat, *Jour. of Geophys. Res.*, Vol. 97, No.4, 1992, pp. 5623-5635.

20) List, J.H.: Breakpoint-forced and Bound Long Waves in the Nearshore : A model comparison, *Proc. of 23rd Coastal Eng. Conf.*, ASCE, 1992, pp. 860-873.

21) Yamamoto, Y. and Tanimoto, K.: Effect of Seabed Slope on Infragravity Waves Caused by Wave Grouping, *Journal of Coastal Engineering*, JSCE, Vol.41, 1994, pp.81-85. (in Japanese)

22) Goda, Y.: Deformation of Irregular Waves due to Depth-Controlled Wave Breaking, *Report of the Port and Harbour Research Institute*, Vol.14, No.3, 1975, pp.59-106. (in Japanese)

23) Yamaguchi, M. and Hatada, Y.: Statistical Characteristics of Infragravity Waves Associated with Waves, *Proc. 30th Coastal Engineering Conference*, JSCE, 1983, pp.148-152. (in Japanese)

24) Hirose, M. and Hashimoto, N.: Characteristics of Infragravity Waves at Tagonoura Port, *Proc. 30 th Coastal Engineering Conference*, JSCE, 1983, pp.163-167. (in Japanese)

25) Katoh, K.: Infragravity Waves and the Formation of Multi-stage Sandbars, *Proc. 31st Coastal Engineering Conference*, JSCE, 1984, pp.441-445. (in Japanese)

26) Katoh, K., Nakamura, S. and Ikeda, N.: Field Observation on the Relation between Wave Coupling and Infragravity Waves, *Journal of Coastal Engineering*, JSCE, Vol.37, 1990, pp.101~105. (in Japanese)

27) Iwata, K., Sawaragi, T. and Nobuta, W.: Run-up Height and Run-up Period of Irregular Waves on a Gentle Slope, *Proc. 28th Coastal Engineering Conference*, JSCE, 1981, pp.330-334. (in Japanese)

28) Takada, A.: Wave Run-up and Overtopping, *Proc. 13th Summer Workshop on Hydraulic Engineering*, B-2, 1977. (in Japanese)

29) Hunt, I. A.: Design of Seawalls and Breakwaters, *Proc. ASCE*, Vol.85, No.WW3, 1959, pp.123-152.

30) Saville, T. Jr. ： Wave Run-up on Composite Slopes, *Proc. 6th Coastal Engineering Conference,* ASCE, 1958, pp.691-699.

31) Nakamura, M., Sasaki, Y. and Yamada, J.: Study on Run-up Height of Waves in Composite Cross Section, *Proc. 19th Coastal Engineering Conference*, JSCE, 1972, pp.309-312. (in Japanese)

32) Yamamoto, Y.: On Wave Run-up Height of Waves after Wave Breaking for Complex Beach Cross-Sections, *Proc. Civil Engineering in the Ocean*, JSCE, Vol. 4, 1988, pp.295-299. (in Japanese)

33) Sunamura, T. and Horikawa, K.: Two-dimensional Beach Transformation due to Waves, *Proc. 14th Inter Conf. on Coastal Engineering*, ASCE, 1974, pp.920-938.

34) Yamamoto, Y., Yamaji, T. and Asano, G.: Importance of Wave Grouping in Wave Overtopping Calculations and Its Engineering Evaluation Method, *Journal of Coastal Engineering*, JSCE, Vol. 43, 1996, pp.741-744. (in Japanese)

35) Savage, R. P.: *Laboratory Data on Wave Run-up on Roughened and Permeable Slopes*, U.S. Army Corps of Engrs., Beach Erosion Board, Tech. Memo., No.109, 1959, 28p.

36) Mase, H.: Random Wave Runup Height on Gentle Slope, *Jour. Waterway, Port and Ocean Engineering*, ASCE, Vol.115, No.5, 1989, pp.649-661.

37) Van der Meer, J. W. and Stam C.M.: Wave Runup on Smooth and Rock Slopes of Coastal Structures, *Jour. Waterway, Port and Ocean Engineering*, ASCE, Vol.118, No.5, 1992, pp.534-550.

38) De Waal, J. P. and Van der Meer, J. W.: Wave Run-up and Overtopping on Coastal Structures, *Proc. 23rd Inter Conf. on Coastal Engineering*, ASCE, 1992, pp.1758-1771.

39) Hosoi, M. and Shuto, N.: On Wave Run-up Height When Waves Enter Obliquely to the Dike, *Proc. 9th Coastal Engineering Conference*, JSCE, 1962, pp.149-152. (in Japanese)

40) Sibul, J.O. and Tickner, E.G.: *A Model Study of the Run-up of Wind Generated Waves on Levees with Slope of 1:3 and 1:6*, U.S. Army Corps of Engrs., Beach Erosion Board, Tech. Memo., No.67, 1955, 19p.

41) Yamamoto, Y., Vu, T.C., Asano, G. and Arimura, J.: Study on Evaluation Method of Wave Overtopping for a Local Coastal Dike, *Journal of Coastal Engineering*, JSCE, Vol. 46, 1999, pp.761-765. (in Japanese)

42) Goda, Y., Kishira, Y. and Kamiyama, Y.: Laboratory Investigation on the Overtopping Rate of Seawalls by Irregular Waves, *Report of the Port and Habour Research Institute*, Vol.14, No.4, 1975, pp.3-44. (in Japanese)

43) Goda, Y. and Kishira, Y.: *Experiments on Irregular Wave Overtopping Charcteristics of Seawalls of Low Crest Types*, Technical Note of the Port and Harbour Research Institute, No.242, 1976, 28p. (in Japanese)

44) Yamamoto, Y. and Horikawa, K.: Proposal of a New Formula for Calculating Wave Overtopping Rate, *Proc. Civil Engineering in the Ocean*, JSCE, Vol. 7, 1991, pp.25-30. (in Japanese)

45) Van der Meer, J. W. and Janssen, W.: Wave Run-up and Wave Overtopping at Dikes, *Wave Forces on Inclined and Vertical Wall Structures*, Kobayashi & Demirbilek, ASCE, 1995, pp.1-27.

46) Isobe, M., et al. 37: *Coastal Technology Library No.39 CADMAS-SURF/3D Research and Development of a Numerical Wave Tank*, Coastal Development Institute of Technology, 2010, 235p. (in Japanese)

3.WAVE RUN-UP, WAVE OVERTOPPING AND WAVE FORCES

47） Watanabe, A. and Kawahara, T.: Relation between Spectrum of Irregular Waves and Wave Height Period Distribution, *Proc. 31st Coastal Engineering Conference*, JSCE, 1984, pp.153-157. (in Japanese)

48） EurOtop: *Manual on Wave Overtopping of Sea Defences and Related Structures, an Overtopping Manual Largely Besed on European Research, but for Worldwide Application*, Second Edition, 2018, 304p. http://www.overtopping-manual.com/assets/downloads/EurOtop_II_2018_Final_version.pdf

49) Vu, T.C., Tanimoto, K. and Yamamoto, Y.: Numerical Simulation of Cross-Shore Beach Profile Change, *Journal of Coastal Engineering*, JSCE, Vol.46, 1999, pp.611-615. (in Japanese)

50） Sato, S. and Suzuki, H.: Evaluation Method of Bottom Velocity Fluctuation Waveforms in Wave Breaking Zone, *Journal of Coastal Engineering*, JSCE, Vol.37, 1990, pp.51-56. (in Japanese)

51） Okayasu, A., Suzuki, Y. and Hanada, M.: Field Observation and Numerical Simulation of Infragravity Waves in a Breaking Wave Band, *Journal of Coastal Engineering*, JSCE, Vol.45, 1998, pp.286-290. (in Japanese)

52） Horikawa Kiyoshi: *Coastal Engineering, an Introduction to Ocean Engineering*, University of Tokyou Press, 1978, 402p.

53） Bagnold, R.A.: Interim report on wave-pressure research, *Jour. Inst. Civil Engrs.*, Vol.12, 1939, pp.202-226.

54） Sainflou, G.: Essai sur les digues maritimes Verticales, *Annales Ponts et Chaussées*, Vol.98, No.4, 1928.

55） Goda, Y.: The Design of Uplight Breakwater, *Design and Reliability of Coastal Structures, Proceedings of the Short Course on Design and Reliability of Coastal Structures*, attached to the 23rd International Conference on Coastal Engineering, ASCE, 1992, pp.19-1 – 19-22.

56） Hiroi, I.: On a Method of Estimating the Force of Waves, *J. College of Eng.*, Tokyo Imperial University, Vol. 10, No. 1, 1919, pp.1-19.

57） Goda, Y.: A New Method of Wave Pressure Calculation for the Design of Composite Breakwaters, *Report of the Port and Habour Research Institute*, Vol.12, No.3, 1973, pp.31-69. (in Japanese)

58） Tanimoto, K., Moto, K., Ishizuka, S. and Goda, Y.: Study on Calculation Formula of Design Wave Forces of Breakwaters, *Proc. 23rd Coastal Engineering Conference*, JSCE, 1976, pp.11-16. (in Japanese)

59） Takahashi, S., Tanimoto, K. and Shimosako, K.: Wave and Block Forces on a Caisson Covered with Wave Dissipating Blocks, *Report of the Port and Habour research Institute*, Vol.29, No.1, 1990, pp.53-75. (in Japanese)

60) Tanimoto, K. and Kimura, K.: *A Hydraulic Experimental Study on Trapezoidal Caisson Breakwaters*, Technical Note of the Port and Habour Research Institute, No.528, 1985, 27p. (in Japanese)

61） PROVERBS (Oumeraci, H., Allsop, N.W.H., De Groot, M.B., Crouch, R.S., Vrijling, J.K. and Kortenhaus, A): *Probabilistic Design Tools for Vertical Breakwaters*, final report (MAS3 - CT95 – 0041), TU Delft Research Repository, 1999. https://repository-tudelft.nl/record/uuid:ce9ec158-78b0-484a-9344-d55c326b4ede

62） Minikin, R.R.: *Winds, Waves and Maritime Structures*, Griffin, London, 1950, pp.38-39.

63） Tanimoto, K.: On Wave Forces Acting on Composite Breakwaters, *Proceedings of the Port and Harbor Research Institute*, 1976, pp.1-26. (in Japanese)

64） Iribarren, R.: *A Formula for the Calculation of Rock-fill Dikes*, Technical Report, HE-116-295, Fluid Mech. Lab., Univ. of California, 1938, translated by D. Heinrich in 1948.

65） Hudson, R.Y.: Laboratory Investigation of Rubble Mound Breakwaters, *Jour. of Waterways and Harbor Div.*, Vol.85, WW3, 1959, pp.93-121.

66） Van der Meer, J.W.: *Rock Slopes and Gravel Beaches Under Wave Attack*, Dissertation of Ph.D., Delft Univ. of Technology, Pub. No. 396, Delft, The Netherlands, 1988, 152p.

67） Engineering Research Development Center: *Coastal Engineering Manual*, U.S. Army Corps of Engineers, 2002. https://www.publications.usace.army.mil/USACE-Publications/Engineer-Manuals /u43544q/636F617374616C20656E67696E656572696E67206D616E75616C/

68） Van der Meer, J.W.: Stability of Cubes, Tetrapodes and Accropode, *Proc. of the Breakwaters '88 Conference*; *Design of Breakwaters*, Institution of Civil Engineers, Thomas Telford, London, UK, 1988, pp.71-80.

69） d'Angremond, K., van der Meer, J.W., and van Nes, C.P.: Stresses in Tetrapod Armour Units Induced by Wave Action, *Proc. of the 24th International Coastal Engineering Conference*, ASCE, Vol.2, 1994, pp.1713-1726.

70） Kashima, R., Sakakiyama, T., Shimizu, T., Sekimoto, T., Kunisu, H., and Kyoya, O.: Evaluation Formula of Deformation of Wave Dissipating Block Coverings against Irregular Waves, *Journal of Coastal Engineering*, JSCE, Vol.40, 1993, pp.795-799. (in Japanese)

71） Hanzawa, M., Sato, H., Takayama, T., Takahashi, S. and Tanimoto, K.: Study on Stability Evaluation Formula of Wave Dissipating Blocks, *Journal of Coastal Engineering*, JSCE, Vol.42, 1995, pp. 886-890. (in Japanese)

72） Brebner, A. and Donnelly, D.: Laboratory Study of Rubble Foundation for Vertical Breakwater, *Proc. 8th Coastal Engineering Conference*, New Mexico City, 1962, pp. 408-429.

73） Tanimoto, K., Yagyu, T., Muranaga, T., Shibata, K. and Goda, Y.: Stability of Armor Units for Foundation Mounds of Composite Breakwaters by Irregular Wave Tests, *Report of the Port and Harbour Research Institute*, Vol.21, No.3, 1982, pp.3-42. (in Japanese)

74） Takahashi, S., Kimura, K. and Tanimoto, K.: Stability of Armor Units of Composite Breakwater Mound against Oblique Waves, *Report of the Port and Habour Research Institute*, Vol.29, No.2, 1990, pp.3-36. (in Japanese)

3.WAVE RUN-UP, WAVE OVERTOPPING AND WAVE FORCES

75) Uda, T., Omata, A. and Saito, T.: *Stability Formula of Armor of Artificial Reef*, Technical Note of Pablic Works Research Institute, No.2893, 1990, 48p. (in Japanese)

76) Seacoast Office, River Bureau, Ministry of Land, Infrastructure and Transport, and Coast Division, River Department, National Institute of Land and Infrastructure Management: *Guide to the Design of Artificial Reefs*, National Association of Sea Coast, 2004, 95p. (in Japanese)

77) Morison, J.R., O'Brien, M.P., Johnson, J.W. and Schaaf, S.A.: The Force Exerted by Surface Waves on Piles, *Petroleum Trans.*, AIME Vol.189, 1950, pp.149-157.

78) MacCamy, R.C. and Fuchs, R.A.: *Wave Forces on Piles: A Diffraction Theory*, U.S. Army Corps of Engrs., Beach Erosion Board, Tech. Memo., No.69, 1954, 17p.

Chapter:4

CURRENTS IN THE SEA

4. CURRENTS IN THE SEA

Currents in the sea area include the following.

(a) **Nearshore currents** (wave-induced currents): currents generated nearshore by wave breaking in nearshore areas.

(b) **Drift currents** (wind-driven currents): the movement of surface and near-surface seawater by the action of prevailing wind blowing for a sufficiently long duration on the sea surface.

(c) **Thermohaline currents** (a kind of density currents): currents generated by spatial differences in temperature and salinity of ocean water.

(d) **Tidal currents:** currents generated by the gravitational force of the moon and sun.

Moreover, the drift and thermohaline currents of the larger scale offshore are collectively referred to as **ocean currents**, and the nearshore, drift, thermohaline, and tidal currents of the smaller scale near the shore are collectively referred to as **coastal currents**.

All the above-mentioned currents are influenced by the Coriolis force, resulting from the rotation of the earth. However, the influence of the Coriolis force is strong for large-scale currents, such as drift currents, thermohaline currents, and tidal currents, but weak for small-scale currents, such as nearshore currents.

4.1 Nearshore Currents

1) Radiation Stress

In fluids in motion, the momentum flux through a surface (the momentum transported per unit time and unit width of a surface) is the integration of the sum of the water pressure acting on the target surface and the momentum of the fluid passing through the target surface from the bottom to the water surface minus the hydrostatic pressure (the hydrostatic pressure must be subtracted from the total water pressure because it acts on both sides of the surface and cancels out).

As shown in **Fig. 4.1**, when the x axis is set in the on-offshore direction, perpendicular to the shoreline, and the y axis in the longshore direction, parallel to the shoreline, and the water velocity in the on-offshore direction is denoted by u and in the longshore direction by v, when waves propagate from offshore to the shore, the momentum flux M_{xx} in x direction through a surface perpendicular to the x axis is expressed by Eq. (4.1).

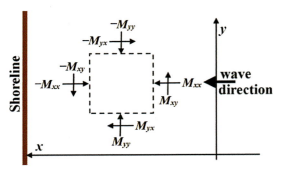

Fig. 4.1 Definition of momentum flux of waves.

$$M_{xx} = \int_{-h}^{\eta}(p+\rho u \times u)dz - \int_{-h}^{\eta}\rho gz\,dz \qquad (4.1)$$

where h is the water depth from the sea bottom to the still water surface, η is the water surface elevation on the still water surface, p is the water pressure, ρ is the density of seawater, z is the vertical coordinate with the origin at the still water surface (positive upward), and g is the gravitational acceleration.

The momentum flux averaged over one wave period S_{xx} is expressed by Eq. (4.2).

$$S_{xx} = \frac{1}{T}\int_0^T \int_{-h}^{\eta}\left(p + \rho u^2\right)dzdt - \frac{1}{2}\rho gh^2 \tag{4.2}$$

The averaged momentum flux S_{xx} is the excess momentum of water motion under waves and thus is an additional force acting on the target surface of unit width. This additional force generated by surface gravity waves is called **radiation stress**.

Similarly, the excess momentum flux M_{yy} in the y direction is given by Eq. (4.3), and the radiation stress S_{yy}, averaged over one wave period, is given by Eq. (4.4).

$$M_{yy} = \int_{-h}^{\eta}\left(p + \rho v \times v\right)dz - \int_{-h}^{\eta}\rho gz dz \tag{4.3}$$

$$S_{yy} = \frac{1}{T}\int_0^T \int_{-h}^{\eta}\left(p + \rho v^2\right)dzdt - \frac{1}{2}\rho gh^2 \tag{4.4}$$

The radiation stress S_{xy}, obtained by averaging the momentum flux in the y direction on a surface perpendicular to the x axis over one wave period, is expressed by Eq. (4.5). Since the momentum parallel to the target plane is considered, the water pressure, which is the stress normal to the surface, is neglected.

$$S_{xy} = \frac{1}{T}\int_0^T \int_{-h}^{\eta}\left(\rho u \times v\right)dzdt \tag{4.5}$$

Similarly, the radiation stress S_{yx}, obtained by averaging the momentum flux in x direction on a surface perpendicular to the y axis over one wave period, is expressed by Eq. (4.6).

$$S_{yx} = \frac{1}{T}\int_0^T \int_{-h}^{\eta}\left(\rho v \times u\right)dzdt \tag{4.6}$$

Assuming that waves propagate perpendicular to the shoreline, i.e., where the water particle velocity u is in the x direction and the water particle velocity v in the y direction is zero, and using the small amplitude wave theory, Eq (4.2) and Eqs. (4.4) to (4.6) lead to Eq. (4.7) in tensor representation.

$$S_{ij} = \begin{pmatrix} S_{xx} & S_{xy} \\ S_{yx} & S_{yy} \end{pmatrix} = \frac{1}{8}\rho gH^2 \begin{pmatrix} 2n - \dfrac{1}{2} & 0 \\ 0 & n - \dfrac{1}{2} \end{pmatrix} \tag{4.7}$$

where i and j are subscripts representing the plane and direction of interest respectively, H is the incident wave height, and n is the group velocity ratio (C_g/C) defined by Eqs. (2.1.50), (2.1.51).

Furthermore, since $n = 1/2$ in deep water and $n = 1$ in very shallow water, the following equations for S_{xx} are obtained.

$$\text{the deep water :} \quad S_{xx} = \frac{1}{16}\rho gH^2 \tag{4.8}$$

$$\text{the shallow water :} \quad S_{xx} = \frac{3}{16}\rho gH^2 \tag{4.9}$$

4. CURRENTS IN THE SEA

For shallow water, the finite amplitude wave theory should be used from the viewpoint of computational accuracy, but from the viewpoint of simplicity of handling, the small amplitude wave theory was used to obtain the analytical solution.

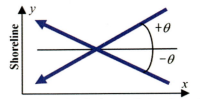

Fig. 4.2 Definition of incident angles of waves.

As shown in **Fig. 4.2**, the radiation stress of a wave with incident angle θ can be calculated using Eq. (4.10).

$$S_{ij} = \begin{pmatrix} S_{xx} & S_{xy} \\ S_{yx} & S_{yy} \end{pmatrix} = \frac{1}{8}\rho g H^2 \begin{pmatrix} n(1+\cos^2\theta)-\frac{1}{2} & \frac{1}{2}n\sin 2\theta \\ \frac{1}{2}n\sin 2\theta & n(1+\sin^2\theta)-\frac{1}{2} \end{pmatrix} \quad (4.10)$$

Longuet-Higgins and Stewart [1] were the first to propose radiation stress as the external force for solving the nearshore seawater motion equation averaged over one wave period. Their theory allows us to understand the mean water level change and the nearshore current due to wave breaking.

2) Average Water Level Inside and Outside the Breaking Zone

According to Longuet-Higgins and Stewart [1], when incoming waves are at right angles to the shoreline, the equation of momentum averaged over one wave period in the on-offshore direction (x direction) for the steady state is expressed as follows:

$$\rho g (h+\overline{\eta})\frac{\partial \overline{\eta}}{\partial x} + \frac{\partial S_{xx}}{\partial x} = 0 \quad (4.11)$$

where $\overline{\eta}$ is the water surface elevation averaged over one wave period from the still water surface.

Outside the wave breaking zone, the second term on the left side of Eq. (4.11) becomes positive because Eq. (4.8) is valid, and the incident wave height increases as the wave approaches the wave breaking depth from offshore due to shoaling deformation. As a result, the average water surface elevation decreases. This phenomenon is called **wave setdown**. Solving Eq. (4.11) using the offshore boundary condition of $\overline{\eta} \to 0$ at $h \to \infty$ under the approximation of $h+\overline{\eta} \approx h$, the amount of wave setdown is obtained as follows:

$$\overline{\eta} = -\frac{H^2}{8}\frac{\kappa}{\sinh 2\kappa h} \quad (4.12)$$

where κ is the wave number.

Eq. (4.12) indicates that the absolute value of $\overline{\eta}$ becomes larger as the wave approaches the wave breaking depth h_b, as shown in **Fig. 4.3**.

Inside the wave breaking zone, the incident wave height can be approximated by $H = r(h+\overline{\eta})$ using the wave height-to-depth ratio r at the wave breaking limit, which means that the incident wave

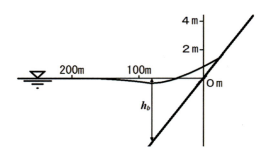

Fig. 4.3 Illustration of wave setup and setdown.

154

4.1 Nearshore Currents

height decreases as the wave approaches the shoreline. Therefore, substituting Eq. (4.9) in Eq. (4.11), the second term on the left-hand side becomes negative under this condition, indicating that $\bar{\eta}$ increases towards the shore. This phenomenon is called **wave setup**. The amount of wave setup can be obtained by solving Eq. (4.11) with the boundary conditions $\bar{\eta} = \bar{\eta}_b$ at the wave breaking depth h_b as follows:

$$\bar{\eta} = \bar{\eta}_b + \frac{1}{1+\frac{8}{3r^2}}(h_b - h) \qquad (4.13)$$

Eq. (4.13) shows that $\bar{\eta}$ is $\bar{\eta}_b$ lower than the still water surface at the wave breaking location, becomes larger as the wave approaches the shoreline from the wave breaking depth h_b, and becomes higher than the still water surface on the shoreline, as shown in **Fig. 4.3**.

3) Planar Structure of Nearshore Currents

If the shoreward seawater transport caused by the change in radiation stress due to wave breaking is balanced by gravity-driven return flow, no further rise in water surface elevation will occur. In this case, the surface seawater is transported onshore while the bottom seawater is transported offshore. The seaward flow of bottom seawater under waves is called **undertow**. However, if incoming waves obliquely approach the shore, water masses in the swash zone where the wave runup on the beach cannot return offshore due to the subsequent waves that come in one after another but will move alongshore in the direction of the incoming waves. This is called **longshore currents**.

Due to the irregularity of coastal topography, incoming waves are refracted and diffracted to concentrate in certain areas. This makes wave setup at some places where waves concentrate higher than that in other places where waves diverge. Due to different surface levels, seawater at high wave setup areas will flow to low wave setup areas, and ultimately return to the sea as narrow water jets. These are called **rip currents**. The currents moving to the shore are called **shoreward currents**. The cell structure of the nearshore currents, consisting of shoreward currents, longshore currents, and rip currents, is shown in **Fig. 4.4**.

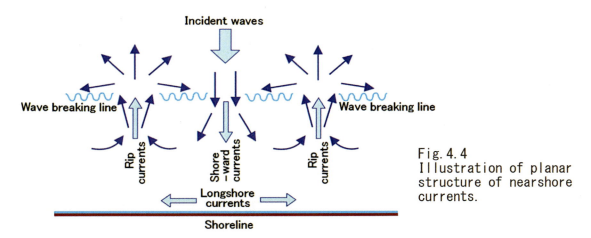

Fig. 4.4 Illustration of planar structure of nearshore currents.

155

4.CURRENTS IN THE SEA

The greater the angle of oblique wave incidence, the greater the component of longshore current (in **Fig. 4.4**, e.g., the rightward longshore current is greater in the case that the wave incidence is from the left). In the wave breaking zone and its outer vicinity, the strong longshore currents and rip currents are connected to form a meandering river-like current. (refer to Harris [2])

The important components of nearshore, longshore, and rip currents are explained in more detail below.

(1) Longshore Currents

According to Longuet-Higgins [3], the velocity $V(x)$ of the longshore current at a distance x offshore from the shoreline can be obtained from the balance between the activation force per-unit-area τ_g due to radiation stress S_{xy} expressed in Eq. (4.14) and the seabed friction resistance force per-unit-area τ_s in the longshore direction defined in Eq. (4.15), given in Eq. (4.16).

$$\tau_g = -\frac{\partial S_{xy}}{\partial x} = \frac{5\rho\gamma^2 \tan\beta}{16} \frac{\sin\theta_o}{C_o}\left[gh(x)\right]^{3/2} \tag{4.14}$$

$$\tau_s = \frac{2f_r}{\pi}\rho \times u_{max} \times V(x) = \frac{f_r}{\pi}\rho \times \gamma\sqrt{gh(x)} \times V(x) \tag{4.15}$$

$$V(x) = \frac{5\pi}{16}g\frac{\gamma \tan\beta}{f_r}\frac{\sin\theta_o}{C_o}h(x) \tag{4.16}$$

where ρ is the density of seawater, γ is the wave height-to-depth ratio (0.7-0.8) at the wave breaking limit, $\tan\beta$ is the seabed slope, θ_o is the incident angle of offshore waves from the normal to the shoreline, C_o is the wave velocity of offshore waves, g is the gravitational acceleration, $h(x)$ is the water depth at distance x offshore, f_r is the friction coefficient when the bottom friction force is defined in Eq. (4.15), u_{max} is the amplitude of the bottom velocity of water particles due to waves.

The velocity distribution given by Eq. (4.16) is proportional to the water depth up to the wave breaking location and becomes zero offshore of the wave breaking location. This equation corresponds to the case $P = 0$ in **Fig. 4.5**, and $X^* = x/x_b$ and $V^* = V(x)/V(x_b)$, using x_b as the distance from the shoreline to the wave breaking depth, $X^* = 1.0$ corresponds to the distance to the wave breaking depth, and V^* at this location is 1.0. If the horizontal diffusion force per unit area τ_m defined in Eq. (4.17) is considered in Eq. (4.16), the velocity of the longshore current V becomes as in Eqs. (4.18) and (4.19), resulting in a smooth velocity distribution seen in **Fig. 4.5** when $P > 0$.

$$\tau_m = \frac{\partial}{\partial x}\left(K_D h(x)\frac{\partial V(x)}{\partial x}\right) = \frac{\partial}{\partial x}\left(N\rho x\sqrt{gh(x)}h(x)\frac{\partial V(x)}{\partial x}\right) \tag{4.17}$$

$$P\left(= \pi\frac{N\tan\beta}{rf_r}\right) \neq \frac{2}{5}: \quad V^* = \begin{cases} AX^* + C_1\left(X^*\right)^{P_1}: & 0 \leq X^* \leq 1 \\ C_2\left(X^*\right)^{P_2}: & 1 \leq X^* \end{cases} \tag{4.18}$$

$$P\left(= \pi\frac{N\tan\beta}{rf_r}\right) = \frac{2}{5}: \quad V^* = \begin{cases} \frac{10}{49}X^* - \frac{5}{7}X^*\log_e X^*: & 0 \leq X^* \leq 1 \\ \frac{10}{49}\left(X^*\right)^{-5/2}: & 1 \leq X^* \end{cases} \tag{4.19}$$

4.1 Nearshore Currents

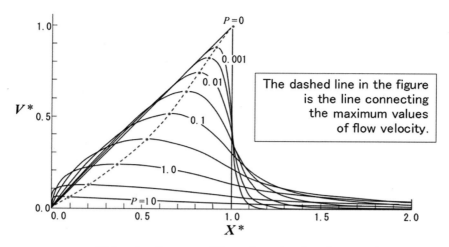

Using the distance offshore from the shoreline x and
the distance from the shoreline to the wave breaking depth x_b,
$X^* = x/x_b$, $V^* = V(x)/V(x_b)$.

Fig. 4.5 Off-shore distribution of the velocities $V(x)$ of longshore currents in the case of regular waves.

where K_D is the horizontal diffusion coefficient, and N is the dimensionless constant. If the distance x is measured offshore from the shoreline and x_b is the distance between the shoreline and the wave breaking depth,

$$X^* = \frac{x}{x_b}, \quad V^* = \frac{V(x)}{V(x_b)}, \quad A = \frac{1}{1-(5P/2)}, \quad C_1 = \frac{P_2 - 1}{P_1 - P_2}A, \quad C_2 = \frac{P_1 - 1}{P_1 - P_2}A$$

$$P_1 = -\frac{3}{4} + \sqrt{\frac{9}{16} + \frac{1}{P}}, \quad P_2 = -\frac{3}{4} - \sqrt{\frac{9}{16} + \frac{1}{P}}$$

Goda and Watanabe [4] used an irregular wave breaking model to obtain a velocity distribution equation for longshore currents at a plane beach. In their equation, a smooth velocity distribution is obtained even when the horizontal diffusion force is set to zero. If the mean value of the offshore distribution of longshore current velocity $\overline{V}$ is required, Eq. (4.20) obtained by Kormer [5] from field observations and hydraulic model experimental data can be used.

$$\overline{V} = 2.7 u_{b\max} \sin\theta_b \cos\theta_b \tag{4.20}$$

where u_{bmax} is the amplitude of the bottom velocity of the water particles during wave breaking and θ_b is the incident angle of the breaking wave from the normal to the shoreline.

(2) Rip Currents

The characteristics of rip currents are summarized from reports (e.g. refer to Horikawa [6]) of numerous field surveys as follows:

(a) If wave height distribution in the longshore direction is not constant, the probability that the water mass transported to the shore returns to the offshore as rip currents (as shown in **Fig. 4.4**)

4.CURRENTS IN THE SEA

is higher than the probability that it returns to the offshore as an undertow with a uniform flow rate in the longshore direction.

(b) Rip currents are more likely to appear on beaches with relatively gentle slopes. Moreover, the higher the incident wave height, the higher the rip current velocity (maximum, around 5 knots ≒2.5 m/s), and the longer the distance to the offshore edge of the rip current. However, current velocity can fluctuate due to the influence of infragravity waves.

(c) When the incident wave height is small, many weak rip currents are generated, but when the incident wave height is large, strong rip currents are generated at intervals of several hundred meters.

Bowen and Inman [7], [8] were the first to present a rip current generation mechanism. According to them, **edge waves** (waves that travel along the shoreline whose amplitude decreases exponentially in the offshore direction) interfere with incoming waves, resulting in periodic wave height changes in the longshore direction, which in turn generates rip currents at low wave heights.

Hino [9], [10] presented a rip current generation mechanism due to hydrodynamic instability. When waves approach at right angles to the shoreline on a uniformly sloping beach, wave setdown and setup are expected to occur uniformly in the longshore direction. But, in reality, the slightest disturbance to the hydrodynamic system can cause instability. In addition, currents can cause sediment movement and instability due to bathymetry change. Based on these ideas, Hino created a vertically integrated 2D numerical model of nearshore currents and seabed topography changes and showed that the optimum interval between longshore rip currents is about four times the width of the wave breaking zone.

Horikawa et al [11] classified the regions where the above mechanisms dominate as shown in **Table 4.1** and **Fig. 4.6** using the **surf similarity parameter** ξ_o [see the Iribarren number in Eq. (2.2.37)] for offshore waves shown in Eq. (4.21), and clarified the physical characteristics of each region.

$$\xi_o = \frac{\tan \beta}{\sqrt{H_o / L_o}} \tag{4.21}$$

The '**infragravity**' column in **Table 4.1** and **Fig. 4.6** corresponds to a region where long-period gravity waves dominate and where the seabed slope is very gentle relative to the wave steepness.

From **Fig. 4.6**, the ratio of the occurrence interval of rip currents (Y_r) to the wave breaking zone width (x_b) can be summarized as follows:

(a) Infragravity region ($0.23 \geq \xi_o$): $Y_r / x_b \approx 157 \times (\xi_o)^2$.

(b) Unstable region ($1 \geq \xi_o \geq 0.23$): $Y_r / x_b \approx 4$.

(c) Edge wave region ($\xi_o \geq 1$):   If the number of offshore modes of edge waves (the number of lines of nodes parallel to the shore) = 1, $Y_r / x_b \approx \pi$.

4.1 Nearshore Currents

When the number of offshore modes = 2, $Y_r / x_b \approx 2\pi/3$.

Table 4.1 Three theories of nearshore current generation from Horikawa et al [11].

Theory (Mechanism)	INFRAGRAVITY	INSTABILITY	EDGE WAVE
Proponent	Sasaki, T. (1974)	Hino, M. (1973)	Bowen & Inman (1969)
Coverage by ξ_o	$0.23 \geq \xi_o$	$1 \geq \xi_o \geq 0.23$	$\xi_o \geq 1$
Feature — Wave breaking type	Spilling	Spilling – Plunging	Plunging
Feature — Surf	Always present	Present or absent	Always none
Feature — Number of waves in the wave breaking zone	3 or more waves	1 - 3 waves	1 or less waves
Feature — Reflection coefficient γ of incident waves	$\gamma < 10^{-2}$	$\gamma \sim 10^{-2}$	$\gamma > 10^{-1}$
Feature — Incident wave characteristics	Wind waves and swells	Swell	
Explanation	When waves with large wave steepness come to a beach with small $\tan \beta$.	When swell comes to a beach with $\tan \beta$ between 1/20 and 1/40.	When waves with small wave steepness come to a beach with large $\tan \beta$. Pure nearshore currents are unlikely to occur.

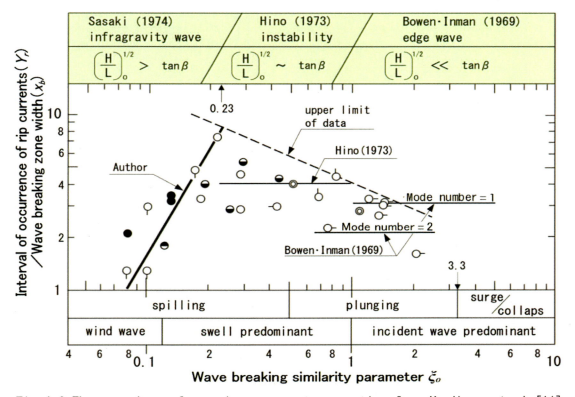

Fig. 4.6 Three regions of nearshore current generation from Horikawa et al [11].

4.CURRENTS IN THE SEA

4) Numerical Simulation Models of Nearshore Currents

When predicting the distribution of nearshore currents in complex coastal topography, hydraulic model experiments in a wave tank are considered, but this time-consuming and labor-intensive while requiring large equipment. Numerical models are therefore commonly used to simulate nearshore currents in recent years.

(1) Numerical Model with Wave Period-Averaged Momentum Equations

The wave field is calculated using the numerical model introduced in Chapter 2, Section 4, to obtain the radiation stress. Next, the nearshore currents are calculated by entering the radiation stress as an external force into the continuity equation and the equations of momentum, averaged over a wave period as shown in Eq. (4.22). The first and second equations in Eq. (4.22) are the cross-shore and longshore momentum equations, and the third is the continuity equation. The first term on the left-hand side of the momentum equations is the time variation of momentum, the second and third terms are the spatial variation of momentum, the fourth term is the pressure term, the fifth and sixth terms are the horizontal diffusion terms, the seventh and eighth terms are the force terms due to radiation stress, and the ninth term is the bottom friction term.

$$
\left.
\begin{aligned}
&\frac{\partial \rho d U}{\partial t}+\frac{\partial}{\partial x}\left(\rho d U^{2}\right)+\frac{\partial}{\partial y}\left(\rho d U V\right)+\rho g d \frac{\partial \bar{\eta}}{\partial x}-\frac{\partial}{\partial x}\left(K_{D} d \frac{\partial U}{\partial x}\right)-\frac{\partial}{\partial y}\left(K_{D} d \frac{\partial U}{\partial y}\right)+\left(\frac{\partial S_{xx}}{\partial x}+\frac{\partial S_{yx}}{\partial y}+\tau_{bx}\right)=0 \\
&\frac{\partial \rho d V}{\partial t}+\frac{\partial}{\partial x}\left(\rho d V U\right)+\frac{\partial}{\partial y}\left(\rho d V^{2}\right)+\rho g d \frac{\partial \bar{\eta}}{\partial y}-\frac{\partial}{\partial x}\left(K_{D} d \frac{\partial V}{\partial x}\right)-\frac{\partial}{\partial y}\left(K_{D} d \frac{\partial V}{\partial y}\right)+\left(\frac{\partial S_{xy}}{\partial x}+\frac{\partial S_{yy}}{\partial y}+\tau_{by}\right)=0 \\
&\qquad\qquad\qquad \frac{\partial \rho d}{\partial t}+\frac{\partial \rho d U}{\partial x}+\frac{\partial \rho d V}{\partial y}=0
\end{aligned}
\right\} \quad (4.22)
$$

where ρ is the density of seawater; d is the total water depth (the water depth h below the still water surface + the averaged water surface elevation $\bar{\eta}$ above the still water surface); U and V are the velocity of the nearshore current in the offshore direction (x direction) and the alongshore direction (y direction); t is the time; g is the gravity acceleration; K_D is the horizontal diffusion coefficient; S_{xx}, S_{yx}, S_{xy}, and S_{yy} are radiation stress; τ_{bx} and τ_{by} are the seabed friction force per unit area.

(2) Numerical Model with Nonlinear Equations of Motion

Because nonlinear equations of motion in two planar dimensions, such as the Boussinesq equation, can accurately calculate the momentum in shallow water areas, waves and nearshore currents can be calculated simultaneously. Therefore, with the addition of run-up boundary conditions and sediment transport equations, they can become effective numerical models for calculating wave run-up and predicting topographic changes including run-up areas.

As a typical example, the numerical model of Sato and Kabiling [12] is introduced here. The basic equations consist of the continuity equation and the Boussinesq equations of motion on a horizontal bed, considering wave energy loss due to wave breaking, shown in Eq. (4.23).

160

4.1 Nearshore Currents

$$\frac{\partial q_x}{\partial t}+\frac{\partial}{\partial x}\left(\frac{q_x^2}{d}\right)+\frac{\partial}{\partial y}\left(\frac{q_x q_y}{d}\right)+gd\frac{\partial \eta}{\partial x}-\frac{h^2}{3}\left(\frac{\partial^3 q_x}{\partial x^2 \partial t}+\frac{\partial^3 q_y}{\partial x \partial y \partial t}\right)-M_{Dx}+\frac{f_w}{2d^2}q_x\sqrt{q_x^2+q_y^2}=0$$

$$\frac{\partial q_y}{\partial t}+\frac{\partial}{\partial x}\left(\frac{q_x q_y}{d}\right)+\frac{\partial}{\partial y}\left(\frac{q_y^2}{d}\right)+gd\frac{\partial \eta}{\partial y}-\frac{h^2}{3}\left(\frac{\partial^3 q_x}{\partial x \partial y \partial t}+\frac{\partial^3 q_y}{\partial y^2 \partial t}\right)-M_{Dy}+\frac{f_w}{2d^2}q_y\sqrt{q_x^2+q_y^2}=0 \qquad (4.23)$$

$$\frac{\partial \eta}{\partial t}+\frac{\partial q_x}{\partial x}+\frac{\partial q_y}{\partial y}=0$$

where q_x and q_y are the flow rates per unit width defined by $q_x=\int_{-h}^{\eta}u\,dz$ and $q_y=\int_{-h}^{\eta}v\,dz$ using the velocities u in the x-direction and v in the y-direction, d is the total depth ($=h+\eta$), h is the water depth below the still water surface, η is the water surface elevation above the still water surface, g is the gravity acceleration, f_w is the sea bottom friction coefficient, t is the time, x is the coordinate in the off-shore direction and y in the longshore direction. M_{Dx} and M_{Dy} are the diffusion terms of the momentum due to breaking waves, and are expressed as follows:

$$M_{Dx}=\nu_t\left(\frac{\partial^2 q_x}{\partial x^2}+\frac{\partial^2 q_x}{\partial y^2}\right), \qquad M_{Dy}=\nu_t\left(\frac{\partial^2 q_y}{\partial x^2}+\frac{\partial^2 q_y}{\partial y^2}\right)$$

Inside the wave breaking zone:

$$\nu_t=\frac{a_D g \overline{d}\tan\beta}{\omega^2}\sqrt{\frac{g}{\overline{d}}\frac{\hat{q}-0.135\sqrt{g\overline{d}^3}}{0.4(0.57+5.3\tan\beta)\sqrt{g\overline{d}^3}-0.135\sqrt{g\overline{d}^3}}} \qquad (4.24)$$

Outside the wave breaking zone:

$$\nu_t=\frac{0.0008}{\tan\beta}\sqrt{g\overline{d}^3}$$

where ν_t is the eddy kinematic viscosity; a_D is the empirical coefficient that is 0 outside the wave breaking zone, increases linearly to 2.5 within a wavelength across the wave breaking position, and has a constant value of 2.5 inside the wave breaking zone; $\overline{d}$ is the mean water depth; $\tan\beta$ is the sea bottom slope; and ω is the angular frequency of the fundamental frequency component. $\hat{q}$ is the amplitude of the flow rate per unit width, which is obtained from the variation of the flow rate per unit width in the previous one period.

Vu et al. [13] improved the numerical model of Sato and Kabiling [12], and Yamamoto et al. [14], [15] validated the numerical model of Vu et al. [13] on several coasts. The detailed structure and usage of the numerical model of Vu et al. [13] are described in Section 2 of the Appendix.

4. CURRENTS IN THE SEA

4.2 Drift Currents

When a prevailing wind blows for a sufficiently long duration on a sea surface, the wind-induced drag force on the sea surface generates currents on and near the sea surface. These are called **drift currents** and are a type of ocean and coastal currents. Nansen, F. [16], during his exploration of the Arctic Ocean from 1893 to 1896, noticed that ice flows drifted at an angle of 30 to 40 degrees to the wind direction. As it was assumed that icebergs moved with the same velocity as the drift currents, this means that the drift currents have a direction which has an angle of 30 to 40 degrees to the wind direction. The phenomenon that the ocean drift currents had a direction that formed an angle with the prevailing wind direction was shown by Ekman [17] to be caused by the Coriolis force due to the Earth's rotation.

When the equation of motion with shear and Coriolis forces balanced in the steady state shown in Eq. (4.25) is solved under the boundary conditions shown in Eq. (4.26), the solution shown in Eq. (4.27) can be obtained.

$$v_t \frac{d^2 u}{dz^2} + f_{cl} v = 0, \quad v_t \frac{d^2 v}{dz^2} + f_{cl} u = 0 \right\} \tag{4.25}$$

$$\left. \begin{array}{l} at\ z = 0 : \quad \rho v_t \dfrac{du}{dz} = 0,\ \rho v_t \dfrac{dv}{dz} = \tau \\ at\ z = -\infty : \quad u = v = 0 \end{array} \right\} \tag{4.26}$$

$$u = V_o e^{\pi z / \delta} \sin\left(\frac{\pi}{4} - \frac{\pi z}{\delta}\right),\ v = V_o e^{\pi z / \delta} \cos\left(\frac{\pi}{4} - \frac{\pi z}{\delta}\right),\ V_o = \frac{\tau}{\sqrt{\rho \varepsilon f_{cl}}},\ \delta = \pi \sqrt{\frac{2\varepsilon}{\rho f_{cl}}} \right\} \tag{4.27}$$

where v_t is the constant eddy kinematic viscosity in the vertical direction [unit: m²/s]; u and v are velocities in the x and y directions; z is the vertical coordinate with the positive direction above the sea surface; f_{cl} is the Coriolis coefficient (= $2\omega\sin\varphi$, ω is the angular velocity due to the rotation of the earth, and φ is the latitude); ρ is the density of seawater; ε is the constant eddy viscosity in the vertical direction [unit: kg /m/s]; V_o is the velocity of the drift current on the sea surface; and δ is the depth to which the flow direction is opposite to the wind direction and is called the **friction depth** [unit: m].

The characteristics of the drift current in the Northern Hemisphere are shown in Eq. (4.27), where the velocity is V_o on the sea surface and the flow direction is shifted 45° to the right (45° to the left in the Southern Hemisphere) relative to the wind direction. Then, in the downward direction, the flow velocity decreases while the flow direction deviates more and more from the wind direction, as shown in **Fig. 4.7**.

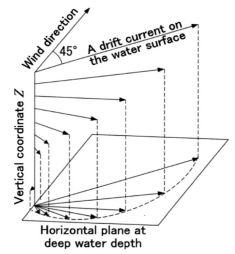

Fig. 4.7 Vertical distribution of flow velocity in a drift current.

4.2 Drift Currents

The friction depth ranges from 10m to 100m, and the velocity at the friction depth is about 4% of the surface flow. This characteristic of the drift current is called the **Ekman spiral**. As a result, seawater is transported to the right of the wind direction in the Northern Hemisphere (to the left of the wind direction in the Southern Hemisphere). This phenomenon is called **Ekman transport**.

What is the Coriolis force?

The **"Coriolis Force"** is a force exerted on an object moving in a rotational coordinate at right angles to the direction of motion with a magnitude proportional to the velocity of the object. In the Northern Hemisphere, an object moving from a low latitude to a high latitude is subjected to this force in eastern direction, and an object moving from a high latitude to a low latitude is subjected to this force in western direction.

Due to the Coriolis force, winds within a typhoon in the Northern Hemisphere do not move straight toward the center of the typhoon. The wind direction distribution viewed from directly above becomes a counterclockwise vortex. Also, if a missile is fired to the north in the Northern Hemisphere, it will always be slightly off to the east of the strait line drawing from its origin position to the North.

The magnitude F_{cl} of the Coriolis force acting on an object of mass m and moving velocity v at the latitude φ can be expressed as follows:

$$F_{cl} = m \times f_{cl} v, \qquad f_{cl} = 2\omega \sin \varphi \qquad \text{(g.1)}$$

where f_{cl} is the Coriolis coefficient, and ω is the angular velocity of the earth's rotation.

Let us now compare the effect of the Coriolis force on a ball thrown at the North Pole, where the Coriolis force is greatest, with a mean velocity of 10.0 m/s for a distance of 100 m and a mean velocity of 50.0 m/s for a distance of 50,000 m.

Since the angular velocity ω of the earth rotation is 0.0000729 rad/s, the acceleration due to the Coriolis force for each case is as follows:

For the mean velocity of 10.0 m/s, the acceleration $f_{cl} \times v = 2\omega \sin \varphi \times v \approx 0.00146$ m/s^2.

For the mean velocity of 50.0 m/s, the acceleration $f_{cl} \times v \approx 0.00729$ m/s^2.

Since each time t for the ball to pass through is 10.0s and 1,000s, each displacement distance x for what can be regarded as a motion of uniform acceleration is as follows:

For the distance of 100 m at $v = 10.0$ m/s: $x = 0.5 \times 0.00146$ m/s$^2 \times t^2 \approx 0.0729$ m $= 7.29$ cm.

For the distance of 50,000 m at $v = 50.0$ m/s: $x = 0.5 \times 0.00729$ m/s$^2 \times t^2 \approx 3,650$ m.

From both comparisons, the effect of the Coriolis force can be neglected for normal velocities and distances, but its effect cannot be neglected for high velocities and long distances.

163

4.CURRENTS IN THE SEA

When the water depth is shallower than the friction depth, solving for zero velocity at $z = -h$ as a boundary condition yields the answer shown in Eq. (4.28).

$$
\left.
\begin{aligned}
u &= V_A \sinh\frac{\pi}{\delta}(h+z)\cos\frac{\pi}{\delta}(h+z) - V_B \cosh\frac{\pi}{\delta}(h+z)\sin\frac{\pi}{\delta}(h+z) \\[2mm]
v &= V_A \cosh\frac{\pi}{\delta}(h+z)\sin\frac{\pi}{\delta}(h+z) + V_B \sinh\frac{\pi}{\delta}(h+z)\cos\frac{\pi}{\delta}(h+z) \\[2mm]
V_A &= \frac{\tau}{\varepsilon}\frac{\delta}{\pi}\frac{\cosh\left(\dfrac{\pi}{\delta}h\right)\cos\left(\dfrac{\pi}{\delta}h\right) + \sinh\left(\dfrac{\pi}{\delta}h\right)\sin\left(\dfrac{\pi}{\delta}h\right)}{\cosh\left(2\dfrac{\pi}{\delta}h\right) + \cos\left(2\dfrac{\pi}{\delta}h\right)} \\[2mm]
V_B &= \frac{\tau}{\varepsilon}\frac{\delta}{\pi}\frac{\cosh\left(\dfrac{\pi}{\delta}h\right)\cos\left(\dfrac{\pi}{\delta}h\right) - \sinh\left(\dfrac{\pi}{\delta}h\right)\sin\left(\dfrac{\pi}{\delta}h\right)}{\cosh\left(2\dfrac{\pi}{\delta}h\right) + \cos\left(2\dfrac{\pi}{\delta}h\right)}
\end{aligned}
\right\}
\qquad (4.28)
$$

4.3 Ocean Currents

Ocean currents (average velocities of about 2 m/s to 0.5 m/s ≈ 4 knots to 1 knot, average widths of 200 km to several 10 km, and average depths of 1000 m to several 100 m) are caused constantly by tangential forces applied on the sea surface due to continuous wind and the homogenizing effects of the differences in water temperature and salinity. They are strongly influenced by the Coriolis force and the flow regime is generally determined on a global scale. There are large-scale clockwise circulations [(1), (7) → (8) → (9) → (10)] between low and mid-latitudes in the northern hemisphere of the Atlantic and Pacific Oceans, and large-scale counterclockwise circulations [(2), (4), etc.] between low and mid-latitudes in the southern hemisphere of the Atlantic and Pacific Oceans, and in the Indian Ocean, as shown in **Fig. 4.8**. In addition, there is an eastward current (3) with a large flow rate around Antarctica.

These circulation currents on the ocean surfaces are mainly generated by tangential forces from large-scale ocean winds. The Coriolis force is weaker at low latitudes and stronger at mid-latitudes, resulting in a phenomenon called **westward intensification**, in which the western current is stronger than the eastern current in each ocean. For example, the Kuroshio Current of (9) is stronger than the California Current of (7). In addition, an eastward flowing current of (6) (flow velocities are about 1.5 m/s to 0.5 m/s, average depth is about 200 m to 100 m) is clearly observed in the equatorial windless zone (3 to 8 degrees north latitude) between the North and South Pacific Great Circulation Currents because the trade winds blowing from east to west raise the water level on the west coast.

(1) North Atlantic Current
(2) South Atlantic Current
(3) Antarctic Current
(4) South Indian Current
(5) South Equatorial Current
(6) Equatorial Countercurrent
(7) California Current
(8) North Equatorial Current
(9) Kuroshio Current
(10) North Pacific Current
(11) Oyashio Current

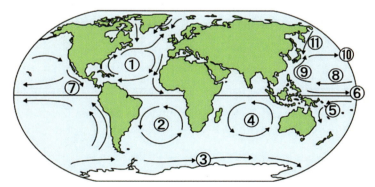

Fig. 4.8 Illustration of the world's major ocean currents. (corrected the ocean current map on the JMA website [18])

Ocean currents flowing from low latitudes with high water temperatures to the polar regions are called **warm currents** [e.g., (9)] because they cool themselves while emitting heat to the atmosphere above them. On the other hand, ocean currents flowing from high latitudes to low latitudes, where the water temperature is lower, are called **cold currents** [e.g., (7) and (11)] because they move while absorbing heat from the atmosphere. Cold currents are naturally rich in nutrients and plankton. The

4.CURRENTS IN THE SEA

areas where warm and cold currents meet, called **current-rips**, can easily attract fish from warm currents as well as from cold currents, making them good fishing grounds.

Furthermore, when a full-scale numerical simulation of ocean currents is required, the following numerical model consisting of the equations of motion with the Coriolis force and continuity, the equation of state for seawater density, and the transport equations for water temperature and salinity are used to obtain the distributions of water level, current velocity, seawater density, water temperature, and salinity within an area by considering tangential forces due to ocean winds, seawater density, water temperature, and salinity as boundary conditions,.

$$
\left.
\begin{aligned}
&\frac{\partial u}{\partial t} + u\frac{\partial u}{\partial x} + v\frac{\partial u}{\partial y} + w\frac{\partial u}{\partial z} - f_{cl}v = -\frac{1}{\rho_w}\frac{\partial p}{\partial x} + \frac{\partial}{\partial z}\left(K_m\frac{\partial u}{\partial z}\right) + F_u \\[6pt]
&\frac{\partial v}{\partial t} + u\frac{\partial v}{\partial x} + v\frac{\partial v}{\partial y} + w\frac{\partial v}{\partial z} + f_{cl}u = -\frac{1}{\rho_w}\frac{\partial p}{\partial y} + \frac{\partial}{\partial z}\left(K_m\frac{\partial v}{\partial z}\right) + F_v \\[6pt]
&\frac{\partial w}{\partial t} + u\frac{\partial w}{\partial x} + v\frac{\partial w}{\partial y} + w\frac{\partial w}{\partial z} = -\frac{1}{\rho_w}\frac{\partial q}{\partial z} + \frac{\partial}{\partial z}\left(K_m\frac{\partial w}{\partial z}\right) + F_w \\[6pt]
&\frac{\partial u}{\partial x} + \frac{\partial v}{\partial y} + \frac{\partial w}{\partial z} = 0, \qquad \rho = \rho(T,S,p) \\[6pt]
&\frac{\partial T}{\partial t} + u\frac{\partial T}{\partial x} + v\frac{\partial T}{\partial y} + w\frac{\partial T}{\partial z} = \frac{\partial}{\partial z}\left(K_h\frac{\partial T}{\partial z}\right) + F_T \\[6pt]
&\frac{\partial S}{\partial t} + u\frac{\partial S}{\partial x} + v\frac{\partial S}{\partial y} + w\frac{\partial S}{\partial z} = \frac{\partial}{\partial z}\left(K_h\frac{\partial S}{\partial z}\right) + F_S
\end{aligned}
\right\}
\tag{4.29}
$$

where t is the time; x and y are the horizontal Cartesian coordinates; z is the vertical coordinate; u, v, and w are the x, y, and z components of the flow velocity; f_{cl} is the Coriolis coefficient; ρ_w is the reference density of seawater; p is the total pressure ($=p_s+p_a+q$, p_s is the hydrostatic pressure, p_a is the atmospheric pressure, q is the nonhydrostatic pressure of water); K_m is the vertical eddy kinematic viscosity; F_u, F_v, and F_w are horizontal diffusion terms; ρ is the seawater density; T is the water temperature; S is the salinity; K_h is the vertical diffusion coefficient; F_T is the horizontal thermal diffusion term; and F_S is the horizontal salt diffusion term.

4.4 Tidal Currents

The change in the composite vector of the gravitational force of the moon or sun and the Earth's centrifugal force is the driving force that produces the tidal water level fluctuations in the ocean. This composite vector is called **tide generating force** or **tidal force**. The variation of tidal level due to the tidal force also changes with time due to the orbits of the Earth and the moon. The time variation of the tide level can be regarded as a long wave. The resonance corresponding to the coastal topography caused by this long wave also causes a change in the tide level. The changes in the tide level related to the tidal force are called **astronomical tides**. Sea water level can also change due to low-pressure systems and strong winds created by typhoons. They are called **meteorological tides**. These tidal level changes (long-period rise and fall of the seawater surface) are collectively called **tides**, and the phenomenon in which seawater moves (flows) from a position where the sea level falls to a position where the sea level rises due to tidal phenomena is called a **tidal current**.

1) Tides

The moon revolves around the Earth in a circular orbit. The Earth also revolves around the common centre of gravity of the Earth and the moon, as shown in **Fig. 4.9**. The centrifugal force due to the Earth's rotational motion has the same direction and magnitude at any point on the Earth, as shown by the blue vectors in **Fig. 4.10**.

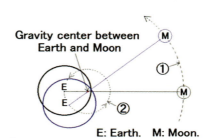

E: Earth. M: Moon.
①: Moon's orbit from gravity centre.
②: Earth's orbit from gravity centre.

Fig. 4.9 Explanation of the rotational motion of the earth and the moon.

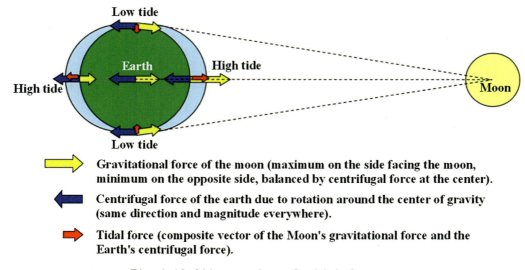

⇨ Gravitational force of the moon (maximum on the side facing the moon, minimum on the opposite side, balanced by centrifugal force at the center).

⇦ Centrifugal force of the earth due to rotation around the center of gravity (same direction and magnitude everywhere).

⇨ Tidal force (composite vector of the Moon's gravitational force and the Earth's centrifugal force).

Fig. 4.10 Illustration of tidal force.

4.CURRENTS IN THE SEA

On the other hand, the gravitational force of the moon is largest on the side of the Earth facing the moon and smallest on the opposite side, as shown by the yellow vectors in **Fig. 4.10**. The centrifugal force generated by the Earth's spinning is shown by the blue vectors. The resultant force of the centrifugal force of the Earth and the gravitational force of the moon is shown by the red vectors. The ocean surface (shown in light blue in this figure) at the side facing the moon and on the opposite side is raised by the outward resultant force vectors from the Earth's center, resulting in a **high tide**. On the other hand, at the face of the Earth in the direction perpendicular to the direction facing the moon, the ocean surface is lowered by the resultant force vectors pointing toward the Earth's center, resulting in a **low tide**. Therefore, this resultant force is called the **tidal force**.

Because the moon orbits the earth in about 29.5 days, the interval between two consecutive times at which the moon is visible to one point on the Earth is not exactly 24 hours, but 24 hours and 50 mins. Therefore, a tide, which is the change in tide level from one high tide to the next high tide or from one low tide to the next low tide, occurs twice in about 24 hours and 50 minutes. The phenomenon that there are two high tides and two low tides in a day is called a **semi-diurnal tide**. The phase of the tidal cycle from one low tide to the next high tide is called a **flood tide**, while that from one high tide to the next low tide is called an **ebb tide**.

In addition, the sun is the only entity other than the moon that exerts a non-negligible tidal force on the Earth (tidal forces by other stars are negligibly small). At the time of the new moon and full moon, when the tidal forces of the moon and the sun coincide, the tides are the largest and are called the **spring tide**. During the first quarter and last quarter moon, when both tidal forces are at their highest offsetting, the tides are the lowest and are called the **neap tide**.

The above is the concept of tidal phenomena explained based on the **equilibrium theory of tides** published by Newton [19] in 1687. However, this theory alone cannot accurately explain the actual tidal motion. For example, the difference between high and low tides according to the equilibrium theory of tides is a few tens of centimeters, but it is often as large as 15 meters in the Bay of Fundy in eastern Canada, 8 to 12 meters in the Bristol Channel in the UK, 7 to 10 meters on the west coast of the Korean Peninsula, and 5 meters in Ariake Bay in Japan. Therefore, Kelvin [20] considered that tidal phenomena are the one in which long waves generated by changes in tidal forces enter bays or inland seas and resonate, causing a large rise and fall of the sea level, and published the **dynamical theory of tides** in 1879, in which the Coriolis force is considered in the equation of motion of long waves. Thus, because tides can be considered as stationary waves caused by resonance phenomena, although the fundamental period of tidal fluctuations is locked in at about 12 hours and 50 minutes when the moon is overhead, it is not always a high tide, but can be a low tide or an intermediate tide, depending on the location. In addition, because tidal changes are delayed by the inertia of seawater and the effects of seafloor friction, high and low tides occur one to three days later than at the time of the new and full moon or of the first quarter and last quarter moons.

As a popular belief, it is said that the difference between high and low tides in spring and autumn equinoxes is larger because the sun and moon are on the equator, and the opposite is true in the summer and winter solstices. However, this does not hold in areas located at mid-latitudes, and the difference between high and low tides during the spring and autumn equinoxes is smaller than the annual average.

2) Harmonic Analysis of Tides

The decomposition of the tide level η (the height from the reference level to the changing sea surface) into N tidal components called **partial tides,** which has often been used in tidal analyses based on the assumption of the **equilibrium theory of tides**. This is called **harmonic analysis of tides**. If the coefficients of Eq. (4.30) are obtained from long-term tide observation data, it is easy to predict the change of the tide level in the observed sea area by Eq (4.30).

$$\eta = H_0 + \sum_{n=1}^{N} f_n H_n \cos\left(V_n + u_n + \sigma_n t - \varphi_n\right) \tag{4.30}$$

where H_0 is the height from the reference plane to the mean sea level during the observation period, H_n is the amplitude of each partial tide, f_n is the modification factor of each amplitude, V_n+u_n is a constant angle that varies with each partial tide, σ_n is the angular frequency of each partial tide, t is the elapsed time, and φ_n is the phase angle of each partial tide.

The amplitudes and phase angles of partial tidal waves are harmonic constants specific to the observation location and are determined by the least squares method or Fourier analysis. The other parameters are given by the observation location and the observation start time. For the actual calculation method, the reader is referred to literature such as Murakami [21].

Table 4.2 lists the names and symbols of major partial tides, and their angular velocities, periods, and relative maximum values of tidal forces.

The **cardinal datum level** (C.D.L.) used in charts and tide tables is the mean tide level minus the sum of the three tidal components M_2, S_2, K_1, and O_1 ($H_m+H_s+H_k+H_o$).

3) Prediction of Tidal Currents

If the prediction of tides and tidal currents in complex topographies is desired, a numerical model based on the equations of motion that take into account the Coriolis force as an external force is used to simulate the distribution of water levels and current velocities within an area by inputting changes in tidal levels over time from the boundaries of the sea area.

To reduce the computational costs, as shown in Eq. (4.31), this numerical model solves the nonlinear long wave equations obtained from the Navier Stokes equation of motion by using the unit-width flow rates q_x and q_y, which are obtained by integrating the offshore (x-direction) velocity u_x and the alongshore (y-direction) velocity u_y along the vertical coordinate z from the water depth h below the still water surface to the water surface elevation η above the still water surface.

4.CURRENTS IN THE SEA

Table 4.2 Example of major partial tides.

Names of partial tides	Symbols of partial tides	Angular velocities (degree/hr)	Periods (hr: min)	Relative maximum values of tidal forces
Main lunar semidiurnal tide	M_2	28.98	12: 25	0.4543
Main solar semidiurnal tide	S_2	30.00	12: 00	0.2113
Main lunar elliptical tide	N_2	28.44	12: 39	0.08796
Sun-Moon composite semidiurnal tide	K_2	30.08	11: 58	0.05752
Sun-Moon Composite Diurnal Tide	K_1	15.04	23: 56	0.2652
Main Lunar Diurnal Tide	O_1	13.94	25: 49	0.1886
Main Solar Diurnal Tide	P_1	14.96	24: 04	0.08775
Lunar Semimonthly Tide	M_f	1.098	327: 5	0.07827

$$\left.\begin{array}{l} \dfrac{\partial q_x}{\partial t}+\dfrac{\partial}{\partial x}\left(\dfrac{q_x^2}{d}\right)+\dfrac{\partial}{\partial y}\left(\dfrac{q_xq_y}{d}\right)+gd\dfrac{\partial\eta}{\partial x}-v_L\left(\dfrac{\partial^2 q_x}{\partial x^2}+\dfrac{\partial^2 q_x}{\partial y^2}\right)+\dfrac{f_b}{d^2}q_x\sqrt{q_x^2+q_y^2}-f_{cl}q_y=0 \\[3mm] \dfrac{\partial q_y}{\partial t}+\dfrac{\partial}{\partial x}\left(\dfrac{q_yq_x}{d}\right)+\dfrac{\partial}{\partial y}\left(\dfrac{q_y^2}{d}\right)+gd\dfrac{\partial\eta}{\partial y}-v_L\left(\dfrac{\partial^2 q_y}{\partial x^2}+\dfrac{\partial^2 q_y}{\partial y^2}\right)+\dfrac{f_b}{d^2}q_y\sqrt{q_x^2+q_y^2}+f_{cl}q_x=0 \\[3mm] \dfrac{\partial q_x}{\partial x}+\dfrac{\partial q_y}{\partial y}+\dfrac{\partial\eta}{\partial t}=0 \\[3mm] q_x=\int_{-h}^{\eta}u_x dz, \qquad q_y=\int_{-h}^{\eta}u_y dz \end{array}\right\} \tag{4.31}$$

where t is the time, d is the water depth (= $\eta+h$), g is the acceleration of gravity, v_L is the horizontal eddy kinematic viscosity, f_b is the friction coefficient at the sea bottom, and f_{cl} is the Coriolis coefficient.

The first and second equations in Eq. (4.31) are the nonlinear long wave equations in the cross-shore and longshore directions respectively, and the third equation is the continuity equation for seawater expressing the conservation law of mass. For the actual simulation method of tidal currents, please refer to literature such as Awaji et al [22].

Tidal currents can also be decomposed into partial currents by harmonic analysis. If the velocity of the tidal current is divided into two horizontal directions and the velocity in each direction is expressed by U, the decomposition is as follows:

$$U=U_0+\sum_{n=1}^{N}f_n U_n\cos\left(V_n+u_n+\sigma_n t-\varphi_n\right) \tag{4.32}$$

where U_0 is the velocity of the constant current component, U_n is the amplitude of each partial current, f_n is the modification factor of each amplitude, V_n+u_n is a constant angle that varies with each partial current, σ_n is the angular frequency of each partial current, t is the elapsed time, and φ_n is the phase angle of each partial current.

170

List of References in Chapter 4

1) Longuet-Higgins, M.S. and Stewart, R.W.: Radiation Stresses in Water Waves; a Physical Discussion, with Applications, *Deep-Sea Res.*, Vol. II, No. 4, 1964, pp.529-562.

2) Harris, T.F.W.: Nearshore Circulations -Field Observation and Experimental investigations of an underlying cause in wave tanks-, *Symposium on Coastal Eng.*, South Africa, 1969.

3) Longuet-Higgins, M.S.: Longshore Current Generated by Obliquely Incident Sea Waves, *Jour. Geophys. Res.*, Vol. 75, No. 33, 1970, pp.6778-6801.

4) Goda, Y. and Watanabe, N.: On the Introduction of an Irregular Wave Model to a Longshore Current Velocity Formula, *Journal of Coastal Engineering*, Vol. 37, JSCE, 1990, pp.210-214. (in Japanese)

5) Komar, D.P.: Beach Processes and Sedimentation, Chapt. 7, *Nearshore Currents*, Prentice-Hall Inc., New Jerssey, 1976, pp.168-202.

6) Horikawa, K.: *Coastal Engineering*, University of Tokyo Press, 1978, 402p.

7) Bowen, A.J.: Rip Currents, Part I Theoretical Investigations, *Jour. Geophys. Res.*, Vol. 74, No. 23, 1969, pp.5467-5478.

8) Bowen, A.J. and Inman, D.L.: Rip Currents, Part II Laboratory and Field Observations, *Jour. Geophys. Res.*, Vol. 74, No. 23, 1969, pp.5479-5490.

9) Hino, M.: Generation Theory of Nearshore Current System (3) -Simplified Theory-, *Proc. of the 20th Coastal Engineering Conference*, JSCE, 1973, pp.339-343. (in Japanese)

10) Hino, M.: Generation Theory of Nearshore Current System, *Proceedings of JSCE*, No.225, 1974, pp.17-29. (in Japanese)

11) Horikawa, K., Sasaki, T. Hotta, S. and Sakuramoto, H.: Study on Nearshore Currents (3rd Report) -Scale of Nearshore Current System-, *Proc. of the 22nd Coastal Engineering Conference*, JSCE, 1975, pp.127-134. (in Japanese)

12) Sato, S. and Kabiling, M.: Numerical Computation of Waves, Nearshore Currents and Beach Deformation Using Boussinesq Equation, *Journal of Coastal Engineering*, Vol. 40, JSCE, 1993, pp.386-390. (in Japanese)

13) Vu, T.C., Yamamoto, Y., Tanimoto, K., and Arimura, J.: Simulation on wave dynamics and scouring near coastal structures by a numerical model, *Proc. 28th Inter. Confer. on Coastal Engineering*, ASCE, 2002, pp.1817-1829.

14) Yamamoto, Y., Charusrojtanadech, N. and Sirikaew, U.: Topographical Change Prediction of the Beach or the Seabed in the Front of a Coastal Structure, *Proc. 22nd Inter. Offshore and Polar Engineering Conference*, 2012, pp.1488-1495.

4.CURRENTS IN THE SEA

15) Charusrojthanadech, N., Rattanarama, P. and Yamamoto, Y.: Examination of Coastal Erosion Prevention in the Back of the Gulf of Thailand, *Proc. 23rd Inter. Offshore and Polar Engineering Conference*, 2013, pp.1355-1362.

16) Beaufort Gyre Exploration Project: *History -Nansen and Drift of the Fram (1893-1896)-*, Woods Hole Oceanographic Institution. https://www.whoi.edu/beaufortgyre/history/history_fram.html

17) Ekman, V.W.: On the Influence of the Earth's Rotation on Ocean-Currents, *Arkiv för Matematik, Astronomi och Fysik*, 2, 1905 (Published in his native Norwegian language in 1902), pp,1-52. https://empslocal.ex.ac.uk/people/staff/gv219/classics.d/Ekman05.pdf

18) Japan Meteorological Agency: Global Distribution Map of Ocean Currents, https://www.data.jma.go.jp/gmd/kaiyou/data/db/kaikyo/knowledge/kairyu.html.

19) Newton, Isaac：*Philosophiæ Naturalis Principia Mathematica*, Jussu Societatis Regiæ ac typis Josephi Streater, prostant venales apud Sam. Smith, 1687, 510p.

20) Thomson, W. (Lord Kelvin): On gravitational oscillations of rotating water, *Proc. Roy. Soc. Edinburgh*, Vol.10, 1879, pp.92-101.

21) Murakami, K.: *The Harmonic Analysis of Tides and Tidal Currents by Least Square Method and Its Accuracy*, Technical Note of the Port and Harbour Research Institute, No. 369, 1981, 38p. (in Japanese)

22) Awaji, T., Imawaki, N. and Sato, S.: On Numerical Experiments of Coastal Tides, *Bulletin on Coastal Oceanography*, Vol. 23, No.1, the Oceanographic Society of Japan, 1985, pp.35-48. (in Japanese)

Chapter:5

STORM SURGES AND TSUNAMIS

5. STORM SURGES AND TSUNAMIS

When a strong low-pressure system approaches the coast, the water surface will be elevated due to the low-pressure system. This marine phenomenon is called a **storm surge**. Where populations and assets are concentrated in low-lying coastal areas, there is a high probability that a major disaster will occur due to flood inundation, making storm surge preparedness important.

If there is an oceanic trench or trough off the coast, it is possible for a **tsunami** to be generated by a submarine earthquake caused by crustal movement. If a tsunami surges over coastal dikes and seawalls, it is likely to cause extensive damage. Even if there is no mass overtopping, there is a high potential of damage to beach users and vessels in ports and harbors. Therefore, tsunami preparedness will also be important.

This chapter describes these storm surges and tsunamis.

5.1 Storm Surges

1) Generation of Storm Surges

Air, heated by the sun in the low-latitude tropics, rises to a certain altitude where its buoyancy equals that of the surrounding air. This air then moves to the mid-latitude zone and falls to the surface. Rising air creates a low-pressure area near the ground or ocean surface, causing air from surrounding higher-pressure areas to converge. The low-pressure zone in the low latitudes is called the **Intertropical Convergence Zone**, and the area of falling air in the mid-latitudes becomes the high-pressure zone. Since the near surface air in the tropics usually has high humidity, the cooling process accompanying warm air ascending facilitates the water vapor in the air to condense and release the latent heat. The released latent heat fuels ascending air for further ascending. If moisture is abundant, this process may create a cumulus cloud, which later can become a cumulonimbus cloud accompanying a thunderstorm and rainfall.

On a warm ocean surface, a continuous supply of moisture from evaporated surface water and heat may make the near surface air buoyant and ascend. Similar to the previous case, the condensation of water vapor in the ascending air releases the latent heat, which in turn fuels further ascending of the air. Ascending air creates a low air pressure area near the water surface. Under favourable conditions, i.e. the sea surface temperatures exceeding 26.5°C, instability of the atmosphere in the troposphere, and weak vertical wind shear, the low-pressure area near the water surface can become an air converging centre. Air from surrounding high air pressure flows to the centre of low air pressure and supplies it with more moisture. If the area is in the 5° to 30° latitude zone, the Coriolis effect is sufficiently strong. In this case, under the action of the Coriolis force, the convergence air does not directly flow to the low air pressure centre but deflects to the right in the Northern Hemisphere in an anticlockwise spiral, as shown in **Fig. 5.1(a)**.

174

In the Southern Hemisphere, the airflow deflects to the left in a clockwise spiral. This spiral circulation continues to transport warm and humid air to the low air pressure centre and further fuels the ascending air to form a thick and rotating cylinder of cumulonimbus, called a tropical depression.

As shown in **Fig. 5.1(b)**, strong winds are generated when the surrounding hot and humid air moves into the updraft (=low-pressure) area. A well-developed tropical depression is called a **tropical cyclone**. In the case that its maximum wind speed exceeds certain thresholds, it is called by different names in different regions, as follows:

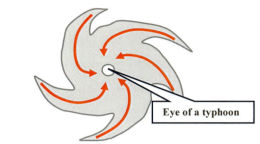

(a) Cross-sectional view at the bottom.

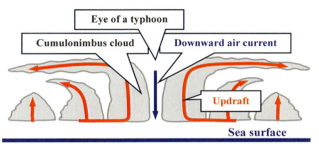

(b) Side view (The ratio of vertical to horizontal of this illustration is greater than the actual ratio. Typhoon heights are around 10 km - 15 km and the diameters of wind fields with speeds above 15 m/s are around 200 km - 2000 km).

Fig. 5.1 Structure of a tropical cyclone.

(a) Hurricane

In the North Atlantic Ocean and the Northeast Pacific Ocean, when the maximum average wind speed over one minute in a tropical cyclone reaches or exceeds 17.2 m/s, the term **tropical storm** is used. When the maximum average speed over one minute reaches or exceeds 32.7 m/s, the term **hurricane** is used. This term was derived from the name of the Mayan mythological god "Huracan," who was believed to cause floods, storms, and rain.

(b) Cyclonic Storm or Tropical Cyclone

In the North Indian Ocean, when the maximum average wind speed over ten minutes in a tropical cyclone reaches or exceeds 17.2 m/s, the term **cyclonic storm** is used.

In the South Indian Ocean, when the maximum average wind speed over ten minutes in a tropical cyclone reaches or exceeds 17.2 m/s, the term **tropical storm** is used, when the same maximum speed reaches or exceeds 32.7 m/s, the term **tropical cyclone** is used.

In the South Pacific Ocean, when the maximum average wind speed over ten minutes in a tropical cyclone reaches or exceeds 17.2 m/s, the term **tropical cyclone** is used.

In addition, the term **cyclone** was suggested by meteorologist Henry Piddington based on the Greek word "Kyklos," meaning a snake coiled on its tail.

5. STORM SURGES AND TSUNAMIS

(c) Typhoon

In the Northwest Pacific Ocean, when the maximum average wind speed over ten minutes in a tropical cyclone reaches or exceeds 17.2 m/s, the term **typhoon** is used. On the other hand, according to the World Meteorological Organization's international classification, a tropical cyclone with a maximum 10-minute average speed of 17.2 m/s or more is called a **tropical storm**. One with the same maximum speed of 32.7 m/s or more is called a **typhoon**. As for the origin of this term, there are several theories including the Arabic word "tufan" meaning storm, the Greek god of wind "typhon", or the Chinese word for strong wind.

A well-developed tropical cyclone generates large waves due to strong winds and forces water surface to rise several tens of centimeters to one meter near its centre due to low pressure. When a tropical cyclone approaches a coast, large waves, and high tidal levels compounded by strong wind cause a typical storm surge disaster.

2) Reality of Storm Surge Disasters

The strongest hurricane in recorded history was "Patricia" in the US and Mexico in October 2015 (a minimum central pressure of 872 hPa, where a maximum average wind speed of over one minute of 95.8m/s). The worst recorded hurricane disaster was caused by "Katrina" in the US in August 2005 (the minimum central pressure of 902 hPa, the maximum average wind speed over one minute of 77.8m/s). According to the observed data, the maximum tide level was MLLW+4.04 m, but was absent before the peak, and was estimated to be around MLLW+8.5 m based on trace data (see **photos 5.1**). The number of dead or missing persons because of flooding and strong winds was 2,541, and the total damages were 108 billion US dollars.

Photo 5.1 Hurricane Katrina under landing to Louisiana in the United States from the Caribbean Sea in August 2005.
(from NASA/Goddard Space Flight Center Scientific Visualization Studio)

5.1 Storm Surges

As for cyclonic storms or tropical cyclones, the strongest in recorded history was the severe tropical cyclone "Mahina" which occurred in Australia in March 1899 (the central pressure 880 hPa). The worst human loss was caused by the extremely severe cyclonic storm "Bhola" in Pakistan (the minimum central pressure 966 hPa) in November 1970 with 300,000 to 500,000 persons were dead or missing and total damages 570 million US dollars in 2019 equivalent. The worst damage was caused by the extremely severe cyclonic storm "Nargis" in Myanmar (the minimum central pressure 962 hPa) in April 2008 with total damages of 11 billion US dollars and 138,366 people were dead or missing.

In the case of typhoons, the strongest was "Tip" occurred in the Philippines (Japanese name is "Showa 54 Nen Typhoon No.20") in October 1979, with a minimum central pressure of 870 hPa and a maximum wind speed of 72 m/s. The worst typhoon disaster was caused by "Vera" in Japan (Japanese name is "Typhoon Ise-wan") in September 1959, with a central pressure of 895 hPa. 5098 people were dead or missing, 40838 houses were destroyed or washed away, and 7576 ships were damaged.

In the Northwest Pacific Ocean, since many typhoons traverse across the Japanese islands in summer and autumn, they cause storm surge disasters in Japan .

Super typhoons such as "Eiso no Kaze" in September 989, "Kouan no Yeki" in August 1281, "Typhoon Siebold" in September 1828, and "Great Wind of Ansei 3 Nen" in September 1856 caused tremendous storm surge damage in Japan. However, Since 1956 when the Coastal Act was enacted, the number of people killed or missing and the amount of damage have both decreased significantly due to the active implementation of coastal protection projects by the Japanese Government.

3) Methods for Calculating Tidal Levels during Storm Surges

The total water level at high tide is a composite of the astronomical tide level due to the tidal force of the moon and sun, and the seawater level departure from normal (= storm surge) caused by tropical cyclones. Therefore, the total water level can be calculated by adding the storm surge calculated by the following methods to the astronomical tide level, obtained from the harmonic analysis of tide level data excluding obvious storm surge.

(1) Empirical Formula for Calculating the Maximum Storm Surge

The Maximum tidal height η_{max} during a storm surge is obtained from Eq. (5.1), which adds the surge of seawater level due to the lowest air pressure, water level rise due to strong wind, and the water level rise due to wave setup or surf beat.

$$\eta_{\max} = a\Delta p + bU_{\max}^2 \cos\theta_w + \eta_s \tag{5.1}$$

where a is the coefficient of 1 cm rise in tide level for a pressure drop of 1 hPa, Δp is the difference between one atmospheric pressure (1013 hPa) and the lowest pressure at the tropical cyclone centre, b is a topography-dependent coefficient (0.002 to 0.2) that is larger for shallow coasts and bays where strong winds blow-in easily, U_{max} is the maximum wind speed, θ_w is the angle of inclination of the

177

5. STORM SURGES AND TSUNAMIS

wind from the normal to the shoreline (0 degrees if the wind direction is perpendicular to the shoreline), and η_s is the surge of seawater level due to wave setup or surf beat, which is given by Eq. (5.2) based on Yamamoto and Tanimoto's Eq. (3.10).

$$\eta_s = \frac{0.033\left(\tan\beta\right)^{1/6}}{\sqrt{\dfrac{H_{o1/3}}{L_{o1/3}}\left(1+\dfrac{h}{H_{o1/3}}\right)}} \times H_{o1/3} \quad \left[\text{range of application: } 1/10 \geq \tan\alpha \geq 1/70\right] \tag{5.2}$$

where $\tan\beta$ is the average seabed slope within the wave breaking zone, $H_{o1/3}$ and $L_{o1/3}$ are the significant wave height and wavelength of the offshore waves, and h is the water depth at the location of interest.

(2) Numerical Model for the Prediction of Storm Surges

In the case of time-varying predictions of the horizontal distribution of seawater motion during a tropical cyclone, a horizontal 2D numerical model consisting of Eqs. (5.3) and (5.4) can be used to calculate the horizontal distributions of water surface elevation and current velocity towards land by inputting the time-varying tide level from the offshore boundary and the time-varying pressure and wind velocity distributions from the sea surface.

The q_x and q_y in Eq. (5.3) are the flow rates per unit width in the offshore (x-direction) and longshore (y-direction) directions. They are determined by integrating the velocity u_x in the offshore direction and u_y in the longshore direction along the vertical coordinate z from the water depth h below the still water surface to the water surface elevation η above the still water surface.

The first and second in Eq. (5.4) are the equations of motion in the offshore and longshore directions, and the third is the continuity equation for seawater implying a conservation law of mass. The first term on the left-hand side of the two equations of motion is the acceleration term, the second and third are advection terms representing the spatial variation of the unit-width flow rates, the fourth is the pressure term representing the movement of seawater from a higher pressure position to a lower pressure position, the fifth is the sea surface friction force due to wind, the sixth is the sea bottom friction force due to currents, and the seventh is the Coriolis force due to the rotation of the earth. In these equations of motion, the viscous terms between water particles and due to eddies (turbulence) are considered relatively small and ignored.

$$q_x = \int_{-h}^{\eta} u_x\, dz, \qquad q_y = \int_{-h}^{\eta} u_y\, dz \tag{5.3}$$

$$\left.
\begin{aligned}
&\frac{\partial q_x}{\partial t} + \frac{\partial}{\partial x}\left(\frac{q_x^2}{d}\right) + \frac{\partial}{\partial y}\left(\frac{q_x q_y}{d}\right) + gd\frac{\partial(\eta-\eta_o)}{\partial x} - \frac{\rho_a}{\rho_w} f_s w_x \sqrt{w_x^2+w_y^2} + \frac{f_b}{d^2} q_x \sqrt{q_x^2+q_y^2} - f_{cl} q_y = 0 \\
&\frac{\partial q_y}{\partial t} + \frac{\partial}{\partial x}\left(\frac{q_y q_x}{d}\right) + \frac{\partial}{\partial y}\left(\frac{q_y^2}{d}\right) + gd\frac{\partial(\eta-\eta_o)}{\partial y} - \frac{\rho_a}{\rho_w} f_s w_y \sqrt{w_x^2+w_y^2} + \frac{f_b}{d^2} q_y \sqrt{q_x^2+q_y^2} + f_{cl} q_x = 0 \\
&\qquad\qquad\qquad\qquad \frac{\partial q_x}{\partial x} + \frac{\partial q_y}{\partial y} + \frac{\partial \eta}{\partial t} = 0
\end{aligned}
\right\} \tag{5.4}$$

5.1 Storm Surges

where u_x and u_y are respective flow velocities in x and y directions, t is the time, d is the total depth (= $\eta + h$), g is the acceleration of gravity, η_o is the elevation in water level due to low pressure, ρ_a is the density of air, ρ_w is the density of seawater, w_x and w_y are the wind speeds offshore and in the coastal direction (values at 10 m above sea level), f_s and f_b are the friction coefficients at sea level and the sea bed, f_{cl} is the Coriolis coefficient.

Determining wind velocity and atmospheric pressure distribution in a typhoon is one of the most important tasks for storm surge modeling. There are several analytical storm models that are widely used in the world, such as the Fujita model [1], and the Holland model [2]. Because these two models are quite simple and easy to use, a summary of the distribution of atmospheric pressure and wind speed according to these two models is provided below.

Fujita [1] suggested that the distribution of atmospheric pressure and wind velocity at an elevation 10 m above sea surface as functions of a distance from the typhoon centre and other typhoon parameters as in Eqs. (5.5) and (5.6).

$$P(r) = \begin{cases} P_\infty - \dfrac{P_\infty - P_c}{\sqrt{1 + 2(r/R)^2}} & (0 \leq r \leq 2R) \\[4mm] P_\infty - \dfrac{P_\infty - P_c}{1 + r/R} & (r > 2R) \end{cases} \tag{5.5}$$

$$V(r) = \begin{cases} \left[\dfrac{f_{cl}^2 r^2}{4} + \dfrac{(P_\infty - P_c)}{\rho_a} \dfrac{2r^2}{R^2} \left(1 + 2\dfrac{r^2}{R^2}\right)^{-3/2} \right]^{1/2} - \dfrac{f_{cl} r}{2} & (0 < r \leq 2R) \\[4mm] \left[\dfrac{f_{cl}^2 r^2}{4} + \dfrac{(P_\infty - P_c)}{\rho_a} \dfrac{r}{R} \left(1 + \dfrac{r}{R}\right)^{-2} \right]^{1/2} - \dfrac{f_{cl} r}{2} & (2R < r < \infty) \end{cases} \tag{5.6}$$

Holland [2] proposed a typhoon model with atmospheric pressure and wind velocity near the sea surface during the typhoon respectively calculated by using Eqs. (5.7) and (5.8).

$$P(r) = P_c + (P_\infty - P_c) \times \exp\left[-\left(\dfrac{R}{r}\right)^B \right] \tag{5.7}$$

$$V(r) = \left\{ \dfrac{B}{\rho_a} \left(\dfrac{R}{r}\right)^B (P_\infty - P_c) \times \exp\left[-\left(\dfrac{R}{r}\right)^B \right] + \left(\dfrac{f_{cl} r}{2}\right)^2 \right\}^{1/2} - \dfrac{f_{cl} r}{2} \tag{5.8}$$

Where $P(r)$ is the atmospheric pressure near the sea surface at a distance r from the typhoon centre, P_∞ and P_c are respectively the atmospheric pressures of the ambient air, and at the centre of the typhoon, R is the radius of maximum wind speed, $V(r)$ is the typhoon wind velocity at r, ρ_a is the air density, f_{cl} is the Coriolis parameter, and B is the typhoon shape parameter which has a value ranging from 1 to 2.5 (refer to Holland [2]).

It should be noted that the water level due to storm surge calculated using Eq. (5.4) does not account for wave setup, even though the value of wave setup during a typhoon can be very significant. The reasons for this are as follows. Simulating nearshore currents and wave setup needs a very fine

179

5. STORM SURGES AND TSUNAMIS

horizontal grid to accurately calculate wave breaking process. However, as the spatial scale of a typhoon is several hundred kilometers, computation with a very fine grid to accurately detect the wave breaking is not possible. Therefore, when simulations of storm surge are executed by using a numerical model based on Eq. (5.4), it is necessary to add the values of typhoon wave setup.

4) Secondary Undulation and Continental Shelf Waves

When high waves with wave grouping characteristics caused by typhoons continue to enter harbors or bays, if the period of the long-period wave components of the wave grouping coincide with or equal an integer number of the natural periods of harbors or bays, resonance can occur and long-period waves are amplified within the harbors and the bays. This phenomenon is called **secondary undulation** in contrast to the main undulation of tides and storm surges, and can cause damage to ships and port facilities.

When typhoon-force winds continue to blow over bays and lakes, the water masses on the windward side move to the leeward side. When the typhoons move directly above and the strong winds suddenly stop, free oscillations occur with the natural periods of the bays and the lakes. When they occur in the bays, they are called **secondary undulation**, and when they occur in the lakes, they are called **seiche**.

These oscillatory phenomena can be reproduced by inputting the water level fluctuations of the wave group from the offshore side or the wind speed changes of strong winds for the entire water area to the numerical model based on Eq. (5.4).

Furthermore, the approximate period (T_1) of the fundamental mode of the secondary undulation shown in **Fig. 5.2** can be calculated using Eq. (5.9), which is obtained from "wave velocity of long waves = wavelength/period", and the approximate period (T_2) of the fundamental mode of the seiche shown in **Fig. 5.3** can be calculated using Eq. (5.10), which is obtained from the same relationship.

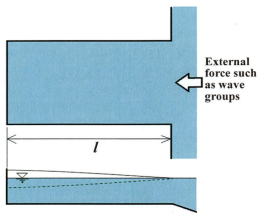

$$T_1 = \frac{4l}{\sqrt{gh}} \qquad (5.9)$$

$$T_2 = \frac{2l}{\sqrt{gh}} \qquad (5.10)$$

where l is the mean length of the harbor or the lake in the direction of interest, g is the acceleration of gravity, and h is the mean water depth.

when the right end is the opening of the harbor and the left end is the back of the harbor, the opening is the node of the water level fluctuation and the back is the loop, as in the two different curves.

Fig. 5.2 Illustration of the basic mode of the secondary undulation.

The secondary undulation and the seiche can also occur for integer fractions of the natural period of the port or the lake, the general equations for the approximate periods are as follows.

$$T_1 = \frac{4l_x}{(2m_x - 1)\sqrt{gh}} \quad (5.11)$$

$$T_2 = \frac{2}{\sqrt{\frac{m_x^2}{l_x^2} + \frac{m_y^2}{l_y^2}}\sqrt{gh}} \quad (5.12)$$

Where m_x and l_x are the number of vibration modes and the mean length in the principal axis of the harbor or the lake, m_y and l_y are the number of vibration modes and the mean length in the direction orthogonal to the principal axis of the lake, $m_x = m_y = 1$ being the fundamental mode and $m_x = m$ or $m_y = m$ being the $1/m$ mode in each direction.

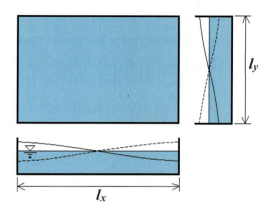

when both ends are lake banks, as in the two types of curves, the two banks are the loop of the water level fluctuation and the centre is the node.

Fig. 5.3 Illustration of the basic mode of the seiche.

Further, long-period waves excited near the coast by typhoons may propagate slowly along the continental shelf to the right in the Northern Hemisphere and to the left in the Southern Hemisphere. These waves are called **continental shelf waves** and require attention because they can cause flood damage due to rising sea water levels a few days after the passage of a typhoon.

5. STORM SURGES AND TSUNAMIS

5.2 Tsunamis
1) Generation of Tsunamis

The original meaning of tsunami was a wave that comes to a harbor or inlet in Japan. Waves caused by earthquakes were called seismic tsunamis and those caused by typhoons were called wind tsunamis. Currently, in the world, long-period seawater waves caused by undersea earthquakes, volcanic activity, and landslides are collectively called **tsunamis**. The English literature scholar Patrick Lafcadio Hearn (known in Japan as Koizumi Yagumo) introduced the Japanese word **tsunami** to the world.

The main types of earthquakes that cause tsunamis are **plate-boundary earthquakes** (also called **subduction zone earthquakes**, or **ocean-trench earthquakes**) and **outer-rise earthquakes**.

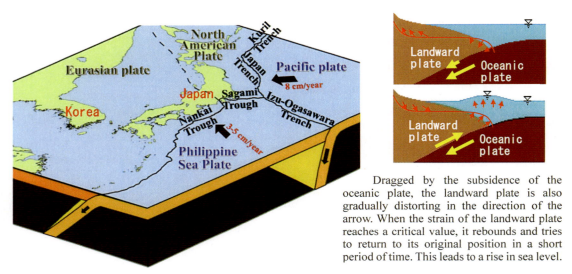

Dragged by the subsidence of the oceanic plate, the landward plate is also gradually distorting in the direction of the arrow. When the strain of the landward plate reaches a critical value, it rebounds and tries to return to its original position in a short period of time. This leads to a rise in sea level.

Fig. 5.4 Explanation of plate boundary earthquakes and tsunamis caused by landward plate rebounds. (modified the illustration of the plates on the JMA website [3])

The crust of the Earth's surface, dragged by the movement of the mantle, is slowly moving in several plates. As shown on the left-hand side of **Fig. 5.4**, in the area around Japan, landward plates are the North American plate including the northern part of the Japanese islands, and the Eurasian plate including the southern part of the Japanese islands. The oceanic plates are the Pacific plate and the Philippine Sea plate. As the mantle beneath the oceanic plates cools during movement and subducts beneath the landward plates, it attempts to drag the oceanic plates with it. Therefore, the Pacific plate is subducting under the North American plate and the Philippine Sea plate is subducting under the Eurasian plate. Moreover, as shown on the right side of **Fig. 5.4**, dragged by the subsidence of the oceanic plates, the landward plates are also gradually distorting. When the strains of the landward plates exceed a certain limit, the edges of the landward plates rebound upwards, and the seawater masses on the edges move upwards. This is the mechanism of plate-boundary earthquakes and the resulting tsunamis.

It should be noted that when a large plate-boundary earthquake occurs, the strain at the centre of the fault is released, but a new strain tends to accumulate at the edges. A new large earthquake can occur at the plate boundary next to this fault in a few years or decades. For example, in July 869, the Jogan Earthquake (Magnitude 8.3 or higher) occurred near Japan Trench, and 18 years later, in August 887, the Ninna Earthquake (Magnitude 8-8.5) occurred near Nankai Trough.

In March 2011, **2011 off the Pacific coast of Tohoku Earthquake** (Magnitude 9.0-9.1, see Photo 5.2) occurred near the Japan Trench, and the Japanese Government announced that there was a 70%-80% probability of a super earthquake occurring in the vicinity of the Nankai Trough within the next 30 years. Shishikura et al [4] also pointed out the existence of consolidated mega earthquakes with a 500 – 1000 year cycle in the Japan Trench and a 400 – 600 year cycle in the Nankai Trough from geological surveys. Azuma et al [5] showed the possibility of a megathrust earthquake with an interval of more than 1000 years in Shizuoka Prefecture (northern part of the Nankai Trough) from geological surveys, and Matsuoka and Okamura [6] also showed the possibility that a megatsunami greater than recorded in Kochi Prefecture (southern

Photo 5.2 Flooding on the Iwanuma coast caused by the megatsunami resulting from the 2011 off the Pacific coast of Tohoku Earthquake. (from Mainichi Shimbun Photo Archive)

Magnitude (M) is an index value expressing the energy scale of an earthquake in logarithmic scale. The magnitude calculated from the maximum amplitude of the seismic waveform can be estimated immediately after the event, but this method underestimates magnitudes above 8. Therefore, **Moment Magnitude (M_w)** calculated from the fault moment by multiplying the fault area by the displacement is now used.

Hence, in the case of megathrust earthquakes, the magnitude value based on the maximum amplitude is announced immediately after the event but is often subsequently revised upwards using Moment Magnitude.

5. STORM SURGES AND TSUNAMIS

part of the Nankai Trough) occurred approximately 2000 years ago from geological surveys.

If a Nankai Trough Boundary Type Megathrust Earthquake occurs then, there is a possibility that the damage would be greater than that caused by 2011 Tohoku Earthquake because this trough is closer to the coastline than the Japan Trench and the land areas near this trough have more population and assets than the land areas near the Japan Trench. Therefore, proactive disaster mitigation efforts are required.

Furthermore, in the case that a reverse fault-type earthquake occurs at a plate boundary, the resistance to subduction of the oceanic plate is lost, and a normal fault-type earthquake, presumably due to the tensile force acting in the shallow section of the oceanic plate bend called the outer rise, can occur later in the shallow section of the outer rise. This is called an **outer-rise earthquake**. In this type of earthquake, the shaking on land is not severe because the earthquake occurs far away from land, but the tsunami tends to be large because of the large vertical displacement of the fault. The Showa Sanriku earthquake occurred in March 1933 (Magnitude 8.4) after the Meiji Sanriku earthquake of June 1896 (Magnitude 8.2-8.5, the maximum tsunami run-up height 38.2m) is said to be a typical example. Several outer-rise earthquakes have been observed in the aftershocks of the 2011 Tohoku earthquake.

However, an **intraplate earthquake** (also called an **epicentral earthquake**, or an **inland earthquake**), which is well known together with the plate-boundary earthquake, is caused by the accumulation of tectonic strain inside the landward plate, which results in faulting. This type of earthquake tends to occur in a land part on the landward plate, so the damage caused by this earthquake itself is likely to be significant. But a huge tsunami is unlikely to occur.

In addition, long-period oscillations of 5 to 200 s are likely to occur in the case of megathrust earthquakes (M > 9) and earthquakes passing through sedimentary basins. For example, Suzuki [7] reported the occurrence of a resonance phenomenon with periods of 1 to 2 min (also called a seiche) at Lake Nishiko, Yamanashi Prefecture, during the 2011 off the Pacific coast of the Tohoku earthquake.

Volcanic activities (eruptions and mountain collapses) and landslides can also generate tsunamis. Major examples are as follows:

(a) The eruption of Santorini in Greece around 1628 BC caused an unusual tsunami with an estimated maximum height of about 90 m.

(b) The May 1792 eruption of Shimabara Unzen-dake and the collapse of Mayuyama caused a huge tsunami (estimated tsunami height at the coastline is 4 - 9 m and the maximum run-up height is 57 m) known as the "Disaster in Shimabara, Nuisance in Higo".

(c) The April 1815 eruption of Mount Tambora and the August 1883 eruption of Krakatau volcano in Indonesia also caused huge tsunamis.

(d) The collapse of a mountain body deep in Lituya Bay by an earthquake in Alaska in July 1958 caused an unusual tsunami with a maximum tsunami height of 524 m.

(e) The earthquake in April 1771 (Magnitude 7.4-8.7) around the Yaeyama islands in Japan caused an unusual tsunami with a maximum tsunami height of 85.4 m on Ishigaki Island, which has been assumed to have been caused by an undersea landslide.

Even though the maximum height of this type of tsunami may be abnormally high compared to the area of influence, the current tsunami forecasting system for earthquakes cannot correctly predict this type of tsunami.

2) Reality of Tsunami Disasters

Subduction zones at the plate boundaries shown in **Fig. 5.5** are called **trenches** when the depth is more than 6000 m, and **troughs** when the depth is less than 6000 m. The trenches and troughs at the Pacific plate boundary are summarised below:

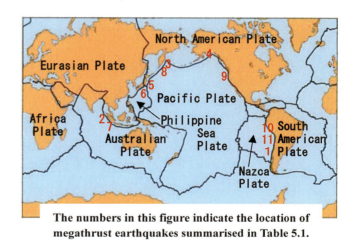

The numbers in this figure indicate the location of megathrust earthquakes summarised in Table 5.1.

Fig. 5.5 Illustration of the location of Tectonic plates and megathrust earthquakes.

(a) The boundary between the Pacific plate and the North American plate: Japan Trench (off eastern Japan), Kuril-Kamchatka Trench (off Russian Far East), Aleutian Trench (off the coast of Alaska, USA), Cascade Subduction Zone (off the west coast of the USA).

(b) Boundary between the Nazca plate and the South American plate: Peru-Chile Trench (off the west coast of South America).

(c) The boundary between the Pacific plate and the Australian plate: Bougainville Trench and Tonga Trench (off South Pacific islands such as New Guinea and Solomon Islands).

(d) The boundary between the Australian plate and the Eurasian plate: Java Trench (off Indonesia).

(e) The boundary between the Philippine Sea plate and the Eurasian plate: Nankai Trough (off Tokai and western Japan).

Around these trenches and troughs, megatsunamis caused by megathrust earthquakes with magnitudes of around 9.0 are listed in **Table 5.1**, and megatsunamis caused by great earthquakes of magnitude 8 class are frequent.

5. STORM SURGES AND TSUNAMIS

Table 5.1 List of megathrust earthquakes of around Moment Magnitude 9.0 since 1700 that caused megatsunamis. (Selected from Wikipedia's "Super Giant Earthquakes" [8], "Chronology of earthquakes" [9], etc.)

Name of each megathrust earthquake	Epicentre	Moment Magnitude	Date of occurrence
Chilean Valdivia Earthquake	1 Off the coast of Chile, South America	9.2 - 9.5	22 May 1960
Sumatra Earthquake	2 Off the coast of Indonesia	9.1 – 9.3	26 December 2004
Kamchatka Peninsula Earthquake	3 Off Kamchatka Peninsula	9.0 – 9.3	18 October 1737
Great Alaska Earthquake	4 Off the coast of Alaska, USA	9.1 – 9.2	27 March 1964
Off the Pacific coast of Tohoku Earthquake	5 Off the east coast of Japan	9.0 – 9.1	11 Mar 2011
Hōei Earthquake	6 Off the Tokai and western Japan	8.7 – 9.3	28 October 1707
South Sumatra Earthquake	7 Off the coast of Indonesia	8.8 – 9.2	25 November 1833
Severo-Kurilsk Earthquake	8 Off Kamchatka Peninsula	9.0	5 November 1952
Cascade Earthquake	9 Off the West Coast of US	8.7 – 9.2	26 January 1700
Chilean Arica Earthquake	10 Off the coast of Chile, South America	8.5 – 9.3	13 August 1868
Chilean Iquique Earthquake	11 Off the coast of Chile, South America	8.9	9 May 1877

The damage caused by the two recent megathrust earthquakes and tsunamis are summarised next.

(a) Tohoku Earthquake (2011)

This earthquake was a megathrust earthquake of moment magnitude 9.0-9.1 that occurred on the 11 March 2011 near the Japan Trench in the Pacific Ocean (the fault area is about 500 km north-south × 200 km east-west, with an average displacement distance of about 62 meters). This earthquake generated a megatsunami, with tsunami heights near the coastline reaching 6-22 m in Iwate Prefecture to northern Miyagi Prefecture and 5-13 m in the Sendai Bay to southern Ibaraki Prefecture, run-up heights of 40-20 m on the Sanriku coast, and inundation areas up to 6 km inland in the low-lying areas of the Sendai Bay to northern Fukushima Prefecture, causing extensive damage along the Pacific coast of eastern Japan. The total damage amounted to 16.9 trillion yen and the number of dead and missing reached 22,312, the majority of whom were tsunami victims. The tsunami also caused a core meltdown, radioactive material release and contamination accident at the Fukushima Daiichi Nuclear Power Station, forcing the evacuation of more than 100,000 residents and causing an electricity crisis in the summer of the same year. Furthermore, land subsidence of up to 120 cm occurred along the Pacific coast of the Tohoku region, making Ishinomaki City and other areas more prone to storm surge damage .

5.2 Tsunamis

(b) Sumatra Earthquake (2004)

Sumatra earthquake that occurred off the north-western coast of Sumatra in western Indonesia on 26 December 2004 was a megathrust earthquake of moment magnitude 9.1-9.3 (the fault area is about 400 km north-south × 150 km east-west, with a maximum displacement of about 20 meters). The tsunami generated by this earthquake spread across the Indian Ocean, reaching around 10 meters or more on some western coastlines of Indonesia and Thailand, and some eastern coastlines of India and Sri Lanka, with a run-up height of 34 meters in northern Sumatra. Most of the damage was caused by this megatsunami, which left 227,898 people dead or missing and caused total damage of more than USD 4.7 billion.

3) Evaluation Methods for Various Tsunami Parameters
(1) Propagation Calculation Methods Based on Formulae
(a) Propagation Velocities of Tsunamis

The propagation velocity v of a tsunami is obtained by the following approximate equation based on the linear long wave equation of motion.

$$v \cong \sqrt{gd} \qquad (5.13)$$

where g is the acceleration of gravity and d is the total water depth (= the depth below the still water surface h + the water surface elevation above the still water surface η).

The relationship between the total water depth and the propagation velocity according to this equation is shown in **Fig. 5.6**. The figure shows that, with a total water depth of 10 m, the propagation velocity is approximately 10 m/s, which is as fast as that of an Olympic runner. Therefore, on flat areas where there is no shelter of sufficient height near-by, it may not be possible to evacuate people if evacuation process was started after the tsunami was visible.

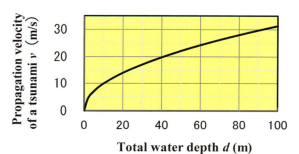

Fig. 5.6 Relationship between the total water depth and the propagation velocity of a tsunami.

(b) Tsunami Deformation

As the tsunami approaches land and the water depth becomes shallower, the propagation velocity slows down but the amount of water transported in unit time changes only slightly due to the conservation law of mass. Therefore, the shallower the water depth and slower the propagation velocity, the higher the water surface elevation becomes and the more forward tilting of the wave occurs.

The theoretical Eq. (5.14) based on the conservation law of wave energy transport between two points can be used to calculate the tsunami wave height due to this shallow water deformation.

5. STORM SURGES AND TSUNAMIS

Furthermore, the tsunami wave height can be even higher in bays where the water depth is not only shallower but also the propagation space is narrower. For this calculation, the theoretical Eq. (5.15) based on the conservation law of wave energy transport between two cross-sections can be used. This equation is called **Green's low**.

$$\frac{H_i}{H_o} = \left(\frac{h_o}{h_i}\right)^{1/4} \quad (5.14), \qquad \frac{H_i}{H_o} = \left(\frac{h_o}{h_i}\right)^{1/4}\left(\frac{b_o}{b_i}\right)^{1/2} \quad (5.15)$$

where H_i, h_i, and b_i are the wave heights of the tsunami, depths, and bay widths on the shore side, respectively; and H_o, h_o, and b_o are those on the offshore side, respectively.

Graphs of Eqs. (5.14) and (5.15) are shown in **Fig. 5.7**. Wave height increases by a factor of 1.4 when the water depth is reduced to 1/4, wave height increases by a factor of 2.0 when the bay width is reduced to 1/4, and wave height increases by a factor of 2.8 when both conditions are combined.

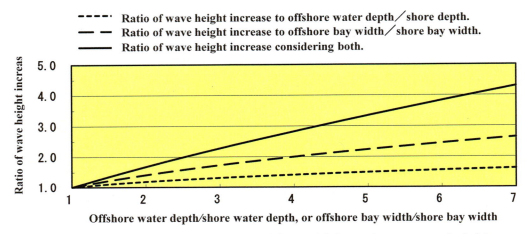

Fig. 5.7 Effect of depth ratio and bay width ratio on wave heights.

Furthermore, refraction and diffraction phenomena occur in both tsunamis and ordinary waves. In the case of ordinary waves, because the wavelength is tens to hundreds of meters, refraction due to changes in water depth within an area of similar lengths and diffraction due to obstacles of similar lengths are observed. On the other hand, in the case of tsunamis, because the wavelength is several hundred times longer than that of ordinary waves, these phenomena are difficult to observe within a sea area of tens to hundreds of meters, or in obstacles of similar length.

(c) Tsunami Heights at the Coastline

The maximum tsunami height H_{max}[m] at the shoreline can be calculated from Eq. (5.16) of Abe [10], using the moment magnitude M_w. The average tsunami height H_{mean}[m] over several tens-of kilometers section of the coastline can be calculated from Eq. (5.17) of Abe [11] using the moment magnitude M_w and the propagation distance Δ [km] from the epicentre to the target coast. These equations were derived using past tsunami data.

$$H_{max} = 10^{(0.5M_w - 3.3)} \quad (5.16), \qquad H_{mean} = 10^{(M_w - \log\Delta - 5.55)} \quad (5.17)$$

The results of various tsunami height calculations using these equations are shown in **Fig. 5.8**. The thick solid line in the figure based on Eq. (5.16) shows that tsunami damage is almost negligible for Mw < 6.0 but for Mw > 7.0, the maximum tsunami height at the coastline is likely to be more than about 2 m, so people should not approach the coastline.

However, the average tsunami height at Mw = 9.0 and $\Delta \leq 100$ km in the figure is abnormally high. This value is considered to be unreliable as it is in an area outside where tsunami data are available, and hence should be used with caution.

Table 5.2 shows a comparison of the measured tsunami heights from Tohoku Earthquake (field survey data from the Coastal Engineering Committee of JSCE, supplemented by field survey data from the author in April 2011) with the calculated maximum and average tsunami heights using Eqs. (5.16) and (5.17). It can be seen that Eqs. (5.16) and (5.17) are useful for obtaining rough estimates of tsunami heights.

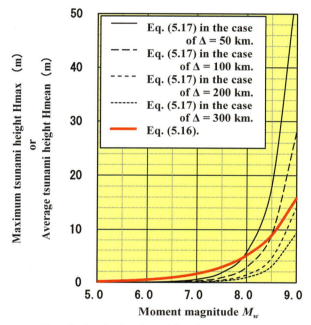

Fig. 5.8 Relationship between moment magnitude and tsunami height on the coastline.

Table 5.2 List of assessed tsunami heights at the coastline from the 2011 Tohoku Earthquake (M_w = 9.0 - 9.1).

Coastal zone name	Propagation distance Δ [km]	Measured tsunami height [m]	Maximum tsunami height H_{max} [m]	Averaged tsunami height H_{mean} [m]
South part of Iwate Prefecture	150	7〜22	15.8	18.8
North part of Miyagi Prefecture	150〜200	6〜18	〃	14.1〜18.8
South part of Miyagi Prefecture	250	6〜13	〃	11.3
North part of Fukushima Pre.	300	6〜11	〃	9.4

(2) Methods for Calculating Tsunami Run-up Heights
(a) Inundation Distance on a Flat Coast

If the coast is flat, the average inundation depth can be estimated from " (tsunami height at the coastline - target ground level) / 2". Then, the average propagation velocity can be obtained by

5. STORM SURGES AND TSUNAMIS

substituting this average inundation depth into the total water depth in Eq. (5.13). So, the inundation distance from the coastline to the inland area can be roughly estimated from this average velocity multiplied by the run-up duration and 2/3.

For example, the tsunami heights in the southern part of Miyagi Prefecture, Japan were 6-13 m, the ground height was approximately 5 m, the run-up time was approximately 20 minutes, and the inundation distances were 2-5 km. Applying these conditions to this calculation method, the average inundation depths are (6m-5m)/2 = 0.5m to (13m-5m)/2 = 4m, and the estimated inundation distances are 1.8km - 5km.

(b) Tsunami Run-up Height on a Uniform Slope

As a formula for calculating the run-up height R on a uniformly sloping beach, Yamamoto et al [12] obtained the Eq. (5.18) by applying the Bernoulli's theorem, which considers the water head loss between the shoreline and the run-up tip. As shown in **Fig. 5.9**, the run-up heights calculated by this equation increase with the steepness of the ground slope and are generally consistent with the measured values from the 2004 Sumatra earthquake tsunami and the 2011 Tohoku earthquake tsunami. **Fig. 5.9** also includes the calculation results using Eq. (5.19) proposed by Freeman and Mehaute [13] for reference.

$$R = \left(\frac{1 + \frac{2}{F_r^2}}{1 + \frac{f_b \times F_r^2}{4 \sin \beta}} \right) \frac{\left(F_r \sqrt{gH_{max}} \right)^2}{2g} \quad (5.18), \quad R = \left(\frac{(1 + \frac{1}{F_r})(1 + \frac{2}{F_r})}{1 + \frac{f_b \times F_r^2}{\tan \beta}} \right) \frac{\left(F_r \sqrt{gH_{max}} \right)^2}{2g} \quad (5.19)$$

where F_r is the Froude number, which ranges from 1.1 to 1.5 for tsunami run-up; β is the average slope angle of the ground slope; f_b is the friction coefficient of the slope; and H_{max} is the tsunami height on the shoreline.

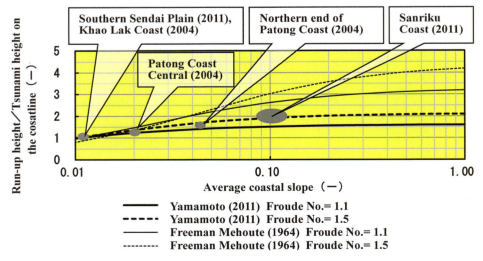

Fig. 5.9 Relationship between average coastal slope and tsunami run-up height.

Note that the equations based on hydraulic model experiments with conventional wave generators do not fully satisfy the similarity laws for megatsunamis because conventional wave generators cannot produce megatsunamis with relatively large wave heights and considerably long periods. Therefore, such equations should not be used except for small tsunamis with wave steepness that can be reproduced in hydraulic model experiments.

(c) Peak Depth and Velocity of Return Flow

In the case of megatsunamis, the flow rate and hydrodynamic force during the return flow are very large. There have been many cases where people and vehicles were swept offshore with the return flow. Also, sea defenses such as revetments can be easily damaged during the return flow as a result of not having support from the seaward side. Therefore, the return flow should not be neglected. Return flow behaviour can be calculated using the numerical model for run-up described in the next section, but if the slope is uniform and the run-up height can be easily obtained, the peak water depth h_r and flow velocity u_r, which are necessary for stability studies during return flow, can be obtained from the Eq. (5.20) of Yamamoto et al [14]. The upper equation is derived from the energy conservation law using the Darcy-Weisbach equation for energy loss, with the water surface shape approximated by a parabolic distribution between the upstream tip on the slope and the target location, while the lower equation is derived from the balance between the slope direction component of the water mass weight on the slope and the slope friction force.

$$\left. \begin{array}{l} h_r = \dfrac{(R-z)f_b/3}{\sin\beta + f_b} \\ u_r = \sqrt{\dfrac{2\sin\beta}{f_b}}\sqrt{gh_r} \end{array} \right\} \quad (5.20)$$

where R is the run-up height, z is the target ground height, f_b is the coefficient of friction on the slope surface, and β is the slope angle of the ground slope.

For the cases where the Froude number is 1.1, the friction coefficient of the slope surface is 0.02, the crown height of a bank is 5 m and the ground slopes are 1/10 and 1/50, the run-up height R for each tsunami height is obtained from Eq. (5.18). Then, the peak depth h_r and velocity u_r of the return flow obtained from Eq. (5.20) are shown in **Fig. 5.10**.

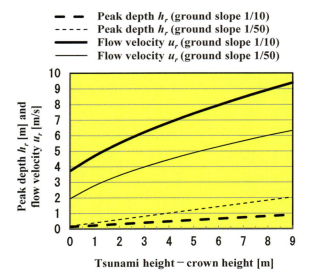

Fig. 5.10 Relationship between "tsunami height − crown height" and peak depth or velocity of a return flow.

5. STORM SURGES AND TSUNAMIS

(3) Numerical Model for Simulating Tsunamis

Authentic numerical simulations of tsunamis are considered to have started with Isozaki and Unoki [15] and Ito et al [16] in 1964. In this section, numerical models of tsunami propagation and run-up on an arbitrary seabed and land topography are presented.

(a) Fault Model

When the strain accumulated in a plate reaches a critical value, a fault is generated and the plates on both sides are rapidly displaced along it. When this displacement occurs within the seabed, vertical uplift/subsidence occurs on the seabed surface, and this uplift/subsidence directly causes vertical fluctuations in sea level and generates a tsunami.

This fault movement can be modelled as shown in **Fig. 5.11**. During an earthquake, a rectangular fault plane of length L and width W is considered to be displaced by a slip amount D.

The direction of slip on the fault plane is called a **rake angle** and is denoted by λ. The intersection of the horizontal plane and the fault plane is usually expressed by the horizontal angle φ from the north and is called a **strike**. The vertical inclination angle of the fault plane is called a **dip** and is denoted by δ. These six parameters are called fault parameters and are determined to match the seismic waves.

Given a magnitude M, the length L, width W and slip amount D of the fault can be determined to a first approximation from the following equations.

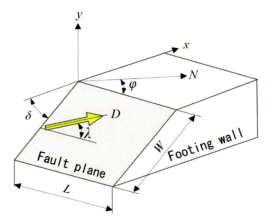

Fig. 5.11 Illustration of a typical fault model.

$$\log_{10} L = 0.5M + 1.1, \quad W = 0.5L, \quad \log_{10} D = 0.5M - 3.2 \tag{5.21}$$

For reference, some existing fault model setups are shown in **Table 5.3**.

Given these fault parameters, the vertical displacement distribution of the seabed near the fault is obtained by the method of Mansinha and Smylie [18] or Okada [19]. This vertical displacement is given as the sea level data and is used as the initial condition for the numerical tsunami simulation. Actual fault movement has a duration of a few seconds to 100 seconds, but in numerical simulations, it is assumed to be completed instantaneously.

5.2 Tsunamis

Table 5.3 Examples of fault models (from Goto and Sato [17]).

Names of earthquakes		1896. Meiji Sanriku Earthquake	1933. Showa Sanriku Oki Earthquake	1968. Tokachi Offshore Earthquake	1960. Chile Earthquake
Names of fault models		Aida (1977) MJ-6	Aida (1977) SY-3	Aida (1978) D2	Kanamori
Length of the fault L [km]		210	185	150	800
Width of the fault W [km]		50	50	100	200
Depth to top edge [km]		1	1	1	53
Dip of the fault δ [degree]		20	45	20	10
Strike of the fault φ [degree]		S 66° W	N 90° W	S 66° W	S 80° E
Slip	Vertical displacement [m], normal fault is +	-10.6	6.6	-2.5	-24.0
	Lateral displacement [m], left displacement +	-6.7	0.0	-3.2	0.0

Plate boundaries have regions where the plates are stable and slide relatively smooth, and regions where they stick together and are difficult to slide. Strain accumulates in these sticking regions. When the plates reach their bearing capacity limit, they slip at once, resulting in plate boundary earthquakes. This sticking zone is called an **asperity**. The Nankai Trough Large Earthquake Model Study Group in Japan used the following definition of asperity when applying the fault model.

① Strong motion generation area: A term used in fault models for evaluating seismic intensity distribution, meaning a region within the fault plane that generates strong seabed motion. Other areas of the fault plane were called background areas.

② Large slip area: A term used in fault models for evaluating tsunamis, meaning a large-slip region in the fault plane. The region of particularly large slip within the large-slip zone is called the super-large-slip zone. Other areas of the fault plane were called tsunami background areas.

(b) Numerical Model for Simulating Behavior of Tsunamis

Tsunami propagation in the sea is generally calculated by inputting the water level variation based on the fault model from the offshore side and solving the nonlinear long wave equations and the continuity equation for seawater to obtain the flow rates per unit width q_x and q_y in the x and y directions and the water surface elevation η. The nonlinear long-wave equations and the continuity equation for seawater are given below.

$$\frac{\partial q_x}{\partial t} + \frac{\partial}{\partial x}\left(\frac{q_x^2}{d}\right) + \frac{\partial}{\partial y}\left(\frac{q_x q_y}{d}\right) + gd\frac{\partial \eta}{\partial x} - \frac{\partial}{\partial x}\left(dv_t\frac{\partial(q_x/d)}{\partial x}\right) - \frac{\partial}{\partial y}\left(dv_t\frac{\partial(q_y/d)}{\partial y}\right)$$

$$+\frac{f_b}{d^2}q_x\sqrt{q_x^2 + q_y^2} = 0 \tag{5.22}$$

5. STORM SURGES AND TSUNAMIS

$$\frac{\partial q_y}{\partial t} + \frac{\partial}{\partial x}\left(\frac{q_y q_x}{d}\right) + \frac{\partial}{\partial y}\left(\frac{q_y^2}{d}\right) + gd\frac{\partial \eta}{\partial y} - \frac{\partial}{\partial y}\left(dv_t \frac{\partial(q_y/d)}{\partial y}\right) - \frac{\partial}{\partial x}\left(dv_t \frac{\partial(q_x/d)}{\partial x}\right)$$

$$+ \frac{f_b}{d^2} q_y \sqrt{q_x^2 + q_y^2} = 0 \tag{5.23}$$

$$\frac{\partial q_x}{\partial x} + \frac{\partial q_y}{\partial y} + \frac{\partial \eta}{\partial t} = 0 \tag{5.24}$$

where d is the total water depth (= depth below still water surface h + elevation above still water surface η), g is the acceleration of gravity, v_t is the eddy viscosity coefficient of seawater, and f_b is the friction coefficient of the seabed.

For specific tsunami simulation methods using this numerical model and the fault model, readers are referred to Goto and Sato [17] and the Earthquake Research Committee of the Headquarters for Earthquake Research Promotion in MEXT [20].

(c) Numerical Model for Simulating Tsunami Run-up

For a tsunami prediction, it is important to clarify the mechanism of tsunami generation and set up an accurate fault model. However, to obtain highly accurate calculation results, a computer with large memory and computing capacity is required to perform propagation and run-up calculations from offshore faulting locations to inland areas with successively finer calculation grids. Therefore, when performing a large number of numerical experiments on a PC to confirm the effectiveness of proposed tsunami countermeasures such as coastal dikes, it is efficient to compare the effectiveness of various countermeasures by inputting time series data of the expected maximum water surface elevation from water depths relatively close to the coast, rather than from the tsunami generation area far offshore.

If the objective is to calculate the landward inundation from a tsunami, water level change data obtained from observation records or existing wide-area numerical calculations are input at the offshore boundary of the shallow-water area, and the nonlinear long-wave equations and continuity equation of seawater, which take into account the non-inundated areas scattered onshore, are solved to obtain the inundation depth and velocity on the land.

Practical numerical models include the following:

(i) The **STOC** model by Tomita and Kakinuma [21], which enables storm surge and tsunami simulations in three dimensions. the calculation software for tsunami simulations is publicly available and downloadable from https://www.pari.go.jp/unit/tsunamitakashio/open-software/t-stoc/download/.

(ii) The model by Yamamoto [22] and Vu et al [23], which enables storm surge and tsunami calculations in two dimensions. The detailed structure and usage of this model is described in Section 3 of the Appendix, and the non-linear long wave equations and continuity equation of seawater used in this model are shown below.

$$\frac{\partial q_x}{\partial t} + \frac{1}{S}\frac{\partial}{\partial x}\left(\frac{Sq_x^2}{d}\right) + \frac{1}{S}\frac{\partial}{\partial y}\left(\frac{Sq_xq_y}{d}\right) + gd\frac{\partial \eta}{\partial x} - \frac{1}{S}\frac{\partial}{\partial x}\left(dSv_t\frac{\partial(q_x/d)}{\partial x}\right)$$
$$-\frac{1}{S}\frac{\partial}{\partial y}\left(dSv_t\frac{\partial(q_y/d)}{\partial y}\right) + \frac{f_d}{d^2}q_x\sqrt{q_x^2+q_y^2} = 0 \quad (5.25)$$

$$\frac{\partial q_y}{\partial t} + \frac{1}{S}\frac{\partial}{\partial x}\left(\frac{Sq_yq_x}{d}\right) + \frac{1}{S}\frac{\partial}{\partial y}\left(\frac{Sq_y^2}{d}\right) + gd\frac{\partial \eta}{\partial y} - \frac{1}{S}\frac{\partial}{\partial y}\left(dSv_t\frac{\partial(q_y/d)}{\partial y}\right)$$
$$-\frac{1}{S}\frac{\partial}{\partial x}\left(dSv_t\frac{\partial(q_x/d)}{\partial x}\right) + \frac{f_d}{d^2}q_y\sqrt{q_x^2+q_y^2} = 0 \quad (5.26)$$

$$\frac{\partial f_y q_x}{\partial x} + \frac{\partial f_x q_y}{\partial y} + \frac{\partial S\eta}{\partial t} = 0 \quad (5.27)$$

where q_x and q_y are respectively the flow rates per unit width in two horizontal directions (x, y), d is the total water depth (= depth below still water surface h + elevation above still water surface η), g is the acceleration of gravity, v_t is the eddy viscosity coefficient of seawater, and f_d is the drag coefficient due to structures or trees.

In a horizontal 2D control volume, the ratio of the inundated area to the total area of the control volume, S is considered. Furthermore, the first and second terms on the left-hand side of the continuity equation consider the inflow and outflow in the x and y directions within the control volume. Therefore, the ratios f_y and f_x of the length of the inundated part to the length of one side parallel to the y and x directions in the control volume, rather than the ratio S of the inundated area can be used.

4) Tsunami Forces
(1) Formulae for Calculating Tsunami Forces at Sea Area

The main formulae to calculate tsunami force in sea areas include those proposed by Fukui et al. [24], Tanimoto et al. [25], Matsutomi [26], Mizutani and Imamura [27], [28] and Ikeno et al. [29], [30], [31], all based on hydraulic model experiments.

When a tsunami with a relatively large wave steepness ($H/L>0.0001$) propagates in shallow water, a phenomenon called **soliton fission** is likely to occur, in which the leading edge of the tsunami splits to form an undular bore shown in **Fig. 5.12**. When soliton fission occurs, the wave height increases and the tsunami force becomes stronger. Therefore,

Fig. 5.12 Illustration of soliton fission.

the formulae they proposed must be differentiated by soliton fission/non-fission. The relationship between tsunami heights in the sea area (= tsunami wave heights) and tsunami force per unit width by using the formulae they proposed is shown in **Fig. 5.13**. In the formulae of Fukui et al. and Mizutani

5. STORM SURGES AND TSUNAMIS

& Imamura, it is necessary to set a target water depth. Thus, in the calculations for this figure, "target water depth = twice the tsunami wave height" was used.

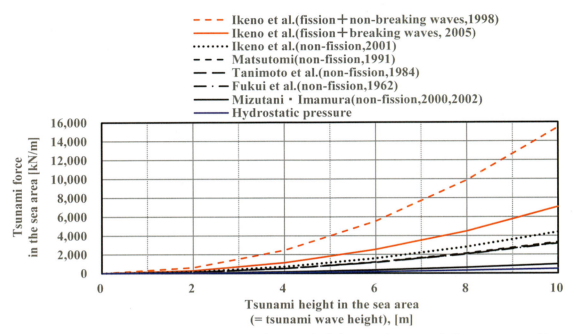

Fig. 5.13 Relationship between tsunami wave heights and tsunami force per unit width using various formulae at sea area.

From this figure, the formula of Ikeno et al. 1998 [29] should be used in the case of soliton fission and no wave breaking. The formula of Ikeno et al. 2005 [31] should be used in the case of soliton fission and wave breaking, while the formula of Ikeno et al. 2001 [30] can be used in the case of no soliton fission (older equations are known to easily underestimate actual measured values).

The formulae of Ikeno et al [29], [31], [30] are summarised below.

(a) Formula of Ikeno et al [29] (1998)

In the case of soliton fission and non-breaking waves, the formula for tsunami pressure (= tsunami force per unit area) $p(z)$ [N/m²] where the vertical distribution is triangular for $z=6H_I$ to $3H_I$ and constant for $z<3H_I$ is given in Eq. (5.28).

$$\left. \begin{array}{l} \text{For } 6 \geq (z/H_I) \geq 3, \quad p(z) = 3.5\rho g H_I \left[6 - (z/H_I)\right]/3 \\ \text{For } 3 \geq (z/H_I), \qquad p(z) = 3.5\rho g H_I \\ \text{Height to the top of tsunami pressure distribution } \eta_{max} = 6H_I \end{array} \right\} \quad (5.28)$$

196

where $\eta_{\max}$ is the height [m] to the top of tsunami pressure distribution, H_I is the tsunami height in the sea area (= tsunami wave height) [m], ρ is the density of seawater, g is the gravitational acceleration, and z is the vertical coordinate with the origin at the still water surface and positive upwards.

(b) Formula of Ikeno et al[31] (2005)

In the case of soliton fission and breaking waves, the formula for the tsunami pressure $p(z)$ [N/m²] where the vertical distribution is triangular for $z=3H_I$ to $2H_I$, trapezoidal for $z=2H_I$ to H_I and constant for $z< H_I$ is given in Eq. (5.29). P_{max} should be used differently depending on the relative water depth h/L.

$$
\left.
\begin{aligned}
&\text{For } 3 \geq \frac{z}{H_I} \geq 2, \quad p(z) = 1.2 p_{\max}\left(3 - \frac{z}{H_I}\right) \\[2mm]
&\text{For } 2 \geq \frac{z}{H_I} \geq 1, \quad p(z) = 0.4 p_{\max}\left(1 + \frac{z}{H_I}\right) \\[2mm]
&\text{For } 1 \geq \frac{z}{H_I}, \qquad p(z) = 0.8 p_{\max} \\[2mm]
&\text{Height to the top of tsunami pressure distribution } \eta_{\max} = 3H_I \\[2mm]
&\text{In the case of } \frac{h}{L} \leq 0.001, \qquad p_{\max} = 2.0\rho g H_I \\[2mm]
&\text{In the case of } 0.001 \leq \frac{h}{L} \leq 0.005, \quad p_{\max} = \left(250\frac{h}{L} + 1.75\right)\rho g H_I \\[2mm]
&\text{In the case of } 0.005 \leq \frac{h}{L}, \qquad p_{\max} = 3.0\rho g H_I
\end{aligned}
\right\} \quad (5.29)
$$

where h is the water depth below the still water surface and L is the wavelength of the bore part of the tsunami at depth h.

(c) Formula of Ikeno et al[30] (2001)

When no soliton fission, the formula for the tsunami pressure $p(z)$ [N/m²] where the vertical distribution is triangular for $z=3H_I$ to 0, trapezoidal for $z=0$ to $-0.5H_I$ and constant for $z<-0.5H_I$, is given in Eq. (5.30).

$$
\left.
\begin{aligned}
&\text{For } 3 \geq \frac{z}{H_I} \geq 0, \qquad p(z) = \rho g H_I\left(3 - \frac{z}{H_I}\right) \\[2mm]
&\text{For } 0 \geq \frac{z}{H_I} \geq -0.5, \quad p(z) = \rho g H_I\left(3 + 1.6\frac{z}{H_I}\right) \\[2mm]
&\text{For } -0.5 \geq \frac{z}{H_I}, \qquad p(z) = 2.2\rho g H_I \\[2mm]
&\text{Height to the top of tsunami pressure distribution } \eta_{\max} = 3H_I
\end{aligned}
\right\} \quad (5.30)
$$

5. STORM SURGES AND TSUNAMIS

(2) Formulae for Calculating Tsunami Forces on Land

The main formulae for calculating the tsunami force in land areas include those proposed by Asakura et al. [32], Iizuka and Matsutomi [33], Ikeno et al. [34] and Yasuda et al. [35], which are based on hydraulic model experiments. These formulae must also be differentiated by soliton fission/non-fission. The tsunami force per unit width is calculated for different tsunami heights above the ground (i.e. inundation depths) using each formula, as shown in **Fig. 5.14**. The Iizuka and Matsutomi formula requires the drag coefficient and the depth of action. The drag coefficient = 2 and the depth of action = 2 × the tsunami height above the ground (assuming perfect reflection at the vertical wall) is used.

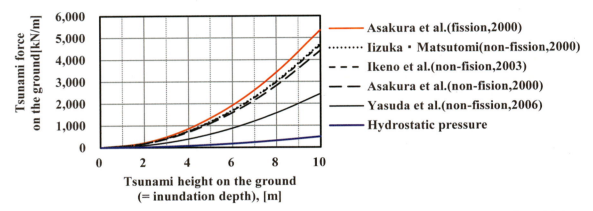

Fig. 5.14 Relationship between tsunami heights and tsunami force per unit width using various formulae on the ground.

From this figure, the formula of soliton fission by Asakura et al [32] can be used for the tsunami force with soliton fission, while the formulae of non-fission by Iizuka and Matsutomi [33], Ikeno et al [34], and Asakura et al [32] can be used for the tsunami force without soliton fission.

(a) Formula of Asakura et al [32] (2000)

In the case of soliton fission, the formula for the tsunami pressure $p(z)$ [N/m²] is given in Eq. (5.31). This vertical distribution is triangular for $z=3.0h_I$ to $0.8h_I$ and trapezoidal for $z<0.8h_I$.

$$\left. \begin{array}{l} \text{For } 3.0 \geq \dfrac{z}{h_I} \geq 0.8, \quad p(z) = \rho g h_I \left(3 - \dfrac{z}{h_I}\right) \\[6pt] \text{For } 0.8 \geq \dfrac{z}{h_I} \geq 0, \quad p(z) = \rho g h_I \left(5.4 - 4\dfrac{z}{h_I}\right) \\[6pt] \text{Height to the top of tsunami pressure distribution } h_{\max} = 3.0 h_I \\[6pt] \text{Tsunami force per unit width } F = 5.46 \rho g h_I^2 \end{array} \right\} \quad (5.31)$$

where h_{max} is the height of the uppermost point where the tsunami pressure acts [m], h_I is the tsunami height on the ground (= inundation depth) [m], F is the tsunami force per unit width on the ground [kN/m], ρ is the density of seawater, g is the gravitational acceleration, and z is the vertical coordinate with the origin at the ground surface and positive in the upward direction.

The formula for the tsunami pressure $p(z)$ [N/m^2] without soliton fission is shown in Eq. (5.32). This distribution is triangular at z=3.0h_I to 0.

$$
\left.
\begin{aligned}
&\text{For } 3.0 \geq \frac{z}{h_I} \geq 0, \quad p(z) = \rho g h_I \left(3 - \frac{z}{h_I}\right) \\
&\text{Height to the top of tsunami pressure distribution } h_{max} = 3.0 h_I \\
&\text{Tsunami force per unit width } F = 4.5 \rho g h_I^2
\end{aligned}
\right\} \tag{5.32}
$$

(b) Formula of Iizuka and Matsutomi [33] (2000)

Iizuka and Matsutomi proposed Eq. (5.33) using the drag formula and tsunami force per unit width F = tsunami pressure $\times$ depth of action. Here, using a Froude number of 1.1, the velocity of a run-up wave = 1.1 $\times$ the propagation velocity of a long wave is assumed.

$$
F = \frac{C_D}{2} \rho u^2 h_f = \frac{C_D}{2} \rho \left(1.1\sqrt{gh_f}\right)^2 h_f \fallingdotseq 0.61 C_D \rho g h_f^2 \tag{5.33}
$$

where C_D is the drag coefficient, ρ is the density of seawater, u is the velocity of the run-up wave, h_f is the inundation depth in front of the target wall on the ground, g is the gravitational acceleration.

(c) Formula of Ikeno et al [34] (2003)

When there is no soliton fission, the formula for the tsunami pressure $p(z)$ [N/m^2] is given in Eq. (5.34).

$$
\left.
\begin{aligned}
&\text{For } 3.0 \geq \frac{z}{h_I} \geq 0.5, \quad p(z) = \rho g h_I \left(3 - \frac{z}{h_I}\right); \quad \text{For } 0.5 \geq \frac{z}{h_I} \geq 0, \quad p(z) = \rho g h_I \left(4 - 3\frac{z}{h_I}\right) \\
&\text{For } 0 \geq \frac{z}{h_I} \geq -0.5, \quad p(z) = \rho g h_I \left(4 + 3.6\frac{z}{h_I}\right); \quad \text{For } -0.5 \geq \frac{z}{h_I}, \quad p(z) = 2.2 \rho g h_I \\
&\text{Height to the top of tsunami pressure distribution } h_{max} = 3.0 h_I \\
&\text{Tsunami force per unit width } F = 4.75 \rho g h_I^2
\end{aligned}
\right\} \tag{5.34}
$$

where z is the vertical coordinate with the origin at the inundation surface, h_I is the tsunami height on the ground.

The vertical distribution is triangular for z=3.0h_I to 0.5h_I, trapezoidal for z<0.5h_I to 0, inverse trapezoidal for z=0 to $-0.5h_I$, and constant for z<$-0.5h_I$.

5. STORM SURGES AND TSUNAMIS

(3) Calculation Method of Tsunami Forces for Complex Sections Using CADMAS-SURF

CADMAS-SURF, introduced in Chapter 3, is a numerical model that can predict the velocity, pressure, and water surface distributions even when the topography and structure cross-sections are relatively complex. This model can be used to calculate tsunami forces. However, it is not easy to set up appropriate input data that can be expected to provide high prediction accuracy. Therefore, it is advisable to prepare validation data for a representative case from hydraulic model experiments, confirm the appropriate data setting method through validation calculations using this data, and then carry out comparative simulations for multiple cases.

For specific implementation methods, see "Research and development of CADMAS-SURF/3D numerical wave tank" by Isobe et al [36]. The source programme can be downloaded from the website of the Coastal Development Institute of Technology (https://www.cdit.or.jp/program/cadmas-download.html).

5) Tsunami Debris Impact Forces

In the case of megatsunamis, drifted objects such as driftwood, containers, vehicles, and ships are generated in large quantities over a wide area of the coastline and the secondary damage caused by these objects cannot be ignored. To construct facilities to prevent drift generation, control drift movement, and improve the ability of surrounding structures to withstand the impact force of drifting debris, it is necessary to accurately estimate the impact force caused by drifting debris. Yamamoto et al [37] selected the following reliable impact force formulae for the main drifting objects (driftwood, containers, vehicles, and vessels) by checking whether the values calculated using each formula are within a reasonable existence range and by comparing the values calculated using each formula with large-scale experimental data.

(1) For Driftwood

Eq. (5.35) of Matsutomi [38], Eq. (5.36) of US Federal Emergency Management Agency (FEMA) & US National Oceanic and Atmospheric Administration (NOAA) [39] and Eq. (5.37) of American Society of Civil Engineers (ASCE) [40] are highly reliable.

(2) For Containers

Eq. (5.35) of Matsutomi [38], Eq. (5.36) of FEMA & NOAA [39], Eq. (5.37) of ASCE [40], Eq. (5.38) of Arikawa et al. [41] and Eq. (5.39) of Ikeno et al. [42] are highly reliable.

(3) For Motor Vehicles

Eq. (5.36) of FEMA & NOAA [39] and Eq. (5.37) of ASCE [40] are highly reliable.

5.2 Tsunamis

(4) For Ships

Eq. (5.40) of Mizutani [43], Eq. (5.36) of FEMA & NOAA [39] and Eq. (5.37) of ASCE [40] are highly reliable.

FEMA & NOAA and ASCE can be applied to all the above drift types because they propose relatively accurate methods for evaluating the stiffness between the impacting object and the impacted object. In addition to those presented here, many other formulae of impact force have been proposed. These formulae are supposed in good agreement with the measurement data obtained by the proponents. However, when compared with the measurement data collected by others, some disagreements were observed. In many cases, this is because the ability to correctly assess the difference in stiffness between the impacting and impacted objects has not been built in, even though different stiffnesses of the impacting and impacted objects have different effects on the reduction of impact forces due to deformation. Therefore, the use of formulae that do not consider the stiffness or variables related to the stiffness (yield stress, Young's modulus, time of action of the impact force, etc.) should be avoided.

When using the formulae presented here, it is necessary to properly set the stiffness or variables related to the stiffness, so it is recommended to refer to Kaida and Kihara [44], Yamamoto et al. [37], as well as FEMA & NOAA [39] or ASCE [40], as necessary.

Eq. (5.35) of Matsutomi [38], Eq. (5.36) of FEMA & NOAA [39], Eq. (5.37) of ASCE [40], Eq. (5.38) of Arikawa et al. [41], Eq. (5.39) of Ikeno et al. [42] and Eq. (5.40) of Mizutani [43] are explained below.

(a) Formula of Matsutomi [38]

$$\frac{F_{ik}}{rD^2 L} = 1.6 C_{MA} \left(\frac{V}{\sqrt{gD}} \right)^{1.2} \left(\frac{\sigma_f}{rL} \right)^{0.4} \tag{5.35}$$

where F_{ik} is the impact force [kN], C_{MA} is the apparent mass coefficient (1.7 for stepped waves and surges, 1.9 for steady flow), V is the impact velocity [m/s], g is the acceleration of gravity, D is the cross-sectional diameter of the drifting object, σ_f is the yield stress [kN/m^2] of the drifting object, r is the unit volume weight [kN/m^3] of the drifting object, and L is the length of the drifting object.

(b) Formula of FEMA & NOAA [39]

$$F_i = 1.3 u_{max} \sqrt{m_d k (1 + C)}, \quad u_{max} = \sqrt{2 g R \left(1 - \frac{z}{R} \right)} \tag{5.36}$$

where F_i is the impact force [N]; 1.3 is the severity factor for hazard category IV; u_{max} is the maximum velocity of the fluid transporting the drifting object; m_d is the mass of the drifting object; k is the

5. STORM SURGES AND TSUNAMIS

effective axial stiffness between the drifting object and the drifted object (2.4×10^6 N/m for driftwood, 6.5×10^8 N/m for a 40-ft container empty, 1.5×10^9 N/m for a 20-ft container empty); C is the additional mass coefficient (0.0 for the longitudinal impact of driftwood, 0.2 for the longitudinal impact of a 40-ft container, 1.0 for transverse impact of a 40-ft container, 0.3 for longitudinal impact of a 20-ft container, and 1.0 for transverse impact of a 20-ft container); R is the maximum run-up height $\times$ 1.3; and z is the target ground height above the water surface.

(c) Formula of ASCE [40]

$$F_i = R_{\max} I_{TSU} C_o u_{\max} \sqrt{k m_d} \qquad (5.37)$$

where F_i is the impact force [N]; $R_{\max}$ is the dynamic response ratio to the impact load ($0.0 - 1.8$, depending on the impact action time); I_{TSU} is the importance factor for a tsunami ($1.0 \sim 1.25$); C_o is the direction coefficient ($=0.65$); $u_{\max}$ is the maximum flow velocity; k is the smaller between the effective stiffness of the drift and the transverse stiffness of the target structure; and m_d is the mass of the drifting object. Here, the effective stiffness of driftwood can be obtained using $k = EA/L$ (E is the longitudinal modulus of elasticity of a log, A and L are respectively the cross-sectional area and length of the driftwood).

(d) Formula of Arikawa et al. [41]

$$\left. F_{ik} = \gamma_p \chi^{2/5} \left(\frac{5}{4} m_{12} \right)^{3/5} V^{6/5}, \quad \chi = \frac{4\sqrt{a}}{3\pi} \frac{1}{k_1 + k_2}, \quad k_i = \frac{1 - v_i^2}{\pi E_i} \right\} \qquad (5.38)$$

$$\text{For driftwood, } m_{12} = C_{MA} m_1; \quad \text{For a container, } m_{12} = \frac{m_1 m_2}{m_1 + m_2} \right|$$

where F_{ik} is the impact force [kN], γ_p is the energy attenuation effect due to plasticity (0.25), V is the impact velocity [m/s], a is half the radius of the impact surface [m], v_i is Poisson's ratio, E_i is Young's modulus [kN/m²], C_{MA} is the apparent mass coefficient, m_i is the mass [t], and the subscript i for m and k denotes the impacting object 1 and the impacted object 2.

(e) Formula of Ikeno et al. [42]

$$F_i = k \left(C_{MA} m_d \right)^{0.6} V^{1.2} D^{0.2} E^{0.4}, \quad E = \frac{E_1 \times E_2}{E_1 \left(1 - v_2^2 \right) + E_2 \left(1 - v_1^2 \right)} \right\} \qquad (5.39)$$

where F_i is the impact force in elastic impact [N]; k is 0.243; C_{MA} is the apparent mass coefficient (2.0 for a cylindrical timber impacting longitudinally, 1.0 for a container); m_d is the mass of the drifting object [t]; V is the impact velocity [m/s]; D is the representative size of the impact cross-section [m]; and E is the representative stiffness (the subscripts 1 and 2 of Young's modulus and Poisson's ratio denote the impacting and impacted objects).

202

5.2 Tsunamis

(f) Formula of Mizutani [43]

$$F_{ik} = 2\frac{WV_x}{gt_d}$$
(5.40)

where F_{ik} is the impact force [kN], W is the small vessel weight [kN], V_x is the drifting speed of the small vessel [m/s], and t_d is the action time of the impact force.

The action time t_d of the impact force is obtained by the following equation.

$$t_d = 2\sqrt{\frac{m_d}{k}}$$
(5.41)

However, the t_d of a loaded container can be obtained by the following equation, where $m_{contents}$ is 50% of the maximum capacity mass of the container.

$$t_d = \frac{m_d + m_{contents}}{\sqrt{km_d}}$$
(5.42)

5. STORM SURGES AND TSUNAMIS

List of References in Chapter 5

1) Fujita, T.: Pressure Distribution within Typhoon. *Geophys. Mag.*, 23, 1952, pp.437–451.

2) Holland, G.J.: An analytic model of the wind and pressure profiles in hurricanes. *Monthly Weather Review*, 108, 1980, pp.1212-1218.

3) Japan Meteorological Agency: Schematic Diagram of the Plates near Japan, https://www.data.jma.go.jp/eqev/data/nteq/nteq.html.

4) Shishikura, M., Fujiwara, O., Sawai, Y., Fujino, S. and Namegaya, Y.: Forecasting Multi-segment Mega Earthquakes by Coastal Paleoseismological Survey, *Chishitsu News*, No.663, 2009, pp.23-28. (in Japanese)

5) Azuma, T., Ota, Y., Ishikawa, M. and Taniguchi, K.: Late Quaternary Coastal Tectonics and Development of Marine Terraces in Omaezaki, Pacific Coast of Central Japan, *the Quaternary Research*, Vol.44, No.3, 2005, pp.169-176. (in Japanese)

6) Matsuoka, H. and Okamura, M.: The 2000 Years Ago Tsunami Event in the Kaniga-ike Pond Innermost the Tosa Bay, *Japan Geoscience Union Meeting 2011*, SSS035-P02, 2011. (in Japanese)

7) Suzuki, T.: Seismic Seiche Occurred at Saiko Lake due to the 2011 Off the Pacific Coast of Tohoku Earthquake, *Journal of JSCE*, A1(Structural and earthquake engineering), Vol.68, No.4, 2012, pp. I_152-I_160. (in Japanese)

8) Wikipedia, the free encyclopedia: *Super Giant Earthquakes*, https://ja.wikipedia.org/wiki/超巨大地震. (in Japanese)

9) Wikipedia, the free encyclopedia: *Chronology of Earthquakes*, https://ja.wikipedia.org/wiki/地震の年表. (in Japanese)

10) Abe, K.: Quantification of Historical Tsunamis by the *M*t Scale, *Earthquakes*, 2nd series, Vol.52, 1999, pp.369-377. (in Japanese)

11) Abe, K.: Physical size of tsunamigenic earthquakes of the northwestern Pacific, *Phys. Earth Planet. Inter.*, Vol.27, Issue 3, 1981, pp.194-205.

12) Yamamoto, Y., Charusrojthanadech, N. and Nariyoshi, K.: Proposal of Rational Evaluation Methods of Structure Damage by Tsunami, *Journal of JSCE*, B2(Coastal Engineering), Vol.67, No.1, 2011, pp.72-91. (in Japanese)

13) Freeman, J. C. and Mehaute, B. L.: Wave breakers on a beach and surges on a dry bed, *Proc. ASCE*, Vol. 90, No.HY2, 1964, pp.187-216.

14) Yamamoto, Y., Nariyoshi, K. and Vu, T.C.: Improvement of Prediction Method of Coastal Scour and Erosion due to Tsunami Back-flow, *Journal of JSCE*, B2(Coastal Engineering), Vol.65, No.1, 2009, pp.511-515. (in Japanese)

15) Isozaki, I. and Unoki, S.: The numerical computation of the tsunami in Tokyo Bay caused by the

Chilean Earthquake in May, 1960, *Study on Oceanography*, Dedicated to Prof. Hidaka in Commemoration of his Sixties Birthday, 1964, pp.389-402.

16) Ito, Y., Toki, S. and Morihira, M.: *Digital Computation on the Effect of Breakwaters Against Long-period Waves (2nd Report)*, Report of Port and Harbour Technical Research Institute, Vol.3, No.7, 1964, 123p. (in Japanese)

17) Goto, C. and Sato, K.: Development of Tsunami Numerical Simulation System for Sanriku Coast in Japan, *Report of the Port and Harbour Research Institute*, Vol.32, No.2, 1993, pp.3-44. (in Japanese)

18) Mansinha, L. and Smylie, D.: The displacement fields of inclined faults, *Bull., Seismological Society of America*, Vol.61, No.5, 1971, pp.1433-1440.

19) Okada, Y.: Surface deformation due to shear and tensile faults in a half-space, *Bull., Seismological Society of America*, Vol.75, No.4, 1985, pp.1135-1154.

20) Earthquake Research Committee in the Headquarters for Earthquake Research Promotion: *Tsunami Prediction Method for Earthquakes with Characterized Source Faults (Tsunami Recipe)*, the Headquarters for Earthquake Research Promotion, 2017, 35p. (in Japanese)

21) Tomita, T. and Kakinuma, T.: Storm Surge and Tsunami Simulator in Oceans and Coastal Areas (STOC), *Report of the Port and Harbour Research Institute*, Vol.44, No.2, 2005, pp.83-98. (in Japanese)

22) Yamamoto, Y.: Design Process of Coastal Facilities for Disaster Prevention, *Proc. Schl. Eng. Tokai Univ.*, Ser. E, Vol. 31, 2006, pp.11-19.

23) Vu, T.C., Yamamoto, Y. and Charusrojthanadech, N.: Improvement of Prediction Methods of Coastal Scour and Erosion due to Tsunami Back-flow, *Proc. 20th International Offshore and Polar Engineering Conference*, 2010, pp.1053-1060.

24) Fukui, Y., Shiraishi, H., Nakamura, M. and Sasaki, Y.: Tsunami Study (II) - Effects of Stepped Wave Tsunamis on Coastal Dikes -, *Proceedings of the 9th Coastal Engineering Conference*, 1962, pp.50-54. (in Japanese)

25) Tanimoto, K., Tsuruya, H. and Nakano, S.: Tsunami Forces and Causes of Damage to Reclaimed Seawalls in the 1983 Central Japan Sea Earthquake Tsunami, *Proceedings of the 31st Coastal Engineering Conference*, 1984, pp.257-261. (in Japanese)

26) Matsutomi, H.: Pressure Distribution and Total Wave Force during Breaking Step-wave Collisions, *Journal of Coastal Engineering*, Vol. 38, 1991, pp. 626-630. (in Japanese)

27) Mizutani, S. and Imamura, F.: Experiments on Step-wave Force Acting on Structures, *Journal of Coastal Engineering*, Vol. 47, 2000, pp.946-950. (in Japanese)

28) Mizutani, S. and Imamura, F.: Proposal for a Design External Force Calculation Flow Which Takes into Account the Impulsivity and Overtopping of Tsunami Bores, *Journal of Coastal Engineering*, Vol.49, 2002, pp.731-735. (in Japanese)

5. STORM SURGES AND TSUNAMIS

29) Ikeno, M., Matsuyama, M. and Tanaka, H.: Experimental Study on the Deformation of Soliton Fission Tsunamis on the Continental Shelf and Their Design Wave Pressures for Breakwaters, *Journal of Coastal Engineering*, Vol.45, 1998, pp.366-370. (in Japanese)

30) Ikeno, M., Mori, N. and Tanaka, H.: Experimental Study on Wave Force due to Breaking Stepped-wave Tsunamis and the Behaviour・Impact Force of Drifting Objects, *Journal of Coastal Engineering*, Vol.48, 2001, pp.846-850. (in Japanese)

31) Ikeno, M, Matsuyama, M., Sakakiyama, T. and Yanagisawa, S: Experimental Study on the Evaluation of Wave Force Acting on Breakwaters due to Tsunamis with Soliton Fission and Breaking Waves, *Journal of Coastal Engineering*, Vol.52, 2005, pp.751-755. (in Japanese)

32) Asakura, R., Iwase K., Ikey, T., Takao, M, Kaneto, T., Fujii, N. and Oomori, M.: Experimental Study on Wave Force Caused by Tsunamis Overtopping Seawalls, *Journal of Coastal Engineering*, Vol.47, 2000, pp.911-915. (in Japanese)

33) Iizuka, H. and Matsutomi, H.: Damage Estimation for Tsunami Inundation Flows, *Journal of Coastal Engineering*, Vol.47, 2000, pp.381-385. (in Japanese)

34) Ikeno, M. and Tanaka, H.: Experimental Study on Land-run Tsunamis and the Impact Force of Drifting Objects, *Journal of Coastal Engineering*, Vol.50, 2003, pp.721-725. (in Japanese)

35) Yasuda, T., Takayama, T. and Yamamoto, H.: Experimental Study on the Deformation and Wave Force Characteristics of Soliton Fission Tsunamis, *Journal of Coastal Engineering*, Vol.53, 2006, pp.256-260. (in Japanese)

36) Isobe, M., et al. 37: *Coastal Technology Library No.39 CADMAS-SURF/3D Research and Development of a Numerical Wave Tank*, Coastal Development Institute of Technology, 2010, 235p. (in Japanese)

37) Yamamoto, Y., Kozono, Y., Mas, E., Murase, F., Nishioka, Y., Okinaga, T. and Takeda, M.: Applicability of Calculation Formulae of Impact Force by Tsunami Driftage, *J. Mar. Sci. Eng.*, 9, 493, 2021. https://doi.org/10.3390/jmse9050493

38) Matsutomi, H.: A Practical Formula for Estimating Impulsive Force due to Driftwood and Variation Features of the Impulsive Force, *Journal of JSCE*, No.621/II-47, 1999, pp.111-127. (in Japanese)

39) FEMA and NOAA: *Guidelines for Design of Structures for Vertical Evacuation from Tsunamis*, Second Edition, 2012, 174p.

40) ASCE: *Standard 7, Chapter 6, Tsunami Loads and Effects*, ASCE, Virginia, WA, USA, 2015.

41) Arikawa, T., Ohtsubo, D., Nakano, F., Shimosako, K. and Ishikawa, N.: Large Model Tests of Drifting Container Impact Force due to Surge Front Tsunami, *Journal of Coastal Engineering*, Vol.54, 2007, pp.846-850. (in Japanese)

42) Ikeno, M., Kihara, N. and Takabatake, D.: Simple and Practical Estimation of Movement Possibility and Collision Force of Debris due to Tsunami, *Journal of JSCE*, B2(Coastal

Engineering), Vol.69, No.2, 2013, pp. I_861-I_865. (in Japanese)

43) Mizutani, N., Usami, A. and Koike, T.: Experimental Study on Behavior of Drifting Boats due to Tsunami and Their Collision Forces, *Journal of Civil Engineering in the Ocean*, Vol.23, 2007, pp.63-68. (in Japanese)

44) Kaida, H. and Kihara, N.: *Evaluation Technologies for the Impact Assessment of Tsunami Debris on Nuclear Power Plants – Review of Current Status and Discussion on the Application -*, Report of Central Research Institute of Electric Power Industry, O16010, 2017, 45p. (in Japanese)

Chapter:6

COASTAL TOPOGRAPHIC CHANGE

6. COASTAL TOPOGRAPHIC CHANGE

In this chapter, the stability conditions of the beach cross-sections are discussed first, which is followed by the evaluation methods of sand movement. Next, the current knowledge on numerical simulation models of beach topography change is introduced (models based on equations for calculating sediment transport rates by only waves, models based on equations for calculating sediment transport by waves and nearshore current, and models using equations of motion for drift sand). Then, the causes and measures of coastal erosion are discussed. Furthermore, scouring in front of coastal dikes and seawalls, suction and outflow of backfill materials from these structures, and wind-blown sand are discussed.

6.1 Beach Profile Change
1) Beach Cross-section Topography

Fig. 6.1 shows a typical beach cross-section topography. The **coastline**, the boundary between land and sea, is set at the water's edge where no more seawater can reach at high tide. In practice, it is not desirable to set a coastline on sandy beaches, whose topography may change daily; thus, the coastline is set at the front of sea cliffs or coastal embankments. Additionally, a waterline at mean low tide is called a **low tide shoreline**, and a waterline at the mean high tide is called a **high tide shoreline**. The area from the coastline to the low tide shoreline is called the **beach**. The area from the low tide shoreline to the tip of the normal wave run-up from the high tide shoreline is called the **foreshore** or **beach face**. The area from the tip of the normal wave run-up at high tide to the coastline is called the **backshore** or **back beach**.

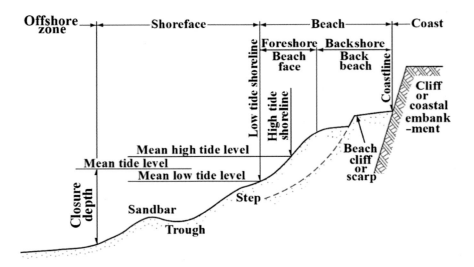

Fig. 6.1 General cross-section of a beach (from the JSCE Hydrodynamic Formulae, Part 5, Chapter 6, "Drifting Sediment and Coastal Processes").

The shoreface is the area with a relatively steep seabed gradient from the low tide shoreline to closure depth (at the position slightly offshore from the wave breaking positions and at the offshore edge of the area where significant wave-induced topographic changes occur). The area outside the shoreface until the continental shelf edge is called the **offshore zone**.

2) Beach Cross-sectional Topographic Change

Between the beach and the shoreface, sand from the foreshore moves to the shoreface during high waves and a sandbar called a **longshore bar** develops (the depression on the landward side of the longshore bar, which is a convex feature on the shoreface, is called a **trough**). The beach with this cross-sectional condition is called a **stormy beach** or **stormy seashore**.

During calm waves, sand from the longshore bar moves and deposits at the foreshore. This deposited terraced topography is called a **step**, further developed and raised topography is called a **berm**, and the beach with this cross-sectional condition is called a **normal beach** or **normal seashore**.

Johnson [1] proposed offshore wave steepness $H_o/L_o \geq 0.025 - 0.030$ as a rough threshold for transforming from a normal beach to a storm beach. However, in addition to wave steepness, sediment grain size and beach slope are also important parameters for this transition. Iwagaki and Noda [2], [3] found the threshold line between stormy and normal beaches shown in **Fig. 6.2**, with the wave steepness of offshore waves on the vertical axis and "offshore wave height / sediment median grain size d_{50}" on the horizontal axis, by using previous studies and their experimental data.

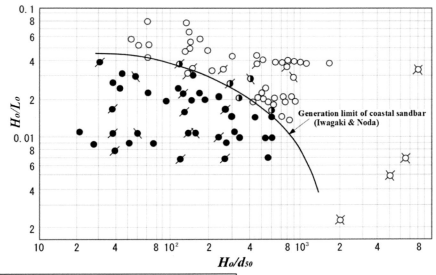

Fig. 6.2 Generation limit of stormy beach according to Iwagaki and Noda [2].

6. COASTAL TOPOGRAPHIC CHANGE

Sunamura and Horikawa [4] classified sandy beach cross-sections into three types as shown in **Fig. 6.3** and Eq. (6.1), with the wave steepness of offshore waves on the vertical axis and a non-dimensional parameter consisting of the beach slope $\tan\beta$ and the representative grain size d on the horizontal axis, by using data from previous studies and their experiments.

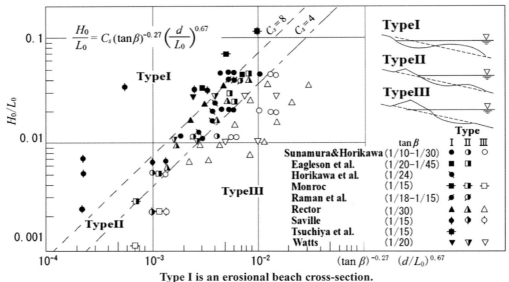

Type I is an erosional beach cross-section.
Type II is a neutral beach cross-section.
Type III is a sedimentary beach cross-section.

Fig. 6.3 Classification of three types of beach cross-sections by Sunamura and Horikawa [4].

$$\left.\begin{array}{l} \text{Type I (erosional beach):} \quad \dfrac{H_o}{L_o} \geq 8 \times (\tan\beta)^{-0.27} \left(\dfrac{d}{L_o}\right)^{0.67} \\[2mm] \text{Type II (neutral beach):} \quad 8 \times (\tan\beta)^{-0.27}\left(\dfrac{d}{L_o}\right)^{0.67} \geq \dfrac{H_o}{L_o} \geq 4 \times (\tan\beta)^{-0.27}\left(\dfrac{d}{L_o}\right)^{0.67} \\[2mm] \text{Type III (sedimentary beach):} \quad \dfrac{H_o}{L_o} \leq 4 \times (\tan\beta)^{-0.27}\left(\dfrac{d}{L_o}\right)^{0.67} \end{array}\right\} \quad (6.1)$$

In **Fig. 6.2**, the reason for the threshold line could be determined without using the coastal slope as a parameter on the horizontal axis is because the coastal slope is a function of wave parameters and sediment grain size, as shown e.g. in Eq. (6.3), when soil adhesion can be ignored.

The dimensionless parameters on the horizontal axis of **Fig. 6.3** did not fully satisfy the similarity law for sand grain size [although the definition of beach slope may differ between research papers as noted in Section 6.1, 3)], Therefore, the coefficients '8' and '4' in **Fig. 6.3** and Eq. (6.1) had to be changed to '18' and '9' respectively. As can be understood from this, in addition to the similarity law for external forces (the Froude number), it is also necessary to satisfy the similarity law for bottom grain size in the study of topographic change phenomena.

In addition to wave parameters and water depth, indicators that have a significant influence on the movement of bottom sediments include the following:

(a) Specific gravity (γ_s) or density (ρ_s) of sediment:

The higher the specific gravity of the sediment, the less likely it is to move. Since an actual sediment layer consists of sediment particles and air/water-filled voids, the dry density should be used if the effect of weight is to be investigated for the water-saturated sediment layer. In order to make the sediment layer less mobile, it can be compacted to increase the **dry density**, other than replacing them with sediments of a higher specific gravity. Here, to expect a sufficient compaction effect, it is necessary to increase the uniformity coefficient.

(b) Median grain size (d_{50}):

The external force to move a sediment is proportional to the cross-sectional area, i.e. the square of the sediment grain size, while the self-weight as a resisting force is proportional to the cube of the sediment grain size. Therefore, the larger the median grain size, the less likely it is forced to move.

(c) Uniformity coefficient (U):

The ratio of the grain size of the 10% mass passage percentage (d_{10}) to the 60% mass passage percentage (d_{60}) in the grain size distribution curve is called the uniformity coefficient. The closer the value of the ratio d_{60}/d_{10} is to 1, the more uniform in size the sand is. The larger the uniformity coefficient, the wider the range of grain sizes, the better the interlocking, and the relatively less mobile the bottom sediment.

(d) Surface roughness:

The rougher the surface of the individual sediment particles, the lesser the slippage between particles, and therefore the sediment is less mobile.

(e) Soil cohesion:

If the sediment layer is composed of fine particles, the adhesive force, which is the resistance to sediment migration, is likely to increase.

---

6. COASTAL TOPOGRAPHIC CHANGE

【Shields' law of similarity】

When conducting movable-bed experiments, Froude's law and Reynolds' law cannot be satisfied simultaneously if the gravitational acceleration and the viscosity coefficient cannot be changed. Therefore, the scale ratio between the laboratory and prototype on model sizes and external forces is determined by Froude's law, while the scale ratio between the laboratory and prototype on the sediment grain size and external force is often determined by a similarity law based on the Shields number (ratio of the shear force acting on the sediment particles to their weight in water), as a dimensionless indicator governing sediment transport.

In the field and in laboratory experiments, the density of water ρ and the density of sediment particles ρ_s will stay the same, but the coefficient of friction f' is a function of sediment particle size d, etc. The Shields number θ_s is expressed as follows.

$$\theta_s = \frac{\dfrac{f'}{2}\rho u^2 d^2}{(\rho_s - \rho)gd^3} = \frac{\dfrac{f'}{2}\rho u^2}{(\rho_s - \rho)gd} = \text{dimensionless constant} \times \frac{f'u^2}{gd} \tag{h.1}$$

where u is the flow velocity and g is the acceleration of gravity.

Assuming that the prototype Shields number (subscript p) is the same as that in laboratory experiment (subscript m), the gravitational acceleration is the same in the field and in the experiment, and that $f_m'/f_p' \fallingdotseq (L_p/L_m)^{1/2}$ from the existing experimental data on the friction coefficient (L is the dimension of length), and as according to Froude's law, $T_m/T_p = \sqrt{L_m/L_p}$, the following Eq. (h.2) is obtained.

$$\frac{d_m}{d_p} = \frac{u_m^2 f_m'}{u_p^2 f_p'} = \left(\frac{L_m^2 T_p^2}{T_m^2 L_p^2}\right)\left(\frac{L_p}{L_m}\right)^{\!1/2} = \left(\frac{L_m^2 L_p}{L_p^2 L_m}\right)\left(\frac{L_p}{L_m}\right)^{\!1/2} = \left(\frac{L_m}{L_p}\right)^{\!1/2} \tag{h.2}$$

It can be seen from Eq. (h.2) that if the ratio between the prototype friction coefficient and the experimental friction coefficient is $f_m'/f_p' \fallingdotseq (L_p/L_m)^{1/2}$, the model grain size will be the power of one-half of that of the prototype. However, the formulation of the part corresponding to the above friction coefficient ratio is not simple, as can be inferred from the fact that bottom sediment movement in a wave field cannot be fully explained by shear force alone.

For laboratory experiments on topography change due to waves, Ito (1986) in his PhD dissertation proposed a method to determine a scale ratio between the physical model and prototype by using Froude's law and the sediment grain size through an evaluation chart made from a comparison between large-scale and small-scale experiments.

Ito, Masahiro: Study on the characteristics of two-dimensional beach deformation and its similarity law, PhD Dissertation, Kyoto University, 1986, 172p. https://repository.kulib. kyoto-u.ac.jp/dspace/handle/2433/74694

6.1 Beach Profile Change

3) Methods for Setting Beach Stability Cross-sections

It is very costly and time-consuming to carry out large-scale experiments to determine the required functional beach cross-sections (beach nourishment cross-sections) from hydraulic model experiments to avoid flaws due to not satisfying similarity requirements. To reduce this burden, numerical models that can predict beach deformation in the wave run-up area have emerged, but these models can only be handled by experts with sufficient knowledge and experience. Therefore, it is also important to introduce accurate calculation formulae that can be used by ordinary engineers.

However, the old empirical equations for determining stable beach cross-sections (beach slopes and sediment grain sizes) in the on-off-shore direction under wave action are regarded by many engineers as inaccurate. Yamamoto and Iwasaki [5] therefore investigated the accuracy of the main empirical equations using existing field observation data and nationwide coastal survey data.

Since "the improved virtual slope (tan α)" defined by using **Fig. 3.8** and Eq. (3.17) captures the characteristics of the beach cross-section well, the accuracy of the calculated wave run-up height by this definition of virtual slope is adequate for a beach steepened by erosion. Therefore, the improved virtual slope is an important indicator when considering the effect of reducing wave overtopping on sandy beaches. **Fig. 6.4** shows the correlation between the measured values as the improved virtual slopes obtained from existing field observation data and the calculated values using main empirical equations, and **Fig. 6.5** shows the correlation between the measured values as the improved virtual slopes based on field data collected under calm wave conditions and large wave conditions at main coasts in Japan and the calculated values using main empirical equations.

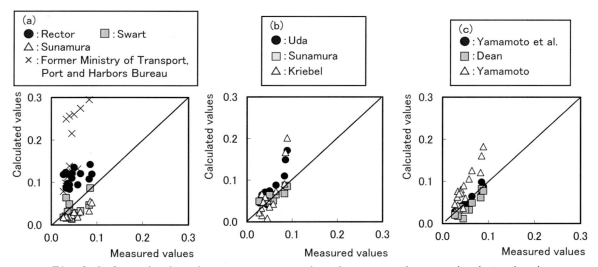

Fig. 6.4 Correlation between measured values as "improved virtual slopes (tan α)" based on existing field observations and calculated values using main empirical equations for the beach slope.

6. COASTAL TOPOGRAPHIC CHANGE

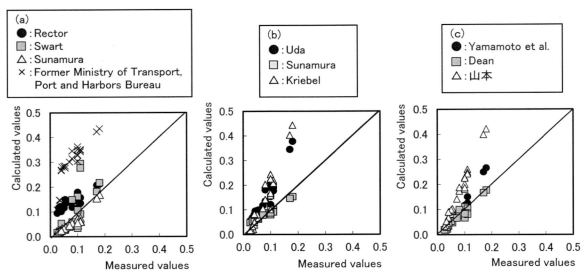

Fig. 6.5 Correlation between measured values as "improved virtual slopes (tanα)" based on coastal surveys in Japan and calculated values using main empirical equations for the beach slope.

In these figures, the equation of Rector [6] is based on small-scale experiments and therefore overestimates. The equations of Swart [7], [8] are based on large-scale experiments, so the calculated values are comparable to the measured values, but the correlation is not good due to the small number of experimental data used to propose the equations. The equation of Sunamura [9] underestimates the improved virtual slopes because it was proposed using seabed slopes between the coastline and 20 m water depth. The equation of the Port and Harbour Bureau of the former Ministry of Transport of Japan [10] overestimates them because it was designed to give upper limits. The equation of Uda [11] was proposed using the relatively steep foreshore slopes rather than the improved virtual slopes, so the calculation results is about 1.5 times the measured values. The equation of Sunamura [12] is an improved version of Sunamura's equation described above and therefore shows a relatively good correlation. The equations of Kriebel et al. [13] and Dean [15] focus on the cross section below the water surface, but the mean slopes of the former correspond to the foreshore slopes, so the calculated results are about 1.5 times the measured values, while the mean slopes of the latter show a good correlation. The equation of Yamamoto et al. [14] was proposed using the improved virtual slopes and therefore shows a relatively good correlation. The equation of Yamamoto [16] was proposed using the foreshore slopes, so the calculated values are about 1.5 times the measured values.

In the remaining equations excluding Kriebel et al., Dean, and Yamamoto's equations, the median grain size of the bottom sediment is explicitly included. So, "the median grain size (d_{50})" of the bottom sediment can be obtained by giving the beach slope. The equations of Yamamoto et al, Sunamura, and Uda, which have relatively good accuracy, were used to determine the median grain

6.1 Beach Profile Change

size of the bottom sediment, and the correlation with the measured values is shown in **Fig. 6.6**. The equation of Yamamoto et al. [14] has the best correlation, followed by the equation of Sunamura [12].

The above can be summarised as follows:

(1) To obtain the beach slope (= improved virtual slope) for the overtopping reduction study, the equations of Dean [15], Yamamoto et al [14], and Sunamura [12] can be used. To obtain the foreshore slope, the equations of Yamamoto [16], Kriebel et al [13], and Uda [11] can be used.

(2) The equation of Yamamoto et al [14] and Sunamura [12] can be used for the median grain size of the bottom sediment.

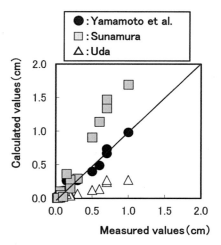

Fig. 6.6
Correlation between measured values from coastal surveys and calculated values based on empirical equations for "sediment grain size (d_{50})".

In addition to the beach slope (or foreshore slope) and the median grain size, a minimum beach width is required for the design of the beach cross-section, which should be determined from the calculation of the wave run-up height or wave overtopping rate to obtain the required wave overtopping reduction effect or from the constraints on beach use.

Specific equations for calculation methods with good correlation with measured values (Uda [11], Sunamura [12], Kriebel et al. [13], Yamamoto et al. [14], Dean [15], and Yamamoto [16]) are given below.

(a) Equation of Uda [11]

Uda proposed Eq. (6.2) for calculating the foreshore slope $\tan\alpha_f$ using field observation data.

$$\tan\alpha_f = 4.5\left(\frac{d_{50}}{H_o}\right)^{0.5} \tag{6.2}$$

where d_{50} is the median grain size, and H_o is the offshore wave height.

(b) Equation of Sunamura [12]

Sunamura proposed Eq. (6.3) for calculating the beach slope $\tan\alpha$ between the wave breaking depth and the wave run-up height, using previous field observation data.

$$\tan\alpha = 0.25\left(\frac{d_{50}}{H_o}\right)^{0.25}\left(\frac{H_o}{L_o}\right)^{-0.15} \tag{6.3}$$

where L_o is the offshore wavelength.

6. COASTAL TOPOGRAPHIC CHANGE

(c) Equation of Yamamoto et al [14]

Based on Eq. (6.4) proposed by Sunamura [17] for the on-offshore sediment transport rate per unit width and unit time q_y [m³/s/m], Yamamoto et al. proposed Eq. (6.5) for the improved virtual slope $\tan\alpha$ and median grain size d_{50}.

$$
\left.
\begin{aligned}
&\frac{q_y}{w_s d_{50}} = K U_r^{0.2} \phi \left(\phi - 0.13 U_r \right) \\[2mm]
&U_r = g H_s \frac{T^2}{\eta_s^2}, \quad \phi = \frac{H_s^2}{\left(\rho_s / \rho - 1 \right) \eta_s d_{50}} \\[2mm]
&H_s = 2.4 \left(\tan\alpha \right)^{0.3} \eta_s, \quad \eta_s = \left(1.63 \tan\alpha + 0.048 \right) H_b, \quad H_b = \left(\tan\beta \right)^{0.2} \left(\frac{H_o}{L_o} \right)^{-0.25} H_o
\end{aligned}
\right\}
\tag{6.4}
$$

where w_s is the settling velocity of the bottom sediment, d_{50} is the median grain size of the bottom sediment, K is the proportionality coefficient, U_r is the **Ursell number**, g is the acceleration of gravity, H_s is the wave height on the shoreline, η_s is the mean water surface elevation above the shoreline, T is the period of the action wave, ϕ is **Hallermeier's non-dimensional number**, ρ_s is the density of the bottom sediment, ρ is the density of the seawater, $\tan\beta$ is the initial sea bottom slope, H_o and L_o are respectively the offshore wave height and wave length.

$$
\left.
\begin{aligned}
&\tan\alpha = \left(0.0864 \times \frac{\rho_s - \rho}{\rho} \times g d_{50} \frac{T^2}{H_b^2} \right)^{2/3} \\[2mm]
&d_{50} = \frac{H_s^2}{0.13 U_r \times \left(\rho_s / \rho - 1 \right) \eta_s}, \quad \eta_s = \frac{H_s}{2.4 \left(\tan\alpha \right)^{0.3}}, \quad H_s = 1.9 \left(\tan\alpha \right)^{0.9} H_b
\end{aligned}
\right\}
\tag{6.5}
$$

where H_b is the wave breaking height.

(d) Equations of Kriebel et al [13] and Dean [15]

The distribution of the water depth h from the shore to the offshore is approximated by the **Bruun rule** shown in Eq. (6.6), and for the shape factor A_e of the equilibrium beach cross-section, Kriebel et al. proposed Eq. (6.7) and Dean proposed Eq. (6.8).

$$
h = A_e \times y^{2/3}
\tag{6.6}
$$

$$
A_e = 2.25 \left(\frac{w_s^2}{g} \right)^{1/3} \quad \left[\text{The unit of } w_s \text{ is "m/s", the unit of } g \text{ is "m/s}^2\text{"} \right]
\tag{6.7}
$$

$$
A_e = 0.067 w_s^{0.44} \quad \left[\text{The unit of } w_s \text{ is "cm/s", the unit is } A_e \text{ is "m}^{1/3}\text{"} \right]
\tag{6.8}
$$

where y is the offshore distance, w_s is the settling velocity of the bottom sediment, g is the gravitational acceleration.

The settling velocity w_s is obtained by substituting the median grain size d_{50} into Eq. (6.9) and so on.

$$w_s = \sqrt{1.333\left(\frac{\rho_s}{\rho} - 1\right)\frac{g d_{50}}{C_D}} \tag{6.9}$$

where C_D is the drag coefficient in turbulent conditions.

From Eqs. (6.6) to (6.8) above, the mean beach slope $\tan\alpha$ is defined as in Eq. (6.10), using the critical depth of sediment movement h_c and the offshore distance W, which is defined by using h_c (the critical depth) and W_r (the offshore distance to this critical depth), shown in Eq. (6.11).

$$\tan\alpha = \frac{h_c}{W} = \frac{5}{3}\frac{A_e^{3/2}}{h_c^{1/2}} \tag{6.10}$$

$$\left.\begin{array}{l} h_c = A_e W_r^{\frac{2}{3}} \\[2mm] W = \dfrac{\int_0^{W_r} A_e y^{\frac{2}{3}}dy}{h_c} = \dfrac{\frac{3}{5}A_e W_r^{\frac{5}{3}}}{h_c} = \dfrac{3}{5}W_r = \dfrac{3}{5}\left(\dfrac{h_c}{A_e}\right)^{\frac{3}{2}} \end{array}\right\} \tag{6.11}$$

(e) Equation of Yamamoto [16]

Yamamoto proposed Eq. (6.12) for calculating the foreshore slope $\tan\alpha_f$ using data from small to large scale hydraulic model experiments.

$$\left.\begin{array}{l} \tan\alpha_f = \sqrt{\dfrac{w_s}{\sqrt{gH_b}}\dfrac{\tan\beta}{\sqrt{H_b/L_o}}} \\[3mm] H_b = \left(\tan\beta\right)^{0.2}\left(\dfrac{H_o}{L_o}\right)^{-0.25}H_o \end{array}\right\} \tag{6.12}$$

where w_s is the settling velocity of the bottom sediment [m/s], H_b is the wave breaking height of the incident wave [m], and $\tan\beta$ is the initial seabed slope.

219

6. COASTAL TOPOGRAPHIC CHANGE

6.2 Drifting Sand

1) Patterns of Sediment Transport

Sediment produced in mountain areas is transported to the sea through river channels. Sediment flowing through river channels is called **flowing sand**, while sediment drifting by waves in the sea is called **drifting sand**.

The modes of **sediment transport** from the river channels to the coastal zone can generally be classified as follows.

(1) Bed Load

Bed Load is the sediment that moves near the bottom of a river or a sea by **sliding**, **rolling**, and **saltation**, and is an important material for the geometry change of the riverbed or seabed. The grain size of the sediment that can be transported as the bed load varies with the strength of the current, ranging from sand to silt when the current is weak, and from stone to gravel when the current is strong.

(2) Suspended Load

Suspended Load is the sediment that is carried in suspension in the water away from the riverbed or seabed, and contributes to sedimentation by settling on the riverbed or seabed when the flow is weak and contributes to erosion by floating again when the flow is strong. The grain size of the sediment that moves as a suspended load varies with the strength of the flow, from silt when the flow is weak to gravel or sand when the flow is strong.

(3) Wash Load

Wash Load is the fine sediment, usually consisting of clay or silt, that flows through water without ever being replaced by sediment from the riverbed or seabed during its movement along the river or coastal zone. It does not contribute to the geomorphic change of the target river or coastal area and is therefore ignored in geomorphic change predictions.

2) Critical Depth for Sediment Movement

As the water depth becomes shallower from the offshore zone to the foreshore, the bottom flow velocity becomes relatively higher, and the bottom shear force increases, so that the sediment begins to move. Water depth at this limit is called the **limit depth for sediment movement** or the **critical depth for sediment movement** and ranges from a few metres to within 100 metres.

Studies on the critical depth for sediment movement began with Sato and Kishi [18] and have been carried out by Ishihara and Sawaragi [19], Sato and Tanaka [20], Sato [21], Horikawa and Watanabe [22], etc. Four classifications can be made by substituting data from hydraulic model experiments and field observations into the following equation, which is obtained from the balance between the frictional force acting on sediment particles and resistance force by their weight.

$$\sinh\left(\frac{2\pi h_c}{L}\right) = A \left(\frac{L_o}{d_{50}}\right)^B \left(\frac{H_o}{L_o}\right)\left(\frac{H}{H_o}\right) \quad (6.13)$$

where h_c is the critical depth for sediment movement, H and L are respectively the wave height and wavelength at the critical depth, H_o and L_o are the wave height and wavelength of offshore waves, d_{50} is the median grain size of bottom sediment, and A and B are respectively the coefficient and exponent summarised in **Table 6.1**.

The critical depth for sediment movement can be classified into the following four categories according to the movement state of the bottom sediment:

(a) **Initial movement limit depth**: The depth at which some of the bottom sediment begins to move.

(b) **General movement limit depth**: The limit depth at which most of the surface particles of the bottom sediment move.

(c) **Surface movement limit depth**: The limit depth at which the surface particles of bottom sediment move collectively in the direction of waves.

(d) **Complete movement limit depth**: The limit depth at which bathymetric changes are clearly visible and significant sand movement occurs.

Table 6.1 Coefficients and exponents for each movement state in Eq. (6.13)

Threshold for sediment movement	Proponents	Coefficient A	Index B
Initial movement limit	Ishihara & Sawaragi [19]	5.85	1/4
General movement limit	Sato & Tanaka [20]	1.77	1/3
Surface layer movement limit	Sato & Tanaka [20]	0.741	1/3
Complete movement limit	Sato [21]	0.417	1/3

3) Equations for Calculating Bed Load Transport Rates

In **Fig. 6.7**, if the fluid force, which is the summation of the drag force on the front of the sediment particle and the shear force on the top of the sediment particle, is greater than the frictional resistance force due to the dead weight of the sediment particle, the particle will move on the seabed. The fluid force F that moves the sediment particle can be expressed as

$$F = \frac{C_D + f'}{2} \rho u_b^2 \times \frac{\pi d^2}{4} \quad (6.14)$$

where C_D is the drag coefficient, f' is the coefficient of friction between the water and the sediment particle, ρ is the water density, u_b is the mean velocity near the bottom and d is the diameter of the sediment particle.

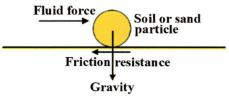

Fig. 6.7 Force acting on sediment particles.

6. COASTAL TOPOGRAPHIC CHANGE

On the other hand, the weight W of the sediment particle in water, which is the resistance to movement, can be expressed using the density of the sediment particle ρ_s and the acceleration of gravity g as follows, assuming the particle is spherical.

$$W = (\rho_s - \rho) g \frac{\pi d^3}{6} \tag{6.15}$$

The ratio of F (the fluid force that moves the sediment particle) to W (the weight of bottom sediment in water) is an important indicator of the degree of sediment particle movement. If we consider the movement of sediment particles on the surface of a sand layer on the seabed, the main fluid force is the shear force acting on the sediment particle area proportional to d^2. Therefore, the ratio of this to the weight of the sediment particle volume proportional to d^3 in water is considered the important indicator of sediment movement and is called **Shields number**.

Shields number is expressed as follows:

$$\theta_s = \frac{\tau d^2}{(\rho_s - \rho) g d^3} = \frac{\frac{f'}{2} \rho u_b^2 d^2}{(\rho_s - \rho) g d^3} = \frac{\frac{\tau}{\rho}}{\left(\frac{\rho_s}{\rho} - 1\right) g d} = \frac{u_*^2}{\left(\frac{\rho_s}{\rho} - 1\right) g d} \tag{6.16}$$

where τ is the shear stress, u_* is the friction velocity.

The power (energy per unit time) required to move sediment as the bed load can be expressed as "shear force × velocity of movement". Therefore, the formula for calculating the bedload transport rate is proportional to "$\tau \times u_b$", or a function of this. From Eq. (6.16), it can be said that Shields number is proportional to "τ" or "u_b^2", so the bed load transport rate per unit time and unit width of the seabed, q_b [m³/s/m], can be considered as proportional to the 1.5 power of Shields number.

There are many examples (e.g. Meyer - Peter and Müller [23], Ashida and Michiue [24]) that use Eq. (6.17) as the basic formula for calculating the bed load transport rate q_b [m³/s/m] in one directional flow.

$$\frac{q_b}{\sqrt{(\rho_s/\rho - 1) g d^3}} = \begin{cases} C_1 [\theta_s - \theta_c]^{1.5} & (\theta_s \geq \theta_c) \\ 0 & (\theta_s < \theta_c) \end{cases} \tag{6.17}$$

where C_1 is a coefficient related to sediment properties, which can be obtained from experiments; θ_c is a physical quantity corresponding to the frictional resistance force, which also can be obtained from experiments. This is called the **critical Shields number**.

In the case of the bed load transport rate q_b [m³/s/m] in oscillatory currents in coastal areas, it is necessary to devise a way to reverse the direction of the drifted sediment if the flow direction is reversed. Hence, examples include Eq. (6.18) of Watanabe et al [25], which uses shear force and velocity instead of Shields number in Eq. (6.16), and Eqs. (6.19) and (6.20) of Ribberink [26], which utilise the absolute value of Shields number instead of Eqs. (6.16) and (6.17).

222

6.2 Drifting Sand

$$q_{bc} = A_c \frac{(\tau - \tau_{cr})u_{bc}}{\rho_w g}, \qquad q_{bw} = A_w \frac{(\tau - \tau_{cr})\hat{u}_{bw}}{\rho_w g} F_D$$

$$F_D = \tanh\left(k_d \frac{\Pi_c - \Pi}{\Pi_c}\right), \qquad \Pi = \frac{u_b^{\,2}}{(\rho_s/\rho_w - 1)gd}\frac{h}{L_o} \tag{6.18}$$

where q_{bc} and q_{bw} are the bed load transport rates by the flow and waves respectively, A_c and A_w are dimensionless coefficients ($= 0.1$ to 1) for the flow and waves respectively, τ is the maximum frictional stress on the bottom under the coexistence of flow and waves, τ_{cr} is the movement-limiting frictional stress on the bottom material, u_{bc} is the near-bottom velocity of the flow, and $\hat{u}_{bw}$ is the amplitude of the bottom orbit velocity of the waves. It is straightforward to determine the drifting direction of sand by the waves from the residual amount of drifting sand during one cycle although this process is complicated. Therefore, the amplitude of the orbital velocity is used to determine the drifting sand rate and the direction discrimination function F_D given in equation (6.18) is used to determine the direction of drifting sand. In that, if Π is smaller than Πc then, the direction is shoreward, if larger, the direction is offshore. k_d is a correction factor, usually taken as 1. Π_c is the value of Π when the net amount of drifting sand is zero, h is the water depth below the still water surface and L_o is the wavelength of the offshore waves.

$$\frac{q_b}{\sqrt{(\rho_s/\rho - 1)gd_{50}^{\,3}}} = \begin{cases} C_2 \left[\left|\theta_s'(t)\right| - \theta_c\right]^{1.65} \dfrac{\theta_s'(t)}{\left|\theta_s'(t)\right|} & \left(\left|\theta_s'(t)\right| \ge \theta_c\right) \\[2mm] 0 & \left(\left|\theta_s'(t)\right| < \theta_c\right) \end{cases} \tag{6.19}$$

$$\theta_s'(t) = \frac{\dfrac{f_w'}{2}\rho|u_b(t)|u_b(t)}{(\rho_s - \rho)gd_{50}} \tag{6.20}$$

where d_{50} is the median grain size of bed load; C_2 is the coefficient of bed load transport rate ($C_2 \fallingdotseq 11$ in the case of sand), which is determined from verification calculations in each coastal area; $\theta_s'(t)$ is the Shields number in oscillatory flow defined by Eq. (6.20); θ_c is the critical Shields number in oscillatory flow which can be calculated from the equation of van Rijn [27]; f_w' is the friction coefficient in the oscillatory flow field; and $u_b(t)$ is the oscillatory flow velocity over the seabed boundary layer (the layer from the seabed surface where the flow velocity is distinctly reduced due to frictional resistance by the seabed surface).

The equation of Watanabe et al [25] gives good a correlation with measured bed load transport rates, although it is difficult to set the coefficients. On the other hand, the equation of Ribberink [26] is easy to use because there is no need to change the coefficient $C_2 \approx 11$ for sand and no other ambiguous coefficients to set.

In addition to the above, various other equations are available for calculating the bed load transport rate. Readers can refer in "Nearshore Dynamics and Coastal Process" of Horikawa [28] or "Collection of Hydraulic Formulae" of the Editorial Sub-Committee on Hydraulic Formulae [29].

6. COASTAL TOPOGRAPHIC CHANGE

【Physical and transport properties of sediment】

① If the effect of external forces and sediment grain size is correctly considered in equations for calculating bed load transport rates, there is no need to change the bed load transport rate coefficient for different sediment grain sizes. Unfortunately, however, there are no such equations available, and the bed load transport rate coefficient needs to be adjusted if the grain size of the bottom sediment changes significantly.

② Because the interlocking effect increases as the uniformity coefficient of the bottom sediment increases, the bed load transport rate in unidirectional flow is reduced to some extent, but equations that can take this effect into account are not available. On the other hand, the effect of larger uniformity coefficients is thought to gradually weaken as classification of the bottom sediment progresses in oscillatory flows.

③ If well compacted and the dry density is greater, compared to the loose density case, the initial outflow rate is less severe, but outflow will gradually become similar to the loose density case.

④ Sassa and Watabe (2007) noted that on dried-out sandy flats, the suction pressure of porewater, which is repeatedly developed and dissipated by the tides, tightens the pore spaces within bottom sediment and increases its strength (see paper [a]).
Shirozu et al. (2018) introduced a correction factor for the specific gravity of bottom sediment in water in the equation for calculating the bed load transport rate, and increased this correction factor in a timely manner so that the suppression effect of suction pressure on the bed load transport rate could be considered, thereby improving the reproducibility of the coastal topographical change prediction model (see paper [b]).

[a] Sassa, S. and Watabe, Y.: Role of Suction Dynamics in Evolution of Intertidal Sandy Flats: Field Evidence, Experiments, and Theoretical Model, *Journal of Geophysical Research*, Vol.112, F01003, 2007. doi: 10.1029/2006JF000575.
[b] Shirozu, H., Yamamoto, K. and Asai, K.: Effect of Suction Dynamics on Maintenance of Micro Topography in Minamigata, Yamaguchi Bay, *Journal of JSCE B2* (Coastal Engineering), Vol.74, No.2, 2018, pp. I_733-I_738. doi.org/10.2208/kaigan.74.I_733.

4) Equations for Calculating Suspended Load Transport

Transport of suspended load is also important in underwater drifting sand transport, and the transport of suspended load due to turbulence for two-dimensional depth-integrated case is described by the following advection-diffusion equation.

$$\frac{\partial C}{\partial t}+\frac{\partial \tilde{u} C}{\partial x}+\frac{\partial \tilde{u} C}{\partial y}=\frac{\partial}{\partial x}\left(\nu_t \frac{\partial C}{\partial x}\right)+\frac{\partial}{\partial y}\left(\nu_t \frac{\partial C}{\partial y}\right)-C_s+C_{ut} \tag{6.21}$$

where C is the depth-integrated suspended load concentration, $\tilde{u}$ is the depth-averaged velocity (averaged over one cycle for oscillatory flow), ν_t is the eddy kinematic viscosity coefficient, C_s is the amount of suspended load settling to the bottom, and C_{ut} is the amount of suspended load being rolled up from the bottom.

The settling rate C_s and the roll-up rate C_{ut} are evaluated by assuming a vertical distribution of suspended load. If the vertical distribution of suspended load concentration is $C(z)$, C_s and C_{ut} are determined by Eq. (6.22).

$$C_{ut}=-\nu_t \left.\frac{\partial C(z)}{\partial z}\right|_{z=z_a}, \quad C_s=w_s C\left(\frac{w_s}{2}\right) \tag{6.22}$$

where z is the vertical coordinate with the origin at the bottom and w_s is the settling velocity of the sediment particles.

As the concentration of suspended sand in unidirectional flow decreases from the bottom upwards, a reduction function such as the Rouse equation given in Eq. (6.23) can be used to determine vertical distribution.

$$\frac{C(z)}{C_a}=\left(\frac{h-z}{z}\frac{a}{h-a}\right)^b \tag{6.23}$$

where C_a is the suspended load concentration at position a; h is the water depth; a is $0.05\times h$ from the bottom; b is called the Rouse number, and is expressed using Kalman's constant κ (= 0.4) and the friction velocity u_* as follows.

$$b=\frac{w_s}{\kappa u_*} \tag{6.24}$$

For the vertical distribution of suspended load concentration in oscillatory currents in coastal areas, Eq. (6.25), or Eq. (6.26) proposed by Soulsby [30] can be used.

$$\frac{C(z)}{C_o}=e^{-C_1 z} \tag{6.25}$$

where C_o is the reference concentration at the bottom and C_1 is the decay parameter.

$$\frac{C(z)}{C_a}=\left(\frac{z_a}{z}\right)^b \tag{6.26}$$

where b is the Rouse number, and for C_a and z_a, the following equation can be used from the proposal of Zyserman and Fredsøe [31].

6. COASTAL TOPOGRAPHIC CHANGE

$$C_a = \frac{0.331(\theta_s - 0.045)^{1.75}}{1 + 0.720(\theta_s - 0.045)^{1.75}}, \quad z_a = 2d_{50} \quad (6.27)$$

For the settling velocity w_s in the oscillatory flow, the following equation from van Rijn [32] can be used.

$$w_s = \begin{cases} \dfrac{\nu D_*^3}{18 d_{50}} & (D_*^3 \leq 16.187) \\[6pt] \dfrac{10\nu}{d_{50}}\left[\left(1 + 0.01 D_*^3\right)^{1/2} - 1\right] & (16.187 < D_*^3 \leq 16187) \\[6pt] \dfrac{1.1\nu D_*^{1.5}}{d_{50}} & (D_*^3 > 16187) \end{cases} \quad (6.28)$$

where ν is the kinematic viscosity and D_* is a parameter defined by the following equation.

$$D_* = d_{50}\left(\frac{g}{\nu^2}\frac{\rho_s - \rho}{\rho}\right)^{1/3} \quad (6.29)$$

5) Continuity Equation of Drifting Sand

In the small area represented by $\Delta x \times \Delta y$ in **Fig. 6.8** on the seabed surface, if a bed load transport rate q_{bx} enters from the left side in the x direction and the same rate q_{bx} flows out from the right side, no change in ground height is caused by these bed load transport rates even if the bed load transport rates are large.

However, if the increased bed load transport rate $q_{bx} + \Delta q_{bx}$ flows out from the right side, the erosion equivalent to this increase Δq_{bx} occurs within the small area. The total discharge per time Δt through width Δy is $\Delta q_{bx} \times \Delta y \times \Delta t$. If the bed load transport rate q_{by} in the y direction is considered in the same way, the total discharge per time Δt through width Δx is $\Delta q_{by} \times \Delta x \times \Delta t$. On the other hand, if the reduction in ground level per time Δt is $-\Delta \zeta$, the total erosion in the small area due to the total discharge per time Δt in both directions is $-\Delta \zeta \times \Delta x \times \Delta y$, giving the relationship shown in Eq. (6.30).

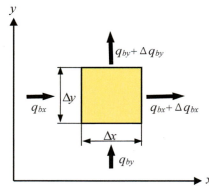

Fig. 6.8 Bed load transport rates in and out of the small area.

$$\Delta q_{bx} \Delta y \Delta t + \Delta q_{by} \Delta x \Delta t = -\Delta \zeta \Delta x \Delta y \quad (6.30)$$

From this equation, Eq. (6.31), which is called the **continuity equation of drifting sand** or the **conservation law of drifting sand**, can be derived.

$$\frac{\partial \zeta}{\partial t} = -\left(\frac{\partial q_{bx}}{\partial x} + \frac{\partial q_{by}}{\partial y}\right) \quad (6.31)$$

In addition, since most of the previously introduced equations for calculating bed load transport rates are treated in terms of the net volume of sand particles, the porosity ε_s in the sand layer must be considered when using the continuity equation of drifting sand for the calculation of topographic change. Therefore, in this case, Eq. (6.32) must be used instead of Eq. (6.31).

$$\frac{\partial \zeta}{\partial t} = -\frac{1}{1-\varepsilon_s}\left(\frac{\partial q_{bx}}{\partial x} + \frac{\partial q_{by}}{\partial y}\right) \qquad (6.32)$$

Furthermore, if we consider the amount of suspended load C_s settling onto the bottom and the amount of suspended load C_{ut} being rolled up from the bottom, these are also treated in terms of the net volume of sand particles, so that the continuity equation of drifting sand is as follows.

$$\frac{\partial \zeta}{\partial t} = -\frac{1}{1-\varepsilon_s}\left(\frac{\partial q_{bx}}{\partial x} + \frac{\partial q_{by}}{\partial y} - C_s + C_{ut}\right) \qquad (6.33)$$

6. COASTAL TOPOGRAPHIC CHANGE

6.3 Coastal Topographic Change Prediction Methods

Numerical models for predicting coastal topographical changes can be classified as follows.

(a) Line Models and Beach Profile Models Based on Wave Parameters

In 1956, Pelnard-Considére [33] proposed a model that predicts shoreline change using the continuity equation of drifting sand. The model calculates the change in shoreline position from the longshore distribution of the amount of sediment moving in the longshore direction (hereafter referred to as the **longshore sediment transport rate**). The longshore transport rate is calculated based on breaking parameters. This model is called the **shoreline change model** or the **one-line model**. Since then, many researchers have improved the shoreline change model, most notably the numerical model 'GENESIS' by Hansen and Kraus [34].

Since Bakker [35] proposed a two-line model (two-isobaths) in 1968, several **multiline models**, which predict changes of multi-isobaths by applying the continuity equation of drifting sand and the on-offshore distribution of longshore sediment transport to each isobath, have been developed. Representative examples include models of Perlena and Deen [36] and Uda and Kawano [37].

Moreover, many models that predict beach profile change using the continuity equation of on-offshore sediment and an equation that directly calculates the amount of sediment moving in the on-offshore direction (hereafter referred to as **on-offshore sediment transport rate**) from wave parameters have also been proposed from around 1980 to 2000. Representative examples are the model of Yamamoto et al. which can be easily incorporated into the shoreline change model and the **SBEACH model** which can predict topographic change in horizontal two-dimensions (x and y coordinates). are introduced in this section.

In 2006, Serizawa et al [38], [39] proposed a numerical model to predict the bathymetry changes on the horizontal x and y coordinates using the continuity equation of drifting sand and an equation that can calculate the sediment transport rates in any direction from wave breaking parameters.

Because the four types of numerical models presented above consider only the sediment transport rates due to waves, they are very computationally efficient, and the coastal topographical change prediction of 10 square kilometres for 10 years can be performed using a typical PC in 2020 in a few hours to a few days (the difference of calculation time depends on the wave field calculation method, and the influence of nearshore currents has been disregarded).

(b) Bathymetric Change Models Based on Waves and Nearshore Currents

Since Wang et al [40] in 1975, many researchers have also developed numerical models that can predict coastal topographical changes by more accurately incorporating the sediment transport mechanism, taking into account not only the sediment transport rates caused by waves but also by

6.3 Coastal Topographic Change Prediction Methods

nearshore currents. These are referred to here as **bathymetric change models based on waves and nearshore currents**.

Full-scale horizontal 2D models of this type appeared around 1980. Due to the calculation of the wave and nearshore current fields and the irregularity of the waves, they are computationally demanding and take several days to calculate the topographic changes during one day in a few square kilometres of the coastal area using a standard PC in 2020.

Numerical models that simultaneously calculate the wave and nearshore current fields using the Navier-Stokes' three-dimensional equations of motion began to appear around 2000. The computational burden of the three-dimensional non-linear equations of motion is so great that it is extremely difficult to use them to model a wide area using a standard PC in 2020, and at least a high-capacity workstation or a supercomputer will be required.

Each model is described in detail in the following.

1) Line Models and Beach Profile Models Based on Wave Parameters

(1) Shoreline Change Model (One Line Model)

When sediment transport is divided into alongshore and on-offshore components, the **on-offshore sediment transport (= cross-shore sediment transport)** moves offshore for high waves during the year and returns to shore for other relatively calm periods (when the wave steepness is smaller than about 0.025), so the offshore and onshore movements are offset when predicting shoreline change during the year. Therefore, unless there are strong currents outside the wave breaking zone and the offshore migrating sand does not escape outside the target beach, the on-offshore sediment transport can be approximately neglected during short timescales.

The above shows that if readers want to predict annual shoreline change over decades rather than short-term erosion due to storms, readers can choose to consider only **longshore sediment transport (= longshore drift or littoral drift)**. Since the longshore sediment transport mainly moves within the wave breaking zone, by studying the relationship between the longshore sediment transport and the power (energy per unit time) of breaking waves using experimental and field observation data, an equation for calculating the longshore sediment transport rates from the wave breaking parameters (breaking wave height, period and direction) can be obtained. If the longshore distribution of wave breaking parameters is obtained from offshore wave parameters, then the longshore distribution of longshore sediment transport rates can also be calculated.

Substituting the longshore change of longshore sediment transport rate Q, dQ/dx, into Eq. (6.34) of the one-dimensional continuity equation of drifting sand, the variation in shoreline length Δy (y: measured in the on-offshore direction from the inland reference line to the shoreline) can be obtained from Eq. (6.35).

6. COASTAL TOPOGRAPHIC CHANGE

$$\frac{\partial A}{\partial t} = D\frac{\partial y}{\partial t} = \frac{\partial Q}{\partial x} \qquad (6.34)$$

$$\therefore \quad \Delta y = \left(\frac{\Delta Q}{\Delta x}\right)\frac{\Delta t}{D} \qquad (6.35)$$

where D is the height obtained from the regression analysis of the eroded area ΔA of the cross-sectional area A of the sandy beach obtained from the bathymetry data and the recession length Δy of the shoreline length y in the upper figure in **Fig. 6.9** and is called **movement height**. Since the longshore direction is the main axis in this model, the x-axis is in the longshore direction and the y-axis in the on-offshore direction.

This numerical prediction model is called the **shoreline change model** or the **one line model** and the basic assumptions of this model are as follows:

(a) Only longshore sediment transport due to waves is considered, while on-offshore sediment transport due to waves and sediment transport due to nearshore currents are ignored.

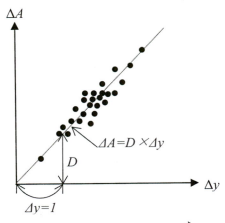
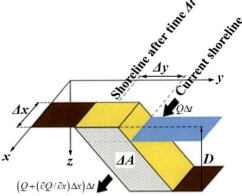

Fig. 6.9 Modelling of beach cross-sectional change.

(b) The on-offshore profile of longshore sediment transport does not change in the longshore direction and the movement height is a constant value (it corresponds to the height between the beach cliff from the high frequency wave breaking depth).

Because on-offshore sediment transport is ignored, this model cannot be used for predicting shoreline change during stormy terms, but while can be used for predicting shoreline change during the long term based on long-term validation simulations.

If the formulae and calculation charts based on the small amplitude wave theory or the parabolic equation which considers only progressive waves are used as the wave equation, the calculation time of wave deformation (by shoaling, refraction, and diffraction), from offshore to nearshore representative points, and form representative points to wave breaking points considering only wave refraction and shoaling, can become very short. Therefore, by continuously providing time series data (several years to several decades) of offshore wave parameters, long-term prediction of shoreline changes over a wide area (several kilometres to several tens of kilometres) becomes possible.

If the amount of beach nourishment and sediment inflow from a river need to be considered in the shoreline change model, the amount of beach nourishment and sediment inflow per Δx and per Δt

6.3 Coastal Topographic Change Prediction Methods

can be expressed as $q[\text{m}^3/\text{m/s}]$ and $(\Delta Q/\Delta x)$ in the right-hand side of Eq. (6.35) can be replaced with $(\Delta Q/\Delta x - q)$.

There have been many models for predicting shoreline change since Pelnard-Considére [33], but the detailed structure and usage method of one example of a shoreline change model that can be applied without difficulty to beaches with a generally straight shoreline is described in Section 1 of the Appendix. The structure of this model is relatively simple, and the source program is available so that the readers can re-create it for themselves relatively easily.

(a) Formula for Calculating Longshore Sediment Transport Rates

The longshore sediment transport rate Q [m³/s] is proportional to the power per unit width of a wave $EC_g\left(=1/8 \times \rho g H_b^2 \times \sqrt{gh_b}\right)$, which according to Komar and Inman [41] is expressed as follows.

$$
\begin{aligned}
Q &= \frac{K_1}{(\rho_s - \rho)g(1-0.4)} \times \left(\frac{1}{8}\rho g H_b^2 \sqrt{gh_b}\right)(\sin\theta_b \cos\theta_b) \\
&= \frac{K_1}{(\rho_s - \rho)g(1-0.4)} \times \left(\frac{1}{8}\rho g H_b^2 \sqrt{gh_b}\right)\left(\frac{1}{2}\sin(2\theta_b)\right)
\end{aligned}
\tag{6.36}
$$

where K_1 is the coefficient of longshore sediment transport (the value used by Komar et al. is 0.77, but is determined from verification simulations to match the measured shoreline change data for each coast); ρ_s is the density of sediment; ρ is the density of seawater; g is the gravitational acceleration; H_b is the wave breaking height (since the transport rate Q is proportional to the power per unit width of a wave EC_g, energy mean wave height, i.e. the squared mean wave height should be used); h_b is the wave breaking depth; and θ_b is the direction of the breaking waves. To obtain the volume of the sand layer, the net sand volume is divided by (1 - 0.4) to account for the porosity of the sand layer (porosity 40% = 0.4).

Furthermore, if the change in wave height due to the diffraction effect of a breakwater cannot be ignored, the following formula for calculating the longshore sediment transport rates by Ozasa and Brampton [42] can be used.

$$
\begin{aligned}
Q &= \frac{K_1}{(\rho_s - \rho)g(1-0.4)} \times \left(\frac{1}{8}\rho g H_b^2 \sqrt{gh_b}\right)\left(\frac{1}{2}\sin(2\theta_b)\right) \\
&- \frac{K_2}{(\rho_s - \rho)g(1-0.4)} \times \left(\frac{1}{8}\rho g H_b^2 \sqrt{gh_b}\right)\left(\frac{dH_b}{dx}\frac{\cos(\theta_b)}{\tan\alpha}\right)
\end{aligned}
\tag{6.37}
$$

where K_2 is the empirical coefficient (the value used by Ozasa and Brampton is $1.62 \times K_1$, but is determined from verification simulations to match the measured shoreline change data for each coast) and $\tan\alpha$ is the mean value of the on-offshore slope within the longshore sediment transport zone.

231

6. COASTAL TOPOGRAPHIC CHANGE

(b) Movement Height

The movement height D can be obtained from the analysis of bathymetric data measured during a period of mild wave conditions every year so that the effect of on-offshore sediment transport is minimised (see the upper graph of **Fig. 6.9**). Moreover, even if the existing bathymetric data are not available, it can be obtained by carrying out two bathymetry surveys, one in areas where shoreline retreat and advance due to longshore sediment transport are occurring and the other in areas where they are not occurring and by comparing these beach profiles.

If bathymetric measurements cannot execute, then the significant limit depth h_c for sediment movement obtained using Eq. (6.38) proposed by Uda et al [43] can be used as the movement height D in the case of general sandy coasts. This equation is the regression equation between the significant limit depth h_c and the incident significant wave height with a 95% non-exceedance probability H_{s95} (typically 1.5 m to 2.5 m in Japan) using measured data.

$$h_c = 3.64 \times H_{s95} \approx D \tag{6.38}$$

(c) Shoreline Change Due to On-offshore Sediment Transport

If the length of shoreline recession due to on-offshore sediment transport immediately after a storm surge is desired in long-term predictions of the shoreline, an equation for obtaining the amount of shoreline recession due to the on-offshore sediment transport rates from wave parameters is required. As an example, based on Eq. (6.4) of the on-offshore sediment transport rate q_y[m³/s/m] per unit width and unit time proposed by Sunamura [17], Eq. (6.39) for calculating the shoreline recession Δy, and the on-offshore sediment transport rate q_y [m³/s/m], proposed by Yamamoto et al [14] using existing field observation data and large-scale hydraulic model experiment data, are presented below.

$$
\left.
\begin{aligned}
&\Delta y = 2.7 \left(\int_0^t q_y dt \right)^{0.5} \\
&q_y = \left(A e^{-Bt/T} \right) U_r^{0.2} \phi (\phi - 0.13 U_r) w_s d_{50} \\
&A = 3.61 \times 10^{-10} \left(d_{50}/H_o \right)^{-1.31}, \qquad B = 4.20 \times 10^{-3} \left(\tan \alpha \right)^{1.57} \\
&\tan \alpha = \tan \alpha_e + \frac{\tan \alpha_0 - \tan \alpha_e}{e^{Bt/T}}, \qquad \tan \alpha_e = \left(0.0864 \frac{\rho_s - \rho}{\rho} g d_{50} \frac{T^2}{H_b^2} \right)^{2/3}
\end{aligned}
\right\} \tag{6.39}
$$

$$\left[\text{applicable range is } 0.017 \le \tan \alpha \le 0.125, \quad 0.00006 \le d_{50}/H_o \le 0.00102 \right]$$

where t is the time elapsed from the start of the action of the target high wave, T is the period of the target high wave, U_r is Ursell number, ϕ is Hallermeier's dimensionless number, w_s is the settling velocity of the bottom sediment, d_{50} is the median grain size of the bottom sediment, H_o is the offshore wave height of the target high wave, $\tan\alpha$ is the average beach slope from wave breaking depth to run-up height after t hours for the target high wave, $\tan\alpha_e$ is the mean beach slope at equilibrium (final) with respect to the target high wave, $\tan\alpha_0$ is the mean beach slope immediately before the onset of

232

the target high wave, ρ_s is the density of bottom sediment, ρ is the density of seawater, g is the gravitational acceleration, and H_b is the breaking wave height.

The Ursell number U_r, Hallermeier's dimensionless number ϕ, and the breaking wave height H_b are obtained from the following equations.

$$\left. \begin{array}{l} U_r = gH_s \dfrac{T^2}{\eta_s^{\,2}}, \quad \phi = \dfrac{H_s^{\,2}}{(\rho_s/\rho - 1)\eta_s d_{50}}, \quad H_b = (\tan\beta)^{0.2}(H_o/L_o)^{-1/4} H_o \\[3mm] H_s = 1.9(\tan\alpha)^{0.9} H_b, \quad \eta_s = \dfrac{H_s}{2.4(\tan\alpha)^{0.3}} \end{array} \right\} \tag{6.40}$$

where H_s is the wave height on the shoreline, η_s is the mean water surface elevation above the shoreline, $\tan\beta$ is the initial seabed slope, and L_o is the offshore wavelength of the target high wave.

(2) Multiline Models

There are models of diverse levels of complexity, from a simple model in which the on-offshore distribution of longshore sediment transport is obtained from the analysis of bathymetric data and incorporated into the shoreline change model so that isobath changes can be represented, to the numerical model of Uda and Kawano [37] in which a continuity equation of drifting sand is established for each isobath so that isobath changes can be predicted without the constraints of simple parallel shifts.

In the numerical model of Uda and Kawano [37], the total longshore sediment transport rate Q is given in Eq. (6.41), the longshore sediment transport rate q_i in each isobath is expressed in Eq. (6.42), and by substituting this q_i into the continuity Eq. (6.43) for drifting sand in each isobath, the variation in isobath displacement Δy_i can be predicted.

$$Q = F_o \left(\tan\theta_b - \frac{\partial y_s}{\partial x} \right) \tag{6.41}$$

where F_o is the power of incoming breaking waves multiplied by a coefficient of longshore sediment transport, θ_b is the direction of breaking waves, x is the alongshore coordinate, and y_s is the shoreline length (the on-offshore length from the reference line parallel to the x-axis to the shoreline).

$$q_i = F_{oi}\left(\tan\theta_b - \frac{\partial y_i}{\partial x} \right), \quad F_{oi} = F_o \times \mu_i, \quad \sum_{i=1}^{n}\mu_i = 1 \right\} \tag{6.42}$$

where μ_i is a weighting factor on the on-offshore distribution of longshore sediment transport rate obtained from the analysis of sediment transport rate in each isobath, and y_i is the on-offshore position of the ith isobath.

$$\Delta y_i = \frac{\Delta t}{h_i}\frac{\partial q_i}{\partial x} \tag{6.43}$$

where Δt is the time taken for the variation Δy_i, and h_i is the height between isobaths.

6. COASTAL TOPOGRAPHIC CHANGE

For the treatment of cases with groynes and breakwaters, see Uda and Kawano [37]; for the treatment of cases with detached breakwaters, see Uda et al [44]; and for the treatment of cases where longshore classification of mixed grain size should be considered, see Uda et al [45].

It should be noted that by substituting Eq. (6.41) for longshore sediment transport rates into the continuity Eq. (6.34) and by setting the wave direction θ_b to zero, the one-dimensional diffusion equation can be yielded. In other words, the line model can be said to be a model for solving the diffusion equation, and when the line model is solved by the finite difference method, the stability condition for the diffusion equation must be satisfied.

(3) SBEACH Model

Larson and Kraus [46], [47] proposed the **SBEACH model** which enables short-term predictions of beach profile change during storm events (cross shore beach change) due to on-offshore sediment transport. Because this model ignores the effect of currents (similar to the line model) and calculates the horizontal two-dimensional distribution of waves by solving the energy equilibrium equation, the topographic changes due to on-offshore sediment transport in horizontal two-dimensions can be calculated within a short time. Therefore, this model can be used to predict beach change due to high waves, using the long-term prediction results generated by the multiline model.

Specifically, the distribution of wave energy transport F is obtained by solving the horizontal two-dimensional steady-state energy equilibrium Eq. (6.44). The Eq. (6.45) for on-offshore sediment transport rate $q(x)$ using the amount F and the continuity Eq. (6.46) for on-offshore sediment transport are used to predict the topographical change in the beach cross-section for each on-offshore cross-section. Here, since the focus is on the on-offshore sediment transport, the x-axis is in the on-offshore direction, and the y-axis is in the alongshore direction, which is the opposite in the case of line models.

$$\left. \frac{\partial (F\cos\theta)}{\partial x} + \frac{\partial (F\sin\theta)}{\partial y} = \frac{K_e}{d}(F - F_s), \quad F_s = \frac{1}{8}\rho g H_s^2 \sqrt{gh} \right\} \tag{6.44}$$

$$\left. \begin{array}{l}
\text{Outer region of wave breaking zone:} \quad q(x) = q_b e^{-\lambda(x - x_b)} \\[2mm]
\text{Wave breaking transition zone:} \quad q(x) = q_p e^{-\lambda_2(x - x_p)} \\[2mm]
\text{Wave breaking zone:} \quad D = \dfrac{\partial F / \partial x}{h}, \quad D_{eq} = \dfrac{5}{24}\rho g^{3/2}\left(\dfrac{H_b}{h_b}\right)^2 A_e^{3/2} \\[4mm]
\text{in the case of } D > \left(D_{eq} - \dfrac{\varepsilon}{K}\dfrac{\partial h}{\partial x}\right), \quad q(x) = K\left(D - D_{eq} + \dfrac{\varepsilon}{K}\dfrac{\partial h}{\partial x}\right) \\[4mm]
\text{in the case of } D \le \left(D_{eq} - \dfrac{\varepsilon}{K}\dfrac{\partial h}{\partial x}\right), \quad q(x) = 0 \\[4mm]
\text{Wave run-up zone:} \quad q(x) = q_z (x - x_r)/(x_z - x_r)
\end{array} \right\} \tag{6.45}$$

234

where F is the wave energy transport; θ is the angle of incident wave; K_e is the wave attenuation coefficient; d is the total water depth; F_s and H_s are the wave energy transport and wave height in the stable state; ρ is the density of sea water; g is the gravitational acceleration; h is the water depth below the still water surface; q_b is the on-offshore sediment transport rate at the wave breaking position; x_b is the wave breaking position; λ is the spatial attenuation coefficient of on-offshore sediment transport; q_p is the on-offshore sediment transport rate at the wave plunge position; x_p is the wave plunge position; λ_2 is also the spatial attenuation coefficient, 0.2-0.5 times λ; H_b is the wave breaking height; h_b is the wave breaking depth; A_e is Dean's shape factor for equilibrium beach section [see Eq. (6.8), m$^{1/3}$]; K is the coefficient of on-offshore sediment transport rate; ε is the coefficient of sediment transport rate dependent on the local seabed slope; q_z is the on-offshore sediment transport rate at the run-up start position; x_z is the run-up start position; and x_r is the run-up end position.

$$\frac{\partial h}{\partial t} = \frac{\partial q(x)}{\partial x} \qquad (6.46)$$

Note that there is no porosity in the continuity Eq. (6.46) for on-offshore sediment transport since the sediment transport rate in Eq. (6.45) includes the amount of porosity in the sand layer.

(4) Model for Predicting Beach Changes Based on Bagnold's Concept

Serizawa et al [38], [39] derived Eq. (6.47) which can obtain the sediment transport in any direction from wave breaking parameters by applying the concept of the equilibrium slope introduced by Inman and Bagnold [48] and the energetics approach of Bagnold [49] based on the coordinates shown in **Fig. 6.10**. They then proposed a beach change prediction model that can obtain the amount of bathymetric change using the horizontal two-dimensional continuity equation for drifting sand shown in Eq. (6.48). This numerical model is also computationally efficient because it does not involve nearshore current calculations, and can be used for the long-term prediction of 3D topographic changes over a wide area.

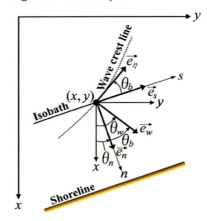

Fig. 6.10
Definition of symbols relating to coordinates.

6. COASTAL TOPOGRAPHIC CHANGE

$$\vec{q} = \frac{K_1 K_\lambda \varepsilon(\zeta)}{(\rho_s - \rho)g(1-0.4)} \frac{\rho g H_b^2 C_{gb}}{8} \cos^2 \theta_b \left(\tan \beta_c \overrightarrow{e_w} - \overrightarrow{\nabla \zeta} + \overrightarrow{C_\lambda} \right)$$

$$\overrightarrow{C_\lambda} = \lambda \left(\tan \phi - \tan \beta_c \right) \sin \theta_b \overrightarrow{e_\eta}, \quad K_\lambda = \left[\frac{\tan \beta_c}{\tan \beta_c + \lambda \left(\tan \phi - \tan \beta_c \right)} \right] \quad (6.47)$$

$$\lambda = \lambda_o \frac{\tan \beta_w}{\tan \beta_c}, \quad \tan \beta_w = \cos \theta_w \frac{\partial \zeta}{\partial x} + \sin \theta_w \frac{\partial \zeta}{\partial y}$$

However, when $\tan \beta_w < 0$, $\tan \beta_w = 0$

$$\frac{\partial \zeta}{\partial t} = -\left(\frac{\partial q_x}{\partial x} + \frac{\partial q_y}{\partial y} \right) \quad (6.48)$$

Where $\vec{q}$ is the vector-displayed sediment transport rate, ρ_s is the density of drift sand, ρ is the density of seawater, g is the gravitational acceleration, 0.4 is the porosity of the soil layer, K_1 is the empirical coefficient of sediment transport determined from verification simulations, $\varepsilon(\zeta)$ is the vertical distribution function of sediment transport based on measured data, ζ is the local bottom height, H_b and C_{gb} is the wave height and group velocity at wave breaking position, θ_b is the intersection angle between the direction perpendicular to the isobath and the direction of breaking waves ($\cos\theta_b \geq 0.5$), $\tan \beta_c$ is the equilibrium seabed slope at which sediment transport ceases, $\overrightarrow{e_w}$ is the unit vector of wave direction at a wave breaking position (= $\cos \theta_w$, $\sin\theta_w$). $\overrightarrow{\nabla \zeta}$ is the slope vector of the terrain (= $\tan \beta \times \cos \theta_n$, $\tan\beta \times \sin\theta_n$), $\tan\beta$ is the seabed slope, λ_o is the coefficient that adjusts λ so that λ is 0 on a horizontal floor and 1 on a steeper slope, ϕ is the internal friction angle of sediment, $\overrightarrow{e_\eta}$ is the unit vector of tangential direction of the wave crest line (= $-\sin\theta_w$, $\cos\theta_w$), $\tan \beta_w$ is the seabed slope in the wave direction, θ_w is the angle from x-direction to the direction of breaking waves, t is the time, q_x and q_y are the sediment transport rates in the x and y directions respectively.

2) Bathymetric Change Models Based on Waves and Nearshore Currents

In the case of bathymetric change models based on waves and nearshore currents, the distributions of sediment transport rates are obtained using the spatio-temporal velocity distributions from the wave and nearshore current fields. Then, they are substituted into the continuity equation for drifting sand to obtain the bathymetric change. A one-dimensional model was first built in the on-offshore direction, which was later expanded to horizontal two-dimensions (x, y Cartesian coordinates). Also, a vertical 2D (x, z Cartesian coordinates) model was built, which was later expanded to three-dimensional models solving the 3D equations of fluid motion. In this way, since the many one-dimensional models and vertical 2D models can be positioned as prototype stages of the target numerical model, some examples of horizontal 2D models and 3D models are introduced here.

6.3 Coastal Topographic Change Prediction Methods

In addition, the horizontal two-dimensional model is sometimes called a three-dimensional topographic change model because it can produce a 3D seafloor change map using a horizontal distribution of bathymetric change.

(1) Horizonal Two-dimensional Seabed Change Models
(a) Case of Separate Model for Wave and Nearshore Current Fields

Since around 1980, numerical models have emerged that can predict bathymetric changes by obtaining the radiation stresses from the wave field calculation (see Section 2.4), substituting them into the equations of momentum averaged over one wave period to calculate the nearshore current fields (see Section 4.1), substituting the velocity distributions of wave fields and nearshore current fields into equations for calculating sediment transport rates (see Section 6.2), and then substituting the distributions of sediment transport rates into the continuity equation for drifting sand.

Eq. (6.49) is the continuity equation for drifting sand corresponding to sediment transport rates which neglect the porosity of the sand layer. On the other hand, if sediment transport rates which consider the porosity of the sand layer are used, the continuity equation which neglects the porosity must be used. However, early numerical models have neglected the effects of suspended load.

$$\frac{\partial \zeta}{\partial t} = -\frac{1}{1-\varepsilon_s}\left(\frac{\partial q_{bx}}{\partial x} + \frac{\partial q_{by}}{\partial y} - C_s + C_{ut}\right) \tag{6.49}$$

where ζ is the local bottom height, q_{bx} and q_{by} are the bed load transport rates in the x and y directions, ε_s is the porosity of the sand layer, C_s is the amount of suspended load settling onto the bottom, and C_{ut} is the amount of suspended load being rolled up from the bottom.

Representative numerical models of this type include Watanabe [50], Yamaguchi et al [51], Yamaguchi and Nishioka [52] and Watanabe et al [25]. However, those models cannot accurately calculate wave run-up due to the limitations of the model used for wave field calculation. Therefore, Shimizu et al. [53] proposed a numerical model that incorporated the calculation method of shoreline changes into a seabed change model separated wave and nearshore current fields, so that the prediction of shoreline change can also be performed accurately.

Based on the numerical model of Roelvink et al [54], The consortium of UNESCO-IHE, Deltares (Delft Hydraulics), Delft University of Technology and the University of Miami have jointly developed a horizontal two-dimensional model for predicting waves, nearshore currents and topographical changes that can consider the effects of irregular waves, long-period gravity waves and their run-up, mixed grain size, and attenuation by seaweeds and mangroves. This numerical model is called the **XBeach.** Initially, XBeach model solved the wave action balance equation at the wave group scale (XBeach- surfbeat). It was later extended to a phase-resolving model (XBeach-X). The structure and usage of XBeach can be found in the "XBeach Technical Reference" by Deltares [55]. The programme can be downloaded from below:

6. COASTAL TOPOGRAPHIC CHANGE

https://oss.deltares.nl/web/xbeach/.

Numerous applications of this model can be found in Pender and Karunarathna [56], Dissanayake et al [57], [58], [59], and many others.

(b) Case of Integrated Model for Wave and Nearshore Current Fields

Since around 1990, Sato and Kabiling [60] and others have proposed numerical models that can simultaneously calculate the velocity distributions of the wave and nearshore current fields [see Eqs. (4.23) and (4.24)] using the Boussinesq equation, which can accurately calculate the momentum in extremely shallow waters, and obtain the horizontal distribution of changes in the seabed by considering bed load and suspended load. A numerical model of this type is also capable of calculating the wave run-up area.

Vu et al [61] proposed a practical numerical model based on the model of Sato and Kabiling [60] with some improvements. The detailed structure and usage method of a modified version of the model of Vu et al [61] are described in Section 2 in the Appendix.

Finite difference methods are usually employed on a staggered grid to solve the Boussinesq equation, where vector quantities are set on the quadrangle grid boundaries and scalar quantities are set at the centre of the quadrangle grid. However, if the staggered grid is not set up carefully, bathymetric changes within the square grid cannot be calculated correctly.

For example, if the ground level in the quadrangle grid between i and $i+1$ is high and the ground level in the surrounding quadrangle grid is low, as shown in **Fig. 6.11**, the correct topographic change mechanism is that in the quadrangle grid between i and $i+1$, since the water depth is shallower than the surrounding area, the velocity is accelerated and causes erosion and the water depth become deeper. Care must be taken to reproduce this when developing models.

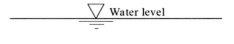
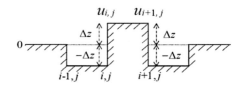

Fig. 6.11 Description of bathymetric change calculations on difference grids.

The model of Vu et al. has been programmed to be able to calculate bathymetric changes using appropriate flow velocities, thus increasing the accuracy of the topographic change calculations without adding an artificial viscosity term to the continuity equation. Furthermore, based on extensive validations using data from various coasts around Japan, Yamamoto et al [62], Charusrojthanadech et al [63], and others have confirmed the good computational accuracy of the improved practical numerical model.

(2) Three-dimensional Models

(a) Models Based on 3D Navier Stokes Equations

For cases where the effects of three-dimensional fluid motion cannot be ignored, Nakamura et al [64] developed a **three-dimensional two-way coupled fluid-structure-sediment interaction model (FSSM model)** using the three-dimensional Navier Stokes equations of motion and a horizontal two-dimensional continuity equation for drifting sand that considers bed load and suspended load.

Although this model is a very good numerical model that incorporates the fluid motion mechanism as precisely as possible, it is hardly in an environment that is freely available to many coastal numerical engineers. Therefore, one solution is to use CADMAS-SURF (capable of numerically calculating three-dimensional fluid motion in porous media [see (4) of 2) in Section 2.4], for which the source programme is publicly available, by incorporating the equations for calculating sediment transport rates and the continuity equation for drifting sand. Suzuki et al [65] confirmed that the vertical two-dimensional CADMAS-SURF/2D model, although not the three-dimensional CADMAS-SURF/3D model, can be used by incorporating the equations for calculating bed load transport rates of Watanabe et al [25] and others.

(b) Quasi-three-dimensional Models

Solving the 3D nonlinear equations of motion is very computationally intensive. Therefore, Kuroiwa et al [66] proposed a numerical model that reduces the computational load by using quasi-three-dimensional equations of motion and also considers mutual interaction between the wave and nearshore current fields to improve calculation accuracy. Specifically, the wave field is calculated using the horizontal 2D energy equilibrium equation, the nearshore current field is calculated using the horizontal 2D nonlinear equations of motion of the nearshore current, the continuity equation is based on the velocities in two horizontal and vertical directions, and the continuity equation based on horizontal inflow and water surface elevation, and the bed load transport rates are calculated using the calculation equations of Watanabe et al [25].

In addition, Deltares (Delft Hydraulics) [67] developed the numerical model **Delft3D** for the prediction of various fluid motions and bathymetric changes using quasi-three-dimensional equations of motion, in collaboration with the Delft University of Technology. The wave field is calculated using the SWAN model. The nearshore current field is calculated using the nonlinear equations of flow in horizontal 2D [Eq. (6.50)] and 3D [Eqs. (6.51)] and the depth average continuity equation [Eq. (6.52)]. The model can also predict the bed load transport rates, the suspended load transport and the bathymetry change.

6. COASTAL TOPOGRAPHIC CHANGE

$$
\begin{aligned}
&\frac{\partial u}{\partial t}+\frac{u}{\sqrt{G_X}}\frac{\partial u}{\partial X}+\frac{v}{\sqrt{G_Y}}\frac{\partial u}{\partial Y}+\frac{w}{h+\eta}\frac{\partial u}{\partial \sigma}-\frac{v^2}{\sqrt{G_X}\sqrt{G_Y}}\frac{\partial\sqrt{G_Y}}{\partial X}+\frac{uv}{\sqrt{G_X}\sqrt{G_Y}}\frac{\partial\sqrt{G_X}}{\partial Y}-f_{cl}v \\
&=-\frac{1}{\rho_w\sqrt{G_X}}P_X+F_X+\frac{1}{\left(h+\eta\right)^2}\frac{\partial}{\partial\sigma}\left(v_V\frac{\partial u}{\partial\sigma}\right)+M_X \\[6pt]
&\frac{\partial v}{\partial t}+\frac{u}{\sqrt{G_X}}\frac{\partial v}{\partial X}+\frac{v}{\sqrt{G_Y}}\frac{\partial v}{\partial Y}+\frac{w}{h+\eta}\frac{\partial v}{\partial \sigma}+\frac{uv}{\sqrt{G_X}\sqrt{G_Y}}\frac{\partial\sqrt{G_{\eta\eta}}}{\partial X}-\frac{u^2}{\sqrt{G_X}\sqrt{G_Y}}\frac{\partial\sqrt{G_X}}{\partial Y}+f_{cl}u \\
&=-\frac{1}{\rho_w\sqrt{G_Y}}P_Y+F_Y+\frac{1}{\left(h+\eta\right)^2}\frac{\partial}{\partial\sigma}\left(v_V\frac{\partial v}{\partial\sigma}\right)+M_Y
\end{aligned}
\right\} \quad (6.50)
$$

$$
\frac{\partial\eta}{\partial t}+\frac{1}{\sqrt{G_X}\sqrt{G_Y}}\frac{\partial\left[(h+\eta)u\sqrt{G_Y}\right]}{\partial X}+\frac{1}{\sqrt{G_X}\sqrt{G_Y}}\frac{\partial\left[(h+\eta)v\sqrt{G_X}\right]}{\partial Y}+\frac{\partial w}{\partial\sigma}=(h+\eta)(q_{in}-q_{out}) \quad (6.51)
$$

$$
\begin{aligned}
&\frac{\partial\eta}{\partial t}+\frac{1}{\sqrt{G_X}\sqrt{G_Y}}\frac{\partial\left[(h+\eta)U\sqrt{G_Y}\right]}{\partial X}+\frac{1}{\sqrt{G_X}\sqrt{G_Y}}\frac{\partial\left[(h+\eta)V\sqrt{G_X}\right]}{\partial Y}=(h+\eta)Q \\[6pt]
&U=\frac{1}{h+\eta}\int_{-h}^{\eta}u\,dz=\int_{-1}^{0}u\,d\sigma, \qquad V=\frac{1}{h+\eta}\int_{-h}^{\eta}v\,dz=\int_{-1}^{0}v\,d\sigma \\[6pt]
&Q=\int_{-1}^{0}(q_{in}-q_{out})\,d\sigma+P-E
\end{aligned}
\right\} \quad (6.52)
$$

where t is the time, X is the longitude coordinate in spherical coordinates [deg], Y is the latitude coordinate in spherical coordinates [deg], u and v are the horizontal flow velocities in X and Y coordinates respectively, w is the vertical flow velocity in the σ coordinate, $\sqrt{G_X}\ (=R\times\cos Y)$ and $\sqrt{G_Y}\ (=R)$ are the coefficient used to transform curvilinear to rectangular coordinates [m], f_{cl} is the Coriolis coefficient [1/s], ρ_w is the reference density of seawater, P_X and P_Y are the gradient hydrostatic pressure in X and Y directions respectively [kg/m^2/s^2], F_X and F_Y are the turbulent momentum flux in X and Y directions respectively [m/s^2] , v_V is the vertical eddy viscosity, M_X and M_Y are the contributions due to external sources or sinks of momentum (external forces by hydraulic structures, discharge or withdrawal of water, wave stresses, etc.) in X and Y directions respectively, $\sigma\ [=(z-\eta)/(h+\eta)]$ is the coordinate converting η to $-h$ into 0 to -1, R is the radius of the Earth [m], z is the vertical coordinate, h is the depth below the still water surface, η is the water surface elevation above the still water surface, q_{in} and q_{out} is the local sources and sinks of water per unit of volume [1/s] respectively, U and V are the depth averaged u and v respectively, Q is the contribution per unit area due to the discharge or withdrawal of water, precipitation and evaporation, P is the non-local source term of precipitation [m/s], and E is the non-local sink term due to evaporation [m/s].

Details of the structure and usage method of this model are summarised in the Delft3D-FLOW User Manual by Deltares [67], and the programme can be downloaded from below.

https://oss.deltares.nl/web/delft3d/

6.3 Coastal Topographic Change Prediction Methods

(c) 3D Models Based on Particle Methods

Above mentioned 3D numerical models, including FSSM model, do not solve the equation of motion for calculating the bed load and suspended load sediment transport, but uses empirical equations that use flow velocity instead. To improve the accuracy of the bathymetric change prediction, a method of solving the equations of motion of sediment transport can be employed. However, as sediments are separate particles with a certain degree of viscosity, the equation of motion of a continuum cannot be used. In this case, the **particle method**, in which the Lagrange-type equation of motion is applied to each sediment particle, can be considered. However, the need to model a very large number of sediment particles increases the computational load so much that even supercomputers cannot handle numerical calculations over large coastal areas. Therefore, it is necessary to increase the virtual sediment particle size to a mass that can be regarded as a single sediment particle. Moreover, since the Lagrangian-type equation of motion is formulated for each of virtual sediment particles, grided approaches, such as finite difference methods, is not necessary.

The main calculation methods that fall into the category of the particle method include a **Distinct Element Method (DEM)** by Cundall and Strack [68], a **Moving Particle Semi-implicit method (MPS method)** by Koshizuka and Oka [69], and a **Smoothed Particle Hydrodynamics method (SPH method)** by Lucy [70].

(i) Distinct Element Method (DEM)

The distinct element method is a numerical method that solves the following equations of motion for each rigid body particle (element).

$$\left. \begin{array}{l} \dfrac{dm_i \vec{u}_i}{dt} = \vec{F}_{flow,i} + \vec{F}_{int,i} + \vec{g}, \qquad \dfrac{dI_i \vec{\omega}_i}{dt} = \vec{T}_{flow,i} + \vec{T}_{int,i} \\[2ex] \vec{F}_{flow,i} = \int_{Vi} \phi_i \rho_i \left(\dfrac{D\vec{u}_i}{Dt} - \vec{g} \right) dV, \quad \vec{T}_{flow,i} = \int_{Vi} \vec{r}_i \times \phi_i \rho_i \left(\dfrac{D\vec{u}_i}{Dt} - \vec{g} \right) dV \end{array} \right\} \tag{6.53}$$

where m_i is the mass of element i, $\vec{u}_i$ is the velocity vector of element i, $\vec{F}_{flow,i}$ is the fluid force vector acting on element i, $\vec{F}_{int,i}$ is the interaction force vector between the elements, $\vec{g}$ is the gravitational acceleration, I_i is the inertia tensor of element i, $\vec{\omega}$ is the angular velocity vector of element i, $\vec{T}_{flow,i}$ is the torque due to fluid force acting on element i, $\vec{T}_{int,i}$ is the torque due to interaction force between the elements, V_i is the calculation range including element i, ϕ_i is the volume occupation of element i, ρ_i is the density of element i, and $\vec{r}_i$ is the relative position vector between the elements.

(ii) Moving Particle Semi-implicit Method (MPS Method)

The MPS method considers a particle interaction model corresponding to the differential

6. COASTAL TOPOGRAPHIC CHANGE

operator, replacing each term in the governing equations with an equivalent particle interaction and calculating the change in variables such as the velocity of each virtual particle based on the particle interaction. The governing equations are the following Navier Stokes equations of Lagrange type and the continuity equations.

$$\left.\begin{array}{ll} \dfrac{D\vec{u}_l}{Dt} = -\dfrac{1}{\rho_l}\nabla p_l + v_l \nabla^2 \vec{u}_l + \vec{g} + \dfrac{\vec{f}_{ls,l}}{\rho_l}, & \dfrac{D\rho_l}{Dt} + \rho_l \nabla \cdot \vec{u}_l = 0 \\[3mm] \dfrac{D\vec{u}_s}{Dt} = -\dfrac{1}{\rho_s}\nabla p_s + v_s \nabla^2 \vec{u}_s + \vec{g} - \dfrac{\vec{f}_{ls,s}}{\rho_s} + \dfrac{\vec{f}_{col}}{\rho_s}, & \dfrac{D\rho_s}{Dt} + \rho_s \nabla \cdot \vec{u}_s = 0 \end{array}\right\} \qquad (6.54)$$

where the subscript l stands for the liquid phase, and the subscript s for the solid phase. $\vec{u}$ is the velocity vector, ρ is the density, p is the pressure, v is the kinematic viscosity coefficient, g is the gravitational acceleration, $\vec{f}_{ls}$ is the interaction vector between the liquid and solid phases, and $\vec{f}_{col}$ is the collision force vector between solid phase particles.

(iii) Smoothed Particle Hydrodynamics Method (SPH Method)

The SPH method considers that the physical quantity possessed by each virtual particle is spread according to a distribution function called a **kernel** and calculates it as a compressible fluid. The governing equations are the equation of state and the conservation law of momentum, which reduces the calculation accuracy for liquid motion. However, because it is solved explicitly, the calculation is relatively rapid. On the other hand, the MPS method uses the particle number density to maintain incompressibility and is solved using a semi-implicit method. Therefore, its computational speed is slower than the SPH method. Therefore, Shao and Lo [71] proposed the **Incompressible Smoothed Particle Hydrodynamics method (ISPH method)**, which incorporates the incompressibility condition of the MPS method into the SPH method to improve computational accuracy.

Shigematsu et al [72], Harada and Goto [73], and others proposed numerical models which solve for the liquid phase by using the three-dimensional Navier Stokes equations of motion of the Euler type, by considering the porosity and the interaction forces between the fluid and the sediment particles, and for the solid phase by using DEM.

Goto et al [74] and others proposed a solid-liquid mixed phase model using the MPS method.

Harada et al [75], [76] and others proposed a solid-liquid mixed phase model based on a hybrid of the MPS method and DEM.

Goto et al [77] proposed a solid-liquid mixed phase model, called a **particle implemented simulator for physical and engineering research (PARISPHERE)**, by hybridising the ISPH method and DEM.

Numerical simulations using the particle method can only be carried out on a small scale, even on supercomputers, if the particle size is set to the actual sediment particle size. If a real coastline of

6.3 Coastal Topographic Change Prediction Methods

about 1 km in the longshore direction is to be calculated, the diameter of the sediment particles will have to be set to about 10 cm, which means that the system is not yet ready for use on a commercial basis. However, if the power of supercomputers increases significantly in future, it may be possible that this approach can be applied to large coastal areas.

6. COASTAL TOPOGRAPHIC CHANGE

6.4 Coastal Erosion

Luijendijk et al [78] found from an analysis of satellite information for the period 1984-2016 that 31% of the ice-free coastline on the Earth was sandy, and that erosion was progressing at rates exceeding 0.5 m/year on 24% of the sandy beaches. Mentaschi et al [79] also revealed that the area of erosion on the Earth was about 28,000 km^2 based on satellite observation data from almost the same period (1984-2015), and that some cases of coastal erosion were caused by natural phenomena, but that anthropogenic cases were very common. Therefore, this section describes the main causes of and basic countermeasures against coastal erosion which have become prominent since the late 20th century.

1) Causes of Coastal Erosion

When waves with high wave steepness strike beaches, sediment near the shoreline is transported in the offshore and longshore directions, causing short-term beach erosion. However, sediment that has moved offshore can return to the shoreline area during calm periods when wave steepness is low (although very fine sediment that is transported too far offshore cannot return). Moreover, even if it is not generally expected that all sediment transported in the longshore and offshore directions will return, most coasts can be considered as dynamically stable in the long term, with a balance between erosion and sedimentation, as large volumes of sediment can be transported to the coasts from mountainous areas and flood plains via rivers. If this natural balance had not been maintained, many sandy beaches would have disappeared a long time ago. However, the balance of this sediment budget (sediment inflow and outflow) was broken by the escalated human socio-economic activities with an interruption of sediment supply to the coast in recent decades, resulting in serious beach erosion at many coasts around the world. Furthermore, global warming has clearly progressed in this century, leading to rising sea levels and atmospheric instability. These phenomena have increased the occurrence frequency of strong winds and high waves, leading to beach sediment instability. The following are the main causes of beach erosion that have intensified since the late 20th century.

(1) Dam Construction and Maintenance Dredging of Reservoirs

Dams have been built upstream of many rivers for important purposes such as flood control, water resource development (securing water for domestic and industrial use, irrigation, or other purposes), and hydropower generation. Although these dams have fully achieved their original construction objectives, they have not only filled reservoirs with water but also with sediment. This leads to a reduction in sediment transport downstream and erosion of the beaches.

Reservoirs need regular dredging to avoid sediment infilling, which reduces their capacity. However, dredged materials were not discharged downstream until recently, in order to avoid contamination of the downstream river water and to avoid unwanted sediment accumulation in areas where it can be problematic to residents or assets.

Coastal erosion due to this has occurred in almost all sandy beaches with river mouths where dams have been constructed upstream of the river. Furthermore, check dams (= sabo dams, sand-trap dams) are important facilities that have been constructed to reduce the occurrence of landslides by controlling rapid sediment discharge, but they also can easily become obstacles for sediment supply to downstream beaches.

(2) Port and Habour Construction, and Maintenance Dredging of Channels

When ports for increasing maritime transport capacity and harbours for promoting fishing activities are constructed on coasts dominated by longshore sediment transport, rapid sedimentation on the updrift side and rapid erosion on the downdrift side occur outside the outer structures (breakwaters) of ports and harbours.

As sedimentation on the upstream side can lead to the burial of the entrances/exits of the ports and harbours, the administrators of ports and harbours carry out maintenance dredging of the navigation channels, including the entrances/exits, as required. However, to avoid artificially discharging dredged sediment downstream, which could pollute seawater or undesirable accumulation of sediment, causing damage to residents and assets downstream, until recently dredged sediment was used only for land reclamation within the areas of the ports and harbours. As a result, the erosion on the downdrift side has occurred on all coasts dominated by longshore sediment transport where the ports and harbours have been built.

(3) Sand and Gravel Mining

As a result of increased demand for aggregate for concrete during the economic growth period since the late 20th century, sand and gravel mining was carried out in many rivers, resulting in rapid erosion of riverbeds and estuaries. Restrictions on sustainable sand and gravel mining in rivers with severe erosion damage have been put in place since around 1990. However, due to the ever-increasing demand for aggregate, marine sand and gravel mining had also become popular, along with mountainous sand and gravel mining, and crushed stone production. Since around 2000, a total ban or restrictions on the amount of marine sand and gravel mining have been imposed on many coasts where coastal erosion and adverse effects on living organisms have become pronounced.

Although at present the erosion has been mitigated on many coasts because sand and gravel mining are now totally banned or heavily regulated on concerned rivers and coasts, in many cases, it takes time to restore an adequate sediment supply because other causes of erosion have not been eliminated.

6. COASTAL TOPOGRAPHIC CHANGE

(4) Installation of Erosion Control Facilities

When wave-absorbing breakwaters are constructed away from a shoreline (**detached breakwaters** or **offshore breakwaters**), the sea area between the land and the breakwaters usually has low wave activity, which is prone to sediment deposition. When shore-normal structures such as **groynes** are constructed between the shoreline and the wave breaking depth of a beach dominated by alongshore sediment transport, sediment deposition occurs on the updrift side of the structure.

However, in the case of detached breakwaters, instead of sedimentation occurring in the landward areas, erosion is more likely to occur on both sides in the longshore direction, especially on the downdrift side of the breakwaters. In the case of groynes, as sedimentation occurs on the updrift side of the longshore sediment transport, erosion occurs on the downdrift side. Therefore, on coasts where these erosion control facilities are installed, if natural or artificial sediment supply is not ensured to match the erosion rate downdrift of the longshore sediment transport of these facilities, then downdrift coastal erosion would steadily progress. For example, on the Pacific side of the Boso Peninsula in Japan, the sea cliffs at Byobugaura and Cape Taito are made of soft rock that can easily collapse due to high waves and an annual shoreline recession of over 1 m had occurred. Therefore, since the 1960s, these sea cliffs have been protected by wave-absorbing breakwaters, and the cliff erosion has greatly been reduced. However, on the Kujukuri Coast, which is located on the downdrift side of the longshore sediment transport of these sea cliffs, coastal erosion is progressing because the amount of sediment supplied by the sea cliff erosion has been greatly reduced.

(5) Reinforcement of Flood Control Facilities

In some cases, coastal erosion is exacerbated by a reduction in sediment supply due to the improvement of river channels and the replacement of river mouths. For example, on the Shimo-Niikawa coast of Toyama Prefecture in Japan, sediment had been supplied to the sea from various parts of the Kurobe Fan due to the inundation of the Kurobe River, but the river channel was fixed and inundation was drastically reduced by channel improvement. However, the sediment supply to the eastern coast away from the Kurobe River mouth was drastically reduced after this intervention. In addition, on the Niigata coast in Japan, the Ohkozu Flood-control Channel, built downstream of the Shinano River, has been very useful in protecting Niigata City at the mouth of the main river from flooding, but the large amount of sediment that had flown into the mouth of the main river has now diverted to the flood-control channel, resulting in significant coastal erosion near the mouth of the main river.

(6) Destruction of Coastal Ecosystems

Coastal wetlands such as mangroves and salt marshes are important coastal ecosystems. Mangrove forests shown in **Photo 6.1** are found along many coasts of low-latitude areas. The mass destruction of mangrove forests to secure charcoal for commercial use and to create aquaculture ponds (e.g. shrimp) has increased since the late 1900s. However, mangrove forests are an important natural wave-attenuation facility. Their removal allows high waves to act directly on bottom sediment, thus causing rapid beach erosion in former mangrove forest areas (Mangrove-covered beaches are clay beaches, and when high waves come into direct action, the surface clay is lost in a short time) and threatening the livelihoods of the local population.

However, it should be noted that a mangrove forest can protect a coast only if waves in the forest are not strong enough to wash away deposited mud at mangrove roots. This happens if a muddy beach exists in front of the mangrove forest which causes sufficient dissipation of wave energy before waves arrive at the forest, or when the mangrove forest is located in a wave shadow area. In an area of high wave exposure, mangrove trees can grow only if there is enough muddy sediment supply to maintain a muddy beach.

Photo 6.1
Example of a mangrove forest with a mixture of various types. (Trees with roots extending like octopus legs, seen from left to centre back, are the brace root type; trees with short bamboo shoot-like root clusters protruding from the ground, seen in the centre front, are the bamboo shoot root type; trees on the far right are the buttress root type)

In addition, the main cause of salt marsh disappearance in the mid-latitudes is land reclamation, and the death of frontal seagrass beds due to water pollution (Eutrophication of seawater) caused by urbanization and sea level rise.

(7) Rising Sea Levels and High Waves Due to Global Warming

Global warming is due to an increase in the concentration of carbon dioxide and other greenhouse gases in the atmosphere with rising seawater temperatures causing the expansion of seawater and the melting of ice in polar regions and glaciers causing sea level rise. Warmer ocean waters also increase the intensity of tropical depressions and typhoons, accompanied by high storm surges, high waves, and strong nearshore currents. The climate change induced sea level rise reduces

6. COASTAL TOPOGRAPHIC CHANGE

beach area, which together with strong nearshore currents and high waves increase sediment instability and intensify coastal erosion.

(8) Human-induced Land Subsidence

Land subsidence in fluvial river delta occurs due mainly to two factors:

(a) Consolidation of shallow Holocene sediments by human loadings, such as buildings, roads, and other infrastructures together with natural loadings, and drainage.

(b) Extraction of groundwater from deep underground aquifers.

Large land subsidence exacerbates erosion. For example, on the Niigata and Kujukuri coasts in Japan, land subsidence caused by large-scale extraction of groundwater for natural gas extraction has exacerbated the damage caused by coastal erosion. In these coasts, areas with annual subsidence of more than 10 cm were widespread during the economic growth period after World War II. This has been reduced in recent years due to the restrictions imposed by the local government on natural gas extraction since 1973, but subsidence continues at around 1 cm/year in some areas of the Niigata coast and around 10 cm/year in some areas of the Kujukuri coast.

(9) Land Subsidence Due to Natural Forces

Plate-boundary earthquakes cause land uplift near landward plate boundaries and subsidence away from landward plate boundaries. In other words, large earthquakes tend to cause coastal areas away from the landward plate boundary to sink. This exacerbates the damage caused by coastal erosion.

【Coastal erosion on the Kaike Coast, Tottori Prefecture, Japan】

The main cause of serious beach erosion on the Kaike Coast facing the Sea of Japan is the decrease in the supply of sediment from the Hino River, which flows into this coast. The dams built upstream of this river have contributed to this decrease in sediment supply, but there is also another unique reason.

In this region of Japan, iron ore had long been extracted by a method known as "kanna nagashi (in Japanese)", in which iron sand is extracted by cutting down iron ore-rich rocks from the mountains and pouring them into channels, where they are separated into sediment and iron ore. The sediment from this process had been supplied in large quantities to the Kaike Coast via the Hino River, but this was rapidly reduced since 1900 due to imports of iron ore from overseas, and the shoreline of the Kaike Coast receded by about 100 m in a 10 km section around the mouth of the Hino River between 1900 and 1970, when full-scale coastal protection projects began.

248

6.4 Coastal Erosion

Uda [80], [81] discussed the actual situations of major coastal erosion in Japan in "Japan's Beach Erosion (Reality and Future Measures)" and summarised measures to deal with major causes of erosion in "Reality and Measures of Coastal Erosion", which should be referred to by those interested in the erosion problem.

2) Basic Approach to Coastal Erosion Control

The common cause of coastal erosion except land subsidence is an imbalance in the sediment budget (inflow and outflow of sediment), resulting in a sediment deficit in the target area. Therefore, the basic idea of coastal erosion control is "how to correct the imbalance of the sediment budget?". However, since the facilities and actions that caused human-induced erosion were built and executed out of necessity, many of the actions have been banned, but many of the facilities cannot be removed, so there are many cases of coastal erosion problems that are difficult to settle.

(1) Case of Existence of Dams or Ports and Harbours

If the role of dams, ports, and harbours has come to an end, they should be removed. However, in most cases, they were built with important missions and are still playing important socio-economic and environmental roles. Therefore, methods to allow downstream flow of sediment should be established. Then, if it takes time for the effects of these measures to become apparent, only then should remedial measures (e.g. beach nourishment and installation of erosion control facilities) be considered.

(2) Case of Sand and Gravel Mining

This case could be largely eliminated by imposing a ban on sand and gravel mining. However, because some time is usually required for beaches to recover after eliminating sand and gravel mining, interim measures such as beach nourishment and construction of temporary sediment control structures will be necessary. If there are still rivers and coasts where gravel extraction continues despite severe beach erosion, the higher authorities should take the lead in shifting the sand and gravel mining only to rivers with high sediment-producing margins and should encourage the enhancement of aggregate recycling systems.

(3) Case of Installation of Coastal Erosion Control Facilities

First, the causes of coastal erosion should be thoroughly investigated. Then, the best approach to reinstate the sediment supply and eliminate coastal erosion should be decided.

However, in cases where the sediment cannot be supplied from the source, such as the relationship between the Kujukuri Coast in the Boso Peninsula of Japan and wave absorbing breakwaters in the sea cliff areas, the prevention of sediment discharge by using erosion control

6. COASTAL TOPOGRAPHIC CHANGE

facilities can only be carried out while considering the overall balance. To minimise the use of the control facilities, limited beach nourishment may be applied if necessary.

(4) Case of Reinforcement of Flood Control Facilities

From the point of view of flood control, it may not be possible to reverse the situation before the reinforcement facilities were put in place. Therefore, a combination of sediment supply (beach nourishment) to sections with inadequate sediment supply and minimal erosion control facilities will be used to combat coastal erosion.

(5) Case of Destruction of Coastal Ecosystems

In the low latitudes, since the mass removal of mangrove forests has brought heavy beach erosion due to the direct action of high waves, it becomes necessary to maintain wave absorbing facilities. However, since the mangrove forested beaches are muddy beaches which are soft ground, wave absorbing breakwaters made of stone or concrete blocks would immediately sink. Therefore, breakwaters made using bamboo wood or reinforced fibre bags filled with silt can be installed to recreate calm areas within which mangroves can be reinstated.

In the mid-latitudes, the disappearance of salt marshes that have not been reclaimed is mainly due to a lack of sediment supply and the disappearance of seagrass communities with the ability to dissipate waves, so in addition to beach nourishment and the minimum necessary wave-dissipating facilities, it is also necessary to consider restoring seagrass communities through improving water quality and planting in the hope that the dwindling small aquatic fauna will recover.

(6) Case of Sea Levels Rise and High Waves Due to Global Warming

Measures against sea level rise and high waves include the strengthening (reinforcement and expansion) of protection functions provided by coastal dikes/seawalls and wave-absorbing facilities.

However, it is crucial to sufficiently reduce the generation of carbon dioxide and other greenhouse gases which is the primary cause. This is achieved by significantly reducing the use of thermal power generation, internal combustion engines, and so on, which consume large quantities of coal, oil and natural gas. To compensate for the energy shortfall caused by this, the use of nuclear power could be considered, but as it is very difficult to ensure safety, it is also extremely important to promote the use of renewable energies, which is explained in Section 8.3.

(7) Case of Land Subsidence

Groundwater extraction could be considerably reduced by imposing restrictions. However, consideration should be given to restoring the groundwater level although full recharge may not be

6.4 Coastal Erosion

possible. It is extremely difficult to backfill other large areas to the original ground level, except in limited areas where this is particularly necessary.

Even in the event of a great earthquake, there is no natural return to the original ground level in a short period and it is extremely difficult to backfill the entire area.

In these cases, the main practical measures would be to raise and widen embankments if there is a significant population and property in the landward area. If there is little population and property in the landward area, relocation to safer locations should be considered.

However, in many instances, coastal erosion occurs as a result of a combination of causes. Therefore, combined measures are often required to mitigate coastal erosion. Specific coastal erosion control facilities and methods are described in Sections 7.1 and 7.2 (in addition, areas where the shoreline protrudes or is discontinuous compared to its surroundings should be eliminated as far as possible, as they are vulnerable to concentrated attack by high waves).

6. COASTAL TOPOGRAPHIC CHANGE

6.5 Scour and Sand Outflow

1) Scour Due to Large Waves

When coastal structures are constructed on silt, sand or gravel beaches, the seabed can be subjected to significant scouring during storms. Even if the structure is designed to withstand wave forces, there is a high probability that the structure will fail due to frontal scour (if readers want to study scour, please refer to Sumer and Fredsøe [82]). Therefore, when designing coastal structures, it is necessary to select a bathymetric change model with high scour prediction accuracy from Section 6.3 "Coastal Topographical Change Prediction Methods" and carry out numerical experiments, or conduct hydraulic model experiments to predict the maximum depth and width of the scour.

However, full-scale three-dimensional numerical models and large hydraulic model experiments are not easily possible. Therefore, Yamamoto et al [83], [84] used the improved numerical model of Vu et al [61] (described in Section 2 of the Appendix), which was found to be highly accurate in predicting scour, to calculate the scour depth in front of a seawall or a breakwater, and prepared **Figs. 6.12(1)** and **6.12(2)** for predicting the maximum scour depth. The data for the maximum scour depth used in the preparation of these calculation charts was obtained by the following procedure.

(a) Three incident wave heights of 3 m, 6 m, and 9 m were set as the incident significant wave heights at the offshore boundary. The corresponding significant wave periods of 8.8s, 12.4s and 15.2s respectively were selected, to satisfy a wave steepness of 0.025, which is neutral to erosion and sedimentation. The main wave direction of the incident waves was set perpendicular to the seawall and the breakwater.

(b) Next, considering a typical fine sand beach (the median grain size 0.2 mm), a beach profile with a plain seabed slope of 1/20 was constructed. To exclude the effects of wave-induced erosion and sedimentation, irregular waves with the above significant wave parameters were applied for more than 12 hours until a stable beach topography for each significant wave was reached.

(c) Then, an upright coastal structure was placed on the stable beach profile, and the beach was subjected to another 12 hrs of the same wave activity. The final maximum scour depth at the front of the coastal structure was determined at the end of 12 hrs. If the scour does not cease after 12 hrs, the maximum scour depth after 24 hours was determined. It is assumed that high waves do not usually last longer than 24 hours.

(d) The scour prediction calculations were repeated for all selected wave conditions and various structure installation depths. When a sufficient number of data was available, **Fig. 6.12(1)** was produced.

(e) The same work was repeated for the case where a 10 m crown width (equivalent to three rows of wave-dissipating blocks) wave-absorbing breakwater with a porosity of 50% was attached to the front of the upright structure, and **Fig. 6.12(2)** was produced.

6.5 Scour and Sand Outflow

It is known from previous studies and surveys that the actual maximum scour depth may be as high as the incident significant wave height; and that during high waves, the liquefaction of submarine ground in the front of the coastal structure is generally advanced at least up to the maximum scour depth, creating a dangerous condition for the coastal structure.

According to a report produced by the Port and Airport Research Institute in Japan using large-scale laboratory experiments on scouring by standing waves, the maximum scour depth tends to appear at the nodes of the standing waves under conditions of perpendicular wave incidence, and at the loops of the standing waves under oblique wave incidence. However, it should be noted that the actual incident waves are irregular, and therefore, these observations may not be valid in real situations.

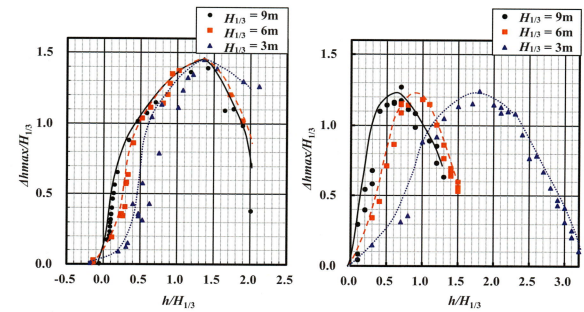

$\Delta hmax$ is the maximum scour depth,
h is the water depth in front of the breakwater,
$H_{1/3}$ is the incident significant wave height.
(1) The sea bottom grain size is 0.2 mm.
(2) The wave steepness is 0.025.
(3) The dominant wave direction is perpendicular to the breakwater front.

Fig. 6.12(1)
Relationship between the maximum scour depth in front of the upright breakwater, the significant offshore wave height and the water depth in front of the breakwater.

$\Delta hmax$ is the maximum scour depth,
h is the water depth in front of the breakwater,
$H_{1/3}$ is the incident significant wave height.
(1) The sea bottom grain size is 0.2 mm.
(2) The wave steepness is 0.025.
(3) The dominant wave direction is perpendicular to the breakwater front.

Fig. 6.12(2)
Relationship between the maximum scour depth in front of the breakwater with wave-dissipating works, the significant offshore wave height and the water depth in front of the breakwater.

The calculated values of the maximum scour depths (after 12-24 hours of wave incidence) in front of a coastal structure at different water depths about the representative significant wave height can be obtained from **Figs. 6.12(1)** and **6.12(2)** for the sediment with a median grain size of 0.2 mm.

6. COASTAL TOPOGRAPHIC CHANGE

However, the amount of scour should decrease as the median grain size of the bottom sediment increases and under oblique wave incidence. Therefore, Yamamoto et al. proposed calculation diagrams for scour reduction factors under different sediment sizes and angles of wave incidence, based on the following procedure. The results are shown in **Figs. 6.13(1)** and **6.13(2)**.

(a) The combinations of significant wave heights and periods of incident irregular waves are the same as **Fig. 6.12**, i.e. 3 m and 8.8 s, 6 m and 12.4 s, and 9 m and 15.2 s. Beaches with five different median grain sizes of 0.2 mm, 0.5 mm, 1.0 mm, 5.0 mm, and 10.0 mm and four different incident wave directions, 0, 20, 40, and 60 degrees to orthogonal direction, are considered.

(b) First, the above irregular waves were applied for more than 12 hours to the beach profile of the 1/20 seabed slope with each median grain size, following the same procedure discussed earlier, to remove the effects of wave-induced erosion and sedimentation, and prepare a stable beach profile.

(c) Then, the coastal structure was placed on the beach for each scenario and waves were run for another 12 or 24hrs. The maximum scour depth at the front of the coastal structure, at the end of this period, was determined

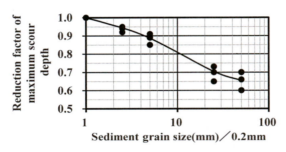

(1) The wave steepness is 0.025.
(2) The dominant wave direction is perpendicular to the front of the breakwater.

Fig. 6.13(1) Reduction factor for the maximum scour depth due to larger median grain size.

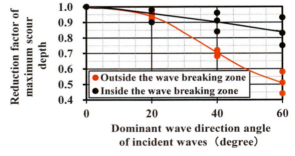

(1) The wave steepness is 0.025.
(2) The dominant wave direction angle of incident waves is the inclination angle from the direction perpendicular to the breakwater front.

Fig. 6.13(2) Reduction factor for the maximum scour depth due to oblique incidence.

The maximum scour depth near the front of the coastal structure can be obtained by multiplying the value read from **Figs. 6.12(1)** or **6.12(2)** by the reduction factor read from **Figs. 6.13(1)** and **6.13(2)**.

Assuming that the scour pit approaches a triangular cross-section with a bottom sediment slope of repose angle (slope of about 30°) at the maximum scour depth, the maximum scour width can be estimated to be $\sqrt{3} \times 2$ times the maximum scour depth.

Furthermore, it should be noted that the scour cross-section surveyed a few days after the end of high waves is likely to be smaller than the scour cross-section during high waves because waves with small wave steepness can encourage sedimentation deposition which can cause backfilling phenomena.

In addition, when examining the stability of the coastal structure during high waves, liquefaction should be assumed to occur in the sediment layer up to the maximum scour depth.

As a reference material, in **Fig. 6.14**, the non-dimensionalised data of the change of scour depth after 12 hours for the three types of significant waves mentioned above, when the coastal structure is located near the wave breaking point are shown in black circles and solid lines. Similar results, when the coastal structure is located offshore from the wave breaking depth, are shown as white circles and dotted lines.

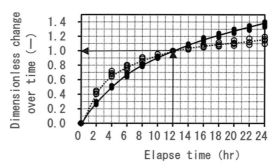

(1) Black circles and solid lines indicate the dimensionless changes over time of maximum scour depth when the breakwater is located near the wave breaking location.
(2) White circles and dotted lines indicate the dimensionless changes over time of maximum scour depth when the breakwater is located offshore from the wave breaking location.

Fig. 6.14 Change in maximum scour depth over time.

2) Outflow (Suction) of Backfill Materials
(1) Case of Dikes or Seawalls Covered with a Single Concrete Slab

As shown in the experiment in **Photo 6.2**, when the frontal scour due to high waves progresses nearer to the lower edge of the front face of a dike or a seawall covered with a single concrete slab, the backfill material is sucked out from the lowest edge and hollowing out progresses due to the maximum suction force becomes greater than the effective suction resistance. As shown in **Fig. 6.15**, the dike or the seawall becomes more fragile even when a wave overtopping rate, which is positively related to wave force, is low. However, sediment supply by backfill material suction weakens the frontal scour. Thus, it is important to study not only the scour caused by high waves but also the prevention of suction. If the frontal scour depth becomes sufficiently deeper than the lowest edge of the front face of the dike or the seawall, the structure will fail more dramatically by total collapse than by suction failure.

Iwasaki et al [85] have shown in a large physical model experiment that even if there are no gaps due to poor construction or ageing of the front cover of the dike, suction can occur from the lowest edge of the front face of the dike.

6. COASTAL TOPOGRAPHIC CHANGE

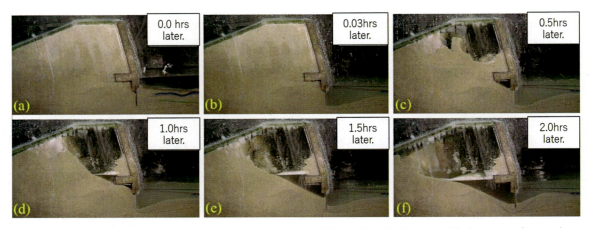

Although it is normal to have a 50 cm thick layer of cobble and crushed stone, this layer was ignored in this experiment in order to clarify the effect of varying the median grain size of the backfill material.

Photo 6.2 Suction experiment at 1/30th scale, 0.2 mm median grain size of backfill material and 14.3 cm incident significant wave height.

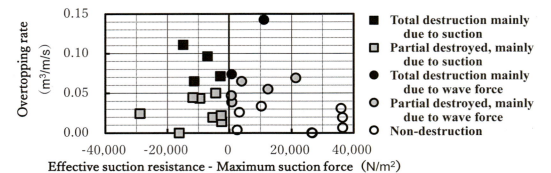

This figure was compiled based on data on damaged cases and healthy dikes and seawalls in Japan, and shows that in the range where the horizontal axis is negative, dikes and seawalls are weakened by suction and fail with small wave overtopping rates.

Fig. 6.15 Relationship between failure degree of dikes and seawalls, wave overtopping rates, and "Effective suction resistance − Maximum suction force".

Maeno et al [86] used a finite element model (FEM) based on poroelastic infiltration theory to represent suction outflow from an upright dike and showed that the suction rate of backfill material decreases with increasing time, with deeper rooting, and with increasing permeability of the ground. Furthermore, Maeno et al [87] reproduced backfill material suction from an upright revetment using a combined Digital Elevation Model and a finite element model (DEM-FEM). Harada et al [88] used a three-dimensional DEM model to reproduce the suction from the same type of revetment. Nakamura et al [89] reproduced the suction from a rubble simple revetment using a FEM model combined with the Volume Of Fluids (VOF) model, and Goto et al [90] reproduced the suction from a concrete square revetment using a numerical prediction model with the MPS method for fluid motion and an elastoplastic model for soil particle motion.

6.5 Scour and Sand Outflow

For backfill material suction from dikes covered on three sides (front side covered with concrete, top and back with concrete or asphalt) and seawalls covered on two sides (front side and top covered in the same way), Yamamoto et al [91] have shown from case studies and stability calculations that damage from frontal scour and suction is predominant if the frontal water depth is shallower than 3 m. Yamamoto et al [92] have shown that the failure of dikes and seawalls can be determined by the wave overtopping rate and the difference between the effective suction resistance and maximum suction force, as shown in **Fig. 6.15**, and that the effective suction resistance increases with the internal friction angle and the grain size of backfill material. Yamamoto et al [91] proposed equations for calculating the effective suction resistance and maximum suction force, considering the effect of the median grain size of backfill material. Ioroi and Yamamoto [93] proposed an equation for calculating the suction rate for sandy backfill material. Yamamoto et al [84] improved this equation so that it can be applied to the damage cases on actual coasts and confirmed that the improved equation reproduces the suction situation well on 19 actual coasts.

The equation of Yamamoto et al [84] is expressed by Eqs. (6.55) – (6.58). The suction rate including porosity q [m³/m/s] per unit time and unit width from the lowest edge of front face of a dike covered on three sides or a seawall covered on two sides is largely governed by the excess pore water pressure within the backfill layer of these structures.

$$
\left.
\begin{array}{l}
\theta_s - \theta_c \geq 0.0: \quad \dfrac{q}{w_s d_{50}} = \beta\left(\theta_s - \theta_c\right)\dfrac{1}{2}\left[1 + \cos\left(\dfrac{\alpha t}{T_{1/3}}\right)\right] \qquad \left[0 \geq \dfrac{\alpha t}{T_{1/3}} \geq \pi\right] \\[4mm]
\alpha = 0.00218 \times e^{-0.038\left(\frac{d_{50}}{0.0002}\right)}, \qquad \beta = 0.056 \times e^{-0.69\left(\frac{d_{50}}{0.0002}\right)^{0.72}} \\[4mm]
\left[0.0002 \text{ is the reference grain size and the unit is 'm'.}\right] \\[2mm]
\left[\text{Scope of application is } 1 \leq \dfrac{d_{50}}{0.0002} \leq 50.\right] \\[4mm]
\theta_s - \theta_c \leq 0.0: \quad \dfrac{q}{w_s d_{50}} = 0.0
\end{array}
\right\} \tag{6.55}
$$

$$
\theta_c = \frac{\tau_r}{(\rho_s - \rho)g d_{50}}, \qquad \tau_r = (\rho_s g d_t - \rho g d_t - P_{ob\,max})\tan\phi \tag{6.56}
$$

$$
\theta_s = \frac{\tau_f}{(\rho_s - \rho)g d_{50}}, \qquad \tau_f = \frac{f}{2}\rho V_{max}^{\,2} \tag{6.57}
$$

$$
\left.
\begin{array}{l}
\dfrac{P_{ob\,max}}{\rho g H_{1/3}} = a \times \left[\tanh\left(\dfrac{0.03 H_{1/3}}{d_t}\right)\right]^b, \qquad V_{max} = \sqrt{\dfrac{2 P_{ob\,max}}{\rho\left[(h/H_{1/3}) + 1.0\right]^{1.11}}} \\[4mm]
a = 0.37\left(0.0002/d_{50}\right)^{0.85} + 0.30, \qquad b = 0.55\left(0.0002/d_{50}\right)^{0.78} + 0.05 \\[4mm]
\left[\text{Scope of application is } 1 \leq \dfrac{d_{50}}{0.0002} \leq 50\right]
\end{array}
\right\} \tag{6.58}
$$

where w_s is the settling velocity of backfill material according to Rubey's formula; d_{50} is the median grain size of backfill material [m] (If there is a layer of cobble/crushed stone in the backfill, a weighted

6. COASTAL TOPOGRAPHIC CHANGE

arithmetic mean of the area ratio between this layer and the sand layer is used to obtain an approximation); t is the time elapsed since the start of suction [s]; $T_{1/3}$ is the incident significant wave period; α is a coefficient that reduces the suction time divided by $T_{1/3}$ to π in order to represent the phenomenon of suction reduction to zero as a curve with a phase angle from 0 to π of the cosine function (hence $t = \pi \times T_{1/3}/\alpha$ is regarded as the suction end time); and β is a proportionality coefficient. θ_c and τ_r are respectively the dimensionless critical suction resistance and the effective suction resistance per unit area during back flow obtained from Eq.(6.56) (if these are negative values, they are considered to make the suction of backfill material more ease); θ_s and τ_f are respectively the dimensionless suction force and the maximum suction force per unit area during back flow obtained from Eq.(6.57); ρ_s is the density of the sand layer (usually around 1.8); ρ is the density of seawater; g is the gravitational acceleration; d_t is the thickness of the sand layer in front of the dike or the seawall (= the height from the ground surface in front of this structure to the lowest edge of the front face of this structure); ϕ is the internal friction angle of the sand layer ($\fallingdotseq 30$ degrees); f is the coefficient of fluid force at the suction layer ($\fallingdotseq 1$); P_{obmax} and V_{max} are respectively the maximum excess pore water pressure (the maximum value of dynamic pore water pressure) and maximum flow velocity at the lowest edge of front face of the dike or the seawall during back flow which can be obtained from Eq. (6.58). Furthermore, h is the water depth at the front of the dike or the seawall up to the suction outlet (= tide level + depth at the lowest edge of the front face of this structure); and $H_{1/3}$ is the incident significant wave height at the front of the dike or the seawall (If the water depth in front of the structure during the targeted stormy weather conditions required to obtain this incident wave height cannot be estimated correctly, the incident wave height at a position offshore about five times of the equivalent offshore wave height can be used).

From Eqs. (6.55) – (6.58), the following can be found.

(a) The suction rate of backfill material increases with the increase of the incident wave height.

(b) Since the suction outlet of backfill material is fixed at the lowest edge of front face of the dike or the seawall, the flow velocity at the suction outlet and the suction rate of backfill material decrease with the increase of water depth at the lowest edge of the front face.

(c) The suction rate of backfill material decreases with the increase of the median grain size of backfill material.

Moreover, the following can also be stated from the wave shoaling phenomenon.

(d) The longer the incident wave period, the relatively larger the incident wave height in front of the dike or the seawall. Therefore, the suction rate of backfill material increases with the increase of the incident wave period.

6.5 Scour and Sand Outflow

Ioroi et al [94] attempted to reproduce the suction outflow process by substituting the excess pore water pressure and flow velocity obtained from CADMAS-SURF (refer to Section 2.4), which can simulate fluid motion in porous media, into the proposed equations for calculating the suction rate. But the accuracy of the reproduction was not good because CADMAS-SURF cannot correctly account for differences in grain size in sand and gravel. They showed from experiments that:

(a) As the uniformity coefficient of backfill material increases, the suction rate decreases.

(b) As the dry density (degree of compaction) increases, the final cumulative volume of suction itself changes slightly, but the time taken to reach the final cumulative volume of suction increases.

Yoshizawa et al [95] proposed equations for correcting excess pore water pressure and flow velocity using CADMAS-SURF to improve the accuracy of reproducing the suction process of backfill material. They also proposed an equation for calculating suction rates that can consider the influence of the uniformity coefficient and the dry density on the backfill material.

Silarom et al [96], [97] proposed a numerical model, which can reproduce the continuous development of a hollow in a dike or a seawall due to suction outflow. Their model uses CADMAS-SURF to determine the maximum pore water pressure p_{CAD} and the maximum flow velocities U_{CAD} and V_{CAD} during the backflow period of each wave in the x-direction (from the front face of the dike or the seawall to the inside of the structure) and the y-direction (alongshore direction), Eq.(6.59) to obtain the suction rate of backfill material (included porosity) q_x and q_y per unit time and unit width, and Eq.(6.60) (a sand continuity equation) to calculate the backfill thicknesses $d_{i,j}$ (where i is the mesh number from the front to the inside of the structure shown in **Fig. 6.16** and j is the mesh number in the longshore direction) in the structure.

Here, the maximum suction force τ_{fx} and τ_{fy}, and the effective suction resistance τ_{rx} and τ_{ry} for each individual incident wave in the x and y directions are obtained from Eq. (6.61). However, Since CADMAS-SURF cannot accurately evaluate the effect of different grain sizes of backfill material on pore water pressure, the maximum pore water pressure p_{CAD} during the back flow period for each wave calculated using CADMAS-SURF must be substituted into Eq. (6.62) derived using some experiment data, to get the correct maximum excess pore water pressure p_{obmax}.

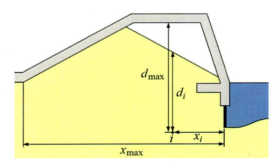

Fig. 6.16 Illustration of main parameters within the cross-section of the structure.

6. COASTAL TOPOGRAPHIC CHANGE

$$
\left.\begin{aligned}
\frac{q_x}{w_s d_{50}} &= \frac{0.321}{e^{0.25(d_{50}/0.0002)}} \times (\theta_{sx} - \theta_{cx}) = \frac{0.321}{e^{0.25(d_{50}/0.0002)}} \times \frac{\tau_{fx} - \tau_{rx}}{(\rho_s - \rho)g d_{50}} \\
\frac{q_y}{w_s d_{50}} &= \frac{0.321}{e^{0.25(d_{50}/0.0002)}} \times (\theta_{sy} - \theta_{cy}) = \frac{0.321}{e^{0.25(d_{50}/0.0002)}} \times \frac{\tau_{fy} - \tau_{ry}}{(\rho_s - \rho)g d_{50}}
\end{aligned}\right\}
\tag{6.59}
$$

$$
\frac{\Delta d_{i,j}}{\Delta t} = \frac{\Delta q_{xi,j}}{\Delta x} + \frac{\Delta q_{yi,j}}{\Delta y} \quad \left(\frac{d_{i+1,j} - d_{i,j}}{\Delta x} \le \tan \alpha_{rp}, \left| \frac{d_{i,j+1} - d_{i,j}}{\Delta y} \right| \le \tan \alpha_{rp} \right)
\tag{6.60}
$$

$$
\left.\begin{aligned}
\tau_{fx} &= \frac{f}{2} \rho U_{CAD}^2, \quad \tau_{rx} = (\rho_s g d_{i-1,j} - \rho g d_{i-1,j} - p_{obmax}) \tan \phi \\
\tau_{fy} &= \frac{f}{2} \rho V_{CAD}^2, \quad \tau_{ry} = (\rho_s g d_{i,j-1} - \rho g d_{i,j-1} - p_{obmax}) \tan \phi
\end{aligned}\right\}
\tag{6.61}
$$

$$
\left.\begin{aligned}
p_{obmax} &= C_x C_d C_{d50} (p_{CAD} - \rho g h) \\
C_x &= 0.00022^{x_i / x_{max}} \\
C_d \le 1: \ C_d &= \exp\left[0.269 - \frac{0.263}{d_{i,j}/d_{max}} + \chi \ln(d_{i,j}/d_{max}) \right] \\
\chi &= 45.0 - 23.7 \exp\left[\frac{-(x_i/x_{max}) - 3.265}{4.953} \right] \\
C_d > 1: \ C_d &= 1 \\
C_{d50} &= 0.65\left(\frac{0.2}{d_{50}} \right)^{0.85} + 0.35
\end{aligned}\right\}
\tag{6.62}
$$

where q_x and q_y are respectively the suction rates of backfill material included porosity in x and y direction per unit width and unit time, θ_x and θ_y are respectively the dimensionless suction force in x and y direction, θ_{cx} and θ_{cy} are respectively the dimensionless critical suction resistance in x and y direction, τ_{fx} and τ_{fy} are respectively the maximum suction force per unit area during backflow for each wave in the x and y directions, and τ_{rx} and τ_{ry} are respectively the effective suction resistance per unit area during backflow for each wave in the x and y directions. $d_{i,j}$ is the backfill thickness from the lowest edge of the front face of the structure at the points i and j, $d_{i+1,j}$ is the backfill thickness from the lowest edge of front face of the structure one mesh further than the point i, $d_{i,j+1}$ is the backfill thickness from the lowest edge of front face of the structure one mesh beyond the point j in the outflow reverse direction, $\Delta q_{xi,j}$ is the difference in the x-directional suction rate between horizontal mesh intervals Δx at the points i and j, $\Delta q_{yi,j}$ is the difference in the y-directional suction rate between horizontal mesh intervals Δy at the points i and j, Δt is the time interval, α_{rp} is the angle of repose of backfill material, $d_{i-1,j}$ is the backfill thickness at the lowest edge of front face of the structure one mesh ahead in the outflow direction (offshore) from the point i, and $d_{i,j-1}$ is the backfill thickness at the lowest edge of front face of the structure one mesh ahead in the outflow direction (lateral) from the point j. C_x, C_d and C_{d50} are correction factors to determine the correct excess pore water pressure in the backfill layer, h is the water depth at the location where the excess pore water pressure in the structure is calculated, x_i is the horizontal distance from the front face of the structure to the point i to be calculated, x_{max} is the total horizontal length from the front face to the back end of the structure, and d_{max} is the total backfill thickness from the lowest edge of front face to the top of the structure.

260

6.5 Scour and Sand Outflow

(2) Case of Dikes or Seawalls Covered with Concrete Blocks

Fukuhara et al [98] conducted physical experiments on a sloping dike or a sloping seawall covered with concrete blocks, as shown in **Photos 6.3** and **6.4**. Here, to clarify the effect of changing the median grain size of backfill material, their experimental models ignored the 50 cm thick layer of cobble and crushed stone, which is usually located inside the front cover of an actual dike or seawall.

Their experiments revealed the following:

(a) When an upright dike or upright seawall with the front face covered with a single concrete slab shown in **Photo 6.2** is subjected to wave action, at first, scour develops at the structure front [(b) of the photo]. When the scour reaches the lowest edge of the structure front, suction begins [(c) of the photo]. the suction stops when the hollow reaches a line extending diagonally up at the slope of repose from the lowest edge of the structure front to the landward side inside the structure [(f) of the photo]. On the other hand, in the case of the concrete block cover shown in **Photos 6.3** and **6.4**, if there is no rigid foot protection in front of the structure, no significant scour occurs and backfill material is sucked out from the gaps between the covered blocks of the structure (not only at the lowest edge of front face). When the blocks fall into the hollow created inside the structure, the backfill material flows out rather than being sucked out until the surface slope of the backfill layer reaches a stable slope for natural beach against wave force.

The scale is 1/25, the crown height is 25 cm, the area ratio of gaps and holes of all blocks to the total surface area is about 10%.

Photo 6.3
Front view of the seawall model covered with perforated concrete blocks.

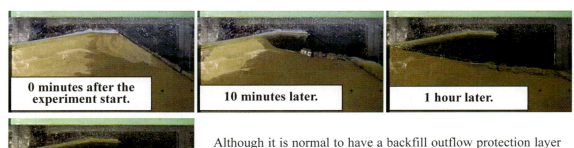

0 minutes after the experiment start.

10 minutes later.

1 hour later.

2 hours later.

Although it is normal to have a backfill outflow protection layer (usually 50 cm thick) made of cobble/crushed stone between the backfill material and the perforated block layer, this layer was ignored in this experiment in order to clarify the effect of varying the median grain size of backfill material.

Photo 6.4 Suction experiment at 1/25th scale, 0.2 mm median grain size of backfill material and 16.0 cm incident significant wave height.

6. COASTAL TOPOGRAPHIC CHANGE

(b) In the case of a sloping dike or a sloping seawall covered with perforated concrete blocks, the outflow outlet is not limited to the lowest edge of the front face of the structure, but extends over the entire gaps between the covered blocks below the water surface on the front surface slope. Therefore, unlike the case of the structure covered with a single concrete slab, the outflow rate (= the suction rate) of backfill material increases with increasing water depth in the front face of the structure. However, the effects of the incident wave height and the period and the grain size of the backfill material on the outflow rate (= the suction rate) are similar to those of the case covered with the single concrete slab. Therefore, the outflow rate for a sloping dike or a sloping seawall covered with perforated concrete blocks can be estimated by replacing Eqs. (6.55) - (6.58) with Eqs. (6.63) – (6.65).

However, the equations proposed by Fukuhara et al. [98] was found to significantly overestimate measured values, so Yamamoto, one of the authors, modified the drag coefficient C_D and the coefficients α and β to match the measured values.

$$\left.\begin{aligned} &\theta_c = \frac{\tau_r}{(\rho_s - \rho)gd_{50}}, && \tau_r = (-P_{ob\max})\tan\phi, && P_{ob\max} = a \times \rho g H_{1/3} \\[2mm] &a = 0.77\left(\frac{0.0002}{d_{50}}\right)^{0.85} + 1.2 && \left[\begin{array}{l} 0.0002 \text{ is the reference grain size and the unit is 'm'.} \\ \text{Scope of application is } 1 \le \dfrac{d_{50}}{0.0002} \le 50. \end{array}\right] \end{aligned}\right\} \quad (6.63)$$

$$\left.\begin{aligned} &\theta_s = \frac{\tau_f}{(\rho_s - \rho)gd_{50}}, && \tau_f = \frac{f}{2}\rho V_{\max}^2, && V_{\max} = \sqrt{\frac{2P_{ob\max}}{\rho C_D}}, && C_D = \left(\frac{h}{H_{1/3}} + 1.0\right)^{1.7} \\ &&&&&& \text{If } C_D < 0.44, \ C_D \text{ is } 0.44 \end{aligned}\right\} \quad (6.64)$$

$$\left.\begin{aligned} &\frac{q}{w_s d_{50}} = \beta(\theta_s - \theta_c)\frac{1}{2}\left[1 + \cos\left(\frac{\alpha t}{T_{1/3}}\right)\right] && \left[0 \ge \frac{\alpha t}{T_{1/3}} \ge \pi\right] \\[2mm] &\alpha = 0.00080 \times e^{-0.00090\left(\frac{d_{50}}{0.0002}\right)}, && \beta = 0.035 \times e^{-0.29\left(\frac{d_{50}}{0.0002}\right)^{0.71}} \\[2mm] &\left[\begin{array}{l} 0.0002 \text{ is the reference grain size and the unit is 'm'.} \\ \text{Scope of application is } 1 \le \dfrac{d_{50}}{0.0002} \le 50. \end{array}\right] \end{aligned}\right\} \quad (6.65)$$

where θ_c and τ_r are respectively the dimensionless liquefaction pressure and the effective liquefaction pressure during the backflow in the backfill layer near the lowest edge in front of the structure, θ_s and τ_f are respectively the dimensionless outflow force and the maximum outflow force per unit area during the backflow in the backfill layer near the lowest edge in front of the structure, $P_{ob\max}$ and $V_{\max}$ are respectively the maximum excess pore water pressure and the maximum flow velocity during the back flow in the backfill layer near the lowest edge in front of the structure, ρ_s is the density of backfill layer [$\approx$ 1800 kg/m³], ρ is the water density, g is the acceleration of gravity, d_{50} is the median grain size of backfill materials [m], ϕ is the internal friction angle [$\approx$ 30°] of the beach sand and backfill material, a is the effect factor of grain size on excess pore water pressure, $H_{1/3}$ is the significant wave height of

6.5 Scour and Sand Outflow

the incident waves considering the setup in front of the structure, f is the outflow force coefficient [≈ 1], C_D is the drag coefficient, h is the water depth at the front of the dike or the seawall, w_s is the settling velocity of backfill material (using Rubey's equation), t is the elapsed time after the start of the outflow, $T_{1/3}$ is the significant period of incident waves, α is the time reduction factor, and β is the outflow adjustment factor.

In the case of perforated block-covered structures, because the layer thickness d_t of the suction-resisting sand layer in front of the structure is zero from the beginning and the hyperbolic tangent function is equal to 1, the coefficient b of the maximum excess pore water pressure $P_{ob\max}$ in Eq. (6.58) can be neglected. The coefficient a can be obtained from Eq. (6.63). Eq. (6.63) means that τ_r is not the resisting force (per unit area) but the force (per unit area) that liquefies backfill material. Next, the Eq. (6.64) for the maximum flow velocity V_{max} was adjusted using experimental data based on Eq. (6.58). As the water depth increases, the frontal incident wave height increases and the outflow rates of backfill material increase. But the increasing rate of the outflow rates is small because the return flow velocity in front of the structure does increase slightly. This mechanism is incorporated in Eq. (6.64). Moreover, the outflow rate including the porosity q [m³/m/s] of backfill material for the structure covered with perforated concrete blocks can be obtained from Eq. (6.65).

It can be seen in **Fig. 6.17,** which shows a comparison between the measured values (included a large-scale experiment data by Noguchi et al [99] and 4 field data in Japan) and the calculated values using Eqs. (6.63) - (6.65), that the accuracy of the equations is very high.

In the case of perforated block-covered sloping structures (the area ratio of gaps and holes of all blocks to the total surface area is about 10%), it is unlikely that significant scour will occur in front of the structures. Therefore, using incident wave heights at an offshore location at a distance equivalent to five times of offshore wave height overestimates the outflow rates easily. Therefore, the incident wave height in front of the structures must be correctly calculated from **Fig. 2.27** (relationship between incident wave height and water depth).

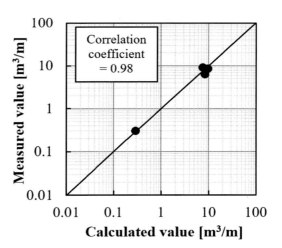

Fig. 6.17
Comparison of measured and calculated final volume of backfill material outflows from dikes and seawalls covered with perforated concrete blocks.

In **Fig. 6.17**, the perforated block-covered sloping structures of the large-scale experiment and the field cases have a backfill outflow protection layer (usually 50 cm thick) made of cobble/crushed

6. COASTAL TOPOGRAPHIC CHANGE

stone between the backfill material and the perforated block layer. In such cases, the mean of the median grain size calculated using the ratios of the cross-sectional areas of the cobble/crushed stone section and the sand section to the cross-sectional area of the backfill layer as weights for weighted arithmetic mean should be used as the median grain size for the calculation of the outflow rate.

Furthermore, if the numerical model CADMAS-SURF (refer to Section 2.4) is modified as follows, the development of hollow in the backfill part can be reproduced as shown in **Fig. 6.18**.

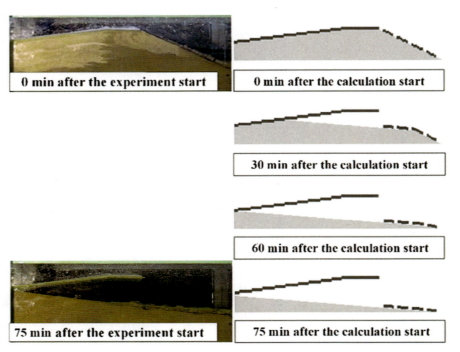

Fig. 6.18 Reproduction of hollowing-out in the backfilled area of the perforated block-covered sloping dike (d_{50} = 0.16mm, the offshore wave height = 16cm). (from Fukuhara et al [98])

(a) Create the batch processing program so that it can intermittently input the positional information of the hollow and the concrete covered blocks that settle with the widening of the hollow (in the case of perforated blocks, they do not move to distant locations due to wave lift, but settle as the surface of the backfill layer settles).

(b) Use Eq. (6.66) to calculate the outflow rate including porosity q [m³/m/s] for perforated block-covered structures using the excess pore water pressure and the outflow velocity at each calculation time Δt during backflow. However, since CADMAS-SURF cannot consider the effect of different grain sizes of sand and gravel, Eq. (6.67) should be applied to the calculated pore water pressure "Calculated p_{ob}" to consider the effect of different grain sizes.

(c) The backfill thickness d_s from the lowest edge of the front face of the structure is determined using Eq. (6.68) (continuity equation for backfill material). However, when the backfill material

6.5 Scour and Sand Outflow

is moved out, the surface slope of the backfill part is maintained with the stable slope of the beach under wave action [e.g. Eq. (6.5)]. The backfill material outflow calculation is stopped when this backfill layer surface connects in line with the outer beach surface of the structure.

$$\left.\frac{q}{w_s d_{50}} = \beta (\theta_s - \theta_c), \qquad \beta = \frac{0.074}{e^{0.11(d_{50}/0.0002)}}\right\} \tag{6.66}$$

$$\left.\theta_s = \frac{0.5 f \rho u_b^2}{(\rho_s - \rho) g d_{50}}, \qquad \theta_c = \frac{-P_{ob} \tan \phi}{(\rho_s - \rho) g d_{50}} \atop \text{Corrected } P_{ob} = \left[0.55 (0.0002/d_{50})^{0.85} + 0.39 \right] \times \text{Calculated } P_{ob}\right\} \tag{6.67}$$

$$\frac{\partial d_s}{\partial t} = \frac{\partial q}{\partial x} \tag{6.68}$$

where u_b and P_{ob} are respectively the flow velocity and the excess pore water pressure during the back flow on the surface of the backfill layer in the structure, x is the coordinate from offshore to shore.

3) Scour Due to Tsunamis

Impact of scour due to great tsunamis caused by plate boundary earthquakes cannot be ignored and, the prediction of scour due to tsunamis is important. Horizontal two-dimensional numerical models for predicting bathymetry changes due to tsunamis include the work of Takahashi et al [100], [101], [102] and Fujii et al [103], while three-dimensional numerical models include the work of Kihara and Matsuyama [104]. The horizontal two-dimensional numerical models of Vu et al [105], Yamamoto et al [106] and Ahmadi et al [107], or a three-dimensional numerical model of Nakamura and Mizutani [108], can be used for the prediction of topographical change in an inundated area on land. When it is important to reproduce complex flows in three dimensions, the model of Nakamura and Mizutani [108] may be used, although it requires a significant amount of computing power due to its three-dimensional nature. On the other hand, when it is desired to understand the horizontal distribution of onshore scour in wide area, the model of Vu et al [105], Yamamoto et al [106] and Ahmadi et al [107], which can significantly reduce the computation capacity, may be used.

The detailed composition, program and usage of Vu, Yamamoto, and Ahmadi's numerical models, which can be run in an ordinary PC, are described in Section 3 of the Appendix. Their numerical model uses Ribberink's Eq. (6.19) to calculate sediment transport rates. However, it should be noted that this equation overestimates the effect of the reduction of sediment transport with increase in sediment size. Therefore, the coefficient in Eq. (6.19) needs to be adjusted. Moreover, since the larger the uniformity coefficient and dry density of sediment, the higher the resistance of the sediment movement, both require a downward adjustment of the coefficient of Eq. (6.19) so that sediment transport rates are lowered. Calculation diagrams for making these adjustments (Figs. A.3.1, A.3.2) are also introduced in Section 3 of the Appendix.

When a tsunami propagates on a step, flow separation may not take place if the flow velocity is slow and the step is low. In such a situation, the scour can be calculated using a normal numerical

6. COASTAL TOPOGRAPHIC CHANGE

model with non-linear equations of motion, but if flow separation occurs at the step position, it is necessary to incorporate a topographic change calculation function into CADMAS-SURF (refer to Section 2. 4), or carry out large scale hydraulic model experiments. However, in situations where these options are not possible or difficult, the following methods equations can be used although they are limited to cases where the bottom sediment is sand.

(1) Equation of Noguchi et al [109]

Noguchi et al. conducted large-scale experiments and noted that a stationary vortex is generated when scour is caused by a separating flow from the crown of a revetment due to a tsunami. They proposed Eq. (6.69) for calculating the maximum scour depth Δh_{max} based on the estimation of the diameter of the vortex. Since the tsunami in their experiments was a solitary wave generated by a wave generator, the wavelength is considerably shorter than that of a great tsunami, and this equation should be regarded as an equation for tsunamis with a period of a few minutes.

$$\Delta h_{max} = 2.1 \times g^{-1/4} q^{1/2} z_f^{1/4} \tag{6.69}$$

where g is the gravitational acceleration, q is the peak flow rate per unit width of falling water [m³/s/m, e.g. Eq. (5.20) can be used], z_f is the height from the still water surface to the top of the revetment.

(2) Equation of Arikawa et al [110]

Arikawa et al. conducted experiments of the same scale as Noguchi et al. using a water circulation device until the scour is stabilized. They found that the maximum scour depth Δh_{max} is about 2.8 times larger than that of Noguchi et al.'s equation. Eq. (6.70) of Arikawa et al. can be used for great tsunamis with a run-up time on the order of 10 minutes.

$$\Delta h_{max} = 5.8 \times g^{-1/4} q^{1/2} z_f^{1/4} \tag{6.70}$$

(3) Equation of Yamamoto et al [111] or Nariyoshi & Yamamoto [112]

Yamamoto et al. proposed empirical equations based on experiments using tsunamic-like water flows in a large tank, assuming that the impulse or kinetic energy of the separating flow from a seawall hitting the ground shown in **Fig. 6.19** is proportional to the scale of the scour. They confirmed that their equations can reproduce scour caused by backflows similar to the scour generated by the 2004 Indian Ocean tsunami.

Nariyoshi and Yamamoto showed that the same empirical equations can be applied to scour caused by overflows on coastal dikes during the 2011 Tohoku Earthquake tsunami.

The maximum scour depth Δh_{max} can be calculated by using Eq. (6.71).

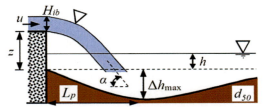

Fig. 6.19 Illustration of scour by a separating flow.

$$
\Delta h_{\max} = \begin{cases} C_{d\Delta h}C_{h\Delta h}C_e \times 0.44 \left(\dfrac{\sum E_y}{\rho g} \right)^{1/3}, & \sum E_y = \dfrac{1}{2}\rho \sum \left(uH_{ib}\Delta t \right)u_y^2 \\[4mm] C_{d\Delta h}C_{h\Delta h}C_f \times 0.15 \left(\dfrac{\sum F_y\Delta t}{\rho \sqrt{gH_{ib}}} \right)^{1/2}, & \sum F_y\Delta t = \rho \sum \left(uH_{ib}\Delta t \right)u_y \end{cases}
$$

$$
C_{d\Delta h} = \exp\left[-0.027\left(\frac{d_{50} - 0.0002}{0.0002} \right) \right] \qquad \begin{bmatrix} 0.0002 \text{ is the reference} \\ \text{grain size, the unit is 'm'.} \end{bmatrix}
$$

$$
C_{h\Delta h} = \begin{cases} \dfrac{1}{2} \times \left[1.0 - \tanh\left(3.7\sqrt{h/z} - 3.0 \right) \right] & \left[(h/z) < 1.0 \right] \\[3mm] \dfrac{1}{2} \times \left[1.0 - \tanh\left(0.7 \right) \right] & \left[(h/z) \geq 1.0 \right] \end{cases}
$$

$$
C_e = 1.11 \times \exp\left[-0.038\left(z/H_{ib} \right) \right], \qquad C_f = 1.04 \times \exp\left[-0.018\left(z/H_{ib} \right) \right]
$$

(6.71)

where $C_{d\Delta h}$ is the correction factor for scour depth on the median grain size d_{50} [the unit is 'm'] of ground sediment, $C_{h\Delta h}$ is the correction factor for scour depth on the frontal water depth h, C_e is the correction factor for the kinetic energy on the crown height z from the frontal ground to the top of the dike or the seawall, C_f is the correction factor for the impulse on the crown height z, ΣE_y is the accumulated kinetic energy, $\Sigma F_y\Delta t$ is the accumulated impulse, ρ is the density of water, g is the gravitational acceleration, H_{ib} and u are respectively the thickness and the horizontal velocity of the water flow on the crown of the structure, Δt is the division time for cumulative calculations, and u_y is the vertical velocity of the water flow jumping from the crown just before it hits the frontal ground.

The horizontal length L_p from the leading edge of the crown of the structure to the maximum scour depth can be calculated by using Eq. (6.72).

$$
L_p = C_{\alpha L}C_{dL}C_{hL}\left(u_{\max} \times t_r \right)
$$

$$
C_{\alpha L} = \begin{cases} 0.9 \times \exp\left[6.3\left(0.67 - \alpha \right) \right] & \left(\alpha \leq 0.67 \right) \\ 0.9 & \left(\alpha > 0.67 \right) \end{cases}
$$

$$
C_{dL} = \exp\left[-0.003\left(\frac{d_{50} - 0.0002}{0.0002} \right) \right] \qquad \begin{bmatrix} 0.0002 \text{ is the reference} \\ \text{grain size, the unit is 'm'.} \end{bmatrix}
$$

$$
C_{hL} = \begin{cases} \dfrac{1}{2} \times \left[1.0 - \tanh\left(2.8\sqrt{h/z} - 3.0 \right) \right] & \left[(h/z) < 1.0 \right] \\[3mm] \dfrac{1}{2} \times \left[1.0 - \tanh\left(-0.2 \right) \right] & \left[(h/z) \geq 1.0 \right] \end{cases}
$$

(6.72)

During run-up: $\quad u_{\max} = F_r\sqrt{gh_{i\max}}$

During back flow: $\quad u_{\max} = \sqrt{\dfrac{2\sin\beta}{f_b}}\sqrt{gh_{b\max}}, \qquad h_{b\max} = \dfrac{(R-z)f_b/3}{f_b + \sin\beta}$

where $C_{\alpha L}$ is the correction factor for horizontal length on the strike angle α [the unit is 'radian'] when the water flow strikes the frontal ground; C_{dL} is the correction factor for horizontal length on the sediment grain size d_{50} [the unit is 'm']; C_{hL} is the correction factor for horizontal length on the frontal water depth; $u_{\max}$ is the peak horizontal velocity of the water flow on the crown of the structure; t_r is the time taken for the water flow to reach the frontal ground from the crown; F_r is the Froude number of the tsunami run-up ($F_r \approx 1.1$ if there are obstacles, $F_r \approx 1.5$ if there is no obstacle); $h_{i\max}$ is the peak thickness of the water flow on the crown during tsunami run-up; β is the average slope angle of the

267

6. COASTAL TOPOGRAPHIC CHANGE

upstream ground; f_b is the friction coefficient on the upstream ground; h_{bmax} is the peak thickness of the water flow on the crown of the seawall or the revetment during tsunami back flow; and R is the tsunami run-up height from the frontal ground [e.g. obtained from Eq. (5.18)].

6.6 Wind-blown Sand

Sand moved by wind is called **wind-blown sand**. Sand on the shoreline is transported inland as wind-blown sand during strong winds, forming sand dunes. On coasts with large dunes, this may sometimes cause disasters, in which small villages can be swallowed up during strong winds. In recent years, this has been a key cause of burial of fishing harbours and the hindrance to the operation of coastal roads.

1）Critical Friction Velocity of Wind-blown Sand

The possibility of the initiation of wind-blown sand can be determined by whether the friction velocity of sand u_* is greater than the critical friction velocity u_{*c}, which is determined from Eq. (6.73). This equation is obtained from the equilibrium between the external force to move the sand grains and the weight of the sand grains in air.

$$u_{*c} = A\sqrt{\frac{\rho_s - \rho_a}{\rho_a} gd} \tag{6.73}$$

where A is the experimental constant (= 0.1 in the case that the sand grain size is uniform between 0.1 and 2.0 mm), ρ_s is the density of sand, ρ_a is the density of air, g is the gravitational acceleration, and d is the sand grain size.

Ishihara and Iwagaki [113] show that Eq. (6.73) and existing experimental data are in good agreement in $0.5 < \sqrt{\rho_s d}$. Moreover, Nakajima [114] showed that the critical friction velocity can be expressed by using Eq. (6.74) from experiments and that the coefficient A_w is about 0.1 when the bottom sediment is dry, and that the coefficient A_w increases relatively rapidly as the water content ratio of the bottom sediment increases. In other words, it can be said that when the bottom sediment contains water, wind-blown sand ceases to occur relatively rapidly.

$$u_{*c} \doteqdot A_w\sqrt{\frac{\rho_s}{\rho_a} gd_{50}} \tag{6.74}$$

where A_w is the experimental constant (= 0.1 in the case that the sand grain is dry and $0.5 < \sqrt{\rho_s d}$) and d_{50} is the median grain size of sand.

2） Equations for Calculating Friction Velocities and Wind-blown Sand Rates

Bagnold [115] proposed Eq. (6.75) as the wind velocity distribution in the presence of wind-blown sand.

$$U(z) = 5.75u_* \log\left(\frac{z}{z_{fp}}\right) + U_{fp} \tag{6.75}$$

where $U(z)$ is the wind velocity [cm/s] at a height z [cm] above the bottom, and for the coefficients z_{fp} and U_{fp} (these are called the focal points). Zingg [116], using wind tunnel experiments, has proposed

6. COASTAL TOPOGRAPHIC CHANGE

Eq. (6.76) which shows a relationship between the grain size d of the sand. Here the units for z_{fp} and d are [mm] and for U_{fp} [cm/s].

$$z_{fp} = 10d, \qquad U_{fp} = 8.8 \times 10^2 d \tag{6.76}$$

The friction velocity u_* for the target sand can be calculated backwards by substituting the coefficients got from the measured grain size of the sand and Eq. (6.76), and the measured wind velocity into Eq. (6.75).

Other empirical equations for friction velocity based on field observations include Eq. (6.77) of Kawamura [117], Eq. (6.78) of Kawata [118], and Eq. (6.79) of Hamada et al [119]. It should be noted that the units for friction velocity and wind velocity above the ground in Eq. (6.79) are [cm/s].

$$u_* \fallingdotseq 0.048 u_{30} \tag{6.77}$$

$$u_* \fallingdotseq 0.053 u_{100} \tag{6.78}$$

$$u_* \fallingdotseq 0.0572 u_{446.5} - 17.1 \tag{6.79}$$

where u_{30} is the wind velocity at a height 30 cm above the ground, u_{100} is the wind velocity at a height 100 cm above the ground and $u_{446.5}$ is the wind velocity at a height 446.5 cm above the ground.

Assuming that the sand transport rate is proportional to the power (= force required to move × movement velocity) and that the force is proportional to the square of the movement velocity, the mass flow rate of wind-blown sand q_m [kg/m/s] per unit width can be considered to be a function of the cube of the friction velocity. Eq. (6.80) of Kawamura [117] and Eq. (6.81) of Bagnold [115] are introduced as representative equations based on this concept.

$$q_m = B_1 \frac{\rho_a}{g} \left(u_* - u_{*c} \right) \left(u_* + u_{*c} \right)^2 \tag{6.80}$$

$$q_m = B_2 \sqrt{\frac{d}{d_o}} \frac{\rho_a}{g} u_*^3 \tag{6.81}$$

where B_1 is a dimensionless coefficient (≈ 2.78) given by experiments or field observations; B_2 is a dimensionless coefficient that depends on the grain size distribution range of the sand, 1.5 for uniform sand, 1.8 for naturally sieved sand, and 2.8 for sand with a wider distribution range; and d_o is the standard sand grain size (= 0.25mm = 0.00025m).

In addition, the equations for calculating the transport rate of wind-blown sand q [m³/m/s] per unit width and unit time are given in Eq. (6.82) for Kawamura [117] and Eq. (6.83) for Bagnold [115].

$$q = B_1 \left(\frac{\rho_a}{\rho_s} \right) \frac{\left(u_* - u_{*c} \right) \left(u_* + u_{*c} \right)^2}{g} \tag{6.82}$$

$$q = B_2 \sqrt{\frac{d}{d_o}} \left(\frac{\rho_a}{\rho_s} \right) \frac{u_*^3}{g} \tag{6.83}$$

List of References in Chapter 6

1） Johnson, J.W.: Scale Effects in Hydraulic Model Involving Wave Motion, *Trans. AGU*, Vol.30, No.4, 1949, pp.517-525.

2） Iwagaki, Y. and Noda, H.: Scale Effects in Hydraulic Models for Beach Process, *Disaster Prevention Research Institute Annual of Kyoto University*, No.4, 1961, pp.210-220. (in Japanese)

3） Iwagaki, Y. and Noda, H.: Laboratory Study of Scale Effects in Two-dimensional Beach Processes, *Proc. 8th Conf. on Coastal Eng.*, ASCE, 1963, pp.194-210.

4） Sunamura, T. and Horikawa, K.: Two-dimensional Beach Transformation Due to Waves, *Proc. 14th Conf. on Coastal Eng.*, ASCE, 1974, pp.920-938.

5） Yamamoto, Y. and Iwasaki, N.: Examination on the Stability of Coastal Protection Facilities Focusing on Beach Topography Change, *Journal of Civil Engineering in the Ocean*, Vol.22, 2006, pp.781-786. (in Japanese)

6） Rector, R. L.: Laboratory Study on Equilibrium Profiles of Beaches, *Tech. Memo.*, No.41, Beach Erosion Board, 1954.

7） Swart, D. H.: Schematization of Onshore-offshore Transport, *Proc. 14th Int. Conf. on Coastal Engrg.*, ASCE, 1974, pp.884-900.

8） Swart, D. H.: Predictive Equations regarding Coastal Transports, *Abstracts of 15th Int. Conf. on Coastal Engrg.*, ASCE, 1976, pp.133-135.

9） Sunamura, T.: Static Relationships among Beach Slope, Sand Size, and Wave Properties, *Geographical Review of Japan*, Vol.48, No.7, 1975, pp.485-489.

10） Liaison Council for Coastal Protection Facility Construction Standards: *Commentary for Revised Coastal Protection Facility Construction Standards*, National Coastal Association of Japan, 1987, 269p. (in Japanese)

11） Uda, T.: *Studies on Nearshore Currents, Sand Drift and Beach Deformation*, Dissertation in Tokyo Institute of Technology, 1982, 144p. (in Japanese)

12） Sunamura, T.: Quantitative Predictions of Beach-face slopes, *Geological Society of America Bulletin*, Vol.95, 1984, pp.242-245.

13） Kriebel, D.L., Kraus, N.C. and Larson, M.: Engineering Method for Predicting Beach Profile Response, *Coastal Sediments '91*, ASCE, 1991, pp.572-587.

14） Yamamoto, Y., Horikawa, K. and Tanimoto, K.: Prediction of shoreline change considering cross-shore sediment transport, *Proc. 25th Int. Conf. On Coastal Engrg.*, ASCE, pp.3405-3418, 1996.

15） Dean, R. G.: *Beach Nourishment Theory and Practice*, World Scientific, 2002, pp.29-34.

16） Yamamoto, K.: *A Study on Beach Morphology and Coastal Processes in Sediment Cells*, Dissertation in University of Tsukuba, 2003, 124p. (in Japanese)

6. COASTAL TOPOGRAPHIC CHANGE

17） Sunamura, T.: Study on the On-offshore Sediment Transport Rate in the Wave Breaking Zone Including the Swash Zone, *Proc. 31st Coastal Engineering Conference*, JSCE, 1984, pp.316-320. (in Japanese)

18） Sato, S. and Kishi, T.: Study on Sand Drift (7), Wave-induced Seabed Shear and Sediment Transport, *Report of the Public Works Research Institute*, Ministry of Construction, No.85-6, 1952. (in Japanese)

19） Ishihara, T. and Sawaragi, T.: On the Critical Velocity, Critical Depth and Transportation Rate of Sediment Movement, *Proc. 7th Coastal Engineering Conference*, JSCE, 1960, pp.47-58. (in Japanese)

20） Sato, S. and Tanaka, N.: Wave-induced Sand Movement on a Horizontal Bed, *Proc. 9th Coastal Engineering Conference*, JSCE, 1962, pp.95-100. (in Japanese)

21） Sato, S.: Drift Sand, *Collection of lectures from the 1966 Summer Workshops on Hydraulic Engineering*, JSCE, 1966, 29p. (in Japanese)

22） Horikawa, K. and Watanabe, A.: A Study of Wave-induced Sand Transport, *Proc. 13rd Coastal Engineering Conference*, JSCE, 1966, pp.126-134. (in Japanese)

23) Meyer-Peter, E. and Müller, R.: Formulas for Bed-load Transport, *Proceedings of 2nd IAHR Meeting*, Stockholm, 1948, pp.39-64.

24) Ashida, K. and Michiue, M.: Study on Hydraulic Resistance and Bed-load Transport Rate in Alluvial Streams, *Proceedings of JSCE*, No.206, 1972, pp.59-69. (in Japanese)

25） Watanabe, A., Maruyama, K., Shimizu, T. and Sakakiyama, T.: Numerical Prediction Model for Three-dimensional Beach Deformation Associated with Installation of Structures, *Proc. 31st Coastal Engineering Conference*, JSCE, 1984, pp.406-410. (in Japanese)

26） Ribberink, J.S.: Bed-load Transport for Steady Flows and Unsteady Oscillatory Flows, *Coastal Engineering*, Vol.34, 1998, pp.59-82.

27） van Rijn, L.C.: *Principles of Sediment Transport in Rivers, Estuaries and Coastal Seas*, Aqua Publications, Amsterdam, The Netherlands, 1993.

28） Horikawa, K. (editer): *Nearshore Dynamics and Coastal Processes*, University of Tokyo Press, 1988, 522p.

29）Editorial Sub-Committee on Hydraulic Formulae in Hydraulic Engineering Committee: *Collection of Hydraulic Formulae*, JSCE, 2019, 927p. (in Japanese)

30） Soulsby, R.: *Dynamics of Marine Sands. A Manual for Practical Application*, Thomas Telford, UK, 1997, 249p.

31） Zyserman, J.A. and Fredsøe, J.: Data Analysis of Bed Concentration of Suspended Sediment, *J. Hydrau. Engg.*, ASCE, 120-9, 1994, pp.1021-1042.

32） Van Rijn, L.C.: Sediment Transport: Part II: Suspended Load Transport, *J. Hydraul. Div.*, ASCE, 110 (HY11), 1984, pp.1613-1641.

33) Pelnard-Considére, R.: Essai de Théorie de l' Évolution des Formes de Rivage en Plages de Sable et de Galets, *Journées de l'Hydrauligue*, 4-1, 1956, pp.289-298.

34) Hanson, H. and Kraus, N.C.: *GENESIS: Generalized Model for Simulating Shoreline Change, Report 1: Technical Reference*, Tech. Rep. CERC-89-19, Department of the Army, Waterways Experiment Station, Corps of Engineers, Vicksburg, MS. 1989, 185p.

35) Bakker, W.T.: The Dynamics of a Coast with Groyne System, *Proc. 11th Coastal Eng. Conf., ASCE*, 1968, pp.492-517.

36) Perlin, M. and Dean, R.G.: *A Numerical Model to Simulate Sediment Transport in the Vicinity of Coastal Struvtures*, U.S. Army Corp of Engrs., C.E.R.C., Miscel. Rep., No.83-10, 1983, 119p.

37) Uda, T. and Kawano, S.: Development of a Predictive Model of Contourline Change Due to Waves, *Journal of JSCE*, No.539/II-35, 1996, pp.121-139. (in Japanese)

38) Serizawa, M., Uda, T., San-nami, T.,and Furuike, K.: A Model for Predicting Beach Changes Based on Bagnold's Concept, *Journal of JSCE(B)*, Vol.62, No.4, 2006, pp.330-347. (in Japanese)

39) Serizawa, M., Uda, T., San-nami, T., and Furuike, K.: Three-dimensional Model for Predicting Beach Changes Based on Bagnold's Concept, *Proc. 30th Inter. Confer. on Coastal Engineering*, ASCE, 2006, pp.3155-3167.

40) Wang, H., Dalrymple, R.A. and Shiau, J.C.: Computer Simulation of Beach Erosion and Profile Modification due to Waves, *Proc. Symp. on Modeling Tecnique*, ASCE, 1975, pp.1369-1384.

41) Komar, P.D. and Inman, D.L.: Longshore Sand Transport on Beaches, *Journal of Geophysical Research*, 75(30), 1970, pp. 5914-5927.

42) Ozasa, H. and Brampton, A.H.: Models for Predicting the Shoreline Evaluation of Beaches Backed by Seawalls, *Report of the Port and Harbour Research institute*, Vol.18, No.4, 1979, pp.77-104. (in Japanese)

43) Uda, T., Serizawa, M., Kumada, T., Karube, R. and Miura, M.: Relations among Wave Climate, Longshore Sand Transport Rate and Closure Depth, *Journal of Civil Engineering in the Ocean*, Vol.18, JSCE, 2002, pp.803-808. (in Japanese)

44) Uda, T., Yamamoto, Y., Itabashi, N. and Yamaji, K.: Field Observation of Movement of Sand Body Due to Waves and Verification of Its Mechanism by Numerical Model, *Journal of JSCE*, No.558/II-38, 1997, pp.113-128. (in Japanese)

45) Uda, T., Aoshima, G., Samejima, T., Yoshioka, A., Furuike, K. and Ishikawa, T.: Application of Model for Predicting both Bathymetric and Grain Size Changes to Shonan Coast, *Journal of Coastal Engineering*, Vol.55, JSCE, 2008, pp.606-610. (in Japanese)

46) Larson, M. and Kraus, N.C.: SBEACH: Numerical Model for Simulating Storm Induced Beach Change, Report 1, Empirical foundation and model development, *Technical Report*, CERC-89-9, 1989.

6. COASTAL TOPOGRAPHIC CHANGE

47） Larson, M. and Kraus, N.C.: SBEACH: Numerical Model for Simulating Storm Induced Beach Change, Report 2, Numerical foundation and model tests, *Technical Report*, CERC-89-9, 1990.

48） Inman, D.L. and Bagnold, R.A.: Littoral Processes, in *The Sea*, M. N. Hill (editor), Vol. 3, New York, Wiley, 1963, pp.529-533.

49） Bagnold, R.A.: Mechanics of Marine Sedimentation, in *The Sea*, M. N. Hill (editor), Vol. 3, New York, Wiley, 1963, pp.507-528.

50） Watanabe, A.: Numerical Simulation of Nearshore Currents and Beach Deformation, *Proc. 28th Coastal Engineering Conference*, JSCE, 1981, pp.285-289. (in Japanese)

51） Yamaguchi, M., Otsu, M. and Nishioka, Y.: Numerical Calculation of Two-dimensional Beach Deformation Due to Unsteady Waves, *Proc. 28th Coastal Engineering Conference*, JSCE, 1981, pp.290-294. (in Japanese)

52） Yamaguchi, M. and Nishioka, Y.: Numerical Method for Three-dimensional Seabed Topography Variation by Detached Breakwater and Jetty Groups, *Proc. 30th Coastal Engineering Conference*, JSCE, 1983, pp.239-243. (in Japanese)

53） Shimizu, T., Kumagai, T., Mimura, N. and Watanabe, A.: Three-dimensional Beach Deformation Long-term Prediction Model Considering Shoreline Change, *Journal of Coastal Engineering*, Vol.41, JSCE, 1994, pp.406-410. (in Japanese)

54） Roelvink, D., Reniers, A., van Dongeren, A., van Thiel de Vries, J., McCall, R. and Lescinski, J.: Modelling Storm Impacts on Beaches, Dunes and Barrier Islands, *Coastal Engineering*, 56(11-12), 2009, pp.1133–1152.

55） Deltares: *XBeach Technical Reference*, Kingsday Release, 2015, 139p.

56） Pender, D. and Karunarathna, H.: A Statistical-process Based Approach for Modelling Beach Profile Variability, *Coastal Engineering*, 81, 2013, pp.19-29. Doi.10.1016/j.coastaleng.2013.06.006

57） Dissanayake, P., Brown, J. and Karunarathna, H.: Modelling Storm Induced Beach/dune Evolution: Sefton Coast, Liverpool Bay, UK, *Marine Geology*, Vol 357, 2014, pp.225-242.

58） Dissanayake, P., Brown, J. and Karunarathna, H.: Comparison of Storm Cluster vs Isolated Event Impacts on Beach/dune Morphodynamics, *Estuarine Coastal and Shelf Science*, 2015. Doi. 10.1016/j.ecss.2015.07.040

59） Dissanayake, P., Brown, J., Wisse, P. and Karunarathna, H.: Effects of Storm Clustering on Beach/dune Evolution, *Marine Geology*, 370, 2015, pp.63-75. Doi.org/10.1016/j.margeo.2015.10.010

60） Sato, S. and Kabiling, M.: Numerical Computation of Waves, Nearshore Currents and Beach Deformation Using the Boussinesq Equation, *Journal of Coastal Engineering*, Vo.40, JSCE, 1993, pp.386-390. (in Japanese)

61） Vu T.C., Yamamoto, Y., Tanimoto, K., and Arimura, J.：Simulation on Wave Dynamics and Scouring near Coastal Structures by a Numerical Model, *Proc. 28th Inter. Confer. on Coastal Engineering*, ASCE, 2002, pp.1817-1829.

62） Yamamoto, Y., Charusrojtanadech, N. and Sirikaew, U.: Topographical Change Prediction of the Beach or the Seabed in the Front of a Coastal Structure, *Proc. 22nd Inter. Offshore and Polar Engineering Conference*, 2012, pp.1488-1495.

63） Charusrojthanadech, N., Rattanarama, P. and Yamamoto, Y.: Examination of Coastal Erosion Prevention in the Back of the Gulf of Thailand, *Proc, 23rd Inter. Offshore and Polar Engineering Conference*, 2013, pp.1355-1362.

64） Nakamura, T., Yim, S.C. and Mizutani, N.: Development of Three-dimensional Coupled Fluid-Structure-Sediment Interaction Model and Its Application to Local Scour around Submerged Object, *Journal of JSCE B2 (Coastal Engineering)*, Vol.66, No.1, 2010, pp.406-410. (in Japanese)

65） Suzuki, K., Osaki, N. and Yamamoto, Y.: On the Estimation of Scour Volume at the Foundations of Breakwaters, *Journal of Coastal Engineering*, Vol.50, 2003, pp.886-890. (in Japanese)

66） Kuroiwa, M., SEIF, A.K., Matsubara, Y., Mase, H. and Zheng, J.: 3D Morphodynamic Model for Considering Wave-current Interaction, *Journal of JSCE B2 (Coastal Engineering)*, Vol.66, No.1, 2010, pp.551-555. (in Japanese)

67） Deltares：*Delft3D-FLOW User Manual*, Version 4.05, 2022, 698p.

68） Cundall, P.A. and Strack, O.D.L.: A Discrete Numerical Model for Granular Assemblies, *Geotechnique*, Vol.29, 1979, pp.47-65.

69） Koshizuka, S. and Oka, Y.: Moving-particle Semi-implicit Method for Fragmentation of Incompressible Fluid, *Nucl. Sci. Eng.*, Vol.123, 1996, pp.421-434.

70） Lucy, L.B.: A Numerical Approach to the Testing of the Fission Hypothesis, *Astron. J.*, Vol.82, 1977, pp.1013-1024.

71） Shao, S. and Lo, E.Y.M.: Incompressible SPH Method for Simulating Newtonian and non-Newtonian Flows with a Free Surface, *Adv. Water Resources*, Vol.26, 2003, pp.787-800.

72） Shigematsu, T., Hirose, M., Nishikiori, Y. and Oda, K.: Development of a Three-dimensional Solid-liquid Multiphase Flow Analysis Model Using the DEM and VOF Methods and Its Application Examples, *Journal of Coastal Engineering*, Vol.48, JSCE, 2001, pp.6-10. (in Japanese)

73） Harada, E. and Gotoh, H.: Numerical Simulation of Vertical Sorting in Sheetflow Sediment Transport by Two-Phase Turbulent Flow Model, *Journal of Coastal Engineering*, Vol.54, JSCE, 2007, pp.476-480. (in Japanese)

6. COASTAL TOPOGRAPHIC CHANGE

74) Gotoh, H., Hayashi M. and Sakai, T.: Numerical Analysis of Wave/Bottom Mud Interaction Using Solid-liquid Two-phase Flow MPS Method, *Journal of Coastal Engineering*, Vol.48, JSCE, 2001, pp.1-5. (in Japanese)

75) Harada, E., Ikari, H., Gotoh, H., Imura, M. and Shimizu, Y.: Numerical Simulation for Sediment Transport in Surf Zone by MPS-DEM Coupling, *Journal of JSCE B2 (Coastal Engineering)*, Vol.72, No.2, 2016, pp. I_583-I_588. (in Japanese)

76) Harada, E., Ikari, H., Gotoh, H., Imura, M. and Shimizu, Y.: Numerical Simulation for Sediment Transport in Surf Zone by 3D DEM-MPS Coupling, *Journal of JSCE B2 (Coastal Engineering)*, Vol.74, No.2, 2018, pp. I_751-I_756. (in Japanese)

77) Gotoh, H., Suzuki, K., Ikari, H., Arikawa, T., Khayyer, A. and Tsuruta, N.: Development of Particle-based Numerical Wave Flume for Multiphase Flow Simulation, *Journal of JSCE B2 (Coastal Engineering)*, Vol.73, No.2, 2017, pp. I_25-I_30. (in Japanese)

78) Luijendijk, A., Hagenaars, G, Ranasinghe, R, Baart, F., Donchyts, G. and Aarninkhof, S.: The State of the World's Beaches, *Nature, Scientific Reports,* 8, Article number 6641, 2018. https://www.nature.com/articles/s41598-018-24630-6

79) Mentaschi, L., Vousdoukas, I.M., Pekel, J.F., Voukouvalas, E. and Feyen, L.: Global Long-term Observations of Coastal Erosion and Accretion, *Nature, Scientific Reports*, 8, Article number 12876, 2018. https://www.nature.com/articles/s41598-018-30904-w

80) Uda, T.: *Japan's Beach Erosion (Reality and Future Measures)*, Advanced Series on Ocean Engineering, Vol.3, World Scientific Publishing Co. Pte Ltd, 2017, 548p. Doi.org/10.1142/10184/May2017

81) Uda, T.: *Reality and Measures of Coastal Erosion*, Sankai-do, 2004, 304p. (in Japanese)

82) Sumer, B.M. and Fredsøe, J.: *The Mechanics of Scour in the Marine Environment*, World Scientific, 2002, 552p. https://doi.org/10.1142/4942

83) Yamamoto, Y., Ioroi, M. and Higa, R.: Damage Prediction of a Dike and a Seawall by Big Waves, *Journal of JSCE B3 (Ocean Development)*, Vol. 68, No.2, 2012, pp. I_882-I_887. (in Japanese)

84) Yamamoto, Y., Ioroi, M. and Oshima, Y.: Destruction Mechanism of a Coastal Dike and a Seawall, and Methods for Estimating the Suction Rate, *Journal of JSCE B2 (Coastal Engineering)*, Vol.71, No.1, 2015, pp.30-41. (in Japanese)

85) Iwasaki, F., Tanaka, S., Sato, S., Nago, H., Maeno, S. and Kotani, Y.: Experimental Study on the Hollowing Occurrence Mechanism of Coastal Embankments, *Journal of Coastal Engineering*, Vol.42, JSCE, 1995, pp.1026-1030. (in Japanese)

86) Maeno, S., Kotani, Y. and Hoshiyama, C.: Study on the Outflow Limit of Backfill Sediment from a Seawall in a Fluctuating Hydraulic Field, *Journal of Coastal Engineering*, Vol.47, JSCE, 2000, pp.926-930. (in Japanese)

87) Maeno, S., Goto, H., Tsubota, Y. and Harada, E.: Flow Analysis of Backfill Sediment from a Seawall in a Fluctuating Water Pressure Field Using DEM-FEM Model, *Journal of Coastal Engineering*, Vol.48, JSCE, 2001, pp.976-980. (in Japanese)

88) Harada, E., Goto, H., Sakai, T. and Tei, T.: 3D Simulation of Wave-induced Cave Growth Process in the Adjacent Layer of Seawall, *Journal of Coastal Engineering*, Vol.50, JSCE, 2003, pp.891-895. (in Japanese)

89) Nakamura, T, Hur D. and Mizutani, M.: Suction Mechanism of Reclaimed Sand Behind a Rubble Seawall, *Journal of JSCE B*, Vol.62, No.1, 2006, pp.150-162. (in Japanese)

90) Gotoh, H., Ikari, H., Komaguchi, T., Mishima, T. and Yodoshi, H.: Numerical Analysis on Cave-Formation Process behind Seawall by Particle Method, *Journal of JSCE B2* (*Coastal Engineering*), Vol.66, No.1, 2010, pp.821-825. (in Japanese)

91) Yamamoto, Y., Kawashima, S. and Fukuhama, M.: Destruction Mechanisms of Coastal Facilities Due to Large Waves and Tsunamis and Application of Destruction Limitation to Actual Coasts, *Journal of Coastal Engineering*, Vol.52, JSCE, 2005, pp.1281-1285. (in Japanese)

92) Yamamoto, Y., Nariyoshi, K. and Higa, R.: Damage Limitation of a Coastal Dike and a Seawall, *Journal of JSCE B3* (*Ocean Development*), Vol.67, No.2, 2011, pp. I_100-I_105. (in Japanese)

93) Ioroi, M. and Yamamoto, Y.: Propose of a Suction Rate Formula of Backfill from the Front Lowest Edge of a Coastal Dike by Big waves, *Journal of JSCE B2* (*Coastal Engineering*), Vol.68, No.2, 2012, pp. I_896-I_900. (in Japanese)

94) Ioroi, M., Yamamoto, Y. and Oshima, Y.: Generalization of a Method for Predicting Suction Rate of Backfilling Materials from the Front Lowest Edge of a Coastal Dike by Big Waves, *Journal of JSCE B2* (*Coastal Engineering*), Vol.70, No.2, 2014, pp. I_1031-I_1035. (in Japanese)

95) Yoshizawa, S., Yamamoto, Y. and Kuisorn W.: Improvement of Methods for Calculating the Suction Rate from a Coastal Dike or a Seawall, *Journal of JSCE B2* (*Coastal Engineering*), Vol.72, No.2, 2016, pp. I_1147-I_1152. (in Japanese)

96) Kornvisith, S., Yamamoto, Y. and Yoshizawa, S.: Numerical Model for Predicting the Sand Outflow Rate of Backfill Materials from a Coastal Dike, *English Journal of JSCE*, Vol.7, 2019, pp.63-71.

97) Kornvisith, S. and Yamamoto Y.: The Reproduction Ability of a Numerical Model for Simulating the Outflow Rate of Backfilling Materials from a Coastal Structure, *J. Mar. Sci. Eng.*, 7, 447, 2019. Doi:10.3390/jmse7120447

98) Fukuhara, R., Yamamoto, Y. and Matsushima, S.: Proposal of Evaluation Methods of the Outflow Rate of Backfill Materials from a Gently Sloping Dike Covered with Blocks, *Journal of JSCE*, Vol.79, No.17, 23-17127, 2023. (in Japanese)

6. COASTAL TOPOGRAPHIC CHANGE

99）Noguchi, K., Tanaka, S., Torii, K. and Sato, S.: Large-scale Physical Model Test of the Deformation of a Gentle Slope Seawall, *Journal of Coastal Engineering*, Vol.47, 2000, pp.756-760. (in Japanese)

100) Takahashi, T., Imamura, F. and Suto, N.: Development of a Tsunami Calculation Method with Sediment Transport, *Journal of Coastal Engineering*, Vol.39, 1992, pp. 231-235. (in Japanese)

101) Takahashi, T., Imamura, F. and Suto, N.: Investigation of the Applicability and Reproducibility of Tsunami Migration Models, *Journal of Coastal Engineering*, Vol.40, 1993, pp.171-175. (in Japanese)

102) Takahashi, T., Suto, N., Imamura, N. and Asai, D.: Development of a Tsunami Movable Bed Model Considering the Amount of Sand Exchanged between Bed and Suspended Load Layers, *Journal of Coastal Engineering*, Vol.46, 1999, pp.606-610. (in Japanese)

103) Fujii, N., Omori, M., Takao, M., Kanayama, S. and Otani, H.: Study on Tsunami Induced Seabed Topography Change, *Journal of Coastal Engineering*, Vol.45, 1998, pp.376-380. (in Japanese)

104) Kihara, N. and Matsuyama, M.: Hydrostatic Three Dimensional Numerical Simulations of Tsunami Scour in Harbor, *Journal of Coastal Engineering*, Vol.54, 2007, pp.516-520. (in Japanese)

105) Vu, T.C., Yamamoto, Y. and Charusrojtanadech, N.: Improvement of Prediction Methods of Coastal Scour and Erosion Due to Tsunami Back-flow, *Proceedings of the Twentieth IOPEC*, 2010, pp.1053-1060.

106) Yamamoto, Y., Hayakawa, M. and Ahmadi, S.M.: Proposal of a Rational Prediction Method of Scour on Land by Huge Tsunami, *Journal of JSCE B2 (Coastal Engineering)*, Vol.75, No.2, 2019, pp. I_697-I_702. (in Japanese)

107) Ahmadi, S.M., Yamamoto, Y. and Vu. T.C.: Rational Evaluation Methods of Topographical Change and Building Destruction in the Inundation Area by a Huge tsunami, *J. Mar. Sci. Eng.*, 8, 762, 2020. Doi:10.3390/jmse810762

108) Nakamura, T. and Mizutani, N.: Sediment Transport Model Considering Pore-water Pressure in Surface Layer of Seabed and Its Application to Local Scouring Due to Tsunami, *Journal of JSCE B2 (Coastal Engineering)*, Vol. 68, No. 2, 2012, pp. I_216-I_220. (in Japanese)

109) Noguchi, K., Sato, S. and Tanaka, S.: Large-scale Model Test of Wave Overtopping and Frontal Scouring of Seawalls Due to Tsunami Run-up, *Journal of Coastal Engineering*, Vol.44, 1997, pp.296-300. (in Japanese)

110) Arikawa, T., Ikeda, T. and Kubota, K.: Experimental Study on Scour Behind Seawall Due to Tsunami Overflow, *Journal of JSCE B2 (Coastal Engineering)*, Vol.70, No.2, 2014, pp. I_926-I_930. (in Japanese)

111) Yamamoto, Y., Charusrojthanadech, N. and Nariyoshi, K.: Proposal of Rational Evaluation Methods of Structure Damage by Tsunami, *Journal of JSCE B2 (Coastal Engineering)*, Vol. 67,

No. 1, 2011, pp.72-91. (in Japanese)

112) Nariyoshi, K. and Yamamoto, Y.: Field Survey and Prediction Method of Topography Change by 2011off the Pacific Coast of Tohoku Earthquake tsunami, *Journal of JSCE A1* (*Structure/ Earthquake Engineering*), Vol. 69, No. 4, 2013, pp. I_559-I_570. (in Japanese)

113) Ishihara, T. and Iwagaki, Y.: On the effect of sand storm in controlling the mouth of the Kiku river, *Disaster Prevention Research Institute*, Kyoto University, 2, 1952, pp.1-32. (in Japanese)

114) Nakashima, Y.: Study on Control of Wind-blown Sand, *Report on the Training Forest*, Faculty of Agriculture, Kyushu University, No. 43, 1979, pp.125-183. (in Japanese)

115) Bagnold, R.A.: *The Physics of Blown Sand and Desert Dunes*, Methuen & Co. Ltd., 1954, 265p.

116) Zingg, A.: Wind-tunnel studies of the movement of sedimentary material, *Proc. 5th Hydraulic Conf.*, Bull. Iowa State University, Studies in Engrg., 34, 1952, pp.111-135.

117) Kawamura, R.: Study on Sand Movement by Wind, *Report of the Institute of Science and Engineering*, University of Tokyo, Vol.5, No.3-4, 1951, pp.95-112. (in Japanese)

118) Kawata, S.: Survey Report on Disaster Prevention Forests, *Reference material for mountain control projects*, Forestry Agency, 1950, pp.1-22. (in Japanese)

119) Hamada, T., Okubo, K. and Hasegawa, N.: *Ishinomaki Port and Watanoha Port Technical Survey Report*, Department of Port and Harbour Physics, Institute of Transportation Technology, 1951, pp.55-58. (in Japanese)

Chapter:7

COASTAL PROTECTION AND
VARIOUS OTHER STRUCTURES

7. COASTAL PROTECTION AND VARIOUS OTHER STRUCTURES

Due to easy access to sea transport, many businesses, residential areas, and port/harbour facilities exist in coastal areas. Therefore, protecting coastal areas is very important to provide safety to assets and facilities against high waves, storm surges, and tsunamis.

This section first presents direct coastal protection facilities such as coastal dikes and seawalls, indirect protection facilities such as detached breakwaters and groynes which reduce coastal erosion, and port/harbour facilities such as breakwaters and quay walls. Then, various types of protection against high waves, storm surges, and tsunamis are explained. Finally, the reliability-based design method is introduced, which is becoming an important design method due to the development of the limit state design method combined with statistical methods.

7.1 Coastal Structures
1) Facilities on Coasts
(1) Coastal Dikes

A structure shown in **Fig. 7.1** in which a bank [(backfilling section) is covered with concrete or other materials to protect it from erosion by high waves, is built along the shoreline and is called a **coastal dike.** Generally, foot protection at the front of the structure is used to prevent scour, water cutoff to prevent seawater seepage to the landward side, and a parapet to reduce wave overtopping is attached to the structure. The standard slope of the front surface is about 0.5:1, but to improve accessibility to the beach, the surface slope may be covered with stepped concrete or sloping concrete blocks with a slope of 2:1 or smaller.

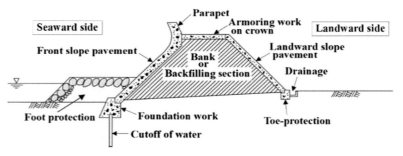

Fig. 7.1 Typical cross-section of a coastal dike. (from "Technical Standards and Commentaries for Coastal Protection Facilities", CDIT [1])

(2) Seawalls

When the ground level at the landside is high and the shoreline is cliffy, a structure similar to that shown in **Fig. 7.2** can be constructed using concrete or other materials to cover the cliff and protect it from erosion by high waves. This type of structure is called a **seawall.** In this case, seawater seepage into the landward side is unlikely to be a problem, so the water cutoff may not be necessary. To improve accessibility, the surface slope may be covered with stepped concrete or concrete blocks with a slope slower than 2:1, similar to the case of a coastal dike.

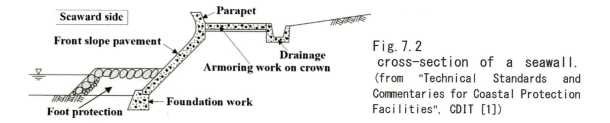

Fig. 7.2 cross-section of a seawall. (from "Technical Standards and Commentaries for Coastal Protection Facilities", CDIT [1])

(3) Wave Absorbing Breakwaters and Wave Absorbing Structures

A stand-alone structure consisting of wave energy absorbing blocks and wave dissipating devices, to reduce wave forces, is called a **wave absorbing breakwater** or **wave dissipating breakwater** as shown in **Photo 7.1**. If the structure is attached to the front of a dike, seawall, or breakwater as shown in **Fig. 7.3**, then it is called a **wave absorbing structure** or **wave dissipating structure**.

Photo 7.1 Wave absorbing breakwater on the Shizuoka coast of Shizuoka Prefecture in Japan.

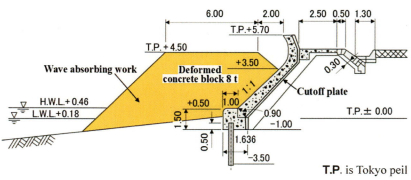

T.P. is Tokyo peil (= Tokyo Bay Medium Tide Level).

Fig. 7.3 Wave absorbing structure and seawall on the Shimo-Niikawa coast of Toyama Prefecture in Japan. (from Kurobe River Office, Ministry of Land, Infrastructure, Transport and Tourism)

7. COASTAL PROTECTION AND VARIOUS OTHER STRUCTURES

(4) Groynes

A **groyne** is a wall-like structure that extends offshore from the beach to prevent beach sediment from being moved along the shore by high waves and nearshore currents. The planform of a groyne is usually I-shaped, inverted L-shaped, or T-shaped (refer to **Fig. 7.4**).

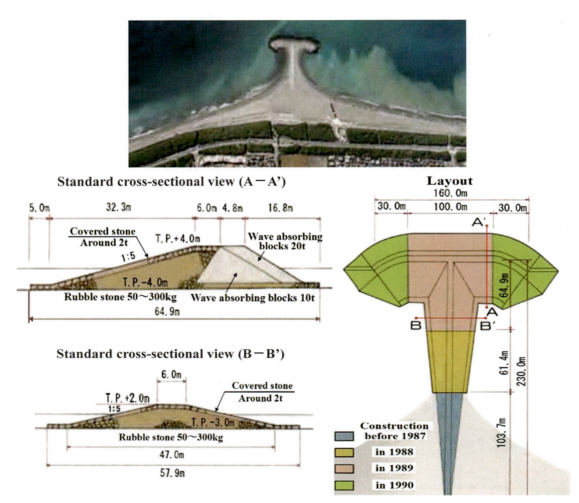

Fig. 7.4 T-shaped groyne at the Chigasaki coast in Kanagawa Prefecture, Japan.
(Source: Erosion Control and Coastal Division, Kanagawa Prefecture)

(5) Detached Breakwaters (Offshore Breakwaters)

A structure in which wave-absorbing blocks are set away from the shore in the form of a wall parallel to the shore to promote sand deposition in the landward calm area created by the blocks as shown in **Fig. 7.5** and **Photo 7.2** is called a **detached breakwater** or **offshore breakwater**. For a long and impermeable breakwater, the supply of sand from the offshore side is prevented. Sand entering the landward calm area behind the breakwater from both ends accumulates only at the ends and cannot move to the inner calm area. Therefore, it is common to use discontinuous breakwaters with wave absorbing blocks. In addition, the foundation of crusher-runs and rubble should always be provided because ordinary wave-absorbing blocks can settle easily due to liquefaction.

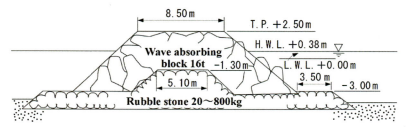

Fig. 7.5 Cross-section of a detached breakwater at the Kaike coast in Tottori Prefecture, Japan. (from Hino River Office, Ministry of Land, Infrastructure, Transport and Tourism)

Beach erosion continued since the Taisho Era due to a lack of sediment supply from rivers caused by the abolition of the iron manufacturing method known as "Kanna Nagashi", and erosion countermeasures were implemented using detached breakwaters.

Photo 7.2 Detached breakwaters at the Kaike coast in Tottori Prefecture, Japan. (from "Coasts and Ports of Japan", the coastal engineering committee of JSCE [2])

(6) Artificial reefs

An **artificial reef** is a submerged breakwater built like a natural coral reef for wave dissipation and sand deposition. In the case of detached breakwaters with a sufficiently high crown height above the water surface, because high wave dissipation efficiency can be expected, the width at the crown is sufficient with three wave dissipating blocks which can be expected to interlock between the blocks. However, in the case of submerged breakwaters, the width at the crown should be around 100 m when the tidal range is about 1 m, and 50 m even when the tidal range is as small as 0.2 m because of their low wave dissipation efficiency. In the artificial reef shown in **Fig. 7.6**, the crown width is significantly reduced by using a special reinforced concrete box with high wave dissipation efficiency.

7. COASTAL PROTECTION AND VARIOUS OTHER STRUCTURES

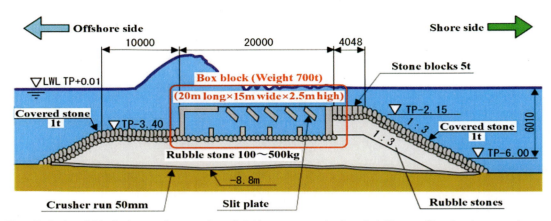

Fig. 7.6 Artificial reef on the Ishikawa coast in Ishikawa Prefecture, Japan.
(from Kanazawa River and National Route Office, Ministry of Land, Infrastructure, Transport and Tourism)

(7) Training Jetties (Training Walls)

A wall-like structure constructed at a river mouth, extending towards the offshore direction to prevent the river mouth from being blocked by the sediment moving along the shore is called a **training jetty** or **training wall.** They can be built along both river banks to block sediment movement (refer to **Photo 7.3**).

Aerial photograph of jetties.

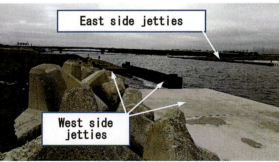
Ground photo of jetties.

Photo 7.3 Training jetties at the Sagami river estuary in Kanagawa Prefecture, Japan.

2) Facilities on Ports and Harbours
(1) Breakwaters

A **breakwater** is a wall-like structure constructed to protect an anchorage from high waves. A breakwater usually consists of a rubble foundation, a main body made of concrete square blocks or concrete caissons, and a wave-dissipating structure to reduce wave force and prevent seaward scour as shown in **Fig. 7.7**.

7.1 Coastal Structures

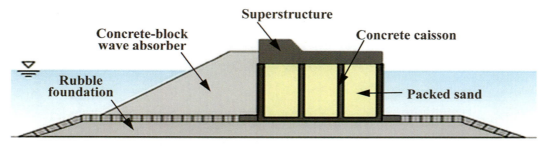

Fig. 7.7 Typical cross section of a breakwater.

(2) Quay Walls and Landing Sites

A structure, as shown in **Photo 7.4**, constructed for mooring vessels, and loading and unloading people and cargo is called a **quay wall** if the structure is large, while a smaller structure is called a **landing site.** The structure can be of the gravity type, constructed with concrete square blocks or caissons, of the sheet pile type, or of the pier type.

Photo 7.4 Landing site at the Ofunato port in Iwate Prefecture, Japan.

(3) Piers

A quay wall or landing site consists of a wall and a reclaimed section along the shore, while a **pier** is a structure with the same purpose as a quay wall or landing site, but with a flat shape projecting out into the sea from the shoreline and built using a piling system like a bridge, as shown in **Photo 7.5**. A pier can be constructed even on soft ground. If the structure is floating, it is called a floating pier.

7. COASTAL PROTECTION AND VARIOUS OTHER STRUCTURES

Photo 7.5 Example of a side view of a pier (Klong Wale Bay in Thailand).

(4) Wharves (Quays)

Mooring facilities consisting of quay walls or landing sites for mooring ships, cargo handling facilities consisting of aprons and gantry cranes, storage facilities for temporarily storing cargo and other goods, harbour roads for cargo transport and passenger facilities are collectively called **wharves** or **quays.**

7.2 Measures against Large Waves

1) Occurrence of Large Waves

Large waves are generated when high-intensity wind blows on the open ocean surface over a prolonged period. Those waves can then be propagated towards the coastlines and can reach significant wave heights $H_{1/3}$ as high as 4 m to 10 m (8 m to 20 m maximum wave heights H_{max}) and significant wave periods $T_{1/3}$ around 8 sec to 16 sec. If strong winds continue to blow in inland seas with a fetch of about 200 km, the significant wave heights and significant wave periods at the coasts can reach around 1.5 m to 4 m (3 m to 8 m at H_{max}) and 4 sec to 8 sec.

The largest recorded significant wave height observed in the North Atlantic in February 2013 was 18.9 m. A giant wave known as a **freak wave** (a high wave that appears once in thousands to tens of thousands of waves and is more than twice the height of a significant wave) with a wave height of 30 m was encountered by the cruise ship Ms Bremen in the South Atlantic Ocean in March 2001.

The wave heights and periods of ocean waves can be easily estimated from the wind field using the calculation charts and formulae of the Significant Wave Method in Chapter 1. If accurate prediction of time and space-varying wave field is necessary, then numerical models based on the spectral method in Chapter 1 should be used.

2) Characteristics of Large Wave Damage

Large waves can cause coastal erosion, flooding due to wave overtopping or overflowing, damage to coastal structures, and destruction to coastal ecosystems. Types of large wave damage include the following.

(1) Destruction to Coastal Structures Due to Strong Wave Force

As wave heights increase, wave forces on coastal structures and wave overtopping can increase (see **Photo 7.6**). The shallower the water depth, the higher the wave height due to wave shoaling, but waves break when the wave height / the water depth exceeds the breaking wave limit (which increases from 0.6 to 1.4 as the seabed slope and offshore wavelength increase).

Photo 7.6
Fuji Coast, Shizuoka Prefecture in Japan, where coastal erosion has greatly reduced the sandy beach, was breached by the strong wave force of Typhoon No. 26 in September 1966. (from "Coasts and Ports of Japan", the coastal engineering committee of JSCE [2])

7. COASTAL PROTECTION AND VARIOUS OTHER STRUCTURES

In other words, the closer the installation depth of the structure is to the wave breaking depth, the stronger the wave force will be. In addition, due to wave refraction, waves are more likely to converge at points where the shoreline protrudes, such as headlands. Also, on beaches with steep bottom slopes, wave forces are more likely to increase. Furthermore, if the sediment supply to the beaches is depleted, the water depths in front of structures may gradually deepen, exposing the structure to larger waves.

(2) Destruction Due to Front Scour and Backfill Material Suction

Even if a structure is designed to withstand wave forces generated by prevailing large waves, repeated occurrence of large waves can cause scouring in front of the structure, thus leading to tilting of the structure and loss of its function. If the structure is a coastal dike or seawall with a backfill constructed of earth, sand, or stones, incident waves can enter the backfill when the front scour progresses to the lowest edge of the front surface of the structure and suck out the backfill material during return flows. This can cause hollowing of the structure, turning it into a "dead body" state on the verge of failure. A structure in this state may fail even under wave forces smaller than the design waves (refer to **Photo 7.7**).

Photo 7.7
On the Hirono coast, Shizuoka Prefecture in Japan, where coastal erosion was in progress, repeated high waves of Typhoon No. 9 in 1997 sucked out the backfill material from the dike, leaving it in a hollowed-out "dead body" state, which was breached by high waves of Typhoon No. 18 in the same year.

(3) Damage Caused by Spray and Wind-blown Sand

When high waves encounter coastal dikes and seawalls, saltwater spray containing sand and gravel particles can be generated. This spray can reach nearby buildings and roads, which is a significant inconvenience or danger to inhabitants and travelers. Also, sand and gravel particles can damage buildings and obstruct traffic if they land on coastal roads. Furthermore, when strong winds are present, the salty droplets can penetrate inland to a considerable extent, causing insulators and power lines to be continuously covered by the droplets, resulting in electrical short circuits and major power outages. There have also been many cases where wind-blown sand generated by strong winds has infiltrated houses, degrading the quality of life and burying fields (refer to **Photo 7.8**).

7.2 Measures against Large Waves

Photo 7.8
The Shizuoka coast, Shizuoka Prefecture in Japan, was damaged by strong waves of Typhoon No. 20 in October 1979, as well as severe traffic disruption caused by spray on the national road along the coast. (from "Coasts and Ports of Japan", the coastal engineering committee of JSCE [2])

3) Concept of High Wave Protection

Disaster prevention and mitigation measures against large waves mean strengthening of coastal, port and harbour facilities to withstand their impacts.

Essentially, the proposal of construction, stability, and component strength of coastal, port, and harbour facilities should be examined against the selected design wave conditions and design high tides during the entire lifetime of the structure.

The official service life of reinforced concrete, which is the main construction material of coastal structures, is 50 years, and under severe service conditions, it reaches its service limit after about 30 years. Therefore, examinations on the function, stability and component strength of the facilities should be carried out for the worst design waves and design high water levels for 30 to 50 years.

Specifically, the external force of the largest hurricane, typhoon or cyclone ever recorded, passing through the area of concern, is considered the worst possible design condition. Furthermore, future projections of sea level rise and giant tropical cyclones due to global warming should also be considered together with potential topographical changes that can take place after extreme typhoons or hurricane attacks.

It is also necessary to assess damage in the event of possible structural failure, formulate plans for the safe evacuation of coastal residents and workers in ports and harbours, and carry out periodic field drills.

In addition, areas where the shoreline protrudes or is discontinuous compared to its surroundings should be eliminated as far as possible because they are vulnerable to concentrated attack from high waves.

(1) Structures Constructed in Sea Areas

Structures installed in sea areas (breakwaters, detached breakwaters, artificial reefs, groynes and training jetties) can be subjected to extreme wave forces that can cause the main body of the structure to slide or overturn, or cause compressive, tensile, bending or shear failure of its components. For those failure modes, the wave pressure distribution acting on the structure due to the design waves

7. COASTAL PROTECTION AND VARIOUS OTHER STRUCTURES

should be determined, and stability and stress/displacement calculations should be carried out. The structure should be designed to maintain its original function with sufficient reliability. Moreover, it should be noted that in the event of repeated high wave action, the bearing capacity of sandy grounds underneath and surrounding the structure in extremely shallow waters can be lost due to liquefaction, and structures with a narrow footprint can easily sink. Waves reflected from the structure and local currents can cause significant scouring around rigid structures. Therefore, foundation works, matting and wave-dampening works should also be installed to prevent settlement and scour.

Actual design should be carried out by the design standards issued by the regulatory authorities. For example, in Japan, "Technical Standards and Commentaries for Coastal Protection Facilities" of CDIT [1] is used for general coasts under the Coast Act, "Technical Standards and Commentaries for Port Facilities" of PHB of MLIT and PARI [3] for ports and harbours under the Port and Harbour Act, and "Reference Book for the Design of Fishing Ports and Fishing Facilities" of Fisheries Agency [4] for fishing ports and fishing areas under the Act on Development of Fishing Ports and Grounds. Internationally, the "Coastal Engineering Manual" by the U. S. Army Corps of Engineers [5] is widely used for designing coastal protection structures.

(2) Structures Located near the Coastline

When coastal dikes and seawalls are constructed in very shallow water, they can be subjected to waves dissipated by depth-limited wave breaking. Quay walls and landing sites in ports and harbours are also usually built in calm areas protected by breakwaters and are therefore subject to small wave forces. It is important to consider the stability of these structures not only against wave forces but also against active earth pressure acting from the landward side of these structures. The specific criteria used for those considerations should be the design standards of the supervising authorities. Since these structures are usually constructed on the land-sea boundary, it is also important to consider measures against wave overtopping by the intended use. Furthermore, if the water depth in front of a coastal dike or seawall is shallower than about 3 m, it is more likely that the structure will fail due to repeated wave scour and backfill material suction (outflow) than lack of stability and strength of the structure against maximum wave force. Therefore, it is important to consider these issues.

As coastal dikes and seawalls are built to withstand a pre-selected design wave force, they may fail in the event of the occurrence of forces exceeding the design condition. Failure due to ageing cannot also be ruled out. In the event of a structural failure, wave damage can reach several tens of metres landward from the shoreline, which can be extremely dangerous for people, buildings, infrastructure, farmlands, and the ecosystems surrounding the damage. Therefore, facility managers need to predict the extent and severity of damage and impact caused by a potential failure of a structure, and municipal mayors need to formulate evacuation plans and publicise them to residents and others to be prepared.

4) Measures against Wave Forces
(1) Calculation of Wave Force Acting on Upright Walls

Various conditions required for wave force calculations can be set as follows.

(a) Design Water Level

The water level used in the design should be as high as possible and **the highest high-water level** (HHWL) is usually used. However, in the case that there are no measured data of sufficient long periods, **the mean monthly-highest water level** (the mean monthly-highest water level within five days of the new moon and full moon) + **the highest high storm tide** (the maximum existing water level rise due to strong winds and low pressure) is used. The highest high storm tide η_M is calculated from the following equation.

$$\eta_M = a\Delta p + bU_{max}^2 \cos\theta_w + \eta_s \tag{7.1}$$

where a is the coefficient of 1 cm rise in tide level for a pressure drop of 1 hPa, Δp is the difference between the atmospheric pressure (1013 hPa) and the minimum pressure at the centre of a low-pressure system, b is a topography-dependent coefficient (0.002 to 0.2) that increases for shallow coasts and bays where strong winds blow in, U_{max} is the maximum wind speed, θ_w is the wind inclination angle from the normal to the coastline (0 degrees if the wind direction is perpendicular to the coastline), and η_s is the amount of water surface elevation due to the surf beat, which can be estimated using Eq. (5.2).

For breakwaters and quay walls in ports and harbours, it is more important that they do not break than that waves do not overflow, so the worst tide level for stability calculations should be adopted.

(b) Design Water Depth

The designed water depth in front of a structure is obtained by considering the designed high tide level and scour depth. If the structure is less important, the scour depth may be determined based on experience, but for more important structures, hydraulic model tests and numerical simulations (for example, refer to **Figs. 6.12** and **6.13**) should be used.

(c) Design Wave Height

The height and period of the maximum significant offshore wave, obtained from wave forecasting or actual wave data over 30 years, are used to determine the design wave height and period.

The wave period is assumed to remain unchanged even if the waves come close to shore, and the wave period set by a coastal manager is used directly as the design wave period. The equivalent offshore wave height H_o' ($= K_r \times K_d \times H_o$) is determined for each coast by considering the refraction

7. COASTAL PROTECTION AND VARIOUS OTHER STRUCTURES

coefficient K_r and diffraction coefficient K_d to the offshore wave height H_o set by the coastal manager. The shoaling coefficient K_s for the design water depth is then obtained and the wave height calculated using this coefficient is used as the design wave height when designing a coastal structure.

When Goda's wave pressure calculation formula [Eq. (3.43) – (3.46)] is used, it is necessary to use the highest wave height H_{max} as the design wave height. If the water depth is expressed as h, the offshore wavelength as L_o and the seabed gradient as $\tan\alpha$, the highest wave height H_{max} is obtained from the following equation using the equivalent offshore wave height and the shoaling coefficient.

$$\left. \begin{aligned} h/L_o \geq 0.2: \quad & H_{max} = 1.8 K_s H_o' \\ h/L_o \leq 0.2: \quad & H_{max} = \min\{(\beta_0^* H_o' + \beta_1^* h), \quad \beta_{max}^* H_o', \quad 1.8 K_s H_o'\} \end{aligned} \right\} \tag{7.2}$$

$$\beta_o^* = 0.052 \left(H_o'/L_o\right)^{-0.38} \exp\left(20 \tan^{1.5}\alpha\right), \quad \beta_1^* = 0.63 \exp\left(3.8 \tan\alpha\right) \tag{7.3}$$

$$\beta_{max}^* = \max\left\{1.65, \quad 0.53\left(H_o'/L_o\right)^{-0.29} \exp\left(2.4\tan\alpha\right)\right\} \tag{7.4}$$

【A calculation example to obtain the design wave height】

Find the equivalent offshore wave height Ho' when an offshore wave height $Ho = 8.82$ m, a refraction coefficient $K_r = 0.850$ and a diffraction coefficient $K_d = 0.800$. Then find the shallow water coefficient K_s for a period $T = 12.0$ s and a water depth $h = 11.0$ m in front of the breakwater. Then determine the design wave height H_{max} under a condition of seabed slope $\tan\alpha = 1/100$.

$$H_o' = K_r \times K_d \times H_o = 0.850 \times 0.800 \times 8.82 \fallingdotseq 6.00\text{m} \tag{i.1}$$

$$L_o = 1.56 \times T^2 = 1.56 \times 12.0^2 \fallingdotseq 225\text{m} \Rightarrow h/L_o = 11.0/225 \fallingdotseq 0.0490 < 1/20 = 0.05$$

Therefore, this wave can be identified as the shallow-water wave

$$K_s = \sqrt{C_o/(2nC)} = \sqrt{1.56 \times 12.0/\left(2 \times 1 \times \sqrt{9.80 \times 11.0}\right)} \fallingdotseq 0.949 \tag{i.2}$$

Because $h/L_o \leq 0.2$, we can use $H_{max} = \min\{(\beta_o^* H_o' + \beta_1^* h), \quad \beta_{max}^* H_o', \quad 1.8 K_s H_o'\}$.

$\beta_o^* = 0.052(H_o'/L_o)^{-0.38} \exp(20 \tan^{1.5}\alpha) = 0.052 \times (6.00/225)^{-0.38} \exp(20/100^{1.5}) \fallingdotseq 0.210$

$\beta_1^* = 0.63 \exp(3.8 \tan\alpha) = 0.63 \times \exp(3.8/100) \fallingdotseq 0.654$

$\beta_{max}^* = \max\{1.65, \quad 0.53(H_o'/L_o)^{-0.29} \exp(2.4\tan\alpha)\}$

$\quad = \max\left\{1.65, \quad 0.53 \times (6.00/225)^{-0.29} \exp(2.4/100)\right\} \fallingdotseq 1.65$

$$\left. \begin{aligned} \therefore H_{max} &= \min\{(0.210 \times 6.00 + 0.654 \times 11.0), \quad 1.65 \times 6.00, \quad 1.8 \times 0.949 \times 6.00\} \\ &= \min(8.46, \ 9.90, \ 10.3) \fallingdotseq 8.46\text{m} \end{aligned} \right\} \tag{i.3}$$

7.2 Measures against Large Waves

(d) Calculation of Wave Force

The open-source software CADMAS-SURF can be used to calculate the wave pressure distribution acting on structures of arbitrary shape and with wave absorbing devices. However, if the structure is a simple upright wall, the pressure distribution can be calculated using Goda's formula, for example, as described in Section 3.2, 1) "Wave Forces Acting on Upright Walls". Then, using the pressure distribution obtained from Goda's formula and others, the total horizontal wave force F_H [N/m] and the total lift force F_U [N/m] can be obtained concerning **Fig. 7.8** as follows.

$$F_H = \frac{p + p_b}{2} \times h' + \frac{p + p \times \frac{\eta' - h_c}{\eta'}}{2} \times h_c \quad (7.5)$$

$$F_U = \frac{p_u \times b}{2} \quad (7.6)$$

where p is the horizontal wave pressure at sea level, p_b is the horizontal wave pressure at the bottom, p_u is the maximum lift pressure, and η' is the height from sea level to the top of the horizontal wave pressure.

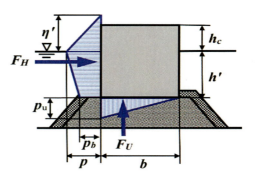

Fig. 7.8 Illustration of wave pressure distribution.

(e) Stability Calculation

As a countermeasure against wave forces, it is necessary to check the stability against slipping and overturning.

For slipping: "the total horizontal wave force × the safety factor (= 1.2) " ≤ "the self-weight taking into account the buoyancy of the structure and the total lift × the friction factor".

For overturning: "the moment of total horizontal wave force × the safety factor (= 1.2) " ≤ "the moment of the self-weight taking into account the buoyancy of the structure and the total lift".

In addition, if there is a possibility that elements of the structure may fail in compression, tension, bending or shear, it is necessary to determine the stresses in each element under the conditions of the design wave force and to check that the values obtained by multiplying them by the safety factor (= 1.2) do not exceed the allowable stress values of each element.

(2) Stability Calculation for Block-Covered Breakwaters

The required mass M of covered stones or concrete blocks against design waves can be obtained from Hudson's Eq. (3.59) and Van der Meer's Eqs. (3.60) - (3.63). The required mass should then be multiplied by a factor of 1.5 if the scattering of the covered material due to uncertainties is to be sufficiently prevented.

7. COASTAL PROTECTION AND VARIOUS OTHER STRUCTURES

【A calculation example to determine the wave forces acting on a breakwater】

Find the total horizontal wave force F_H and the total lift wave force F_U from Goda's equations with $H_{max} = 8.46$ m, $H_o' = 6.00$ m, $T = 12.0$ s, the dominant wave direction angle = 15.0°, the seabed slope $\tan\alpha = 1/100$, the water depth in front of the breakwater $h = 11.0$ m, the depth from the still water surface to the bottom of the concrete blocks $h' = 8.00$ m, the depth from the still water surface to the surface of the foundation mound $d = 6.50$ m, the height from the still water surface to the crown of the breakwater $hc = 4.00$ m, the breakwater width $b = 15.0$ m, and the seawater density ρ = 1030 kg/m³.

$$\eta' = 0.75(1+\cos\theta')H_{max} = 0.75\times[1+\cos(15.0°-15.0°)]\times8.46 \fallingdotseq 12.7 \text{m}$$

where θ' is the most dangerous angle within plus or minus 15.0 degrees of the dominant wave direction angle.

Then, in the case of $h/L_o = 0.0490$, $L = \sqrt{9.80\times11.0}\times12.0 \fallingdotseq 125 \text{m}$.

Therefore, $k = 2\pi/L = 2\times3.14/125 \fallingdotseq 0.0504 \text{m}^{-1}$, and the water depth h_b at a location about 5 times offshore of the significant wave height $H_{1/3}$ from the front of the breakwater,

$$h_b = h + 5K_sH_o'\tan\alpha = 11.0 + 5\times0.949\times6.00/100 \fallingdotseq 11.3 \text{m}$$

$$\alpha_1 = 0.6 + \frac{1}{2}\left\{\frac{2kh}{\sinh 2kh}\right\}^2 = 0.6 + \frac{1}{2}\left\{\frac{2\times0.0504\times11.0}{\sinh(2\times0.0504\times11.0)}\right\}^2 \fallingdotseq 0.937$$

$$\left.\begin{array}{l} \alpha_2 = \min\left\{\frac{h_b-d}{3h_b}\left(\frac{H_{max}}{d}\right)^2, \frac{2d}{H_{max}}\right\} \\[3mm] = \min\left\{\frac{11.3-6.50}{3\times11.3}\left(\frac{8.46}{6.50}\right)^2, \frac{2\times6.50}{8.46}\right\} = \min\{0.239, 1.54\} \fallingdotseq 0.239 \end{array}\right\}$$

$$\alpha_3 = 1 - \frac{h'}{h}\left\{1 - \frac{1}{\cosh kh}\right\} = 1 - \frac{8.00}{11.0}\left\{1 - \frac{1}{\cosh(0.0504\times11.0)}\right\} \fallingdotseq 0.901$$

$$\left.\begin{array}{l} p_1 = \frac{1}{2}(1+\cos\theta')(\alpha_1 + \alpha_2\cos^2\theta')\rho gH_{max} \\[3mm] = \frac{1}{2}(1+1)(0.937+0.239)\times1.03\times9.8\times8.46 = 100.44\cdots \fallingdotseq 100\left[\text{kN/m}^2\right] \end{array}\right\}$$

$$p_2 = \frac{p_1}{\cosh kh} = \frac{100}{\cosh(0.0504\times11.0)} \fallingdotseq 86.8\left[\text{kN/m}^2\right]$$

$$p_3 = \alpha_3 p_1 = 0.901\times100.44\cdots \fallingdotseq 90.5\left[\text{kN/m}^2\right]$$

$$F_H = \frac{1}{2}\left(\frac{12.7-4.00}{12.7}\times100+100\right)\times4.00 + \frac{1}{2}(100+90.5)\times8.00 \fallingdotseq 1102[\text{kN/m}] \quad (j.1)$$

$$\left.\begin{array}{l} F_U = \frac{b}{2}p_u = \frac{b}{2}\times\frac{1}{2}(1+\cos\theta')\alpha_1\alpha_3\rho gH_{max} \\[3mm] = \frac{15.0}{2}\times\frac{1}{2}(1+1)\times0.937\times0.901\times1.03\times9.8\times8.46 \fallingdotseq 541[\text{kN/m}] \end{array}\right\} \quad (j.2)$$

7.2 Measures against Large Waves

The required mass of the foundation mound covering material of a composite breakwater M can be obtained from Eqs. (3.64), (3.65). The required mass of covered material M for a fully submerged breakwater (artificial reef) can be obtained from Eq. (3.66) of Uda et al.

5) Wave Overtopping Measures

To avoid damage due to wave overtopping, the crown height of coastal dikes and seawalls should be higher than the wave run-up height calculated using the design wave height and the design water level.

However, to formulate a coastal protection plan that considers the use of the backland (hinterland), the design of the structure should be such that the wave overtopping rate calculated according to Section 3.1, 4) "Calculation of Wave Overtopping Rate", etc. should not exceed the allowable value for the given usage. The allowable wave overtopping rate depends on the structure type, the type of object used, and the allowable level of damage as shown in Tables 7.1 to 7.4.

For the damage limits of coastal dikes and seawalls, Table 7.1 of the Coastal Engineering Manual by the U. S. Army Corps of Engineers [5] and Table 7.2 of Goda [6] can be used.

For the use limits of backlands, Table 7.3 of the Coastal Engineering Manual by the U. S. Army Corps of Engineers [5] and Table 7.4 of Fukuda et al [7] can be used.

Table 7.1 Damage limit overtopping rates [m³/s/m] according to C.E.M. by U.S. Army Corps of Engineers [5].

Coverage level	Coastal dike	Seawall	Grass covered dike
Concrete covering on three sides (front, crown and back)	0.05	—	0.01
Concrete covering on two sides (front and crown)	0.02	0.2	0.01
Concrete covering on front surface only	Less than 0.002	0.05	Less than 0.001

Table 7.2 Damage limit overtopping rates [m³/s/m] of coastal dikes and seawalls according to Goda [6].

Coverage level	Coastal dike	Seawall
Concrete covering on three sides (front, crown and back)	0.05	—
Concrete covering on two sides (front and crown)	0.02	0.2
Concrete covering on front surface only	Less than 0.005	0.05

Table 7.3 Allowable wave overtopping rates [m³/s/m] based on backland use according to C.E.M. by U.S. Army Corps of Engineers [5].

Target of use	Dangerous	Should be cautious	Safe enough
Pedestrians	Greater than 3×10^{-5}	3×10^{-5} - 4×10^{-6}	4×10^{-6}
Vehicles	Greater than 2×10^{-5}	2×10^{-5} - 1×10^{-6}	1×10^{-6}
Houses	Greater than 3×10^{-5}	3×10^{-5} - 1×10^{-6}	1×10^{-6}

7. COASTAL PROTECTION AND VARIOUS OTHER STRUCTURES

Table 7.4 Allowable wave overtopping rates [m³/s/m] based on backland use according to Fukuda et al [7].

Target of use	50% safe directly behind a dike or a seawall	90% safe directly behind a dike or a seawall
Pedestrians	2×10^{-4}	3×10^{-5}
Bicycles	2×10^{-5}	1×10^{-6}
Houses	7×10^{-5}	1×10^{-6}

6) Measures against Coastal Erosion and Sedimentation in Harbours

Beach erosion is progressing in many parts of the world (refer to Vu et al [8]). If left unprotected, not only will important beaches be lost, but the destructive power of incoming waves will increase due to deeper frontal water depths. Therefore, from the point of view of coastal protection and disaster prevention, it is very important to predict the potential future beach erosion and implement appropriate countermeasures. When constructing breakwaters to protect ports and harbours, including fishing ports, or dredging to maintain channels and anchorage areas, their impact should be predicted in advance and the prediction results should be reflected in the implementation plan so that it is possible to minimize negative impacts on the surrounding beaches.

For the prediction of the impacts of these interventions, numerical models introduced in Section 6.3 can be used. When using these models, it is common to consider the input waves used for calculations as follows.

(a) In the case of long-term forecasts (several decades), multiline or shoreline models can be used to reduce the computational load. If the change of waves over time (especially the change in wave direction) is simple, the mean wave condition ($\bar{H}, \bar{T}, \bar{\theta}$) over the entire year is obtained from Eq. (7.7) and used repeatedly during the prediction period. If the change of waves over time is complex, then the mean wave ($\bar{H}, \bar{T}, \bar{\theta}$) is determined from Eq. (7.7) for each period (season, month, day) and used repeatedly over the prediction period. Furthermore, as more severe condition, if erosion due to high waves during half-day to one-day needs to be considered, the expected high waves should be input just before the output of the long-term forecast results.

(b) In the case of numerical models that also consider nearshore currents, even a one-year prediction can be very computationally demanding. Therefore, the amount of erosion occurred for one year is approximated by inputting high waves that come several times a year, considering that waves during calm periods do not lead to erosion.

$$\bar{H} = \left\{ \frac{\sum_{i=1}^{n} \left(H_i^2 \times T_i \right)}{\sum_{i=1}^{n} T_i} \right\}^{1/2} \quad , \quad \bar{T} = \frac{\sum_{i=1}^{n} T_i}{n}, \quad \bar{\theta} = \frac{\sum_{i=1}^{n} \left(H_i^2 \times T_i \times \theta_i \right)}{\sum_{i=1}^{n} \left(H_i^2 \times T_i \right)} \qquad (7.7)$$

where, H_i, T_i, θ_i are the observed data of the significant wave height, significant period and dominant wave direction respectively; and n is the number of observations for an energy mean processing period.

7.2 Measures against Large Waves

Coastal sediment transport is considered as a continuum where sediment is transported from the mountains to the coast via rivers and then transported in the cross-shore and longshore direction by waves and currents. Therefore, it is important to maintain sediment balance in the coastal system by working with nature rather than going against nature when implementing measures against beach erosion. Specifically, it is necessary to consider how to increase the sediment supply from upstream to the target beach (for example, a sand bypass and other beach nourishment) and how to reduce sediment loss from the target beach (for example, groynes, detached breakwaters, a sand recycle system), regarding the basic ideas summarised in Section 6.4. Attention should also be paid to the scour and the suction (outflow) of backfill material as summarised in Section 6.5.

For the design of facilities that will be required for coastal erosion control, the "Coastal Engineering Manual" by the U. S. Army Corps of Engineers [5] and "Design Manual for Coastal Facilities – 2000" by the Japan Society of Civil Engineers [9] can be consulted. Furthermore, Specific case studies include Kornvisith et al [10] and Promngam et al [11], which have reported examination results of topographical change predictions and comparison of the effectiveness of countermeasures.

(1) Groyne Works

When the dominant wave direction is sufficiently oblique to the shore-normal direction, a significant amount of longshore sediment transport can occur within the surf zone. If a groyne is constructed across the surf zone then, longshore sediment transport will be interrupted. As a result, sedimentation can occur on the upstream side of the groyne while erosion can occur on the downstream side until the shoreline becomes perpendicular to the dominant wave direction, as shown in **Fig. 7.9**. If a second groyne is constructed to the right of the first groyne as in the figure and if the shoreline between the two groynes can become perpendicular to the dominant wave direction, the erosion in this section may cease. If there is no further groyne is constructed, the shoreline to the righthand side the second groyne can continue to erode as shown in the figure unless sediment supply to that section of the beach is restored.

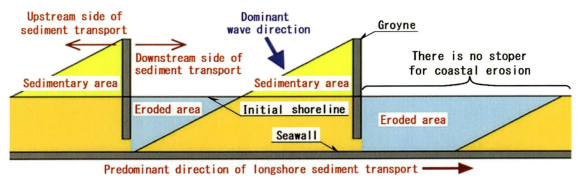

Fig. 7.9 Plane view illustrating the trend of shoreline change due to groynes on a beach dominated by longshore sediment transport.

7. COASTAL PROTECTION AND VARIOUS OTHER STRUCTURES

It is advisable to install a field of groynes at suitable intervals rather than a single groyne to minimize erosion downstream of each groyne. It should be supported by artificial beach nourishment to maintain the shoreline upstream and downstream as well as within the groyne field. In addition, as the construction of a groyne may create turbulent eddies at the tip of the groyne which enhance cross-shore sediment transport and loss of coastal sediment, an overhang with a sufficiently high crown height should be constructed at the upstream side of each groyne, i.e. a T shape groyne, to prevent sediment from being easily washed away.

(2) Detached Breakwater Works

As shown in **Fig. 7.10**, if a detached breakwater is installed in such a way that there is enough calm water region behind the breakwater, sand deposition at the landside of the breakwater can be expected due to the creation of a calm water area and wave diffraction effects at both ends of the breakwater. Furthermore, if the breakwater is permeable sand driven in from the offshore side will also accumulate in the calm area. The tongue-shaped sedimentation area called a **tombolo** will be developed between the shoreline and the breakwater.

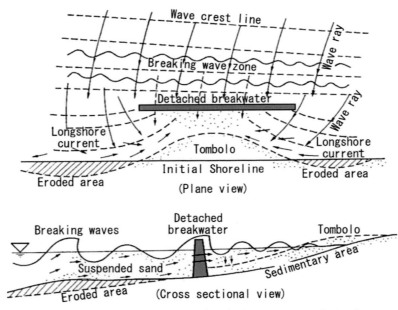

Fig. 7.10 Formation of sedimentary area (tombolo topography) in a calm water behind the landward side of a detached breakwater (from "Technical Standards and Commentaries for Coastal Protection Facilities", CDIT [1])

According to unpublished data on main coasts in Japan by Yamamoto [one of the co-authors], the characteristics of detached breakwaters that can be expected to have a high sedimentation effect are the following:

7.2 Measures against Large Waves

(a) The on-off distance from the initial shoreline before installation to the centre of the detached breakwater should be 1.5 to 2 times the on-off distance from the initial shoreline to the future shoreline to be restored, which is a geometrically reasonable value. However, the installation depth and surrounding seabed topography should be favourable for construction. If the installation depth is close to the breaking depth of high waves, the wave forces will be very large and the stability of the detached breakwater should be carefully considered.

(b) In the case of constructing a group of detached breakwaters, the length of each detached breakwater should be 2 (if some shoreline erosion is tolerated) to 3 (if shoreline erosion is not tolerated) times longer than the gap between breakwaters so that diffraction waves through the gap become weak. If the length of the detached breakwaters is too long, enough sand will not move to the shadow area behind the detached breakwater and a tombolo may not form. In addition, the detached breakwaters should be permeable so that sediment can easily enter from the seaward side and water can be exchanged between the seaward and landward sides to maintain water quality.

(c) The crown height of the detached breakwater should be at least 2 m or more above the still water surface, because, as the wave overtopping rate becomes bigger, the wave-absorbing effect will be significantly reduced, and the sedimentation effect will be diminished. The crown width should be 3 pieces of blocks in a row to make the blocks fully interlock with each other. The evaluation method described in Section 2.3, 3) "Transmission and Reflection of Waves" should be used if the wave transmission coefficient needs to be determined accurately.

When it is necessary to consider landscape and water quality conservation, wide submerged breakwaters (artificial reefs) may be selected. The coastal engineering committee of JSCE [9], and the coast division of NILIM [12] are useful references for the design of wide submerged breakwaters. However, it should be noted that the wave attenuating effect due to the wide crown surface of a submerged breakwater is weaker than that of a normal detached breakwater. In addition, although there will be sedimentation in the calm area behind the landward side of a wide submerged breakwater, a tombolo may not form.

(d) The installation of one large-detached breakwater may lead to excessive erosion downstream of the breakwater. Therefore, it is common to install a group of detached breakwaters with the length of each breakwater being about 100-200 m instead.

(e) If the construction takes a long time, then temporary measures such as beach nourishment should be considered to control erosion during construction. (refer to (1) "Groyne Works").

7. COASTAL PROTECTION AND VARIOUS OTHER STRUCTURES

【Effectiveness of Mangrove Forests in Preventing Erosion】

Mangrove forests native to the coasts in the low latitude zone have been re-evaluated as valuable natural wave-absorbing facilities. As a result, attempts have been made to restore mangrove forests that have been destructed to secure charcoal for commercial use or to create shrimp ponds.

Yamamoto et al [13], using their own experiments and past experimental data, established a diagram showing the relationship between the cross-shore width of mangrove forests and the wave-absorbing effect, as shown in the figure below. They tried to evaluate the erosion prevention effect of mangrove forests using a wave transmission ratio (= transmitted wave height / incident wave height) obtained from this diagram and the numerical simulation model of Vu et al. in 2), (1) of Section 6.3.

Suzuki [14], using the open source software CADMAS-SURF, has also proposed a method that can simulate fluid motion (flow velocity distribution, water surface change related to wave dissipation effect) in vegetated forests, such as mangrove forests.

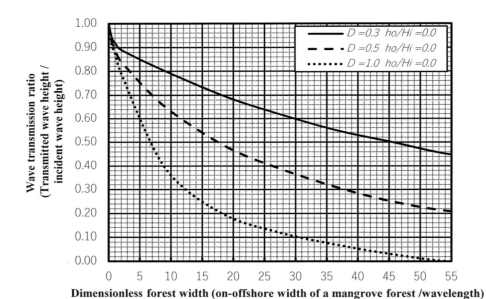

D: Number of main trunks per unit area [No./m^2].

ho/Hi : The height from the still water surface to the top of pillar roots / the incident wave height, where $ho/Hi = 0$ means that almost all of the pillar roots are just submerged in the water.

Fig.k.1 Diagram of the relationship between on-offshore widths of mangrove forests and wave transmission ratios (from Yamamoto et al [13]).

7.2 Measures against Large Waves

(3) Headland Defence Works

As shown in **Fig. 7.11**, the sandy beach between headlands is transformed to an arch shape due to the wave diffraction effect of the two headlands, and the wave refraction effect. A similar method of stabilising a beach by enclosing it within artificial headlands is called **headland defence works**. As shown in **Fig. 7.12**, artificial headlands can also be created using detached breakwaters to form a tombolo in between two breakwaters.

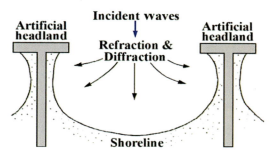

Fig. 7.11 Illustration of headland defence works.

Silvester [15] proposed a method in which the shoreline shape can be determined by the location of both artificial headlands and the direction of incident waves. The reader is referred to the "Design Manual for Coastal Facilities – 2000" of JSCE [9], which explains this in detail. Furthermore, the planning of headland placement on dynamically stable beaches can be assisted by the simulations from the line models described in Section 6.3, 1).

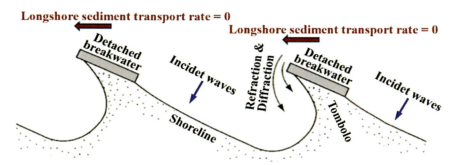

Fig. 7.12 Illustration of the headland control with detached breakwaters.

(4) Artificial Sediment Supply Methods

The erosion control by constructing structures described in (1) to (3) can be ineffective if there is no sufficient sediment supply to the target beach. Therefore, in the event of inadequate natural sediment supply to sustain beach stability, artificial sediment supply methods should be considered.

(a) Sand Discharge Systems from Dammed Reservoirs

One of the major reasons for reducing natural sediment supply to beach systems is building dams across rivers to create reservoirs for irrigation or hydropower generation purposes. Dams block natural sediment flow down rivers while infilling reservoirs and reducing their capacity over time. To avoid this, dams should be equipped with gates and channels dedicated for sediment discharge and the timely release of sediment accumulated in reservoirs. However, it is necessary to investigate the

7. COASTAL PROTECTION AND VARIOUS OTHER STRUCTURES

impact of regulated sediment discharge on the maintenance of river channels downstream, water quality and the impact on the beach, and take measures in advance to ensure that the discharged sediment does not cause any harm to the environment, agriculture and general public. For example, if the sedimentation of organic matter such as fallen leaves is left for a long period, the presence of large quantities of decaying organic matter is a concern and requires particularly careful attention. Furthermore, if the amount of sediment accumulation is large, it may be worth considering the installation of pipelines to transport sediment in the form of mixed sediment water (called **slurry transportation**), with the pumps powered by their own small-scale hydroelectric power generation, which may be cost-effective.

(b) Sand Bypass Method

When a port or harbour is constructed on a coast where littoral drift cannot be ignored, attached breakwaters and other similar structures can interfere with longshore sediment transport. As a result, sedimentation can occur upstream of longshore sediment transport direction while erosion occurs downstream. Furthermore, when the upstream sedimentation becomes saturated, the sediment overflows to the downstream side around the tip of the breakwater, resulting in the burial of the mouth of the port or harbour. To avoid downstream erosion and burial of the mouth, artificial transport of sediment deposited upstream to the downstream side called a **sand bypass method**, can be facilitated.

However, sand transport by a large number of dump trucks would degrade the environment around the road and the total cost of pipeline transport would be similar to or lower than that of dump truck transport if the transport distance is not too long. Therefore, it is worth considering the installation of a pipeline powered by some renewable energy source such as wave or solar power to transport the slurry to the erosion area downstream. **Photo 7.9** shows a sand bypass project using jet pumps and pipelines to transport slurry from the sedimentation area (right) to the eroding beach (left) at Fukuda fishing port in Shizuoka Prefecture, Japan.

(c) Sand Recycling Method

If a sand trap can be installed at the downstream end of a beach where the supply of sediment transport from the upstream side is not sufficient, a **sand recycling method** can be adopted, with the sediment collected at the trap returned upstream and circulated again to maintain dynamic stability. A renewable energy source to power the pumps can be considered as in (b). **Fig. 7.13** shows the sand recycling project using submerged sand pumps and pipelines to transport slurry on the Tottori coast in Tottori Prefecture, Japan.

7.2 Measures against Large Waves

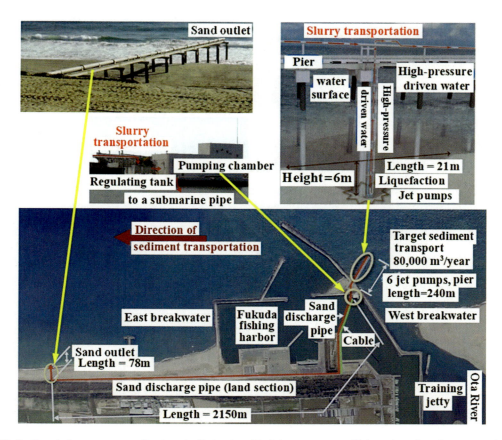

Photo 7.9 Sand bypass project at Fukuda fishing port, Shizuoka Prefecture in Japan (from Shizuoka Prefecture, Fukuroi Civil Engineering Office data).

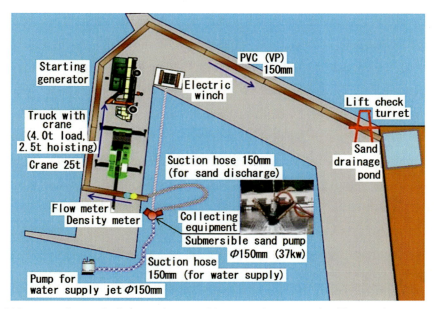

Fig. 7.13 Illustration of the sand recycling project on the Tottori coast, Tottori Prefecture in Japan (from Committee documents by Land Development Department of Tottori Prefecture).

7. COASTAL PROTECTION AND VARIOUS OTHER STRUCTURES

At the Shimizu coast in Shizuoka Prefecture, Japan, an erosion control project is being implemented by combining headland defence works, a sand bypassing and recycling system based on conventional truck transportation.

7) Measures against Scour and Sand Outflow

(1) Breakwaters (Concrete blocks or Caissons without Backfill)

If the frontal scour of breakwaters is left unchecked, the breakwater may overturn and even collapse. Therefore, scour protection measures, such as rooting foundations and mats, should be installed to adequately cover the estimated area of scour. If the scale of the scour is expected to be very large, wave-absorbing breakwaters or wave-absorbing works ("breakwaters" if separate from main structures and independent, or "works" if attached to main structures) should be installed to reduce wave forces.

(2) Coastal Dikes or Seawalls with Backfill

As in the case of breakwaters, the same scour prevention measures are essential, but it is also important to take measures to prevent suction (outflow) induced by scour.

To increase the resistance against suction (outflow) of the backfill materials, determined from Eqs. (6.55) – (6.58) in Chapter 6, it is possible to (i) replace the sand layer with angular, heavy, and large materials, (ii) increase the thickness of the sand layer (as a resistance layer) in front of the structure, and (iii) reduce the excess pore water pressure in the backfill materials.

(i) Some materials that are heavier and have a better interlocking effect than normal sand are iron ore, angular large stones, and reinforced and deformed concrete blocks. Although replacing the sand layer in front of the coastal dike or the seawall with these materials can reduce wave forces on the structure, that means building foot protection or a wave absorber with large stones or concrete blocks. If care is not taken when building these structures, the landscape and the surrounding environment of the front beach can be degraded.

Meanwhile, replacing backfill materials with cobbles is expected to have a sufficient reduction effect while having less impact on the surrounding environment. However, in the case of the coastal dike, if all the backfill materials are replaced by cobbles, the possibility of seawater infiltration to the landward side may increase. Therefore, it is necessary to provide an impermeable layer/wall near the back slope surface.

(ii) In the case of increasing the thickness of the sand layer in front of the structure, in order to ensure that the sand layer is not lost during high waves, it is advantageous to cover the surface of the sand layer with angular large stones or reinforced and deformed concrete blocks, or to install wave-

7.2 Measures against Large Waves

absorbing works in front of it. Therefore, precautions in this case are the same as the first part of (i) and should not lead to the deterioration of the landscape and use of the sandy beach area.

(iii) To reduce the excess pore water pressure in the sand layer, it can be suggested from Eqs. (6.55) – (6.58) in Chapter 6 that the incident significant wave height should be reduced by using wave-absorbing facilities, and the sand layer should be replaced by materials with larger median grain size. In coastal dikes or seawalls covered by a single concrete slab, the front sand layer thickness can be increased by driving sheet piles in front of the dike or seawall to an adequate depth. The rooting depth required to prevent suction (outflow) and the median grain size of the backfill materials can be estimated by substituting Eqs. (6.56) - (6.58) in Chapter 6 into the conditional equation $\tau_r \geq \tau_f$ [the median grain size d_{50} in Eqs. (6.55) - (6.58) is the grain size of the backfill material within the coastal dike and the seawall, as it is assumed that the sand layer in front of these structures is lost due to scour].

The method of increasing the median grain size of the sand layer to reduce the suction (outflow) rate is inherently effective for both the sand layer in front of coastal dikes and seawalls and the backfill material inside them. Kato et al [16] proposed the replacement of the sand layer in front of the dike or seawall with rubble rooting works, while Yamamoto et al [17] proposed the replacement of the backfill material inside them with pebbles. In the former method, care must be taken and suitable temporary protection should be provided to avoid damage from high waves during construction. In the latter case, because the permeability of the seaward part of the backfill is increased, the permeability of the landward part of the backfill must be sufficiently low to prevent seawater from entering the landward side.

Ioroi et al [18] found that increasing the uniformity factor of the backfill material can reduce the amount of suction; and that increasing the dry density by compaction of the backfill material can delay the progress of suction, although the final amount of suction remains unchanged unless the median grain size or uniformity factor is changed. Furthermore, Kuisorn et al [19] proposed a formula for calculating the amount of suction that can take these effects into account.

8) Measures against Splash and Wind-blown Sand

Even if large waves are not strong enough to destroy coastal dikes and seawalls, the spray and wind-blown sand can extend several hundred metres inland, when large waves are accompanied by strong wind. Therefore, coastal managers in high wind areas should not only maintain coastal forests but also attempt to widen them if possible.

Iizuka et al [20] and others have investigated the effect of coastal forests in reducing airborne salinity in a black pine forest on the Yokohama coast in Aomori Prefecture, Japan (as shown in **Fig.**

7. COASTAL PROTECTION AND VARIOUS OTHER STRUCTURES

7.14, the width of the forest from seaward to landward is 204 m, the height of the trees between 44 m from the seaward edge increases from 2 m to 6 m, and the height between 160 m from that point to landward is about 12 m).

Their results are shown in **Fig. 7.14**. According to this figure, when there is a coastal forest with a 200 m width from seaward to landward, the airborne salinity decreases rapidly with the distance from the coastline and approaches zero at the landward edge of the coastal forest. Behind the forest, the airborne salinity increases and returns to the same level as the dotted line of the case without the coastal forest at a distance of about 200 m from the edge of the forest inland. Therefore, Kudo [21] suggested that several coastal forest zones of at least 50 m width to the landward side with a gap of 15 times the height of the trees should be established on the inland side of a coastal forest.

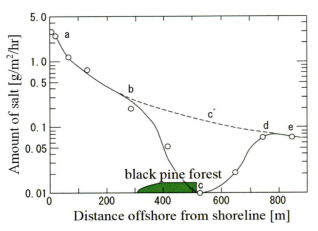

Fig. 7.14
Reduction effect of airborne salinity by a coastal forest. (from Iizuka et al [20])

Tsukamoto et al [22] and others have investigated the effect of coastal forests on reducing the amount of wind-blown sand. **Fig. 7.15** shows the results of a survey by Tsukamoto et al. on the Shonan coast (observation period April 1999 - January 2000, coastal forest width from shore to land side approximately 100 m, average tree height 4.5 m) in Kanagawa Prefecture, Japan. According to this figure, a coastal forest with an average tree height of 4.5 m and a width of at least 50 m to the landward side significantly reduced the amount of wind-blown sand.

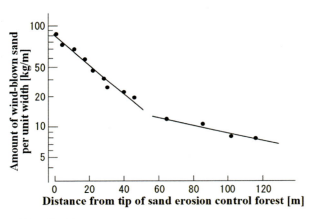

Fig. 7.15
Reduction effect of the amount of wind-blown sand (total amount during the observation period) by a coastal forest. (from Tsukamoto et al [22])

7.3 Storm Surge Protection
1) Characteristics of Storm Surge Damage

As discussed in Section 5.1, the high temperatures and humidity near the centre of a strong low-pressure system such as a hurricane, a cyclone, or a typhoon generate strong updrafts. This generates a low-pressure system at the centre, which in turn generates winds blowing towards the centre; the effect of the Earth's rotation causes the winds to blow counterclockwise in the Northern Hemisphere, as shown in **Fig. 7.16** and clockwise in the Southern Hemisphere. In this figure, the area to the lower right of the cyclone centre is dangerous because the combined velocity of the wind blowing into the centre and the moving velocity of the cyclone becomes greater.

The low atmospheric pressure and the prevailing wind cause the water surface near the centre of the tropical cyclone to rise above the surrounding water level. This is called a **meteorological deviation** or **meteorological anomaly**, also known as a storm surge, which can be evaluated using Eq. (5.1) and is often higher than 1 m.

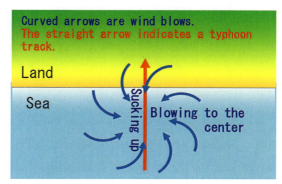

Fig. 7.16 Illustration of sea level elevation due to wind blowing towards the center of a tropical cyclone and sucking up at the center.

Tropical cyclones generate large waves. If meteorological deviations coincide with astronomical high tides, the wave overtopping rates will increase. If the enhanced wave forces lead to the breaching of coastal dikes, large volumes of seawater can enter inland due to overtopping, causing severe damage to buildings, roads, and farmlands in flooded areas, and forcing people and livestock to be evacuated to safer locations. If tropical cyclones strike low-lying coastal areas, typically densely populated, it will cause extensive damage. Global warming may exacerbate the situation due to an increase in tropical cyclone intensity and mean sea level.

The combination of high tides and river floods can cause severe inundation around river mouths. **Photo 7.10** shows the overflow of river banks during Typhoon Vera in 1959 struck,

Photo 7.10 Overflow phenomenon at the Yamazaki River of Nagoya City in Japan caused by Typhoon Vera in September 1959.
(from slide collection "Coasts and Ports in Japan" of JSCE [2])

7. COASTAL PROTECTION AND VARIOUS OTHER STRUCTURES

in which the number of people died or went missing and damaged houses reached 5,098 and 1.2 million, respectively. On coasts where tropical cyclones make landfall, not only water levels will be higher, but strong shoreward winds and waves can also be present, which can cause boats and vessels to wash up on shores, as shown in **Photo 7.11**.

Photo 7.11
Wash-up of the cargo ship "Galatic" on the Fuji Coast of Shizuoka Prefecture in Japan, by Typhoon No. 20 in October 1979.
(from slide collection "Coasts and Ports in Japan" of JSCE [2]).

2) Measures against Storm Surges

Inundation due to storm surges can extend over the entire ground area below the total water level. If coastal dikes protecting the area fail due to external force exceeding the design value or due to the life of the structure exceeding its design life, and if floods of water channels occur simultaneously or accidents prevent the sluice gates from closing, in the zero meter above sea level zone, enormous damage will be caused.

(1) Safety of Residential Areas and Important Facilities

It is necessary to check the safety degree of the functionality, stability, and part strength of coastal dikes and other coast protection structures against the worst design waves and design water levels during the period in which residential properties and other important facilities are expected to continue to present. In practice, the degree of safety is checked using the external force conditions of the strongest cyclone ever recorded (e.g. Hurricane Patricia, Tropical Cyclone Mahina, and Typhoon Tip) approaching the target coast on its worst cyclone track.

However, it is not possible to ignore that a stronger cyclone may strike with a force greater than that ever occurred in the past and that topographical conditions may change due to coastal erosion or human interventions, or both. Therefore, it should be assumed that damage may occur in the event of failing or overtopping of coastal prevention facilities in addition to topographical change, and evacuation plans should be put in place that incorporate regular field training for residents. A system is also needed to regulate the construction of residential buildings and facilities in areas inherently prone to danger such as low-lying areas. A storm surge hazard map required for this purpose can be

7.3 Storm Surge Protection

prepared by referring to the "Tsunami and Storm Surge Hazard Map Manual" by the Cabinet Office, MLIT, and MAFF [23].

The possibility of simultaneous occurrence of storm surge inundation and river flooding is not low; so, evacuation routes, evacuation centers (facilities such as schools and community centers where people take shelter for a certain period of time), and evacuation sites (designated temporary evacuation sites such as parks) must be safe for both.

(2) Safety of Non-critical Areas

In practice, it is necessary to check the safety of the functionality, stability, and part strength of coastal dikes and other structures against the worst design waves and design water level during the service life (30 to 50 years) of reinforced concrete as the main material used in structures.

It is also necessary to establish a system whereby residential developments and the construction of important facilities are not permitted in areas that are topographically vulnerable to severe damage from storm surges and large waves.

(3) Consideration of Global Warming

The Sixth Assessment Report (2022-2023) of the **Intergovernmental Panel on Climate Change (IPCC)** [24] reports the actual situation and future projections of global warming as follows.

(a) Global warming has resulted in a global average temperature rise of 0.8-1.3°C between 1850-1900 and 2010-2019. In addition, a global average sea level rise of 0.15-0.25 m was observed between 1901 and 2018, with 1.3 mm/year for the first 71 years, 1.9 mm/year for the next 36 years and 3.7 mm/year for the last 13 years, indicating an acceleration in the recent years.

(b) The latest climate prediction models predict a global average temperature rise of 3.6°C and a sea-level rise of 0.8 m between 1850-1900 and 2081-2100 if greenhouse gas emissions double between 2015 and 2100.

A temperature rise of 3.6°C is likely to increase the development of cumulonimbus clouds and increase the intensity of tropical cyclones, even if a decrease in the number of cyclones, thus increasing the damage caused by river floods, storm surges, and large waves. A 0.8 m rise in sea level will exacerbate flooding and coastal erosion caused by storm surges and large waves. It is therefore becoming increasingly important to rethink coastal protection measures, as reported by Kato and Tajima [25] and others, and to consider the impact of global warming on the deterioration of the coastal protection environment.

Numerous studies have been carried out around the world examining the impact of worsening external forces due to global warming on coastal disaster prevention. For example, Yamamoto et al [26] found that global warming increases wave overtopping by one order of magnitude and that the severity of the impact will be greater on a beach with a mild slope from the middle of the beach profile

7. COASTAL PROTECTION AND VARIOUS OTHER STRUCTURES

than on a beach with a steep slope, based on the calculation of wave overtopping rates using the following conditions.

(a) the conditions on beach profiles and coastal dikes: (i) the case where a coastal dike with a crown height of MWL+ 9.0 m is placed at a beach height of MWL+ 6.5 m on a composite beach with an upper beach slope of 1:10 and a lower beach slope of 1:50, where the water depth at the boundary between the upper and lower beaches is 7.4 m, and (ii) the case where the same dike is placed at the same beach height on a steep beach with the slope of 1/10.

(b) The conditions on external force: (i) the case where global warming is ignored, and (ii) the case where global warming causes a 65 cm rise in mean water level and a further 3°C rise in temperature, causing a 30% larger pressure decrease at the typhoon center and increase in wave height from 9.4 m to 10.4 m.

Okayasu and Sakai [27], Suh et al [28], and others have studied the effects of global warming on the stability of caisson breakwaters, while Yasuhara et al [29] have noted that rising groundwater levels due to global warming can lead to instability of many structures in coastal areas.

7.4 Tsunami Protection
1) Characteristics of Tsunami Damage

If a tsunami encroaches an unobstructed flat area where the ground level is sufficiently lower than the tsunami height, seawater can propagate landward from the coastline up to about half of the tsunami wavelength, because the water level remains above the mean water level for about half a wavelength. Naturally, if there is a high wall-like obstacle, it will stop there. Because the wavelength of a mega tsunami at the coast may reach about 6 km to 36 km, seawater can propagate several kilometres inland and inundate the land as shown in **Photo 7.12**.

Rivers and roads extending landwards from the seaside can be ideal channels for landward tsunami propagation.

Large quantities of sediment can be transported seaward during the return flow of a tsunami, which can cause a large scale scour on beaches and river mouths.

Furthermore, a large amount of salt contained in seawater can remain in the inundated areas. Unless seawater does not drain out quickly, agricultural land will become barren as shown in the example in **Photo 7.13**.

Photos taken by satellite IKONOS, provided by King Mongkut's University of Technology via the Government of Thailand.

Photo 7.12 Comparison of the situation on the KhaoLak coast in Thailand before and after the 2004 Sumatra Earthquake tsunami struck (Moment Magnitude = Mw = 9.1 - 9.3).

Photo 7.13 Situation after the 2011 Tohoku Earthquake tsunami (Mw = 9.0) inundated the Sendai Plain in Japan.

313

7. COASTAL PROTECTION AND VARIOUS OTHER STRUCTURES

As shown in **Photo 7.14**, the runup height during the 2011 Tohoku tsunami in the sloping area behind Ryogoku coast, Japan, was 1.5 to 2 times higher than that at the coastline. On this coast, the tsunami height at the coastline was about 16 m. The tsunami destroyed all houses around the road and travelled up to a ground level of 30 - 40 m above mean sea level.

As shown in the example of **Photo 7.15**, wooden houses are more likely to be destroyed if the inundation depth is more than 2 m. Brick/mortar and unreinforced concrete buildings are likely to be destroyed if the inundation depth is more than 3 m. Structures built with larger columns, thicker walls, and narrower column spacings have better resistance against tsunami damage.

As shown in the example of **Photo 7.16**, a reinforced concrete structure is unlikely to be destroyed unless the inundation depth exceeds 5 m. However, as shown in **Photo 7.17**, even if the main body of the reinforced concrete structure is not destroyed, overturning or slip failure can still occur if the foundation is not solidly built or the structure is not firmly integrated with its base.

In the case of steel-framed concrete structures, as shown in the example of **Photo 7.18**, the skeleton may remain unbroken even at a flood depth of 7 m. However, the panels forming the walls of the steel-framed concrete structures can be destroyed.

Photo 7.14 Damage situation after the 2011 Tohoku Earthquake tsunami ran up to the Ryogoku coast in Japan.

Photo 7.15 Damage situation after the 2004 Sumatra Earthquake tsunami inundated the Yala coast in Sri Lanka.

Photo 7.16 Damage Situation after the 2004 Sumatra Earthquake tsunami inundated the Tangalle port in Sri Lanka.

7.4 Tsunami Protection

Photo 7.17 Damage situation after the 2011 Tohoku Earthquake tsunami inundated Onagawa Town in Japan.

Photo 7.18 Damage situation after the 2011 Tohoku Earthquake tsunami inundated Yamada Town in Japan.

Evacuation of people will be difficult if the flood depth exceeds 0.5 m. If the flood depth exceeds 1 m, not only fallen trees, pieces of timber, and household goods, but also cars and boats can become drifting debris which can destroy buildings, and potentially threaten lives. **Photos 7.19** and **7.20** show cars that drifted and crashed into buildings due to the fluid force of the 2011 Tohoku Earthquake tsunami. **Photo 7.21** shows some fishing boats drifted ashore in the Kesennuma fishing port area, destroying the surrounding facilities. **Photo 7.22** shows a large fishing boat (330 tonnes) left in the town and **Photo 7.23** shows a steel-framed building deformed by a colliding fishing boat.

Although the seaward side of conventional coastal dikes and seawalls are built to withstand the wave force of the design wave, the landside of those structures is not usually designed to withstand such large forces or large overtopping rates. Therefore, there is a high possibility that

Photo 7.19 A crashed car had left after the 2011 Tohoku Earthquake tsunami inundated Miyako City in Japan.

Photo 7.20 Some crashed cars had left after the 2011 Tohoku Earthquake tsunami inundated Sendai City in Japan.

315

7. COASTAL PROTECTION AND VARIOUS OTHER STRUCTURES

excessive wave overtopping during a large tsunami can damage coastal dikes or seawalls due to excessive scouring at the landward side or due to crown failure, as shown in the example of **Photo 7.24** (a 10 m high tsunami had struck the dike with the crown height of 6 m).

Photo 7.25 shows an example of a breastwork that was heavily scoured around its foundation by the overflow and overturned to the seaside by the backflow during the 2011 Tohoku Earthquake tsunami.

Photo 7.26 shows an example of a block revetment of a road overturned seaward by the backflow of the 2004 Sumatra Earthquake tsunami.

Photo 7.21 Some fishing boats had left in the fishing port area after the 2011 Tohoku Earthquake tsunami inundated Kesennuma City in Japan.

Photo 7.22 A large fishing boat had left in the town after the 2011 Tohoku Earthquake tsunami inundated Kesennuma City in Japan.

Photo 7.23 A deformed building had left after the same earthquake tsunami inundated the Kesennuma fishing port area in Japan.

Photo 7.24 A landward-broken dike and a big scour pond had left after the 2011 Tohoku Earthquake tsunami inundated Yamamoto Town in Japan.

7.4 Tsunami Protection

Photo 7.25 A flood wall overturned seaward when the 2011 Tohoku Earthquake tsunami inundated Hadenya fishing port in Japan.

Photo 7.26 A Block revetment overturned seaward when the 2004 Sumatra Earthquake tsunami inundated Patong Beach of Phuket Island in Thailand.

2) Tsunami Protection Measures

In the case of tsunamis with a return period of more than 500 years, such as the 2011 Tohoku Earthquake tsunami, it would be unrealistic to build all concrete structures to be unbreakable, considering that the service life of a normal concrete structure is around 100 years, it would be more realistic to provide for safe evacuation against tsunamis with such a low recurrence probability. The Basic Disaster Management Plan of Japan was revised in the wake of the 2011 Tohoku Earthquake tsunami as follows:

(a) For **Level 1** tsunamis (the return period of 50 – 200 years), which are likely to occur within the service life of coastal protection facilities and are not as large as a mega tsunami but cause significant damage, the construction of coastal protection facilities should be steadily promoted from the perspective of protecting human life and the property of residents, stabilising local economic activities, and securing efficient production bases.

(b) For **Level 2** tsunamis (the return period of 500 – 2000 years), which are extremely infrequent but can cause extensive damage if they do occur, the highest priority is placed on protecting the lives of residents, and the governments should promote **multiple defenses** by mobilising a combination of hardware and software measures, such as raising residents' disaster prevention awareness, development of early evacuation warning systems, construction of coastal protection facilities and embankment roads with inundation prevention functions, and promotion of land use (included land raising and building restrictions) based on tsunami inundation assumptions and comprehensive measures to reduce damage to Industrial and logistics functions in coastal areas, according to local conditions.

7. COASTAL PROTECTION AND VARIOUS OTHER STRUCTURES

(1) Protective Facility Development Plans and Cost-benefit Ratios

To develop a protection facility development plan, it is necessary to check that the **Cost-Benefit Ratio (B/C)** is greater than 1. Here, the construction and maintenance costs of the protection facilities can be regarded as the cost, and the amount of damage in the absence of the protection facilities can be taken as the benefit. Therefore, it is necessary to prepare a method for evaluating the damage to buildings and other structures from the tsunami inundation depth and velocity.

Suto [30] proposed the tsunami intensity I ($= \log_2 H$, H is the tsunami inundation depth on land or the tsunami height in ports and harbours) and summarised the relationship between this and the damage degree of houses, reinforced concrete buildings, and fishing boats, as shown in **Fig. 7.17**.

Iizuka and Matsutomi [31] investigated the criteria for the destruction to buildings based on the inundation depth, flow velocity, and drag force, as shown in **Table 7.4**. Therefore, by obtaining the inundation distributions from numerical simulations of tsunami inundation and comparing them with this table, it is possible to estimate the degree and the amount of damage due to a tsunami.

Koshimura et al [32], Vescovo et al [33], and some others assumed that the relationship between external force and damage frequency can be expressed as a normal distribution (or lognormal distribution) and applied the method of obtaining the damage probability $P_D(x)$ by the damage function Φ defined by Eq. (7.8) to determine tsunami damage.

$$P_D(x) = \Phi\left(\frac{x - \mu}{\sigma}\right) = \int_{-\infty}^{x} \frac{1}{\sqrt{2\pi}\sigma} \exp\left[-\frac{(t - \mu)^2}{2\sigma^2}\right] dt \qquad (7.8)$$

where x is the external force parameters such as inundation depth, flow velocity, and drag force; μ is the mean value, and σ is the standard deviation.

Tsunami intensity	0	1	2	3	4	5
Tsunami height [m]	1	2	4	8	16	32
Wooden house damage	Partial destruction	Total destruction				
Stone house damage	Withstandable		No documentation	Total destruction		
Damage to reinforced concrete buildings	Withstandable			No documentation		Total destruction
Damage to fishing boats		Damage occurred	50% damage ratio	100% damage ratio		

Fig. 7.17 Relationship between tsunami intensity and damage according to Shuto [30].

Table 7.4 Tsunami destruction criteria according to Iizuka and Matsutomi [31].

House type	Medium damage			Major damage		
	Flood depth [m]	Flow velocity [m/s]	Drag force [kN/m]	Flood depth [m]	Flow velocity [m/s]	Drag force [kN/m]
Reinforced concrete	–	–	–	Above 7.0	Above 9.1	Above 332–603
Concrete block	3.0	6.0	60.7–111	7.0	9.1	332–603
Timber construction	1.5	4.2	15.6–27.4	2.0	4.9	27.4–49.0
Damage degree	Columns survive, part of wall destroyed			Significant parts of walls and columns destroyed or washed away		

7.4 Tsunami Protection

Fig. 2.18 shows the relationship between the tsunami external force and the damage probability calculated by Koshimura et al [32] using Eq. (7.8). Damage data read from satellite images (satellite IKONOS), as an example in the 2004 Sumatra Earthquake tsunami that hit Banda Aceh City of Indonesia. By comparing the obtained diagram of the relationship between the tsunami external force and the damage probability at the target coast with the inundation distributions using numerical simulation results of tsunami inundation, it is possible to estimate the degree and the amount of damage due to tsunamis.

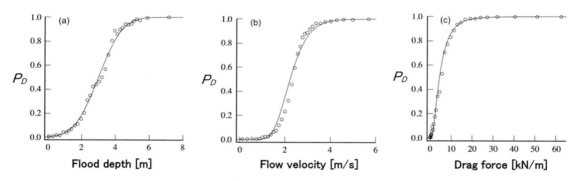

Fig. 7.18 Example of house damage ratio (the damage probability P_D) against (a) Flood depth, (b) Flow velocity, and (c) Drag force according to Koshimura et al [32].

(2) Design of Tsunami Protection Facilities

Tsunami protection facilities should be designed according to the standards of FEMA & NOAA [34], ASCE/SEI [35], PHB of MLIT & PARI [3], and numerous other guidance available in Japan and other developed countries. The experiences of the 2004 Sumatra Earthquake tsunami and the 2011 Tohoku Earthquake tsunami have revealed the importance of developing **resilient facilities** that do not fail even with tsunami overflow.

Conventional coastal dikes and seawalls in Japan were designed only to prevent wave overtopping against giant tsunamis generated by great earthquakes ($8 \leq$ Magnitude < 9). The landward slopes and the landward foundations of those structures were not designed to withstand violent overtopping from mega tsunamis caused by megathrust earthquakes ($9 \leq$ Magnitude). Therefore, the 2011 Tohoku Earthquake tsunami destroyed many coastal facilities in Iwate, Miyagi, and Fukushima Prefectures in Japan due to severe scouring of landward foundations by overtopping, loss of covering material from the crown and landward side, and the loss of backfill material. However, the cost of constructing tsunami protection facilities that will not be overtopped by mega tsunamis will be astronomical. Furthermore, while the return period of a mega tsunami is estimated to be around 1000 years, the service life of a reinforced concrete structure is around 100 years (legally 50 years). Therefore, it is likely that the facilities need to be rebuilt many times before the occurrence of a mega

7. COASTAL PROTECTION AND VARIOUS OTHER STRUCTURES

tsunami. In addition, the construction of large protection facilities will degrade the landscape and accessibility to the sea.

It is important to complete as soon as possible the construction of coastal dikes and seawalls that can withstand wave overtopping from mega tsunamis within a range where the cost-benefit ratio is reasonable. This is because they provide adequate protection against more frequent large tsunamis and, moreover, if the coast dikes and seawalls do not break when a mega tsunami strikes, they can reduce the overtopping rate and the extent of damage, and ensure a longer evacuation time for residents.

Kato et al [36] proposed a resilient coastal dike as shown in **Fig. 7.19** (in addition to the stability against tsunami pressure acting in front of the protection facility, the crown, the landward slope, and the landward foundation have been strengthened). Moreover, because the service life of reinforced concrete is too short compared to the return period of mega tsunamis, embankments shown in **Fig. 7.20** made of **CSG** (a cement mixture that allows for variation in grain size and unit water volume but ensures the necessary strength through constant quality control) and coastal forests (with strong potential for overflow reduction, self-renewal capacity, and prevention of landscape degradation) are favourable in reducing costs and extending the service life.

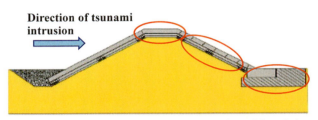

The red circled crown, backslope and backward foundation must be withstood level 2 tsunamis.

Fig. 7.19 Illustration of a resilient dike.

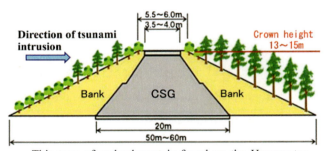

This type of embankment is found on the Hamamatsu coast of Shizuoka Prefecture, the Iwaki coast of Fukushima Prefecture and the southern coast of Miyagi Prefecture.

Fig. 7.20 Illustration of the cross section of a CSG embankment.

Furthermore, although remote management of sluices and floodwall gates is becoming more widespread, it is essential to provide facilities that can drain floodwaters even when sluices are closed, and stairs to evacuate people left behind in seaward spaces even when floodwall gates are closed.

(a) Investigation of the Stability

To check the stability of coastal protection facilities against tsunamis, the formulae introduced in section 5.2, 3), (4) "Tsunami Force" can be used. For the design of breakwaters against tsunami force, MLIT's PHB and PARI [3] recommend a calculation method that combines these formulae.

7.4 Tsunami Protection

Furthermore, Arikawa et al [37] have shown that CADMAS-SURF (see Section 2.4) can be applied to calculate the tsunami force when structure cross-sections and topography are complicated.

For the required mass of the covering material, the following calculation method by Mitsui et al [38], which improved the use of the calculation formula of Isbash [39], can be used.

$$M_s = \frac{\pi \rho_s U^6}{48 g^3 N^6 (\gamma_s - 1)^3 (\cos\alpha - \sin\alpha)^3}, \quad M_b = \frac{\pi \rho_s U^6}{48 g^3 Y^6 (\gamma_s - 1)^3 (\cos\alpha)^3} \quad (7.9)$$

where M_s is the required mass of a stone [t], M_b is the required mass of a concrete block [t], ρ_s is the density of the covering material (e.g. stones) [t/m³], U is the maximum flow velocity at the top of the covering work [m/s], g is the gravitational acceleration, γ_s is the specific gravity of the covering work, α is the slope angle [degrees] in flow direction, and N and Y are Isbash's constants as follows.

For N: 0.86 for exposed stones and 1.20 for embedded stones.

For Y: about 1.5 is appropriate for perforated flat blocks and deformed blocks, because Y decreases from 2.0 to 1.0 as the fall distance of overflow in the water increases from zero.

(b) Investigation of Tsunami Scour

When a giant tsunami overflows a coastal protection structure, there is a significant possibility of structural failure due to scour on the landward side of a coastal dike during tsunami overflow and on the seaward side during tsunami backflow. Therefore, it is necessary to predict the scale of scour and design countermeasures. For the prediction of tsunami scours, the methods introduced in Section 6.5, 3) "Scour Due to Tsunamis" can be used. It is a standard practice in Japan to design foot protection works that can adequately cover the area affected by scouring as a countermeasure.

(3) Safety Studies for Onshore Flooded Areas
(a) Stability Studies of Buildings

In stability studies, tsunami forces should be considered in addition to normal loads such as dead weight and earth pressure, and the safety against overturning, sliding, settlement and member stresses at key locations should be tested. However, it is very labour-intensive and costly to perform this for all the buildings located in the target area. Therefore, it is reasonable to separate buildings into two groups: (a) those that are sufficiently safe and, (b) those that are not sufficiently safe, using the calculation diagrams prepared according to the following concept by Yamamoto et al [40] and Ahmadi et al [41]. Then the safety of group (b) only can be examined in detail.

(i) Even if the actual building has a complex shape, but the basic structure is a portal rigid frame as shown in **Fig. 7.21**, and the building failed when the bending and shear stresses at the bases of the columns of the portal rigid frame exceed the member strength. The strength of the walls of a building is underestimated if they are analysed as cantilever beams. Therefore, they can be considered to be rigidly connected to the ceiling via beams and evaluated as portal rigid frames.

7. COASTAL PROTECTION AND VARIOUS OTHER STRUCTURES

In this case, since the upper load is supported by the columns, only the tsunami force is considered.

(ii) In the case of a tsunami strong enough to fail buildings, the hydrostatic pressure inside and outside the walls of a building is offset because the windows and doors are instantly broken, and seawater enters the building. If the walls are collapsed, the building is considered as partially collapsed. If the pillars are also broken, the building is considered as collapsed.

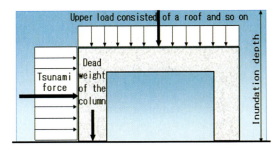

Fig. 7.21 Illustration of the loading to a portal rigid frame.

(iii) The tsunami force is expressed as a function of the inundation depth and velocity using Iizuka and Matsutomi's Eq. (5.33). However, for convenience, the Froude number is fixed at 1.1 and it can be expressed only by the inundation depth.

(iv) Buildings are designed for human use and the dimensions of the main components are standardised based on representative human dimensions. Therefore, combinations of basic dimensions such as column spacing and column height are limited.

Calculation diagrams from Yamamoto et al [40] and Ahmadi et al [41] where the column width and wall thickness at the failure limit can be obtained for an arbitrary flooding height for frequently used combinations of column spacing, column height, etc. can be seen in **Figs. 7.22** to **7.30**. These diagrams plot the damage data from the 2011 Tohoku Earthquake tsunami, the 2004 Sumatra Earthquake tsunami, and the 1993 Hokkaido-Nansei-Oki Earthquake tsunami with circles, where white circles indicate non-destructive cases, grey circles indicate partially destructive cases, black circles indicate destructive cases, and the numbers attached to the circles are the column spacings [unit: m]. In all the diagrams, the curves indicate the failure limits, which are generally located between the white and black circles. These diagrams can be used as a safety guide. For the walls of a wooden house shown in **Fig. 7.30**, although a wide variety of materials exist, to give a rough indication, an average strength was established for a variety of materials. In **Figs. 7.22** to **7.30**, the area above the curve is the safe area.

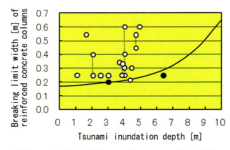

Column height 3.0 m, column spacing 5.0 m, double rebar cross section with seaward and landward arrangement, rebar ratio 0.0365, cover thickness 5 cm, band rebar ignored, concrete compressive bending strength and shear strength 20 N/mm² and 2 N/mm², rebar tensile bending strength 300 N/mm².

Fig. 7.22 Relationship between tsunami inundation depth and the breaking limit width of rein-forced concrete columns in a two-storey building.

7.4 Tsunami Protection

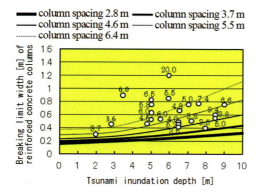

Column height 3.35 m, double rebar cross section with seaward and landward arrangement, rebar ratio 0.05, cover thickness 5 cm, band rebar ignored, concrete compressive bending strength and shear strength 20 N/mm^2 and 2 N/mm^2, steel bar tensile bending strength 300 N/mm^2.

Fig. 7.23 Relationship between tsunami inundation depth and the breaking limit width of reinforced concrete columns in a two-storey building.

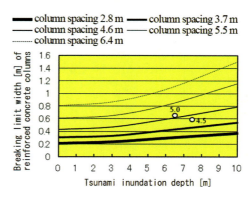

Various conditions are the same as for a two-storey building.

Fig. 7.24 Relationship between tsunami inundation depth and the breaking limit width of reinforced concrete columns in a five-storey building.

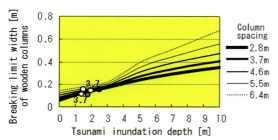

Column height 3.3 m, bending and shear strengths of intermediate timbers 20 N/mm^2 and 2.4 N/mm^2.

Fig. 7.25 Relationship between tsunami inundation depth and the breaking limit width of wooden columns in a one-storey wooden house.

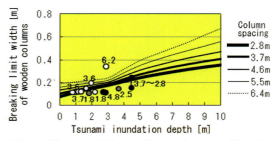

The solid vertical line indicates that the affected cases are distributed between the black circles. Other calculation conditions are the same as for the one-storey case.

Fig. 7.26 Relationship between tsunami inundation depth and the breaking limit width of wooden columns in a two-storey wooden house.

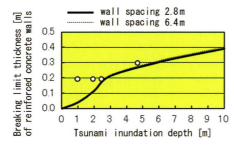

Wall height 3.0 m, single rebar cross section with centreline arrangement, rebar ratio 0.0133, strength of concrete and rebar same as Fig. 7.23.

Fig. 7.27 Relationship between tsunami inundation depth and the breaking limit thickness of reinforced concrete walls.

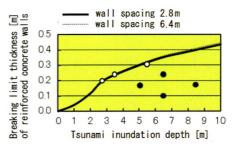

Wall height 3.35 m, other calculation conditions are the same as Fig. 7.28.

Fig. 7.28 Relationship between tsunami inundation depth and the breaking limit thickness of reinforced concrete walls.

7. COASTAL PROTECTION AND VARIOUS OTHER STRUCTURES

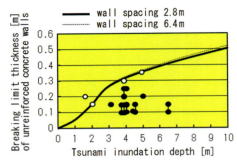

The solid vertical line indicates that the affected cases are distributed between the black circles. wall height 3.0 m, concrete or brick/mortar wall, bending strength 6 N/mm², shear strength 2 N/mm².

Fig. 7.29 Relationship between tsunami inundation depth and the breaking limit thickness of unreinforced concrete walls.

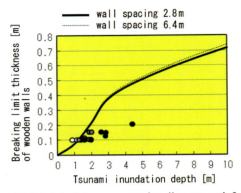

Wall height 3.0 m, average bending strength 3 N/mm², average shear strength 1 N/mm².

Fig. 7.30 Approximate relationship between tsunami inundation depth and the breaking limit thickness of wooden walls.

When using the above calculation diagrams, the following points should be noted.

(i) When calculating the tsunami force using Iizuka and Matsutomi's Eq. (5.33), the Froude number is fixed at 1.1, which may underestimate the tsunami force in the vicinity of the coastline. Conversely, the tsunami force may be overestimated for buildings behind shields.

(ii) It is assumed that the hydrostatic pressure inside and outside the building walls cancels out due to the breaking of windows and doors instantly and water penetrating the building. If there are no or small windows and doors, this assumption will not be valid, and the external force will be underestimated.

(iii) It is a prerequisite that the building columns and walls are integrated with the beams and that the buildings and walls are integrated with the ground. If these integrations are insufficient, the calculation diagrams introduced earlier cannot be used.

(iv) For ground with a low N-value (soil strength by standard penetration test) or sandy ground with a risk of liquefaction, a safety check for the ground is required.

(b) Stability Studies of Block Walls

Yamamoto et al [40] and Ahmadi et al [41] have also proposed charts showing the relationship between the strength of the concrete used and the tsunami height of the unbreakable limit, even for walls made of stacked concrete blocks (10 cm wide × 40 cm long).

The relationship between unbreakable limit depth to tsunami inundation and the concrete bending strength in the case of a wall not reinforced with rebar is shown in **Fig. 7.31**.

The relationship between unbreakable limit depth and the concrete bending strength in the case of the wall reinforced by inserting one rebar (1.3cm diameter) every 40 cm in length is shown in **Fig. 7.32**.

7.4 Tsunami Protection

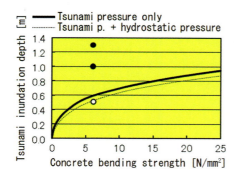

Fig. 7.31 Breaking limit inundation depth of a wall (10 cm thick) made of stacked concrete blocks.

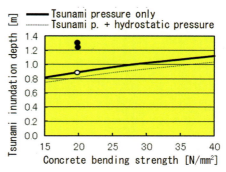

Rebar tensile strength is 15 times greater than concrete compressive strength.

Fig. 7.32 Breaking limit inundation depth of the same type wall (10 cm thick) with rebars for precaution.

In these diagrams, the limit curves are shown for both cases: (a) only the tsunami pressure needs to be considered because the seawater immediately turns around, and (b) for the case where hydrostatic pressure also needs to be considered. The areas below the curves are the safe regions. When calculating the tsunami force using Iizuka and Matsutomi's Eq. (5.33), the Froude number is fixed at 1.1, which may underestimate the tsunami force near the shoreline and overestimate it for buildings behind shields.

(c) Measures against tsunami drifting debris

During a giant tsunami, a large amount of drifting debris is generated and the potential for secondary damage caused by this cannot be ignored. The following measures are necessary to prevent damage caused by drifting debris:

(i) Construction of facilities to prevent the generation of drifting debris in the case of failure.

(ii) Construction of facilities to control the movement of drifting debris or upgrading the surrounding structures to withstand the impact force of drifting debris, if the generation of drifting debris cannot be prevented.

For this purpose, it is necessary to accurately estimate the impact forces caused by drifting debris. Yamamoto et al [42] (see (5) in 3) of Section 5.2) have introduced reliable impact force calculation formulae for each of the main drifting debris types.

(i) **For Driftwood**: Eq. (5.35) of Matsutomi, Eq. (5.36) of US FEMA & NOAA, and Eq. (5.37) of ASCE are highly reliable.

(ii) **For Containers**: Eq. (5.35) of Matsutomi, Eq. (5.36) of FEMA & NOAA, Eq. (5.37) of ASCE, Eq. (5.38) of Arikawa et al. and Eq. (5.39) of Ikeno et al. are highly reliable.

7. COASTAL PROTECTION AND VARIOUS OTHER STRUCTURES

(iii) For Motor Vehicles: Eq. (5.36) of FEMA & NOAA and Eq. (5.37) of ASCE are highly reliable.

(iv) For Ships: Eq. (5.40) of Mizutani, Eq. (5.36) of FEMA & NOAA, and Eq. (5.37) of ASCE are highly reliable.

The reason why FEMA & NOAA and ASCE formulae can be applied to all the above drift types is because the stiffness of the impacting and impacted objects can be assessed with relative accuracy, and the use of formulae that do not consider stiffness or stiffness-related variables (yield stress, Young's modulus, duration of impact force action, etc.) should be avoided.

(4) Soft Measures

As effective measures for giant or mega tsunamis where damage cannot be sufficiently prevented by protective structures (hardware measures), it is extremely important to raise disaster awareness to residents and to develop early warning and evacuation systems.

(a) Development of Tsunami Observation Networks and Early Warning Systems

In Japan, until 2011, a **major tsunami warning** (expected tsunami height $H > 3$ m), **tsunami warning** (3 m $\geq H > 1$ m), or **tsunami advisory** (1 m $\geq H > 0.2$ m) had been announced for the target forecast area within 3 minutes of the earthquake occurrence. The warning was based on the information from tsunami observation facilities located along the coast and the national database. However, when the 2011 Tohoku Earthquake tsunami hit, the tsunami warning was not accurate, so the following improvements were made.

(i) Improvement of Observation Networks

In Japan, the National Research Institute for Earth Science and Disaster Prevention (NIED), which operates the system for Monitoring of Waves on Land and Seafloor (MOWLAS), had constructed the Seafloor observation Network for Earthquakes and Tsunamis along the Japan Trench (S-net) using a large number of seismometers and hydrometers by 2020 and also took over the Dense Oceanfloor Network system for Earthquakes and Tsunamis (DONET) in the vicinity of the eastern Nankai Trough developed by the Japan Agency for Marine-earth Science and Technology (JAMSTEC) in 2016. In addition, NIED had constructed the Nankai trough seafloor observation Network for Earthquakes and Tsunamis (N-net) in the vicinity of the western Nankai Trough by 2024.

At present, NIED continues submarine earthquake and tsunami observation using these networks around the Nankai Trough and the Japan Trench, and timely announces important information through the Japan Meteorological Agency (JMA) and other organisations.

7.4 Tsunami Protection

(ii) Enhancement of Alarm Systems

During the 2011 Tohoku Earthquake tsunami, JMA immediately announced that it was a giant tsunami. However, the wording used in the warning led to the underestimation of the size of the tsunami by residents, and then failed to change their perception. Therefore, JMA changed the notification method as follows:

In the first tsunami warning (first report), the expected tsunami height is announced using the word "huge" in the case of a giant tsunami and "high" in the case of a not giant tsunami, to convey that it is an emergency.

Next, the scale of the earthquake is ascertained with good accuracy within 15 minutes of its occurrence and the expected tsunami height is expressed between "huge" and "high" within five categories: "over 10 m" if it exceeds 10 m, "10 m" if it exceeds 5 m but below 10 m, "5 m" if it exceeds 3 m but below 5 m, "3 m" if it exceeds 1 m but below 3 m, and "1 m" if it exceeds 20 cm but below 1m. If the tsunami height is low, it is expressed as "under observation".

If there is not enough time to issue a tsunami warning through the standard system, a warning is issued by "J-ALERT" (a system that uses satellites to automatically activate municipal disaster prevention administrative radio systems from the national government to transmit information such as warnings and emergency bulletins).

(b) Organisation of Reliable Refuge Processes and Facilities

It is important to raise disaster awareness of residents living in tsunami-prone areas as well as improve evacuation facilities. When robust protective structures against tsunamis are constructed, there is a tendency for a greater number of residents not following evacuation orders when a warning is issued, due to a high sense of security. This can be dangerous, given the recent trend of the occurrence of giant tsunamis.

In addition to strengthening the tsunami warning systems by the government and further strengthening protective facilities and lifelines, it is necessary to disseminate hazard maps that provide information on the route, speed, and maximum reach of inundation, the locations of evacuation centres and evacuation sites, and safe evacuation routes. Moreover, it is also necessary to familiarise residents, school children, members of community associations, and company employees in tsunami inundation hazard areas with the mechanisms of earthquakes and tsunamis and their damage characteristics, and to conduct evacuation drills regularly.

Furthermore, the government should establish the following improvements in tsunami inundation hazard areas:

(i) Designate a sufficient number of earthquake-resistant buildings in the neighbourhood as evacuation buildings (higher than the expected maximum tsunami height, with a structure and sufficient space for evacuees to enter easily).

7. COASTAL PROTECTION AND VARIOUS OTHER STRUCTURES

(ii) Construct the necessary number of evacuation towers or evacuation mounds in areas where there are no suitable earthquake-resistant buildings.

(iii) Seismic reinforcement of roads, bridges, and surrounding structures that serve as evacuation routes. In the maintenance of evacuation routes, consideration should be given to provide accessibility for people with mobility difficulties, and, if necessary, to provide them assistance. Signs should be provided so that people who are not residents can easily understand the dangers and identify evacuation facilities.

(iv) Since some designated hazardous and safe locations are not likely to be in accordance with hazard maps, and some locations are likely to change over time, it is also important to install monitoring centres and surveillance cameras at key locations urgently, and to develop information processing centers and software for smartphones to facilitate the selection and communication of information useful for individual evacuation.

For hazard mapping, the Cabinet Office, MLIT, and MAFF [23] can be consulted. For tsunami evacuation measures, the Intergovernmental Oceanographic Commission in UNESCO [43] and tsunami evacuation manuals of various government agencies can be referred to (the evacuation manual for citizens of Rikuzentakata City [44] in Japan can also be referred to).

Finally, **Fig. 7.33** can be taken as a reference for tsunami evacuation planning. This figure shows that at the seaside where the tsunami is strong, even adult males cannot stand upright if a tsunami inundation depth exceeds 40 cm, while women and children are in danger if a tsunami inundation depth exceeds 20 cm. Therefore, visiting the seaside to observe tsunamis should be avoided. Furthermore, in inland areas where the upwelling of the tsunami is decreasing, people should be evacuated to higher ground before the inundation depth exceeds 50 cm (because at a depth of 0.5 m, the total hydrostatic pressure in the horizontal direction per 1 m width alone is 125 kgf. As a result, the door can't be opened against flood water when the inundation depth exceeds 0.5 m).

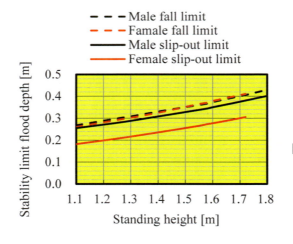

The curves were validated using large scale hydraulic experiment data from the Port and Airport Research Institute.

Fig. 7.33 Stability limit flood depth at which a person can stand on the seashore.
(According to Yamamoto et al [45])

7.5 Performance-based Design and Reliability-based Design

1) Performance-based Design

Performance-based design refers to a design method that specifies the performance required of a structure and objectively confirms that the structure will maintain this performance for its service period. Many national design standards now conform to performance-based design.

Agreement on Technical Barriers to Trade of the World Trade Organisation (WTO), launched in 1995, called for the development of national standards based on international standards so that design standards and procedures for assessing their conformity would not pose unnecessary obstacles to international trade based on ISO 2394 (General principles for the reliability of structures). ISO 2394 established in 1998 is based on the principle of performance specification of facility design.

The following roles are expected of this performance-based design.

(a) To explain and display the performance of structures to the public and give a sense of confidence to them.

(b) To provide freedom in the choice of materials used and structures, and to contribute to the improvement of the productivity of structures.

(c) To enable innovative technologies to be utilised to ensure performance.

(d) To respond to the globalisation of technology, such as the internationalisation of the distribution of construction materials.

The performance-based design of structures is generally classified into the following four hierarchical levels:

(i) Purpose of installation.

(ii) Performance requirements (the performance that the structure must possess to achieve its purpose).

(iii) Performance specifications (specific provisions necessary for the required performance to be met).

(iv) Performance verification (verification that the required performance is satisfied).

2) Reliability-based Structural Design

ISO 2394 is a design standard of performance-specifying and verification-type, which considers the required performance as the limit state of a structure. The reliability-based design based on it is regarded as the best design method to carry out performance design.

In the conventionally used **Allowable Stress Design**, the value obtained by dividing the yield point strength in the stress-strain curve by the material safety factor (usually 1.7) for steel, and the value obtained by dividing the design basis strength by the material safety factor (usually 3) for concrete, are regarded as the allowable stress. The design should be carried out so that the value obtained by multiplying the stress level due to the external force applied by the safety factor (usually

7. COASTAL PROTECTION AND VARIOUS OTHER STRUCTURES

1.2) does not exceed this allowable stress level. The reason for estimating low allowable stresses such as this is that the stress-strain curve of a structural member used in the field may be considerably lower than the stress-strain curve of the member used in the tests. For example, concrete, which constitutes the main part of civil engineering structures, is generally cast in-situ and may have extremely low strength if curing is not carried out correctly.

On the other hand, the research on **Limit State Design**, which allows permanent strain beyond the elastic limit until the ultimate limit state or service limit is reached, started in the former Soviet Union in 1937 and has blossomed in Europe and the USA since the 1960s. Furthermore, since the 1980s, **Reliability-based Design**, which expresses the uncertainty of external forces, used components, etc. in terms of probability distributions (probability density functions) and combines this with Limit State Design to reduce the probability of failure at the ultimate limit or limit of use to less than the allowable probability, has attracted worldwide attention.

Reliability-based design is explained in detail in the "Coastal Engineering Manual" of the U.S. Army Corps of Engineers [5], and is classified into the following three levels according to the level of theory dealing with uncertainty.

(1) Level III Design Method

The Level III design method is a method that expresses the variation of the strength R of a member section of a structure and external force S with a probability density function, searching for the probability to which the structure exceeds a limit state within a durable term on the design, and checking this being below the allowable probability. Therefore, a special knowledge of probability models of external force and resistance force is needed for the probabilistic analysis of the limit state.

The relationship between the probability of a structure exceeding a certain limit state (called the probability of failure) P_f and the allowable probability of failure P_{fa} is expressed in the following equation.

$$P_f \leq P_{fa} \tag{7.10}$$

Furthermore, the margin of safety of the resistance force R against the external force S acting on the structure is expressed in the following equation.

$$Z = R - S \tag{7.11}$$

where the function Z is called the **limit state function**. The probability that $Z < 0$ is the probability exceeding the limit state, i.e. the **probability of failure**.

$$\left. \begin{array}{ll} R - S > 0: & \text{safety} \\ R - S = 0: & \text{limit state} \\ R - S < 0: & \text{failur} \end{array} \right\} \tag{7.12}$$

330

7.5 Performance-based Design and Reliability-based Design

If R and S are random variables, then $R\text{-}S$ is also a random variable and the safety margin is also indicated by a probability value. In other words, the probability of failure P_f can be defined as shown in Eq. (7.13), where the reliability R_f is defined as the extra-event probability of failure.

$$P_f = \text{Prob}[Z = R - S < 0], \quad R_f = \text{Prob}[Z = R - S > 0] = 1 - P_f \quad (7.13)$$

The probability of failure P_f is obtained by integrating the simultaneous probability density function $f_{S,R}(x,y)$ of S and R over the domain ($0 < R < S$) and is expressed as follows:

$$P_f = \text{Prob}[S > R] = \int_0^\infty \int_0^x f_{S,R}(x,y) \, dy \, dx = \int_0^\infty \int_y^\infty f_{S,R}(x,y) \, dx \, dy \quad (7.14)$$

Furthermore, if S and R are independent of each other, since $f_{S,R}(x,y) = f_S(x) \times f_R(y)$, Eq. (7.14) can be expressed as follows:

$$\begin{aligned} P_f &= \int_0^\infty f_S(x) \left\{ \int_0^x f_R(y) \, dy \right\} dx = \int_0^\infty f_S(x) F_R(x) \, dx \\ &\text{or} \\ P_f &= \int_0^\infty f_R(y) \left\{ \int_y^\infty f_S(x) \, dx \right\} dy = \int_0^\infty f_R(y) \{1 - F_S(y)\} \, dy \end{aligned} \quad (7.15)$$

where $F_S(x)$ and $F_R(x)$ are the distribution functions of S and R respectively, and $f_S(x)$ and $f_R(x)$ are the probability density functions of S and R respectively.

If the probability density functions of the external force S and the resisting force R are shown in **Fig. 7.34**, the probability of S being between x and $x+dx$ is $f_S(x)dx$ (yellow coloured area in the figure); and the probability of R being less than this x is given by $F_R(x)$ (grey coloured in the figure). If S and R are independent of each other, the probability of these two events occurring simultaneously can be expressed as the product $f_S(x)dx \times F_R(x)$, which can be integrated over the entire region of S as shown in the first half of Eq. (7.15) to obtain the fracture probability.

Specific examples of this Level III design method on the expected sliding capacity of breakwaters include the work of Shimosako and Takahashi [46], [47] and Goda and Takagi [48].

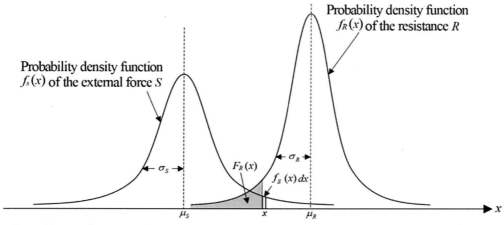

(σ_S and σ_R are the respective standard deviations, μ_S and μ_R are the respective mean values)

Fig. 7.34 Explanation of failure probability in case of level III

7. COASTAL PROTECTION AND VARIOUS OTHER STRUCTURES

(2) Level II Design Method

Level II is a one-step simplified design method of Level III, where the **reliability index (β)** is obtained from the mean and standard deviation of force and strength, assuming that their variation can be expressed in a tractable probability distribution.

Assuming that the probability distributions of the external force S and the resisting force R are independent of each other and normal, the limit state function Z is also normal, and if the normal probability density function $f_Z(x)$ of Z is shown as in **Fig. 7.35**, the failure probability P_f is represented by the grey area of the figure which can be obtained by Eq. (7.16). Therefore, using the function Φ representing the cumulative probability of the standard normal probability distribution, the destruction probability P_f is as in Eq. (7.17).

$$P_f = \int_{-\infty}^{0} f_Z(x)dx \tag{7.16}$$

$$P_f = \Phi(-\mu_Z/\sigma_Z) = \Phi(-\beta) \tag{7.17}$$

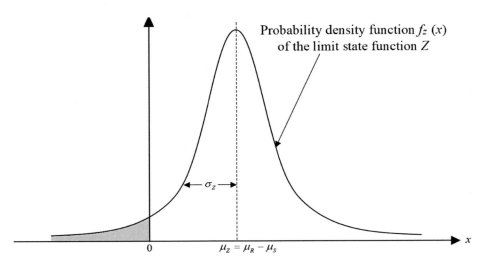

(σ_z is the standard deviation, μ_z is the mean value)

Fig. 7.35 Explanation of failure probability in case of level II

where μ_z is the mean value of Z if R and S are normally distributed ($\mu_z = \mu_R - \mu_S$), σ_z is the standard deviation of Z and $\sigma_Z = \sqrt{\sigma_R^2 + \sigma_S^2}$.

As shown in Eq. (7.17), the probability of failure can be specified by the reliability index β (= μ_z/σ_z). In the actual design, the reliability index should be above the target value as in Eq. (7.18).

$$\beta \geq \beta_T \tag{7.18}$$

Here, β_T is the target reliability index obtained from the allowable probability of failure.

Examples of specific research on design methods at Level II include the work of Toyama [49], who presented a theory for calculating reliability indices, and the work of Burcharth and Sørensen [50] on various types of breakwaters.

7.5 Performance-based Design and Reliability-based Design

Some of the examples proposed by Burcharth and Sørensen [50] are given below. Calculation methods for the stability of various covered blocks and for the stability and scour of breakwater foundations are given in the "Coastal Engineering Manual" by the U. S. Army Corps of Engineers [5].

(a) Sliding Failure of Upright Caisson Breakwaters

The failure function Z for sliding failure can be expressed by the following equation:

$$Z = \left(F_G - U_{Vforce} \times F_U \right) f \Big/ \gamma_z - U_{Hforce} \times F_H \qquad (7.19)$$

where F_G is the caisson weight reduced Buoyancy, F_U is the vertical wave force by Goda formula, f is the friction coefficient, γ_z is the partial safety factor for the resistance, F_H is the horizontal wave force by Goda formula, $U_{Vforce} = 0.77$ is the bias factor applied to the wave force F_U, and $U_{Hforce} = 0.90$ is the bias factor applied to the wave force F_H. The force F_H and F_U are calculated by using the partial safety factor for wave height (γ_H) × the significant wave height (H_s) with return period T.

The failure probability P_f during any reference period of duration T years can be estimated from **Tables 7.5 – 7.8** by using the **variational coefficient of wave data ($\sigma_H{}'$)** defined by the following equation instead of the reliability index β.

$$\sigma_H{}' = \frac{\text{the standard deviation of wave data}}{\text{the mean value of wave data}} \qquad (7.20)$$

In the case measured by accelerometer buoys: $\sigma_H{}' = 0.05$.

In the case where waves are estimated by the SMB method: $\sigma_H{}' = 0.2$.

(b) Overturning Failure of Upright Caisson Breakwaters

The failure function Z for overturning failure can be expressed by the following equation:

$$Z = \left(M_G - U_{Vmoment} \times M_U \right) - U_{Hmoment} \times M_H \qquad (7.21)$$

where M_G is the moment of the caisson weight reduced Buoyancy around the shoreward heel of the target caisson, M_U is the moment of the vertical wave force by Goda formula around the shoreward heel of the target caisson, M_H is the moment of the horizontal wave force by Goda formula around the shoreward heel of the target caisson, $U_{Vmoment} = 0.72$ is the bias factor applied to the moment M_U, and $U_{Hmoment} = 0.81$ is the bias factor applied to the moment M_H. The moments M_H and M_U are calculated by using the partial safety factor for wave height (γ_H) × the significant wave height (H_s) with return period T.

The failure probability P_f during any reference period of duration T years can be estimated from **Tables 7.9 – 7.10** by using the variational coefficient of wave data $\sigma_H{}'$ ($\sigma_H{}' = 0.05$ corresponds to the case measured by accelerometer buoys and 0.2 corresponds to the case where waves are estimated by the SMB method).

7. COASTAL PROTECTION AND VARIOUS OTHER STRUCTURES

Table 7.5 Relationship between σ_H', γ_H, γ_z and P_f in the deep water. (In the case of design without model tests, according to Burcharth and Sørensen [50])

P_f	$\sigma_H' = 0.05$		$\sigma_H' = 0.2$	
	γ_H	γ_z	γ_H	γ_z
0.01	1.4	1.7	1.5	1.7
0.05	1.3	1.4	1.4	1.4
0.10	1.3	1.2	1.4	1.3
0.20	1.2	1.2	1.3	1.2
0.40	1.1	1.0	1.1	1.1

Table 7.6 Relationship between σ_H', γ_H, γ_z and P_f in the deep water. (In the case of wave load based on model tests, according to Burcharth and Sørensen [50])

P_f	$\sigma_H' = 0.05$		$\sigma_H' = 0.2$	
	γ_H	γ_z	γ_H	γ_z
0.01	1.3	1.5	1.4	1.5
0.05	1.2	1.4	1.3	1.4
0.10	1.2	1.2	1.3	1.2
0.20	1.1	1.2	1.2	1.2
0.40	1.0	1.2	1.1	1.0

Table 7.7 Relationship between σ_H', γ_H, γ_z and P_f in the shallow water. (In the case of design without model tests, according to Burcharth and Sørensen [50])

P_f	$\sigma_H' = 0.05$		$\sigma_H' = 0.2$	
	γ_H	γ_z	γ_H	γ_z
0.01	1.3	1.9	1.4	1.9
0.05	1.2	1.6	1.3	1.6
0.10	1.2	1.4	1.3	1.4
0.20	1.1	1.3	1.2	1.3
0.40	1.0	1.2	1.0	1.2

Table 7.8 Relationship between σ_H', γ_H, γ_z and P_f in the shallow water. (In the case of wave load based on model tests, according to Burcharth and Sørensen [50])

P_f	$\sigma_H' = 0.05$		$\sigma_H' = 0.2$	
	γ_H	γ_z	γ_H	γ_z
0.01	1.2	1.6	1.3	1.6
0.05	1.1	1.5	1.2	1.5
0.10	1.1	1.3	1.2	1.3
0.20	1.1	1.2	1.1	1.2
0.40	1.0	1.1	1.0	1.1

Table 7.9 Relationship between σ_H', γ_H and P_f. (In the case of design without model tests, according to Burcharth and Sørensen [50])

P_f	$\sigma_H' = 0.05$	$\sigma_H' = 0.2$
	γ_H	γ_H
0.01	-	-
0.05	2.7	-
0.10	2.0	2.5
0.20	1.6	1.7
0.40	1.2	1.2

Table 7.10 Relationship between σ_H', γ_H and P_f. (In the case of wave load based on model tests, according to Burcharth and Sørensen [50])

P_f	$\sigma_H' = 0.05$	$\sigma_H' = 0.2$
	γ_H	γ_H
0.01	2.1	2.3
0.05	1.7	1.9
0.10	1.4	1.6
0.20	1.3	1.4
0.40	1.1	1.2

(c) Stability of Stones Covered Sloping Breakwaters based on Hudson Formula

The failure function Z of covering stones can be expressed by Eq. (7.22) based on Eq. (3.58):

$$Z = \left(\frac{\rho_s}{\rho} - 1 \right) D \left(K_D \cot \alpha \right)^{1/3} / \gamma_z - \gamma_H H_s \tag{7.22}$$

where ρ is the density of seawater, ρ_s and D are respectively the density and the representative dimension (the cubic root of the volume) of the covering stones, K_D is the empirical stability coefficient, α is the frontal slope angle of a sloping breakwater, γ_z is the partial safety factor for the

7.5 Performance-based Design and Reliability-based Design

resistance, γ_H is the partial safety factor for the wave height, and H_s is the significant wave height with return period T.

The failure probability P_f during any reference period of duration T years can be estimated from **Table 7.11**.

(d) Stability of Tetrapod-Covered Sloping Breakwaters based on Van der Meer Formula

The failure function Z of Tetrapods can be expressed by Eq. (7.23) based on Eq. (3.63):

$$Z = \left(\frac{\rho_s}{\rho} - 1\right) D \left(3.75\frac{N_{od}^{0.5}}{N_w^{0.25}} + 0.85\right)\left(\frac{2\pi H_s}{gT_m^2}\right)^{-0.2} \bigg/ \gamma_z - \gamma_H H_s \qquad (7.23)$$

where ρ is the density of seawater, ρ_s and D are the density and the representative dimension (the cubic root of the volume) of tetrapods, N_{od} is the number of blocks that have moved from within the vertical strip of width D to outside the covering layer, N_w is the number of incident storm waves, and T_m is the mean period, γ_z is the partial safety factor for the resistance, γ_H is the partial safety factor for the wave height, H_s is the significant wave height with return period T.

The failure probability P_f during any reference period of duration T years can be estimated from **Table 7.12**.

Table 7.11 Relationship between σ_H', γ_H, γ_z and P_f. (In the case of design without model tests, according to Burcharth and Sørensen [50])

P_f	$\sigma_H' = 0.05$		$\sigma_H' = 0.2$	
	γ_H	γ_z	γ_H	γ_z
0.01	1.7	1.04	2.0	1.00
0.05	1.4	1.06	1.6	1.02
0.10	1.3	1.04	1.4	1.06
0.20	1.2	1.02	1.3	1.00
0.40	1.0	1.08	1.1	1.00

Table 7.12 Relationship between σ_H', γ_H, γ_z and P_f. (In the case of design without model tests, according to Burcharth and Sørensen [50])

P_f	$\sigma_H' = 0.05$		$\sigma_H' = 0.2$	
	γ_H	γ_z	γ_H	γ_z
0.01	1.7	1.02	1.9	1.04
0.05	1.4	1.06	1.5	1.08
0.10	1.3	1.04	1.4	1.04
0.20	1.2	1.02	1.3	1.00
0.40	1.0	1.08	1.1	1.00

(3) Level I Design Method

Level I is a further simplified version of Level II to maintain continuity with conventional design methods; and is also called the partial safety factor method. This method has been used in several existing design standards as one of the limit state design methods. Although probability calculations are used to assist in setting the coefficients included in the design equation, designers are not required to calculate the probabilities.

7. COASTAL PROTECTION AND VARIOUS OTHER STRUCTURES

List of References in Chapter 7

1） Coastal Development Institute of Technology: *Technical Standards and Commentaries for Coastal Protection Facilities*, Japanese general foundation Coastal Development Institute of Technology, 2018, 351p. (in Japanese)

2） The coastal engineering committee of JSCE: *Coasts and Ports of Japan* (240 slides.), JSCE, 1994.

3） Ports and Harbours Bureau of MLIT, and Port and Airport Research Institute: *Technical Standards and Commentaries for Port and Harbour Facilities in Japan*, the Overseas Coastal Area Development Institute of Japan, 2020, 2181p.

4） Fisheries Agency: *Reference Book for the Design of Fishing Ports and Fishing Facilities*, Fisheries Infrastructure Department of Fisheries Agency, 2015. (in Japanese)
https://www.jfa.maff.go.jp/j/gyoko_gyozyo/g_thema/sub52.html

5） U. S. Army Corps of Engineers: *Coastal Engineering Manual*, U. S. Army, Washington, D.C., 2014.

6） Goda, Y.: Estimation of the Rate of Irregular Wave Overtopping of Seawalls, *Report of the Port and Harbour Research Institute*, Vol. 9, No.4, PARI, 1970, pp.3-41. (in Japanese)

7） Fukuda, N., Uno, T. and Irie, I.: Field Observation on Overtopping of Wave Protection Seawalls (2nd Report), *Proc. 20th Coastal Engineering Conference*, JSCE, 1973, pp.113-118. (in Japanese)

8） Vu, T.C., (convenor and lead member), Bynoe, T., Duong, T.M., Eliot M., Hall, F., Ranasinghe, R., Schipper, M., and Joshua T. Tuhumwire, J.T. (co-lead member): Chapter 13. Changes in erosion and sedimentation, *The Second World Ocean Assessment*, United Nations, 2021, pp.185-200.

9） The coastal engineering committee of JSCE: *Design Manual for Coastal Facilities – 2000*, JSCE, 2000, 586p.

10） Kornvisith, S., Yamamoto, Y. and Charusrojthanadech, N.: Analysis of Coastal Erosion in Khlong Wan Coast, *Proc. 27th International Ocean and Polar Engineering Conference*, ISOPE, 2017, pp.1519-1526.

11） Promngam, A., Charusrojthanadech, N., Maleesee, K. and Yamamoto, Y.: Effect of Jetties in Northern Part of Coastal Change at Chumphon Estuary, *Proc. 28th International Ocean and Polar Engineering Conference*, ISOPE, 2018, pp.1379-1386.

12） The coast division of National Institute for Land and Infrastructure Management: *Guide to the Design of Artificial Reefs* (*revised edition*), National Association of Sea Coast, 2004, 95p. (in Japanese)

13） Yamamoto, Y., Rattanarama, P. and Nopmueng, A.: Evaluation Methods of the Effect for Wave Dissipation and Erosion Prevention by Mangroves, *Journal of JSCE*, B2 (Coastal Engineering), Vol.71, No.2, 2015, pp. I_799-I_804. (in Japanese)

14) Suzuki, T.: *Wave Dissipation over Vegetation Fields*, Delft University of Technology, 2011, 176p.

15) Silvester, R.: *Stabilization of Sedimentary Coastlines*, Nature, No.188, 1960, pp.467-469.

16) Kato, F., Sato, S. and Tanaka, S.: Wave-induced pore water pressure fluctuations in the ground around coastal embankments, *Journal of Coastal Engineering*, Vol. 43., 1996, pp.1011-1015. (in Japanese)

17) Yamamoto, Y., Ioroi, M. and Oshima, Y.: Destruction Mechanism of a Coastal Dike and a Seawall, and Methods for Estimating the Suction Rate, *Journal of JSCE B2 (Coastal Engineering)*, Vol.71, No.1, 2015, pp.30-41. (in Japanese)

18) Ioroi, M., Yamamoto, Y. and Oshima, Y.: Improvement of a Suction Rate Formula from the Bottom of the Sheet Pile of a Coastal Dike, *Proc. 25th International Ocean and Polar Engineering Conference*, ISOPE, 2015, pp.1483-1488.

19) Kuisorn, W., Charusrojthanadech, N. and Yamamoto, Y.: Improvement of Suction Rate Methods of Backfilling Materials from a Coastal Dike or a Seawall, *Proc. 26th International Ocean and Polar Engineering Conference*, ISOPE, 2016, pp.1473-1478.

20) Iizuka, H., Tamate, S., Takakuwa, T. and Sato, T.: Experiment on Model Windbreak (1 st Report): Study on the Effect of Windbreak Forests to Reduce Salt Content in Sea Wind, *Report of Forestry and Forest Products Research Institute*, No. 45, 1950, pp.1-15. (in Japanese)

21) Kudo, T.: *Explanatory Series on the Public Interest Functions of Forests (10) Windbreak function of forests*, General Incorporated Association for Mountain and Water Management in Japan, 1988, 46p. (in Japanese)

22) Tsukamoto, Y., Kosaka, I., Uchiyama, K., Sakatume, S., Sasaki, M. and Satou, K.: Characteristics of Sand Distribution Flying into the Shonan Coastal Pine Forest, *Journal of Japanese Forest Society*, Vol. 83, No. 1, 2001, pp.40-46. (in Japanese)

23) Cabinet Office, MLIT and MAFF: *Tsunami and Storm Surge Hazard Map Manual*, Japan Government, 2004, 128p. https://www.unisdr.org/preventionweb/files/53507_icharmtsunami andstormsurgehazardmap.pdf or https://www.pwri.go.jp/icharm/publication/pdf/2004/tsunami_ and_storm_surge_hazard_map_manual.pdf

24) Intergovernmental Panel on Climate Change (IPCC): *Sixth Assessment report*, the United Nations, 2022~2023. https://www.ipcc.ch.

25) Kato, F. and Tajima, Y.: Coastal adaptation to climate change in Japan: a review, *Coastal Engineering Journal*, Vol.65, JSCE, 2023, pp.597-619. https://doi.org/10.1080/21664250.2023.2259187

26) Yamamoto, Y., Horikawa, K. and Naganuma, Y.: Influence of Global Warming on Wave Overtopping of Coastal Dikes, *Journal of Coastal Engineering*, Vol.39, JSCE, 1992, pp.1036-1040. (in Japanese)

7. COASTAL PROTECTION AND VARIOUS OTHER STRUCTURES

27) Okayasu, A. and Sakai, K.: Effect of sea level rise on sliding distance of a caisson breakwater - Optimization with probabilistic design method, *Proc. 30th International Conference on Coastal Engineering*, 2006, pp.4883-4893. Doi:10.1142/9789812709554_0409

28) Suh, K.D., Kim, S.W., Mori, N. and Mase, H.: Effect of Climate Change on Performance-Based Design of Caisson Breakwaters, *Journal of Waterway, Port, Coastal, and Ocean Engineering*, Vol.138, No.3, ASCE, 2011. https://doi.org/10.1061/(ASCE)WW.1943-5460.0000126

29) Yasuhara, K., Murakami, S., Mimura, N., Komine, H. and Recio, J.: Influence of Global Warming on Coastal Infrastructural Instability, *Sustainability Science*, Vol.2, Springer, 2007, pp.13-25. Doi:10.1007/s11625-006-0015-4

30) Shuto, N.: Tsunami Intensity and Damage, *Tsunami Engineering Technical Report*, No.9, Tohoku University, 1992, pp.101-136.

31) Iizuka, H. and Matsutomi, H.: Damage Assumptions for Tsunami Inundation Flows, *Journal of Coastal Engineering*, Vol. 47, JSCE, 2000, pp.381-385. (in Japanese)

32) Koshimura, S., Namegaya, Y. and Yanagisawa, H.: Fragility Functions for Tsunami Damage Estimation, *Journal of JSCE (B)*, Vol.65, No.4, 2009, pp.320-331. (in Japanese)

33) Vescovo, R., Adriano, B., Mas, E. and Koshimura, S.: Beyond Tsunami Fragility Functions: Experimental Assessment for Building Damage Estimation, *Scientific Reports*, Springer Nature, 2023. https://www.nature.com/articles/s41598-023-41047-y

34) FEMA and NOAA: *Guidelines for Design of Structures for Vertical Evacuation from Tsunamis (Second edition)*, Federal Emergency Management Agency & National Oceanic and Atmospheric Administration, 2012, 174p.

35) ASCE/SEI: *ASCE Standard 7-22 Minimum Design Loads and Associated Criteria for Buildings and other Structures*, ASCE, 2022, 975p.

36) Kato, F., Hatogai, S. and Suwa, Y.: Structures for Coastal Dike with Concrete Armors Resilient to Tsunami Overflow, *Journal of JSCE B2 (Coastal Engineering)*, Vol.69, 2013, pp.1021-1025. (in Japanese)

37) Arikawa, T., Yamada, F. and Akiyama, M.: Investigation of the applicability of a three-dimensional numerical wave tank to tsunami wave force, *Journal of Coastal Engineering*, Vol.52, 2005, pp.46-50. (in Japanese)

38) Mitsui, J., Matsumoto, A. and Hanzawa, M.: Derivation Process of the Isbash Formula and Its Applicability to Tsunami Overtopping Breakwater, *Journal of JSCE B2 (Coastal Engineering)*, Vol.71, No.2, 2015, pp. I_1063-I_1068. (in Japanese)

39) Isbash, S.V.: Construction of Dams by Dumping Stones into Flowing Water, *Sci. Res. Inst. Hydrotech. Leningrad*, 1932, translated by A. Dovjikov, U.S. Army Corps of Engineers, 1935.

40) Yamamoto, Y., Charusrojthanadech, N. and Nariyoshi, K.: Proposal of Rational Evaluation Methods of Structure Damage by Tsunami, *Journal of JSCE B2 (Coastal Engineering)*, Vol.67,

No.1, 2011, pp.72-91. (in Japanese)

41) Ahmadi, S.M., Yamamoto, Y. and Vu, T.C.: Rational Evaluation Methods of Topographical Change and Building Destruction in the Inundation Area by a Huge Tsunami, *Journal of Marine Science and Engineering*, Vol.8, 762, 2020. Doi:10.3390/jmse8100762

42) Yamamoto, Y., Kozono, Y., Mas, E., Murase, F., Nishioka, Y., Okinaga, T. and Takeda, M.: Applicability of Calculation Formulae of Impact Force by Tsunami Driftage, *J. Mar. Sci. Eng.*, 9, 493, 2021. https://doi.org/10.3390/jmse9050493

43) Intergovernmental Oceanographic Commission in UNESCO: *Tsunami Preparedness* (*Information Guide for Disaster Planners*), UNESCO, 2008, 26p.

44) Rikuzentakata City: *Evacuation Manual -A Guide to Evacuation During Times of Disaster-*, Rikuzentakata City, 2021, 26p. https://www.city.rikuzentakata.iwate.jp/material/files/group/61/Evacuation_Manual.pdf

45) Yamamoto, Y., Wuttichan, W. and Arikawa, T.: Improvement of Prediction Method of Coastal Damage due to Tsunami, *Journal of Coastal Engineering*, Vol.55, JSCE, 2008, pp.301-305. (in Japanese)

46) Shimosako, K. and Takahashi, S.: Application of Deformation-based Reliability Design for Coastal Structures -Expected Sliding Distance Method of Composite Breakwaters-, *Coastal Structure '99*, Spain, 1999, pp.363-371.

47) Shimosako, K. and Takahashi, S.: Application of Expected Sliding Distance Method for Composite Breakwaters Design, *Proc. 27th International Conference on Coastal Engineering*, 2000, pp.1885-1898.

48) Goda, Y. and Takagi, H.: A Reliability Design Method of Caisson Breakwaters with Optimal Wave Heights, *Coastal Engineering Journal*, Vol.42, No.4, JSCE, 2000, pp.357-387.

49) Toyama, S.: *The Application of the Reliability Theory to the Breakwater*, Technical Note of the Port and Harbour Research Institute, No.540, Japan, 1985, 49p. (in Japanese)

50) Burcharth, H.F. and Sørensen, J.D.: The PIANC Safety Factor System for Breakwaters, *Coastal Structure '99*, Spain, 1999, pp.1125-1144.

Chapter:8

ENVIRONMENTAL PROTECTION AND WAVE POWER GENERATION

8. ENVIRONMENTAL PROTECTION AND WAVE POWER GENERATION

Nowadays, nutrients and plastic litter are extensively polluting and extremely dangerous to the World's Oceans.

Wastewater from households and factories and drainage from sewage treatment plants flow out into the sea via rivers. According to WWAP [1], more than 80% of wastewater is discharged untreated, making it one of the largest sources of marine pollution. Plastic litter is a major and most dangerous part of marine litter, including plastic, rubber, metal, glass, or wood. Marine litter and nutrients originate mainly from land bases, as seen in WWAP [1], Malone et al [2], Galgani et al [3], and GESAMP [4], [5], with more than one million tons of plastic waste entering the oceans from rivers every year (Lebreton et al [6], Van Emmerick et al [7], Schmidt et al [8]).

Wastewater and drainage discharged into the sea alter the water quality of coastal areas. In addition, illegally dumped solid and liquid wastes, fertiliser, and pesticides from farmlands are washed into the sea via rivers. Ocean currents can spread pollutants in the open seas, some of which can drift onto beaches elsewhere. The former part of this chapter discusses these issues and measures that can be taken to deal with environmental degradation.

In the latter half of this chapter, the current situation of renewable energy, which is necessary to reduce the use of fossil fuels that cause greenhouse gas emissions, is explained and a proposal for accelerating wave power generation is presented.

8.1 Protection of Water Quality in Marine Areas
1) Actual Situation of Water Quality Deterioration

During the period of rapid economic and industrial growth in many countries, large quantities of wastewater from households and factories and drainage from sewage treatment plants are discharged into the sea. Although in open seas, these wastewater and drainage are naturally diluted, still there are harmful effects on marine life.

In semi enclosed water bodies such as gulfs and bays, those pollutants significantly increase nitrogen and phosphorus in the seawater. This leads to eutrophication, which causes a rapid increase in phytoplankton in seawater, leading to the frequent occurrence of **red tides** (see **Photo 8.1**) and **blue tides (anoxia,** see **Photo 8.2**).

The name **red tide** is derived from the red or pink colour of the sea, caused by the abnormal proliferation of plankton. The high incidence of red tides has led to mass mortality of fish and shellfish because the plankton consumes large amounts of oxygen, their attachment to gills prevents fish from breathing, and some plankton produce toxins.

In the case of **blue tides**, there are two types as follows:

(i) When plankton gathers due to eutrophication and other factors, red tides are more likely to occur.

If the constituent plankton is mainly noctoplankton, a luminous phenomenon occurs at night, causing the sea to glow blue-white or other colours. This may also be accompanied by a decrease in the oxygen concentration in the seawater, causing severe hypoxia leading to fish mortality in certain areas of the sea.

(ii) When eutrophication becomes severe, dead phytoplankton that have proliferated abnormally are deposited on the seabed and decomposed by bacteria. This decomposition consumes the oxygen in the bottom layer of seawater, resulting in anaerobic conditions. As a result, anaerobic bacteria increase and sulphate-reducing bacteria among the anaerobic bacteria produce large amounts of hydrogen sulfide. When the lower water containing this large amount of hydrogen sulfide rises due to strong winds, etc., the hydrogen sulfide is oxidised to produce sulphur particulates and sulphur oxides. These fine particles reflect sunlight and turn the seawater blue-white or other colours. In this case, the lack of oxygen in the seawater is severe and leads to mass mortality of fish and shellfish. It is also accompanied by a terrible smell of rotten eggs due to the sulphur content.

Photo 8.1 Red tides in Yokosuka Port, Japan. (Source: https://www.istockphoto.com/jp/)

Photo 8.2 Blue tides in Tokyo Bay, Japan. (Source: https://www.pakutaso.com/20170602177post-12220.html)

To prevent the deterioration of water quality, regulations against the **Potential of hydrogen (PH)**, which is an indicator of the degree of acidity or basicity of a solution, **Chemical Oxygen Demand (COD)**, the amount of oxygen needed to oxidise substances in the water, and **Dissolved oxygen (DO)**, etc., have been used extensively in many countries. However, an excessive eutrophication of seawater means a poor habitat for fish and shellfish. Therefore, effluent regulations for nitrogen and phosphorus have also been implemented. These regulatory thresholds vary from country to country and examples from Japan (Ministry of the Environment, Japan [9]) are presented in **Table 8.1**.

In Japan, the regulations of wastewater and drainage and efforts to improve and disseminate sewage purification technology have halved the amounts of PH, COD, and DO, as well as total nitrogen and total phosphorus in water. Although fish catches, peaked in 1980-1990, have been declining ever since. In recent years, nutrient deficiencies in seawater have been reported [10]. The

8. ENVIRONMENTAL PROTECTION AND WAVE POWER GENERATION

reasons for the continuing decline in catches include the over-harvest of fish and shellfish and the decrease in the number of fishery successors, but it also shows that focusing solely on the regulations of wastewater and drainage (reducing inflow loads) is not an effective countermeasure. It has also been pointed out that coastal reclamation and revetment construction have reduced the number of organisms feeding on plankton and fluidity for water purification in tidal flats and extremely shallow waters, which has led to massive plankton blooms.

Table 8.1 Conservation standards for the water quality environment in marine areas according to Basic Act on the Environment (Act No. 91 of 1993) by Ministry of the Environment in Japan [9].

Item / Class	Water use	Standerd value				
		Hydrogen-ion concentration (pH)	Chemical oxygen demand (COD)	Dissolved oxygen (DO)	Total coliform	N-hexane extract (oil, etc.)
A	Fishery class 1, bathing, conservation of the natural environment, and uses listed in B-C	7.8≤pH≤8.3	≤2 mg/L	≥7.5 mg/L	≤1,000 MPN/100 mL	Not detectable
B	Fishery class 2, industrial water and the uses listed in C	7.8≤pH≤8.3	≤3 mg/L	≥5 mg/L	–	Not detectable
C	Conservation of the environment	7.0≤pH≤8.3	≤8 mg/L	≥2 mg/L	–	

Remark: Total coliform should be 70MPN/100mL or less for the fishery class 1 to cultivate oyster to be eaten raw.

Notes: 1) Conservation of the natural environment: Conservation of sightseeing and other environments.

2) Fishery class 1: For such marine products as red sea bream, yellowtail, and seaweed, and marine products for fishery class 2.
Fishery class 2: Such marine products as mullet and dried seaweed.

3) Conservation of the environment: Limit of not disrupting the day-to-day lives of the population (including things likes walks along the beach).

Item / Class	Water use	Standard value	
		Total nitrogen	Total phosphorus
I	Conservation of the natural environment and uses listed in II-IV (except fishery classes 2 and 3)	≤0.2 mg/L	≤0.02 mg/L
II	Fishery class 1, bathing, and the uses listed in III-IV (excpt fishery classes 2 and 3)	≤0.3 mg/L	≤0.03 mg/L
III	Fishery class 2 and the uses listed in IV (excpt fishery classes 3)	≤0.6 mg/L	≤0.05 mg/L
IV	Fishery class 3, industrial water, and conservation of habitable environments for marine biota	≤1 mg/L	≤0.09 mg/L

Remarks: 1) Standard values are set in terms of annual averages.

2) Standard values are applicable only to marine areas where marine phytoplankton blooms may occur.

Notes: 1) Conservation of the natural environment: Conservation of sightseeing and other environments.

2) Fishery class 1: A large variety of fish, including benthic fish and shellfish, are taken in good balance and stably.
Fishery class 2: Marine products (mainly fish) are taken with the exception of some benthic fish and shellfish.
Fishery class 3: Specific types of marine products highly resistant to pollution mainly taken.

3) Conservation of habitable environments for marine biota: Level where bottom-dwelling organisms can habitat year-round.

Item / Class	Adaptability of the habitat status of aquatic life	Standard value		
		Total zinc	nonylphenol	Linear alkylbenzenesulphonic acid and its salts
Class A organisms	Water areas inhabited by aquatic life	≤0.02 mg/L	≤0.001 mg/L	≤0.01 mg/L
Special class A organisms	Of the water areas inhabited by Class A organisms, those that should be conserved as spawning/rearing areas of aquatic life	≤0.01 mg/L	≤0.0007 mg/L	≤0.006 mg/L

8.1 Protection of Water Quality in Marine Areas

Water quality standards for dangerous substances in the UK Marine Special Areas of Conservation (SACs) are also presented in **Tables 8.2, 8.3** (from Cole et al [11]) for reference.

Oil pollution caused by incidental or inadvertent spills from ships, factories, and pipelines also has significant adverse impacts on water quality. As for the oil pollution of the oceans, international efforts began with the International Convention for the Prevention of Pollution of the Sea by Oil (OILPOL Convention) in 1954. When major oil spills have occurred, convention-based prevention measures have been strengthened. Laws and regulations on oil pollution in domestic waters are also being developed in many countries.

Table 8.2 Water quality standards for List I substances for the UK Marine Special Areas of Conservation (SACs) Project. (according to Cole et al [11])

Parameter	Unit	Water quality standard		Standstill Provision[a]
		Estuary[b]	Marine	
Mercury	μg Hg/l	0.5 DAA	0.3 DAA	yes
Cadmium	μg Cd/l	5 DAA	2.5 DAA	yes
Hexachlorocyclohexane	μg HCH/l	0.02 TAA	0.02 TAA	yes
Carbon tetrachloride	μg CCl₄/l	12 TAA	12 TAA	no
Dichlorodiphenyltrichloroethane				
(all 4 isomers, total DDT)	μg DDT/l	0.025 TAA	0.025 TAA	yes
(para, para-DDT)	μg ppDDT/l	0.01 TAA	0.01 TAA	yes
Pentachlorophenol	μg PCP/l	2 TAA	2 TAA	yes
Total 'drins'	μg/l	0.03 TAA	0.03 TAA	yes
Aldrin	μg/l	0.01 TAA	0.01 TAA	yes
Dieldrin	μg/l	0.01 TAA	0.01 TAA	yes
Endrin	μg/l	0.005 TAA	0.005 TAA	yes
Isodrin	μg/l	0.005TAA	0.005 TAA	yes
Hexachlorobenzene	μg HCB/l	0.03 TAA	0.03 TAA	yes
Hexachlorobutadiene	μg HCBD/l	0.1 TAA	0.1 TAA	yes
Chloroform	μg CHCl₃/l	12 TAA	12 TAA	no
1,2-Dichloroethane (ethylenedichloride)	μg EDC/l	10 TAA	10 TAA	no
Perchloroethylene (tetrachloroethylene)	μg PER/l	10 TAA	10 TAA	no
Trichlorobenzene (all isomers)	μg TCB/l	0.4 TAA	0.4 TAA	yes
Trichloroethylene	μg TRI/l	10 TAA	10 TAA	no

Notes

Substances are listed in the order of publication of Directives.

D: dissolved concentration, ie usually involving filtration through a 0.45-μm membrane filter before analysis.

T: total concentration (ie. without filtration).

AA: standard defined as annual average.

a: Most directives include, in addition to the standards for inland, estuary and marine waters, a provision that the total concentration of the substance in question in sediments and/or shellfish and/or fish must not increase significantly with time (the "standstill" provision).

b: In the UK the standards for estuaries are the same as for marine waters - The Surface Waters (Dangerous Substances) (Classification) Regulations 1989.

8. ENVIRONMENTAL PROTECTION AND WAVE POWER GENERATION

Table 8.3(1) Water quality standards for List II substances for the protection of saltwater life in the UK Marine SACs. (from Cole et al [11])

Parameter	Unit	Water Quality Standard (see footnotes)	Uncertainties in the derivation : Details obtained from the relevant EQS derivation reports
Lead	µg Pb/l	25 AD	The preliminary EQS was multiplied by a factor of 2 to account for overestimation of Pb toxicity in laboratory studies compared to the field environment. The EQS was considered tentative as a result of the paucity of reliable data, in particular for sub-lethal chronic studies with invertebrates and fish, and for field studies.
Chromium	µg Cr/l	15 AD	There were limited data on the sub-lethal effect of Cr and long-term exposure to freshwater and saltwater life. Separate standards for different Chromium valences (Cr(VI) and Cr(III)) were not recommended as a consequence of the lack of data for Cr(III). In addition, a comparison of the toxicities of each oxidation state was not possible. Some data were available that indicated higher sensitivity of some saltwater organisms to low salinities. The EQS was based on data generated at salinities typical of normal seawater. Therefore, further research on the effect of Cr at lower salinities was recommended.
Zinc	µg Zn/l	40 AD	The dataset available for the toxicity of Zn to saltwater life illustrated that at the EQS, adverse effects on algal growth had been reported. However, it was considered that there was currently insufficient evidence to suggest that the EQS would not adequately protect saltwater communities.
Copper	µg Cu/l	5 AD	Further data were considered necessary on the sensitivity of early life stages and life-cycle tests to confirm the sensitivity of saltwater life.
Nickel	µg Ni/l	30AD	Marine algae were reported to be adversely affected by Ni at concentrations as low as 0.6 mg l^{1} which is below the EQS to protect saltwater life. However, it was considered that there was insufficient evidence to justify a lower EQS based solely on results with algae and that further research into this area was desirable. There was also limited evidence to suggest that invertebrates in estuarine systems may be more susceptible to the effects of Ni than invertebrates in marine systems. Thus, an EQS to protect estuarine life may be needed in future when further data become available.
Arsenic	µg As/l	25AD	Based on crab 96 hour LC50, and an extrapolation factor of 10 applied. Standards may need to be more stringent where sensitive algal species are important features of the ecosystem.
Boron	µg B/l	7000 AT	Few data available. However the standard was based on Dab 96 hour LC50, with an extrapolation factor of 10 applied.
Iron	µgFe/l	1000AD	The EQS for the protection of saltwater life was based on observed concentrations and general assessments of water quality. It was recommended, therefore, that the standard should be reviewed as soon as direct observations of water concentrations and biological status become available. Limited data did not allow an assessment of the importance of Fe species.
Vanadium	µgV/l	100 AT	Data on the toxicity of vanadium on saltwater life were limited. As there were limited data for vanadium, it was not possible to recommend standards based on dissolved concentrations or separate standards for migratory fish. With regard to the latter, it may be necessary to base judgement of any risk in applying the EQS on knowledge of local risks and circumstances.
Tributyltin	µg/l	0.002 MT	The standards for TBT weres tentative to reflect a combination of the lack of environmental data, toxicity data or data relating to the behaviour of organotins in the environment.
Triphenyltin (and its derivatives)	µg/l	0.008 MT	The standards for TPT were tentative to reflect a combination of the lack of environmental and toxicity data or data relating to the behaviour of organotins in the environment.
PCSDs	µg/l	0.05 PT	In view of the lack of data for the mothproofing agents, both from laboratory and field studies, the EQSs were reported as tentative values.
Cyfluthrin	µg /l	0.001 PT	In view of the lack of data for the mothproofing agents, both from laboratory and field studies, the EQSs were reported as tentative values.

Notes

Substances are listed in the order of publication of Directives.

A : annual.

D : dissolved concentration, ie usually involving filtration through a 0.45-µm membrane filter before analysis.

T : total concentration (ie. without filtration).

AA : standard defined as annual average.

MAC : maximum concentration.

8.1 Protection of Water Quality in Marine Areas

Table 8.3(2) Water quality standards for List Ⅱ substances. (from Cole et al [11])

Parameter	Unit	Water Quality Standard (see footnotes)	Uncertainties in the derivation : Details obtained from the relevant EQS derivation reports
Sulcofuron	µg /l	25 PT	As a consequence of the general paucity of data for the mothproofing agents, both from laboratory and field studies, the EQSs were reported as tentative values. The data for sulcofuron suggested that embryonic stages for saltwater invertebrates could be more sensitive than freshwater species and, therefore, the EQS for the protection of marine life, derived from the freshwater value, may need to be lower.
Flucofuron	µg /l	1.0 PT	In view of the lack of data for the mothproofing agents, both from laboratory and field studies, the EQSs were based on freshwater values.
Permethrin	µg /l	0.01 PT	In view of the lack of data for the mothproofing agents, both from laboratory and field studies, the EQSs were reported as tentative values.
Atrazine and Simazine	µg /l	2 AA : 10 MAC	The EQSs for the protection of saltwater life were proposed as combined atrazine/simazine to take account of the likely additive effects when present together in the environment.
Azinphos-methyl	µg /l	0.01AA ; 0.04 MAC	In view of the relatively high soil organic carbon sorption coefficient, it is likely that a significant fraction of the pesticide present in the aquatic environment will be adsorbed onto sediments or suspended solids. However, it is likely that this form will be less bioavailable to most aquatic organisms. As the adsorbed pesticide is more persistent than the dissolved fraction, it is possible that levels may build up that are harmful to benthic organisms. Insufficient information on saltwater organisms was available to propose a standard. In view of the paucity of data, the standards to protect freshwater life were adopted to protect saltwater life.
Dichlorvos	µg /l	0.04 AA and 0.6 MAC	Based on data for sensitive crustaceans.
Endosulphan	µg /l	0.003 AA	There is little evidence on the ultimate fate of endosulfan and its metabolites or degradation products in sediments and on any effects on freshwater benthic organisms. Consequently, it is possible that some sediment-dwelling organisms, such as crustaceans, may be at risk.
Fenitrothion	µg /l	0.01 AA : 0.25 MAC	As there were limited data with which to derive EQSs to protect saltwater life, the freshwater values were adopted. However, the annual average for the protection of freshwater life may be unnecessarily stringent in view of the uncertainties associated with the acute toxicity data used in its derivation. The uncertainties exist because the original sources were unavailable for certain studies. Lack of confirmatory data existed in the published literature and data for warm water species were considered in the derivation.
Malathion	µg /l	0.02AA ; 0.5MAC	It was recommended that further investigation for both field and laboratory conditions into the effects of malathion on crustaceans and insects and on UK *Gammarus* species, in particular, should be carried out.
Trifluralin	µg /l	0.1AA : 20 MAC	None mentioned with regard to the annual mean.
4-chloro-3-methyl phenol	µg /l	40 AA : 200 MAC	Insufficient saltwater data were available to propose a standard. Therefore, the standard was based on freshwater value.
2-chlorophenol	µg /l	50 AA : 250 MAC	Insufficient saltwater data were available to propose a standard. Therefore, the standard was based on freshwater value.
2,4-dichlorophenol	µg /l	20 AA : 140 MAC	Insufficient saltwater data were available to propose a standard. Therefore, the standard was based on freshwater value.
2,4D (ester)	µg /l	1 AA : 10 MAC	For the EQS proposed for 2,4-D esters, comparison of the data and derivation of standards were complicated by the number of esters and organisms for which studies were available. In addition, the toxicity of the esters may have been underestimated in some of the studies due to their hydrolysis. There were limited data on the toxicity of 2,4-D ester to saltwater life. Consequently, the freshwater value was adopted until further data become available.
2,4D	µg /l	40 AA : 200 MAC	There were limited data on the toxicity of 2,4-D non-ester to saltwater life. Consequently, the freshwater value was adopted until further data become available.

Notes

Substances are listed in the order of publication of Directives.

A: annual.
D: dissolved concentration, ie usually involving filtration through a 0.45-µm membrane filter before analysis.
T: total concentration (ie. without filtration).
AA: standard defined as annual average.
MAC: maximum concentration.

8. ENVIRONMENTAL PROTECTION AND WAVE POWER GENERATION

Table 8.3(3) Water quality standards for List II substances. (from Cole et al [11])

Parameter	Unit	Water Quality Standard (see footnotes)	Uncertainties in the derivation : Details obtained from the relevant EQS derivation reports
1.1.1 trichloroethane	µg /l	100 AA : 1000 MAC	The 1.1.1-TCA dataset available for freshwater species contained comparatively few studies where test concentrations were measured and, consequently, comparison of studies using measured concentrations vs. those using nominal values indicated that data from the latter type of study could be misleading.
1.1.2-trichloroethane	µg /l	300 AA : 3000 MAC	For 1.1.2-TCA, few data were available on chronic toxicity to freshwater fish. There were limited data on the toxicity of 1.1.2-TCA to saltwater life and, consequently, the EQS to protect freshwater life was adopted.
Bentazone	µg /l	500 AA : 5000 MAC	In view of the relatively high soil organic carbon sorption coefficient, it is likely that a significant fraction of the pesticide present in the aquatic environment will be adsorbed onto sediments or suspended solids. However, it is likely that this form will be less bioavailable to most aquatic organisms. As the adsorbed pesticide is more persistent than the dissolved fraction, it is possible that levels may build up that are harmful to benthic organisms. Insufficient information on saltwater organisms was available to propose a standard. In view of the paucity of data, the standards to protect freshwater life were adopted to protect saltwater life.
Benzene	µg /l	30 AA : 300 MAC	Limited and uncertain chronic data available.
Biphenyl	µg /l	25 AA	The data available for marine organisms were considered inadequate to derive an EQS for the protection of marine life. However, the reported studies for saltwater organisms indicate that the EQS for freshwater life will provide adequate protection.
Chloronitrotoluenes	µg /l	10 AA : 100 MAC	The dataset used to derive the EQS to protect freshwater life was limited. Toxicity data were available for comparatively few species and there was limited information on the bioaccumulation potential of the isomers. There were few chronic studies available to allow the assessment of the long term impact of CNTs. There were no reliable data for the toxicity to or bioaccumulation of CNTs by saltwater species and, therefore, the EQSs proposed for freshwater life were adopted.
Demeton	µg /l	0.5 AA ; 5 MAC	Insufficient saltwater data were available to propose a standard. Therefore, the standard was based on freshwater value.
Dimethoate	µg /l	1 AA	The available data for marine organisms were considered inadequate to derive an EQS for the protection of marine life. Crustaceans were considered to be the most sensitive organisms, but more data are required to confirm this. In view of the uncertainties associated with the marine toxicity dataset, the freshwater EQS was adopted. This was based on the toxicity of dimethoate to insects. Although there are no marine insects, there is some evidence that marine organisms are more sensitive than their freshwater counterparts.
Linuron	µg /l	2 AA	In view of the lack of data for saltwater life, the EQS proposed for the protection of freshwater life was adopted until further data become available.
Mecoprop	µg /l	20 AA : 200 MAC	There were limited data relating to the toxicity of mecoprop to aquatic life. The dataset for saltwater life comprised data for one marine alga, a brackish invertebrate and a brackish fish. Consequently, the freshwater values were adopted until further data become available.
Naphthalene	µg /l	5 AA : 80 MAC	Limited and uncertain chronic data available.
Toluene	µg /l	40 AA : 400 MAC	The dataset used to derive the EQS to protect saltwater life relied on static tests without analysis of exposure concentrations. Consequently, the derived values are considered tentative until further data from flow-though tests with analysed concentrations become available.
Triazophos	µg /l	0.005 AA ; 0.5 MAC	The dataset available for freshwater life was limited to a few studies on algae, crustaceans and fish. No information was available for the target organisms (insects), on different life-stages or on its bioaccumulation in aquatic organisms. There were no data on the toxicity to or bioaccumulation of triazophos in saltwater organisms. Consequently, the EQSs to protect freshwater life were adopted until further data become available.
Xylene	µg /l	30 AA : 300 MAC	Limited information available. Freshwater data used to ' back up' the standards.

Notes

Substances are listed in the order of publication of Directives.

AA : standard defined as annual average.

MAC : maximum concentration.

8.1 Protection of Water Quality in Marine Areas

2) Actions for Water Quality Protection

Regulation of water quality for domestic and industrial wastewater is extremely important, but proper management of nutrients in drainage from sewage treatment plants and factories is becoming increasingly important, as well as the monitoring of nutrients in naturally flowing water in ordinary rivers.

Furthermore, the backfilling of submarine holes created by mining sea gravel and minerals in various shallow water areas is promoted to avoid them becoming anaerobic, which can cause blue tides. Moreover, since tidal flats and extremely shallow waters provide habitats for shellfish and small fish and can be expected to suppress abnormal plankton blooms, their restoration is also promoted.

In case of oil pollution, in addition to legal control, it is essential to develop, improve and promote oil leakage prevention technology, oil fences, oil treatment agents and oil adsorbents, as well as the maintenance of oil recovery vessels and recovery machines. It is also important to predict and evaluate the effects of these measures. It is also necessary to develop, improve and maintain water quality and ecology monitoring systems to collect the necessary data for this purpose and to develop and improve precise data analysis systems together with predictive modelling capabilities.

Examples of useful prediction models include the following:
(a) Sasaki et al [12], [13] have developed a numerical prediction model that combines a three-dimensional inner bay fluid motion model with dissolved oxygen, organic matter and phosphate phosphorus concentration models, and a highly accurate three-dimensional long term fluid motion numerical model.

【Blue Carbon】

The regeneration and creation of seagrass beds, kelp forests and mangrove forests leads to the sequestration and storage of carbon through the action of marine organisms, which can reduce the increase in atmospheric carbon concentration. This carbon sequestered from the atmosphere and stored is called **blue carbon**, and projects to promote blue carbonisation are useful measures for preventing global warming.

The regeneration of mangrove forests on this coast not only prevents coastal erosion based on wave dissipation by the forests but also reduces greenhouse gases through CO_2 sequestration by photosynthesis.

Fig. I.1 Mangrove forest restoration project at the Samut Sakhon coast, Thailand.

8. ENVIRONMENTAL PROTECTION AND WAVE POWER GENERATION

(b) Sasaki et al [14] proposed a simple method for predicting the date, time, and area of occurrence of blue tides without field monitoring, based on the Sasaki et al [12], [13] 3D fluid motion numerical model combined with information from past blue tide observation data.

(c) Shigematsu et al [15] proposed an oxygen consumption model that can estimate the oxygen consumption flux throughout a year by analysing time variation data of the oxygen consumption flux in enclosed sea areas using the chamber method, which is a simple measurement technique.

The open-source Fortran 2003 programming framework FABM (https://github.com/fabm-model) can be used to build an integrated fluid motion/water quality/ecology modelling platform.

8.1 Protection of Water Quality in Marine Areas

【FVCOM and FABM】

Finite Volume Community Ocean Model (FVCOM) is an open-source 3D ocean flow model using the finite-volume method, employing triangular unstructured grids in the horizontal direction and the σ coordinate in the vertical direction. The governing equations are as shown in Eq. (m.1), this model can be used for storm surge and current calculations and can be downloaded from http://fvcom.smast.umassd.edu/.

$$
\left.
\begin{aligned}
&\frac{\partial u}{\partial t} + u\frac{\partial u}{\partial x} + v\frac{\partial u}{\partial y} + w\frac{\partial u}{\partial z} - f_{cl}v = -\frac{1}{\rho_w}\frac{\partial(p_s + p_a)}{\partial x} - \frac{1}{\rho_w}\frac{\partial q}{\partial x} + \frac{\partial}{\partial z}\left(K_m\frac{\partial u}{\partial z}\right) + F_u \\
&\frac{\partial v}{\partial t} + u\frac{\partial v}{\partial x} + v\frac{\partial v}{\partial y} + w\frac{\partial v}{\partial z} + f_{cl}u = -\frac{1}{\rho_w}\frac{\partial(p_s + p_a)}{\partial y} - \frac{1}{\rho_w}\frac{\partial q}{\partial y} + \frac{\partial}{\partial z}\left(K_m\frac{\partial v}{\partial z}\right) + F_v \\
&\frac{\partial w}{\partial t} + u\frac{\partial w}{\partial x} + v\frac{\partial w}{\partial y} + w\frac{\partial w}{\partial z} = -\frac{1}{\rho_w}\frac{\partial q}{\partial z} + \frac{\partial}{\partial z}\left(K_m\frac{\partial w}{\partial z}\right) + F_w \\
&\frac{\partial u}{\partial x} + \frac{\partial v}{\partial y} + \frac{\partial w}{\partial z} = 0, \qquad \rho = \rho(T, S, p) \\
&\frac{\partial T}{\partial t} + u\frac{\partial T}{\partial x} + v\frac{\partial T}{\partial y} + w\frac{\partial T}{\partial z} = \frac{\partial}{\partial z}\left(K_h\frac{\partial T}{\partial z}\right) + F_T, \quad \frac{\partial S}{\partial t} + u\frac{\partial S}{\partial x} + v\frac{\partial S}{\partial y} + w\frac{\partial S}{\partial z} = \frac{\partial}{\partial z}\left(K_h\frac{\partial S}{\partial z}\right) + F_S
\end{aligned}
\right\}
\tag{m.1}
$$

where t is time; x and y are the horizontal Cartesian coordinates; z is the vertical coordinate; u, v, and w are respectively the x, y, and z components of the flow velocity; f_{cl} is the Coriolis coefficient, ρ_w is the reference density of seawater; p_s is the hydrostatic pressure; p_a is the atmospheric pressure; q is the nonhydrostatic pressure of water; K_m is the vertical eddy kinematic viscosity; F_u, F_v, and F_w are horizontal diffusion terms; ρ is the seawater density; T is the water temperature; S is the salinity; K_h is the vertical diffusion coefficient; F_T is the horizontal thermal diffusion term; and F_S is the horizontal salt diffusion term.

Reference: Chen, C, Beardsley, R.C. and Cowles, G.: An Unstructured Grid, Finite-volume Coastal Ocean Model (FVCOM) System, Special Issue Entitled "Advance in Computational Oceanography", *Oceanography*, 19(1), 2006, pp.78-89.

Framework for Aquatic Biogeochemical Models (FABM) is an open-source Fortran 2003 programming framework for linking numerical hydrodynamic and biogeochemical models. FABM can be downloaded from https://github.com/fabm-model. To connect to FABM, models are coded once, and any combination of numerical hydrodynamic and biogeochemical models can be created. Furthermore, FABM enables the optimal utilisation of the expertise of scientists, programmers and end-users by enabling distributed development and user-controlled coupling of biogeochemical models.

Reference: Bruggeman, J. and Bolding, K.: A General Framework for Aquatic Biogeochemical Models, *Environmental Modelling & Software*, Vol.61, 2014, pp. 249–265.
Doi: 10.1016/j.envsoft. 2014.04.002

8. ENVIRONMENTAL PROTECTION AND WAVE POWER GENERATION

8.2 Marine Litter
1) Actual Situation of Drifted Litter

The actual situations of drifted litter obtained from the field surveys carried out by the Ministry of the Environment [16] in Japan at 10 beaches (litter collected per 50 m of beach extension at each beach) in 2019 were as follows:

(a) On a volumetric basis, drifted litter consisted of more man-made objects than natural objects (e.g. driftwood) on seven of ten beaches.

(b) On most beaches, drifted litter had a larger proportion of plastics, mainly fishing nets, ropes, and beverage bottles.

(c) The linguistic description of these bottles drifted ashore showed that more than 50% of the bottles in the western part of the Japanese archipelago were from overseas, as shown in **Fig. 8.1**.

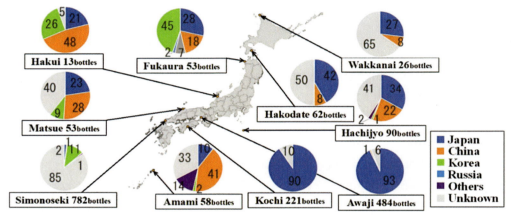

Fig. 8.1 Breakdown of countries of manufacture of beverage bottles within marine litter. (Source: Results of the 2019 Marine Litter Survey by Ministry of the Environment [16])

As mentioned above, it has become common knowledge worldwide that more than half of all drifted litter is plastics, which are less perishable and more prone to drift. The global production of plastics was around 2 million tonnes in 1950 but reached 380 million tonnes in 2015. It is estimated that around 2-3% of this was discharged into the oceans, with the total global discharge in 2015 thought to be between 7.6 and 11.4 million tonnes. If this trend continues, the amount of drifted plastic litter will exceed the total fish catch by around 2050.

Several studies have been conducted to understand the spill situation of this plastic litter. For example, van Sebille et al [17] studied the distribution of drifted plastic litter using data from surface-towing plankton net trawls. **Fig. 8.2** shows their distribution map (blank areas mean no data). According to this figure, high-density areas of plastic litter exist near the centres of oceans, and it is recognised that drifted plastic litter does not stay in the nearshore waters, but spreads to the distant oceans through ocean currents. Jambeck et al [18] also predicted mismanaged plastic waste in 2025 using country-specific mismanaged plastic waste data for 2010, which are summarised in **Table 8.4**. This table shows that mismanaged plastic waste from populous and emerging countries is high.

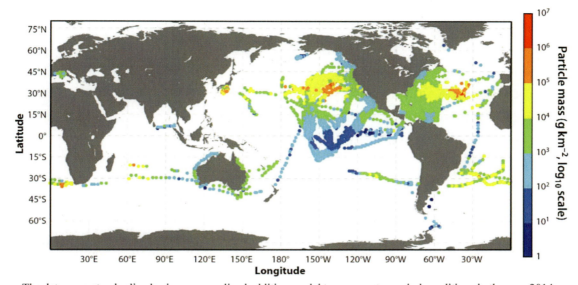

The data were standardized using a generalized additive model to represent no-wind conditions in the year 2014.

Fig. 8.2 particle mass of plastic samples collected from 11,854 surface-towing plankton net trawls. (Source: van Sebille et al [17])

Table 8.4 Country ranking of estimated mass of mismanaged plastic waste by Jambeck et al [18].

Rank	Estimated values in 2010		Predicted values in 2025	
	Country name	MMT/year	Country name	MMT/year
1	People's Republic of China	8.82	People's Republic of China	17.81
2	Indonesia	3.22	Indonesia	7.42
3	Philippines	1.88	Philippines	5.09
4	Vietnam	1.83	Vietnam	4.17
5	Sri Lanka	1.59	India	2.88
6	Thailand	1.03	Nigeria	2.48
7	Egypt	0.97	Bangladesh	2.21
8	Malaysia	0.94	Thailand	2.18
9	Nigeria	0.85	Egypt	1.94
10	Bangladesh	0.79	Sri Kanka	1.92
11	Republic of South Africa	0.63	Malaysia	1.77
12	India	0.60	Pakistan	1.22
13	Algeria	0.52	Myanmar	1.15
14	Turkey	0.49	Algeria	1.02
15	Pakistan	0.48	Brazil	0.95
16	Brazil	0.47	Republic of South Africa	0.84
17	Myanmar	0.46	Turkey	0.79
18	Morocco	0.31	Senegal	0.74
19	North Korea	0.30	Morocco	0.71
20	United States of America	0.28	North Korea	0.61

They assumed the 2010 per capita waste generation rate to be constant until 2025 and predicted the mass of mismanaged plastic waste in 2025 by using the coastal population projection for each country from 2010 to 2025.

8. ENVIRONMENTAL PROTECTION AND WAVE POWER GENERATION

Natural organic materials such as wood have **biodegradability** (the property of being ultimately broken down into carbon dioxide and water by microorganisms in nature). However, plastic is a man-made compound that is not biodegradable. Various organisations have reported increasing numbers of cases of aquatic animals ingesting plastic waste mistakenly as food or becoming entangled in them, resulting in emaciation and death (see Law [19]).

Furthermore, plastics are not biodegradable; but are pulverised into small particles by water and UV radiation. Plastics with a particle size of 5 mm or less shown in **Photo 8.3** are called **microplastics**. Microplasticisation makes their recovery from the oceans more difficult. The fact that plastics adsorb toxic substances such as Poly Chlorinated Bipheny (PCB) and that many of the additives used in manufacturing plastics are harmful to the human body increases the likelihood that microplastics will be ingested and accumulated in the bodies of fish and shellfish, thereby harming the ecosystem and human health.

Photo 8.3 Example of microplastics drifted ashore. (Source: https://www.istockphoto.com/jp/)

In developed countries, even though plastic waste is collected by governments and landfills are properly managed, many microplastics have been detected. Much more, in developing countries, where plastics are left out in the open and landfills are poorly managed, they are practically unregulated.

2) Measures against Drifted Litter

Much of the litter that drifted ashore has been collected by local authorities and sometimes by residents and incinerated at waste incineration plants or on local beaches. However, nowadays open burning is regulated in many countries and the majority of collected litter is either incinerated at waste incineration plants or taken to final disposal sites. However, incineration of salinated refuse has problems such as damaging incinerators and producing dioxins.

Therefore, it is most important to reduce the total amount of litter drifting ashore by involving neighbouring countries in strengthening laws and regulations on the disposability of goods that lead to litter, and by ensuring that litter is collected and reused/recycled more thoroughly. **Table 8.5** shows the results of a study by the Ministry of the Environment of Japan on the current situation and the future potential of reusing and recycling litter drifted ashore. Furthermore, because plastics with biodegradability are being developed, the improvement and expansion of this technology should also be promoted.

8.2 Marine Litter

Table 8.5 Results of the Ministry of the Environment's survey on the reuse and recycling of drifted litter.

Material	Item	Reuse	Recycle
Plastics	Beverage plastic bottles	—	Limited suppliers, but effective use is possible
	Fishing buoys	Buoys in good condition are reused in various places	—
	Ropes, etc.	—	Limited suppliers, but effective use is possible Effective use for coasters, etc. is possible
Styrofoam	Fishing buoys	Buoys in good condition are reused in various places	Styrofoam buoys are reduced in volume with solvents and recycled
Metals	Empty cans, Steel scrap, etc.	—	They can be sold (market value)
Natural drift material	Shrubs, Driftwood	—	They are chipped and recycled Charcoal

To maintain the durability of PCs and household appliances, it is essential to use conventional plastics, but they must be collected after use more efficiently. However, in cases where recovery is difficult, such as fishing nets and fishing lines, or where durability is not required, such as mulch film for agriculture, it is advisable to use biodegradable plastics.

For cases where durability is required but recovery is difficult, technology is being developed to embed plastic-degrading enzymes inside the plastic so that degradation proceeds when the plastic is discharged into the sea. As the appropriate rate of biodegradation depends on the purpose of use, technology is also being developed to change the rate of degradation according to the purpose.

There are three types of biodegradable plastics.

(a) Plastics with significant biodegradability are called **biodegradable plastics**, but this includes plastics derived from petroleum if they are biodegradable.

(b) Plastics made from biomass, which is a renewable resource, are called **biomass plastics**. Bio-polyethylene, which is produced from ethanol obtained from plants such as sugarcane, is included in biomass plastics, but not in biodegradable plastics because it is not biodegradable.

(c) The term **bioplastics** is usually used as a generic term for biodegradable plastics and biomass plastics.

Furthermore, it should be noted that only some biodegradable plastics, such as Poly Hydroxy Butyrate / Hydroxy Hexanoate (PHBH), can be degraded in aquatic environments. The biodegradable plastic Poly Lactic Acid (PLA) is degraded in hot and humid environments, but not in soil or water environments. The same plastic, bio Poly Butylene Succinate (PBS), is degraded in hot and humid environments and soil, but not in aquatic environments.

3) Countermeasures against the Smell of Rotting Seaweed and Seagrass

Foul smells can be emitted due to the decay of large quantities of seaweed (algae growing in marine areas that can be identified by the naked eye) and seagrass (seed plants growing in marine areas) washed onshore during high waves. The chemical spraying method by Yamamoto and Minami

8. ENVIRONMENTAL PROTECTION AND WAVE POWER GENERATION

[20] and Yamamoto and Nariyoshi [21] is introduced below as the study example of a method to reduce odours caused by seaweed and seagrass drifting ashore.

Photo 8.4
Seaweed (mainly, Eisenia bicyclis) drifted ashore in large quantities on the Kamakura coast, Kanagawa Prefecture in Japan, in July 2007.

(1) Study of Chemical Spraying Method by Yamamoto and Minami [20]

The following three deodorants from among the agents that can be expected to reduce the smell of rotting seaweed were selected, because they are low-cost, easy to obtain, and cause little harm to human bodies and the coastal environment.

(a) Powdered calcium: Powdered shells of oysters, scallops, and surf clams which are natural purifiers with strong bactericidal properties.

(b) Effective microorganism solution: a complex of beneficial micro-organisms such as yeast, lactic acid bacteria and natto bacillus that ferment organic matter and purify the environment.

(c) Proteolytic enzyme solution: an enzyme complex that promotes the decomposition of organic matter and the growth and activation of effective bacteria. The product is manufactured by DMT World Trade, which has been verified by the manufacturer to be harmless to humans and animals.

[Preliminary Experiment]

To minimise costs, the above deodorants were diluted 1000, 3000 and 5000 times using tap water. Next, a direct spraying method on the seaweed/seagrass surface was adopted to minimise the time and effort required for treatment. Plastic containers (15 cm wide × 21 cm long × 6 cm deep) containing seaweed/seagrass with strong odours were first placed indoors. Then, the deodorant solutions were sprayed 10 times from 15 cm above the surface by using a small sprayer. As a baseline scenario, one plastic container filled with seaweed and seagrass was left untreated.

In the case where the test specimens with a strong odour were left without any treatment, it took about 12 days to subsite the odour to a state where it could only be smelled when approached up to 50 cm by drying out naturally. However, the case in which proteolytic enzyme solution was sprayed took about half a day, the case in which effective microorganism solution was sprayed took about one day, and the case in which powdered calcium solution was sprayed took about 1.5 days to eliminate odours to the same level. Moreover, after about 2 days, no difference in dilution ratio could be detected, and after about 4 days, in all cases of the deodorant solutions, the foul smells were no longer bothersome even in the immediate vicinity.

Next, natural water was sprayed 2 and 6 hours after the start of the experiment using a sprinkler to mimic rainfall to the extent that it could not be washed away. The deodorising effect disappeared and the deodorising chemical had to be re-sprayed. However, when the same water was sprayed 1 and 3 days later, the deodorising effect recovered to its original level after 2 days, although the odour returned temporarily.

[Main Experiment]

On an actual sandy beach, it was speculated that the deodorant chemical solution would be absorbed by the beach sand and would not have a sufficient deodorant effect. Therefore, large plastic containers (35 cm wide × 45 cm long × 20 cm deep) were filled with a 10 cm thick sand layer (grain size 0.2 mm to 0.66 mm), on which seaweed and seagrass (which had been left in a state with seawater on it until it emits a strong odour) were placed with three times the volume of the preliminary experiment. Then, the deodorant solutions were sprayed 30 times from about 15 cm above the surface by using a small sprayer. In addition, to mimic rainfall, five times the volume of natural water used in the preliminary experiment of was poured over the test specimens using a sprinkler, to investigate the deodorising effect.

In the experimental results shown in **Fig. 8.3**, in the case where the specimen was left without any treatment (blue curve), it took about 4 days to subside the ordour to a state where it could only be smelled when the specimen was 50 cm closer by drying out naturally, which was 1/3 shorter than the 12 days in the preliminary experiment. Compared with the preliminary experiment, the biggest difference in the main experimental method is that the test specimens were placed on a sand layer, but the sand layer itself has odour adsorption performance. Moreover, the sand brought from the beach has natural micro-organisms with an oxidative decomposition function for organic substances, indicating that some deodorising effect by the sand layer can also be expected.

Furthermore, in the summer, all deodorant solutions with a dilution ratio of 5000 times or less, eliminated the odour to the same level as above in about 1.5 days. The odour was not bothersome after 2.5 days even from proximity. Therefore, it can be concluded that it is worth spraying deodorant solutions to prevent bad odours of rotting seaweeds and seagrass.

In addition, the results of watering 2 hours and 6 hours after the start of the experiment, and then 1 day and 3 days after the start of the experiment show that although the bad odour temporarily returned, the deodorant effect soon recovered to the original level by itself. This indicates that even if a small

8. ENVIRONMENTAL PROTECTION AND WAVE POWER GENERATION

amount of rain falls, the effect of all deodorant chemical solutions does not disappear when a sand layer is present.

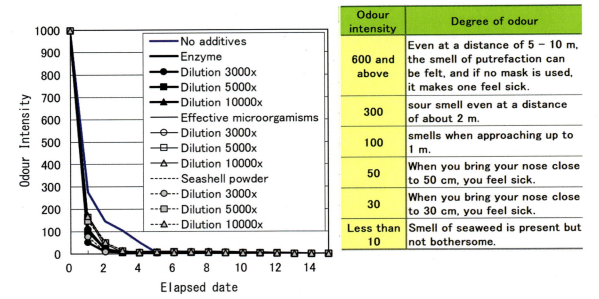

Fig. 8.3 Deodorising effect of three deodorising chemical spray methods.

(2) Continuing Study by Yamamoto and Nariyoshi [21]

They studied the effects of temperature and humidity on the deodorising effect.

Yamamoto and Nariyoshi [21] found that when the daytime temperature is 15°C or higher, all deodorant chemical solutions show a significant deodorising effect. The deodorising effect was slightly higher at higher humidities (75-90%).

When the daytime temperature is below 15°C and the humidity is low (50-65%), the odour is not so severe that it is difficult to recognise the deodorising effect of the chemical solution. On the other hand, when the humidity is higher (70-85%), the odour becomes comparatively worse when left untreated, so the deodorising effect of the chemical solution can be recognised to some extent.

【A Thought on Coastal Litter Recycling】

When Styrofoam is placed in a volume-reducing liquid (e.g. ECOKATON 50), which is mainly composed of citrus oil extracted from grapefruit peels, it is separated into a volume-reducing liquid and a reduced gel-like substance. The volume-reducing liquid can be reused. The gel-like substance can be recycled as a raw material for construction adhesives, polystyrene raw materials, water-repellent paints, firework fuels, solid fuels, etc.

Since the gel-like substance obtained from Styrofoam has the property of bonding and solidifying with other substances, it can be recycled as a pavement material for coastal footpaths by mixing the gel-like substance instead of asphalt and the non-degradable drifted litter instead of aggregate as a new treatment method for non-degradable drifted litter (this gel-like substance has less toxic components than bitumen and is not black). The method to make prototype paving blocks from the gel-like substance and non-degradable drifted litter is described as follows:

(1) Put Styrofoam in a heat-resistant container containing citrus oil to make and accumulate a gel-like substance.
(2) After removing some amount of citrus oil, place the heat-resistant container on a heater, and heat the gel-like substance of the container to 170°C so that the remaining citrus oil is evaporated while adding more Styrofoam to increase the consistency.
(3) After removing the citrus oil until a syrup-like consistency is obtained, pour the gel-like substance into a mould containing the non-degradable litter and stir well (see **Photo n.1**).
(4) Make a mixture consisting of the gel-like substance and the non-degradable litter harden sufficiently by heating it to 170°C and shaping it as a prototype block.

Loading tests were carried out on the sufficiently solidified prototype blocks (see **Photo n.2**), confirming that the compressive strength at 5% deformation was above 1 N/mm^2, which is the acceptable limit for use (see **Fig. n.1**). Moreover, confirm the durability of the prototype blocks, the prototype blocks (20-30 cm in length) were laid on a pathway at Tokai University and examined for damage after two years; as shown in **Photo n.3**, there was no damage.

Photo n.1 the gel-like substance and non-degradable litter in moulds.

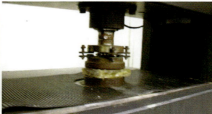

Photo n.2 Loading tests on blocks.

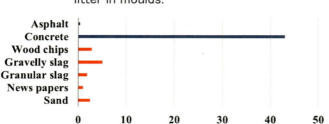

Fig. n.1 Compression test results by aggregate type.
(Strengths of concrete and asphalt are for reference)

Photo n.3 Durability test of blocks.
(Installed in 12/2006, 2 years later)

8. ENVIRONMENTAL PROTECTION AND WAVE POWER GENERATION

8.3 Widespread Use of Wave Power
1) Renewable Energy

By using **Fig. 8.4** and so on, the Intergovernmental Panel on Climate Change (IPCC) has shown that global warming, which causes atmospheric instability [22, 23, 24], sea level rise, and other global environmental degradation [22], is being caused by a rapid increase in emissions of greenhouse gases such as carbon dioxide and methane due to human economic and industrial activities.

Therefore, the **Kyoto Protocol** was adopted at the United Nations Framework Convention on Climate Change in 1997 as an international framework for the regulation of greenhouse gas emissions until 2020 and entered into force in 2005. However, in 2013, the fifth assessment report of IPCC [25] showed that if the world wants to stop global warming, it must achieve the goal of zero addition of greenhouse gases into the atmosphere. Therefore, a stronger post-2020 regulatory framework was adopted at the same convention in Paris in 2015 and entered into force in 2016. This is known as the **Paris Agreement** and requires sufficient efforts to limit the global average temperature increase to 1.5°C, well below the industrial period average of 2°C. For this, the Paris Agreement called on all countries to set emissions targets based on their Nationally Determined Contributions (NDCs) and encourages Parties to provide voluntary support to reduce greenhouse gas emissions.

For this reason, it is extremely important to know from which sources and to what extent greenhouse gases are being emitted, and developed countries are obliged to prepare and submit to the Convention Secretariat an annual list showing the quantities of specific substances emitted or absorbed, from which sources or sinks within a certain period. This data compiling the amount of greenhouse gases emitted or absorbed by each country in a year is known as the **Greenhouse Gas Inventory**.

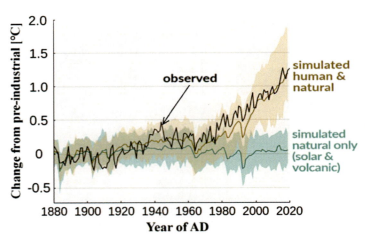

Fig. 8.4 Change in the Earth's surface temperature relative to pre-industrial times (1850 – 1900 average), based on the IPCC's Sixth Assessment Report [22].

A significant reduction in fossil fuel consumption is effective in reducing greenhouse gas emissions. To achieve this, the spread of nuclear power generation based on a significant improvement in safety technology is a possibility, but considering the tragedy that would occur in the event of an accident, we must be cautious at present. The active expansion of **Renewable energy**, which is safe

8.3 Widespread Use of Wave Power

and does not rely on fossil fuels, is required.

The status of the main power generation technologies based on renewable energy is as follows:

(i) Photovoltaic energy: The efficiency of photovoltaic cells is not high at around 20%, but the increase in efficiency and the cost reduction due to the mass production of cells is expected to increase the revenue from the sale of electricity. The current cost of photovoltaic power generation is comparable to thermal power generation's; and may be reduced further in 10 years. However, as the amount of electricity generated depends on solar power, there are some disadvantages, such as the amount of electricity generated varies depending on the weather (cloudy or rainy) and electricity cannot be generated at night.

(ii) Wind power: Due to years of technological development, the efficiency of power generation is not bad at 20-45%. The current cost of power generation is only slightly higher than that of thermal power, but it is expected to be on par with thermal power in 10 years. However, as it relies on wind energy, the amount of electricity generated depends on the frequency and duration of strong winds, and suitable locations for the installation are limited.

(iii) Biomass power: Biomass power generation is fuelled by renewable resources of biological origin other than fossil fuels. This enables the supply of fuels without having to rely on imports, but the generation efficiency is low at around 20% and availability of suitable biomaterial can be limited. Therefore, the cost of biomass power generation tends to be higher than that of fossil fuels.

(iv) Geothermal electric power: In volcanic countries, geothermal reservoirs, in which groundwater is converted into steam by magma heat and accumulated, are found in shallow underground areas. The steam from these reservoirs, which fluctuates little and is not easily exhausted, is used to turn turbines and generate electricity. This offers a relatively stable means of power generation. However, the efficiency of power generation is currently low at 10-20%. Although the total cost of geothermal power generation is comparable to that of fossil fuel power generation, it should be noted that the costs of surveying and construction in volcanic areas are likely to become high, and there are also legal regulations to ensure safety.

(v) Hydroelectric power: Large-scale hydropower generation by water stored in reservoirs behind dams has a very high generation efficiency of 80% and costs as low as fossil fuel power generation. However, such large-scale hydropower generation creates large environmental impacts. Small-scale hydropower, especially cascade hydropower with small reservoirs in

8. ENVIRONMENTAL PROTECTION AND WAVE POWER GENERATION

between two riverbanks, has low power generation efficiency but lower environmental impacts. As power is generated using water stored in reservoirs, it can be a stable source of power supply unless during prolonged periods of drought. The spread of hydropower generation in advanced countries is currently at a standstill because there are no more suitable sites for dams. Therefore, the current situation of hydropower generation is that expensive power generation facilities compared to the amount of electricity generated have been installed in rivers, so it is hoped that the cost of power generation can be reduced through technological development and mass production, as in the case of solar and wind power generation. As with other types of power generation using water currents, there are those using tidal currents and ocean currents. However, they are unlikely to spread widely due to the limited number of locations where permanently fast currents can exist and the high construction costs of construction at sea.

(vi) **Wave power**: This is a method of power generation that converts the kinetic energy of waves, whose energy per unit area is 20-30 times greater than that of sunlight and 5-10 times greater than that of wind, into electricity. Countries facing the sea can easily obtain wave energy. However, the range of wave conditions suitable for power generation can be limited as too large or too small waves are not suitable. In addition, construction and maintenance of facilities in coastal areas is more expensive than on land-based facilities. Therefore, the current efficiency of wave power generation is 10-25% and the cost of power generation is more than double that of fossil fuel power generation. The reasons for the lack of commercialisation of wave power technology can be summarised as follows:

① Expensive due to low power generation efficiency and high investment in power generation and transmission infrastructure. The more complex the mechanisms of power generation and transmission facilities, the more fragile they are by large wave forces.

② Unstable. The power generation capacity of wave energy depends on the wave conditions and becomes very small in weak wave conditions. Therefore, a power source that can be continuously generated is needed to stabilize the grid.

③ Not financially attractive enough to attract investors.

However, efforts are being made to increase the efficiency to more than 40% and the cost to the same level as wind power generation in 10 years.

The development of power generation from various renewable energies has been attempted as described above, and various wave power generation methods are introduced below, finally, a proposal is made for a method of promoting wave power and small-scale hydropower (hydropower without dams).

2) Examples of Wave Power Generation Methods

The types of wave power generation methods can be classified according to the installation method of the power generation facility and the driving/generating method, as follows:

(1) Classification by Installation Methods

(i) **Fixed Type:** Wave power generation facilities are fixed to the coastline area by foundation works. To reduce the construction costs, wave power generation facilities may be installed at wave dissipation works attached to breakwaters.

(ii) **Floating Type:** Power generators are moored to the sea surface or underwater using anchors and wires or chains, such as self-powered navigation beacon buoys. When moored in shallow waters of 20-25 m depth at a separation distance of 500 m or less from the shore, they are called **nearshore devices**; when moored in deeper waters at a greater distance, they are called **offshore devices**.

(2) Classification by Driving/Generation Methods

(i) **Pneumatic Type:** This type converts wave motion into airflow in an air chamber and turns a turbine to generate electricity. In principle, it is the same as wind power generation; and is less likely to break against huge wave forces than the next mechanical type, but its power generation efficiency is lower than that of the mechanical type.

(ii) **Mechanical Type:** This type generates electricity by directly receiving wave energy in a mechanical drive unit. Pendulum and crankshaft types are moving back and forth, buoy types that move up and down, and rotating buoy types. Pendulum and rotating buoy systems are relatively unbreakable against huge wave forces.

(iii) **Overflow Type:** This type generates electricity by turning a turbine with the flow of overtopping water down a channel. In principle, it is the same as hydropower and is less fragile than the mechanical type, but, its efficiency is lower than the mechanical type.

(iv) **Artificial Muscle and Piezoelectric Element Type:** **Electroactive Polymer Artificial Muscle (EPAM)** is a material consisting of a thin polymer membrane made of acrylic or silicon resin sandwiched between stretchable electrodes, where a potential difference is applied between the electrodes. When a potential difference is applied between the electrodes, the electrodes are attracted by electrostatic forces, and the polymer film is stretched thin. Conversely, if the polymer film is stretched by a wave, electricity is generated between the electrodes. A **piezoelectric element** is a material that generates a voltage when pressure is applied and can generate electricity by applying an oscillating force from waves. These materials themselves generate electricity, so no turbines or generators are needed, but the durability of the materials needs to be improved.

8. ENVIRONMENTAL PROTECTION AND WAVE POWER GENERATION

(3) Examples of Wave Power Generation

There are numerous examples of wave power generation around the world, in the following, some basic examples based on installation and drive/power generation methods are introduced.

(i) Example of Pneumatic Drive in a Fixed Installation
Air turbine system due to oscillating water column

As shown in **Fig. 8.5**, the turbine is turned by pressure changes in the air chamber generated by the vertical movement of the waves. The air turbine used is a Wells turbine that rotates in one direction regardless of the positive or negative pressure. This system is not directly subjected to wave forces at the drive unit, which ensures sufficient durability but reduces the efficiency of power generation. If a pressure above the specified pressure occurs in the air chamber, the air is released by a safety valve to protect the equipment and prevent damage or destruction. In the present case, the power generation efficiency was 11.3% for the offshore wave power (approx. 90.54 kW).

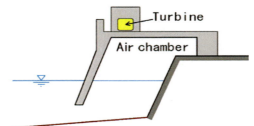

Fig. 8.5 Coastal fixed wave power generation facility by Hotta et al [26]. (offshore Tsuruoka City, Yamagata Prefecture).

(ii) Example of Pneumatic Drive in a Fixed Installation (No.2)
Wave power generation system using a breakwater

The power generation method is the same as in the example (i), by integrating the turbine and generator for power generation into the breakwater as shown in **Fig. 8.6**, the cost of setting up those accommodations is reduced. An air chamber is installed in front of the existing caisson to convert wave energy into wind energy, which turns a turbine to generate electricity. If projecting walls are provided, wave power can be efficiently absorbed over a wide range of wave periods. When the internal pressure of the air chamber exceeds a predetermined value, a safety valve is opened to allow air to escape, thus protecting the turbine and generator.

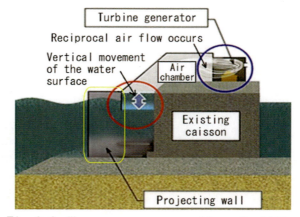

Fig. 8.6 Wave power generation facility fixed to a breakwater at the Port of Sakata by New Energy and Industrial Technology Development Organization (NEDO) in Japan (from Kihara et al [27]).

8.3 Widespread Use of Wave Power

The wave power generation caisson system shown in **Fig. 8.7**, in which a power generation unit is integrated with the breakwater caisson and has a wave dissipating function, is considered the most economical and feasible wave power generation system.

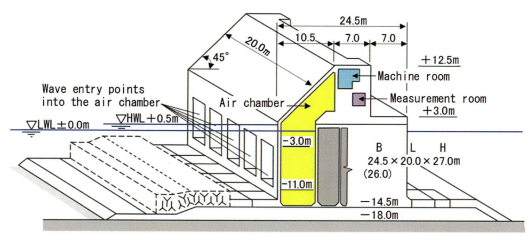

Fig. 8.7 Wave-powered caisson breakwater by the former Japanese Ministry of Transport. (from Takahashi [28])

(iii) Example of Mechanical Drive in a Fixed Installation
   Movable wave-catching plate type

In the power generation system shown in **Fig. 8.8**, a wave plate receives wave energy directly and absorbs it at the hinge, which in turn drives the generator. For a movable wave-catching plate type, three positions: upper, middle, and lower can be considered as the hinge position. Tamaki et al [29] showed from the theoretical examination that the efficiency is lowest with the middle hinge support and maximum with the upper and lower supports. Experimental evidence showed that the efficiency is slightly higher with the lower support than with the upper support.

Compared to the pneumatic type, this system has less variation in power generation performance due to changes in tidal level and can generate power relatively well even when waves are small..

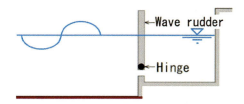

Fig. 8.8 Movable wave-catching plate type by Tamaki et al [29].

(iv) Example of Mechanical Drive in a Fixed Installation (No. 2)
   Top-supported pendulum type (or wave rudder type)

Maruyama et al [30] developed the wave power generator, shown in **Figs. 8.9 and 8.10**, based on the principle of a pendulum-type generator developed by the Muroran Institute of Technology in the 1980s, used existing technology for wind power and commercial products to reduce costs.

365

8. ENVIRONMENTAL PROTECTION AND WAVE POWER GENERATION

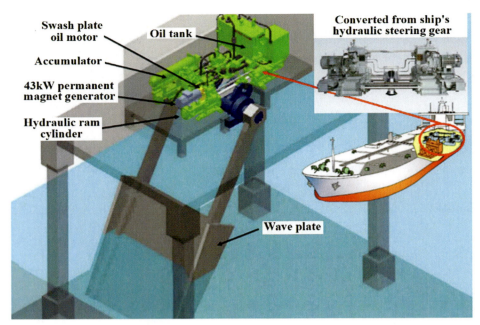

Fig. 8.9 3D view of a pendulum-type wave power generator by Maruyama et al [30]. (Test construction and operation in Kuji City of Iwate Prefecture and Hiratsuka City of Kanagawa Prefecture in Japan)

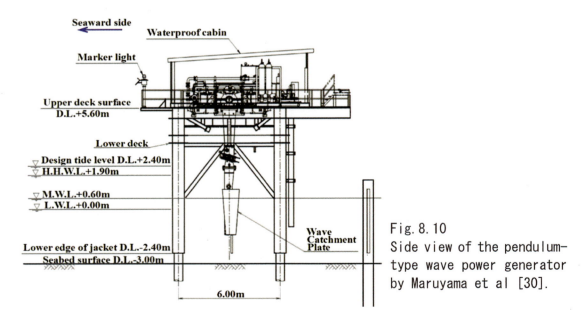

Fig. 8.10 Side view of the pendulum-type wave power generator by Maruyama et al [30].

In front of a breakwater, a steel structure with four pillars is built at a depth of about 6 m at mean tide level, on which a waterproof cabin is placed. A hydraulic power generator (a system in which the reciprocating motion of the wave plate due to waves is converted into oil flow by a ram cylinder, which is then converted into rotational motion by a swash plate oil motor to generate electricity) is installed in the cabin. Moreover, the wave-catching plate of the pendulum-type is hung from the rotating shaft

366

8.3 Widespread Use of Wave Power

of the device, which is designed so that even when 20-year probability waves (in this case, the maximum wave height 5.5 m, the maximum wave period 12 s) act on it, the torque of the hydraulic generator is less than the allowable value (102 t-m). A power generation efficiency of 30% has been achieved, with a target of 50%.

(v) Example of Mechanical Drive in a Fixed Installation (No. 3)
Shaft-type linear motor type

Gomyo et al [31] proposed a new wave power generation system in which a floating body located in a cylinder moves in conjunction with the vertical movement of waves, and a shaft connected to this floating body moves a generator in the slider as shown in **Fig. 8.11**. The shaft-type linear motor consists of a shaft with N- and S-poles arranged sequentially in a cylindrical shape and a coil-wound slider section. The linear motors are not custom-made, as they are converted from those used in other equipment. Although theoretically a sufficiently large output can be expected as regards the amount of power generation and efficiency, there are still issues for practical application.

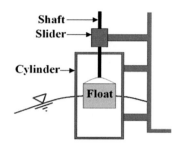

Fig. 8.11
Illustration of a wave power generator using a shaft-type linear motor by Gomyo et al [31].

(vi) Example of Pneumatic Drive in Floating Equipment
Floating and oscillating water column type

According to the White Paper on Renewable Energy Technology by the New Energy and Industrial Technology Development Organization (NEDO) [32], Yoshio Masuda developed the four-valve wave power generator shown in **Fig. 8.12**, after conducting demonstration tests of several types of floating wave power generation from the 1940s.

The reciprocating motion by waves causes a reciprocating motion of the air in the generator cavity, which is regulated into the airflow in the same direction by cleverly arranged valves 1 to 4 in this figure to rotate the turbine. A 500 W generator was used, and 400 W of power generation was recorded in waves of 2 m wave height off Yokosuka City, Kanagawa Prefecture. This generator was adopted

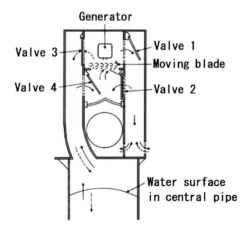

Fig. 8.12
Diagram of the four-sheet valve type wave power generator developed by Masuda.

367

8. ENVIRONMENTAL PROTECTION AND WAVE POWER GENERATION

by the Japan Coast Guard in 1965 to supply power to navigation beacon buoys, as shown in **Photo 8.5,** and the number of units in operation has now increased to more than 1000 in Japan and abroad.

(vii) Example of Pneumatic Drive in Floating Equipment (No. 2)
Oscillating water column type

The Japan Marine Science and Technology Center (JAMSTEC) took the oil crisis in 1973 as an opportunity to start full-scale development of a large-scale wave power generator using the same power generation method as (vi).

First, a floating experimental installation named **KAIMEI** (80 m long, 12 m wide, 5.3 m high, 2.1 m draft, equipped with three to eight impulse turbines of four-valve type, etc., see Ishii et al [33]), shown in **Photo 8.6**, was used to carry out demonstration experiments for the conversion of wave energy into electrical energy from 1976 to 1986 (as an international joint project with six Western countries since 1979) off the coast of Tsuruoka City, Yamagata Prefecture. The initial cost of electricity generation was 360 yen/kWh, but this led to an improvement method that can be expected 50 yen/kWh to generate..

Next, a continuous experimental installation named **Mighty Whale** (50 m long, 30 m wide, 12 m high with a draft of 8 m, three tandem-type Wells turbines, see Hotta et al [34]), shown in **Photo 8.7** and **Fig. 8.13**, was used to continue the technical improvement experiments from 1998 to 2002 offshore of Tokai County, Mie Prefecture.

In the **KAIMEI**, the longitudinal direction of the air chambers was parallel to the direction of wave travel and the air

Photo 8.5
Navigation beacon buoy using Masuda type wave power generator (from Ryokuseisha hormpage. https://www.ryokuseisha.com/product/beacon/index.html)

Photo 8.6
Floating wave power generation test facility **KAIME** by JAMSTEC (off the coast of Tsuruoka City, Yamagata Prefecture, Japan).

Photo 8.7
Floating wave power generation test facility **Mighty Whale** by JAMSTEC (off the coast of Tokai County, Mie Prefecture, Japan).

chambers had no bottom plate, but in the **Mighty Whale**, the longitudinal direction of the air chambers was perpendicular to the direction of wave travel and the air chambers had bottom plates. The **Mighty Whale** has improved the efficiency of wave energy absorption by a factor of three or more.

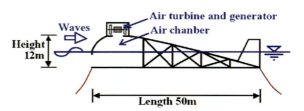

Fig. 8.13 Structural diagram of of **Mighty Whale** of JAMSTEC.

(viii) Example of Mechanical Drive in Floating Equipment
Point absorber wave energy converter (PA-WEC)

A point-absorber wave energy converter, as shown in **Fig. 8.14**, has a relatively simple structure that generates power by transmitting water surface fluctuations by waves into the heave motion of the float section and is largely submerged, offering advantages in terms of reliability and safety. By applying a control force from a linear generator mounted on the spar section, both the motion of the float section and the power output can be increased or decreased. The power output can be increased by optimising this control force, but when multiple PA-WECs are arranged, the hydrodynamic interferences between the PA-WECs must be considered for the optimisation of the control force (refer to Funada et al [35]).

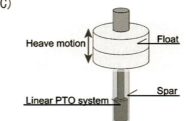

Fig. 8.14 Illustration of a PA-WEC by Funada et al [35].

One power generation device that is becoming commercially viable is the **PowerBuoy** developed by Ocean Power Technologies [36], shown in **Fig. 8.15**. A 40 kW class test unit was operated off the coast of New Jersey, USA, from 2005 to 2008, and has been in full-scale operation since August 2008, withstanding waves of more than 12 m. Moreover, in the same year, **PowerBuoy** was selected for a wave power plant (1.39 MW) of a joint venture between Iberdrola (the largest renewable energy company in the world) and the Spanish Government.

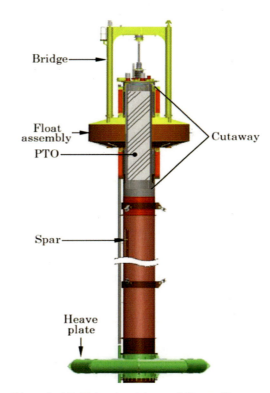

Fig. 8.15 Illustration of **PowerBuoy** developed by Ocean Power Technologies, Inc. [36].

8. ENVIRONMENTAL PROTECTION AND WAVE POWER GENERATION

(ix) Example of Mechanical Drive in Floating Equipment (No. 2)
Gyroscopic wave energy converter

Kanki [37] developed a wave power generator using the gyro effect. A rotating frame keeps turning without collapsing because it has the property of maintaining the direction of the rotational axis constant. Here, any attempt to tilt the rotational axis of a rotating object will result in a repulsive force. In **Fig. 8.16**, when a disc called a flywheel is rotated at high speed by using a motor, if the flywheel is tilted by waves, a rotational force called the gyro-moment is generated. This causes the gimbal, which covers the outside of the flywheel, to rotate, transferring the rotational energy of the gyro-moment to the generator and generating electricity.

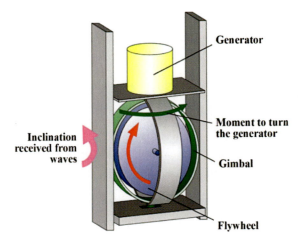

Fig. 8.16 Illustration of a gyroscopic wave energy converter by Kanki [37].

In short, when the rotational axis of a rotating object is tilted by waves, a moment is generated that resists the tilting and allows the object to continue rotating. This is used to generate electricity and is more efficient than common air turbine types. The demonstration experiments in Japan shown in **Fig. 8.17** were executed off the Tottori (Karo) port of Tottori Prefecture in 2004 - 2007 and off the Susami fishing port of Wakayama Prefecture in 2008 - 2010. As of 2010, 45 kW of electricity had been successfully generated.

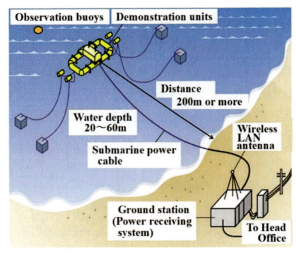

Fig. 8.17 Illustration of the field experiment in Japan by Kanki [37].

(x) Example of Overflow Drive in Floating Equipment
Overflow type wave energy converter (OT-WEC)

Tanaka et al [38] proposed a wave energy converter (overflow type), shown in **Photo 8.8** and **Fig. 8.18**, in which waves are overtopped into a vertical cylinder through a side slope and convergence channels installed near the water surface of the vertical cylinder. The overtopped water falls naturally to rotate a water wheel inside the vertical cylinder. A demonstration test of this system was carried out at the Yokohama port of Kanagawa Prefecture.

Photo 8.8 Demonstration experiment of the OT-WEC by Tanaka et al [38].

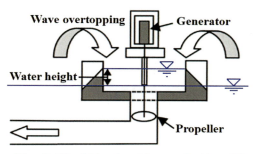

Fig. 8.18 Illustration of the OT-WEC by Tanaka et al [38].

(xi) Example of Artificial Muscle Type in Floating Equipment
Electroactive polymer artificial muscle type (EPAM)

Chiba et al [39], in collaboration with the SRI International (US research institute), have developed a new type of power generation device using an **Electroactive Polymer Artificial Muscle (EPAM)**. As shown in **Fig. 8.19**, the EPAM has a structure in which a thin polymer film made of acrylic or silicon resin is sandwiched between stretchable electrodes. When a potential difference is applied to the upper and lower electrodes, the upper and lower electrodes attract each other due to Coulomb forces, resulting in the polymer film becoming thin and elongating. Conversely, if the polymer membrane is stretched by the force of waves, electricity is generated (see **Fig. 8.20**). Demonstrations conducted in California, USA, have confirmed that 40 kg of EPAM can generate 1 to 2 kWh of electricity, even at an average annual wave height of 1 m.

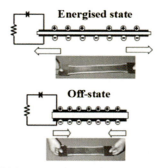

Fig. 8.19 Illustration of elongation by energisation and recovery by cut-off of EPAM by Chiba et al [39].

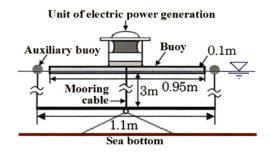

Vertical fluctuations in sea level cause the artificial muscles in the power generating unit to stretch and generate electricity.

Fig. 8.20 Illustration of EPAM wave power generator by Chiba et al [39].

8. ENVIRONMENTAL PROTECTION AND WAVE POWER GENERATION

(xii) Example of Piezoelectric Power Generation in Floating Equipment
    Piezoelectric element type

Mutsuda et al [40] developed a device that generates electricity by deforming piezoelectric elements by wave pressure.

Tokai University conducted model experiments using piezoelectric elements (see **Photo 8.9**) in a wave-making tank shown in **Photos 8.10** and confirmed that a power generation efficiency of around 20% could be achieved, but it was also found that durability improvement is essential for practical use.

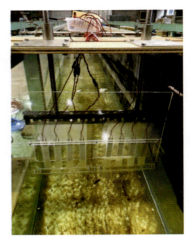

Photo 8.9 Piezoelectric elements used at Tokai University.

Photo 8.10 Device for generating electricity by fixing the top and bottom ends of piezoelectric elements and applying wave pressure to the centre of each element.

3) A Thought for the Spread of Wave Power and Small-scale Hydropower

The United Kingdom is currently the most committed country to wave and tidal power generation. The European Marine Energy Center established in 2003 is the first large-scale experimental facility in the world (refer to EMEC[41]). It was followed by the USA in the second place, Canada in the third, Norway the fourth, Denmark the fifth, and Sweden, Ireland, Japan, China and Australia the sixth place (in the case of Japan, wave power was actively researched in addition to hydropower in the 1900s, but since the government turned towards nuclear power at the end of the 20th century, the research and development in this field has declined since 2000).

To move away from fossil fuels as a measure against global warming and under conditions where concerns about the safety of nuclear power cannot be ruled out, the further spread of renewable energy sources is important, and measures to reduce the cost and increase the efficiency of wave power and small-scale hydropower to the same level as the power generated by mainstream fossil fuels are needed. One of the common reasons why wave power and small-scale hydropower are difficult to popularise, apart from being strongly influenced by natural conditions, is the high cost of production unless large orders of generating equipment are placed. Therefore, because river bed subsidence, estuarine blockage, coastal erosion, and harbour entrance/exit burial are becoming problems around the world, and requiring comprehensive management of the sediment budget in the sediment transport

8.3 Widespread Use of Wave Power

system from the upper reaches of rivers to the coasts, as a catalyst for the widespread use of wave power and small hydropower, the application of its power generation capacity to sediment transport from deposition zones to erosion zones in many sediment transport systems is proposed as following.

In developed countries, dams and harbours have been built at various locations to increase socio-economic activities. Although these facilities have effectively satisfied their construction objectives, they have interfered with the supply of sediment downstream, resulting in serious erosion of the downstream side. There are many cases where sediment dredging (including sand and gravel extraction) was carried out to maintain the function of dams and ports, but the dredged sediment was not discharged downstream, resulting in serious riverbed and coastal erosion. Therefore, it can be proposed that integrated sediment management from the river to the sea should be implemented by artificially moving sediment accumulated in areas where it is not allowed to accumulate to the lower erosion areas. The active use of wave power and small-scale hydropower as a power source for this sediment transport would lead to their widespread use.

(1) Case Study for Small-scale Hydropower

There may be many rivers where the installation of small-scale hydropower plants (plants with an output of 1000 kW or less that do not require dam construction) has not been considered. As small hydropower generation technology is well established, it can be popularised by allocating funds and developing policies that encourage such developments. If the river administrator acts as the project owner and the project is recognised as part of an existing river conservation project, securing costs and land will not be much of a problem at least in Japan, but conflicts with existing water and electricity business rights must be avoided. If it is widely known that the project will lead to the reduction of greenhouse gas emissions, the promotion of small-scale hydropower generation will be accelerated.

(a) Possible Amount of Electricity Generated by Water Turbines

The amount of electricity P_w [kW] generated by a water turbine can be obtained from Eq. (8.1). The lost water depth or lost hydraulic head is determined using the Darcy-Weisbach equation and others. In addition, the flow velocity v [m/s] in an open channel can be obtained from mean velocity formulae such as Manning's equation and it in a pipeline can be obtained from Eq. (8.2).

$$P_w = gQh_e \times \eta \tag{8.1}$$

$$v = \sqrt{\frac{2gh}{1 + f(l/d)}} \tag{8.2}$$

where g is the acceleration of gravity [m/s²], Q is the flow rate of the water supply [m³/s], h_e is the net head [m] considering losses, η is the efficiency, h is the head, f is the friction loss coefficient, l is the length of the pipeline and d is the inner diameter of the pipeline.

373

8. ENVIRONMENTAL PROTECTION AND WAVE POWER GENERATION

As can be understood from Eq. (8.1), it is important to secure a sufficient flow rate and a sufficient net head to generate enough electricity. Therefore, an alluvial fan with a slope of 1/25 to 1/20, where an elevation difference of 4 m to 5 m can be secured between 100 m horizontal distance, on a river where a stable flow rate of more than about 1 m³/s can be expected even during drought periods, is selected as the construction site (**Fig. 8.21**). Small-scale hydropower generation facilities are then installed at two locations on either side of the selected river area. Moreover, in order not to reduce the cross-sectional area of the river, water is drawn in by a channel of about 100 m. A 4 - 5 m drop is secured to act on the turbines. After electricity is generated, water is returned to the river via a channel (discharge channel length: approx. 50 m) to avoid problems relating to water rights.

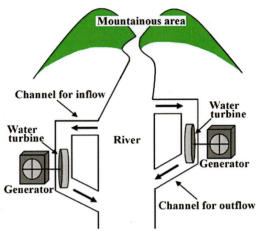

Fig. 8.21 Illustration of small hydropower facilities.

Assuming the channels are pipelines, $h = 4 - 5$ m, $f = 0.02$, $l = 100$ m from the river to the turbines, $d = 0.6$ m, $Q \geq 1$ m³/s (from $v \geq 4$ m/s calculated using Eq.(8.2)) and $\eta = 0.7$, and substituting into Eq. (8.1), the two turbine power stations will generate at least 15 kW of electricity. Assuming an effective operating period of 300 days per year, the annual cumulative power generation would be more than 390 GW×year (= 15kW× 3600s×24hr×300 days). As will be explained in more detail later, more than 180,000 m³ of sediment can be transported with this amount of power generation. In reference, the annual sediment deficit in rivers that leads to downstream coastal erosion is typically between 50,000 m³ and 300,000 m³ per river.

(b) Costs of Hydropower Generation Plants

The construction costs and 50-year operation/repair costs of the above small-scale power generation facilities will be calculated as follows in Japan in 2020. However, no land acquisition costs shall be incurred as the facilities will be located on river management land.

(i) Construction Costs for Two Hydropower Generation Plants
① Water turbine costs: 15,000,000 yen (per site),
② Generator costs: 40,000,000 yen (ditto),
③ Shed and other facilities: 20,000,000 yen (ditto),
④ Construction costs for two locations: 150,000,000 yen {= (①+②+③) ×2}.
The costs of ① to ③ will be reduced if it becomes possible to order large quantities.

8.3 Widespread Use of Wave Power

(ii) Construction Costs of Water Pipes and Electric Lines

⑤ Unit cost of pipe installation: 400,000 yen (per meter),

⑥ Pipe construction cost for the 150 m length: 120,000,000 yen {= (⑤×150 m)×2}.

⑦ Unit cost of wire installation: 25,000 yen (per meter),

⑧ Wire construction cost for the 1,000 m length: 50,000,000 yen {= (⑦×1,000 m)×2}.

(iii) Operation and Repair Costs over 50 Years

⑨ Unit cost of labour and miscellaneous utilities: 8,000,000 yen (per year),

⑩ Unit repair cost: 7,600,000 yen (per year, 2% of construction cost for durable materials,

    10% of the purchase cost for a water turbine and a generator),

⑪ 50-year operation and repair costs: 1,560,000,000 yen {= (⑨+⑩)×50 years ×2}.

From the above ④+⑥+⑧+⑪, the construction costs of the two sites and the 50-year operation and repair costs are estimated to be approximately 1.88 billion yen.

(2) Case Study for Wave Power

Compared with photovoltaic and wind power, wave energy which utilises the kinetic energy of water, is not affected by sunlight hours. Since water has a heavier specific gravity than air, a large amount of power can be generated if energy can be generated efficiently.

Moreover, wave power generation methods with high energy conversion efficiency are continuously being developed in high latitude marine states. For example, a new type of wave power generation system developed by the German company NEMOS [42] in 2019, used the motion of floating bodies caused by waves. This technology is easier to install and less expensive than conventional types, and the construction period can also be shorter. In addition, the size can be adjusted to suit the environment, thus maximising efficiency. Development is currently underway to increase energy conversion efficiency to 70%. If the development of such excellent power generators progresses and they become widespread, large amounts of power generated at low cost can be expected.

(a) Possible Amount of Electricity Generated by Wave Power

The amount of power generated per unit width P_w [kW/m] can be obtained using Eq. (8.3).

$$P_w = \left(\frac{1}{8}\rho g H^2\right) \times C_g \times \eta \tag{8.3}$$

where ρ is the density of seawater [t/m³], H is the incident wave height, C_g is the velocity of the wave group and η is the efficiency.

Using the incident wave height and wave group velocity of offshore waves and with $\eta \approx 0.5$, the following calculation equation for offshore waves is obtained.

$$P_w \approx 0.5 \times H_o^2 \times T \tag{8.4}$$

8. ENVIRONMENTAL PROTECTION AND WAVE POWER GENERATION

where H_o is the offshore wave height and T is the wave period.

Furthermore, using the incident wave height and wave group velocity at the point of wave breaking where the breaking wave height is 0.6 times the water depth, the following calculation equation for breaking waves is obtained.

$$P_w \approx 0.161 \rho g^{1.5} \times H_b^{2.5} \times \eta \qquad (8.5)$$

where H_b is the breaking wave height.

Takahashi and Adachi [43] and Katayama et al [44] have estimated the wave power generation capacity due to offshore waves across Japan using Eq. (8.4). **Table 8.6** gives the power generation capacity calculated from Eq. (8.5) using annual average wave height for 2018 obtained by wave stations of the Port and Airport Research Institute (PARI), Japan.

Table 8.6 Calculated wave power based on wave height data for 2018 of PARI.

Place name	Mean breaking wave height [m]	Electricity generated for one hour [kW×hr/m]	Coast length [km]	Total electricity generated for one hour [GW×hr]
Hokkaido				
Rumoi	1.232	30,927	400	12,371
Setana	1.276	33,763	222	7,495
Tomakomai	0.836	11,731	310	3,637
Subtotal			932	23,503
Sea of Japan side				
Sakata	1.298	35,237	310	10,923
Niigata (offing)	1.023	19,431	150	2,915
Wajima	1.232	30,927	60	1,856
Kanazawa	1.144	25,697	130	3,341
Tottori	1.078	22,149	170	3,765
Hamada	1.177	27,590	175	4,828
Ainoshima (Kyushu)	0.682	7,051	95	670
Subtotal			1090	28,298
Pacific Ocean side				
Shibushiwan (Kyushu)	0.682	7,051	50	353
Murotsu (Shikoku)	0.847	12,121	130	1,576
Kobe	0.308	966	33	32
Shionominaki	1.276	33,763	215	7,259
Kashima	1.397	42,345	95	4,023
Onahama	1.232	30,927	180	5,567
Sendai-shinkou	0.935	15,518	190	2,948
Kamaishi	0.88	13,336	120	1,600
Hachinohe	1.21	29,565	160	4,730
Subtotal			1173	28,088
Okinawa				
Nase	1.221	30,241	100	3,024
Naha	0.946	15,979	30	479
Nakagusuku-wan	1.265	33,040	45	1,487
Subtotal			175	4,990
Grand total				84,879

8.3 Widespread Use of Wave Power

The reason why offshore wave heights were not used in this calculation is that as waves approach the shore from offshore, wave heights are affected by the topography, and therefore the wave heights at the location of a power generation facility are different from the offshore wave heights. Therefore, the actual power generation capacity is also different. In general, the shallower the water depth, the greater the wave height due to wave shoaling. Wave height (and hence wave power generation potential) reduces after wave breaking. Therefore, it is efficient to locate the power generation facility at the wave breaking position. However, as actual waves are irregular, the power generation facility and appliance can be destroyed if they are subjected to waves that are too large. So, the power generation facility should be installed in water depth where all waves stronger than the design strength are broken.

Furthermore, since the power generation value is calculated using the mean wave height, which is smaller than the energy mean wave height, it is considered less likely that the actual amount of electricity generated will be less than the calculated value.

The capacity of electricity generation during one hour over the total length of the coastline of all sites of 3,370 km was approximately 84,879 GW×hr. If the power generation efficiency is 40% then, the total amount of power generated will be 33,952 GW×hr. If the effective operating period is 300 days per year, the annual cumulative electricity generated is approximately 2.44×10^8 GW×year (=33,952GW ×24hr×300day).

The top-supported pendulum-type wave power generation facility of Maruyama et al [30] can be considered an example of a wave power generation facility with high power generation efficiency in operation in Japan. Power generated by one wave power generation station using the approach is 9.63 kW from Eq. (8.5) under conditions of the installation depth 4.0 m, the wave plate width 3.0 m, the average wave breaking height 1.2 m, and the efficiency 0.4. Assuming an effective operating period of 300 days per year, the annual cumulative electricity generation is approximately 250 GW×year (= 9.63kW×3600s×24hr×300day).

(b) Cost of a Wave Power Generation Plant

The construction cost and 50-year operation/repair cost of a top-supported pendulum-type wave power generation facility is as follows in Japan in 2020. However, no land acquisition costs incurred as the facilities will be located on coast management land.

(i) Construction Cost for a Wave Power Generation Plant

① Wave plate cost: 12,000,000 yen,

② Generator cost: 40,000,000 yen,

③ Shed and other facilities: 100,000,000 yen,

8. ENVIRONMENTAL PROTECTION AND WAVE POWER GENERATION

④ Construction cost for one location: 152,000,000 yen {= ①+②+③}.

The costs of ① to ③ will be reduced if it becomes possible to order large quantities.

(ii) Operation and Repair Costs over 50 Years

⑤ Unit cost of labour and miscellaneous utilities: 8,000,000 yen (per year),

⑥ Unit repair cost: 7,200,000 yen (per year, 2% of construction cost for durable materials,

10% of purchase cost for a wave plate and a generator),

⑦ 50-year operation and repair costs: 760,000,000 yen {= (⑤+⑥)×50 years}.

From the above ④+⑦, the construction cost of the one site and the 50-year operation and repair costs are estimated to be approximately 910 million yen.

(3) Proposals for Economical Integrated Sediment Management

Sediment accumulated in reservoirs formed by constructing dams across rivers not only reduces the effective storage capacity and leads to reduced dam functionality, but also accelerates coastal erosion. Therefore, sediment should be artificially discharged downstream. In alluvial fan areas, the flow velocity of the river abruptly decreases, thus accumulating sediment on riverbeds. In the outer bays of meandering channels, the flow velocity is faster, and riverbed erosion occurs. Those fluctuations of the riverbed should be controlled from the point of view of flood hazard reduction. In addition, the safety of coasts is greatly diminished when beach erosion progresses, since coastal beaches are natural wave-absorbing facilities that protect lives and property on land from coastal disasters. Therefore, the supply of sediment to beach erosion sites is also extremely important.

Based on the above, it is important for administrators of rivers and coasts to take comprehensively sediment transport systems from the mountains to the coasts, and to manage sediment transport comprehensively in cooperation with the various agencies concerned.

(a) River Sediment Management Using Hydropower

A water turbine power generation for river sediment management is explained as follows.

Usually, there are several places in a river where sediment from the sedimentation areas should be moved to the erosion areas. Three places are assumed here as shown in **Fig. 8.22**. The sediment from the sedimentation area is suctioned using submersible pumps in a slurry state including water, and discharged onto the erosion area, using power generated by a water turbine. For suction and discharge, several pumps with moving units are used, and these three pumps are considered as one unit.

If a reasonable lift of 4 m, a sediment transport drop of 2 m, a standard transport distance of 700 m, an easily available upper limit of 0.6 m for the inner diameter of the transport pipe, a standard friction loss coefficient of 0.02, a reasonable slurry specific gravity of 1.2 and a standard efficiency of

8.3 Widespread Use of Wave Power

0.7 are used, the flow velocity and flow rate will be 1.27 m/s and 0.36 m³/s respectively. The suction and discharge power required per hour for one unit would be 385.1 MW×hr.

If the amount of sediment to be moved per location is set at 60,000 m³, the amount of sediment to be moved to the three selected sites in one river will be 180,000 m³ (assuming an average value of 50,000 m³ to 300,000 m³ of problematic sediment in one river). Under the above suction and discharge conditions, the time required to move all 180,000 m³ of sediment will be 39.9 days.

Therefore, the annual cumulative electricity required for suction and discharge for these 39.9 days will be approximately 369 GW×year (≈385.1MWh×24hr×39.9day), which is less than the annual cumulative power generation of the two water turbine power stations with a capacity of approximately 390 GW×year mentioned above.

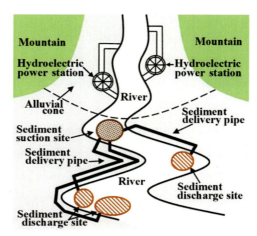

Fig. 8.22
Illustration of a river sediment management system beased on small-scale hydropower.

However, if the power stations are operated continuously for 39.9 days, large suction holes are expected to appear, so the operation should be carried out intermittently, depending on the condition of the suction holes, etc.

The construction cost of one pump suction/discharge unit and the 50-year operation/repair costs will be as follows in Japan in 2020. However, no land acquisition costs incurred as the facilities will be located on river management land.

(i) Construction Cost of One Suction/discharge Unit

① Length: 700m (in one place),
② Unit cost of pipe installation: 500,000 yen (per meter),
③ Cost of suction/discharge pumps (6 units): 15,000,000 yen (per site),
④ Cost of suction/discharge mobiles and installation (two sets): 100,000,000 yen (per site),
  If the suction/discharge pumps are fixed, large holes are created at the suction points and mountains at the discharge points. To prevent this, electrically movable devices are installed.
⑤ Construction cost of control shed etc.: 20,000,000 yen (per site),
⑥ Construction cost of one suction/discharge unit: 1,455,000,000 yen {=(①×②+③+④+⑤)×3}.
  ②～⑤ can be expected to be reduced if large orders can be placed.

8. ENVIRONMENTAL PROTECTION AND WAVE POWER GENERATION

(ii) Operation and Repair Costs over 50 Years

⑦ Unit cost of labour and miscellaneous utilities: 8,000,000 yen (per year and per site),

⑧ Unit repair cost: 18,900,000 yen (per year and per site, 2% of the construction cost for durable materials, 10% of costs for pumps and mobiles),

⑨ 50-year operation and repair costs: 4,035,000,000 yen{(⑦+⑧)×50×3}.

From (⑥+⑨), the cost of one pump suction/discharge unit over 50 years is estimated to be approximately 5.49 billion yen.

On the other hand, if sediment is moved using conventional methods such as power shovels or backhoes and dump trucks to transport 180,000 m³ of sediment in one river by 700 m using 10-ton dump trucks, 120 m³ can be transported per day. If the cost of the dump truck + driver + excavator + operator is 130,000 yen/day (which is underestimated in comparison with the proposal), the cost per year will be 195 million yen (=180,000÷120×130,000 yen), and the total cost for 50 years will be approximately 9.75 billion yen (=195,000,000yen/year ×50years), which is significantly higher than the cost of using water turbines (the construction, operation and repair costs of two hydropower facilities + the construction, operation and repair costs of one suction/discharge unit = 1.88 billion yen + 5.49 billion yen =7.37 billion yen).

(b) Coastal Sediment Management Using Wave Power

A proposal for using wave power generation for coastal sediment management is explained as follows.

On one coast, there may be more than one place where sediment from a sedimentary area should be moved to an erosion area, so two locations are assumed.

Sediment from the sedimentation area is suctioned by submersible pumps in the slurry state including water, and discharged onto the erosion zone, using electricity from wave power generation as shown in **Fig. 8.23**. For suction and discharge, pumps with multiple moving units are used, and two pumps are considered as one unit.

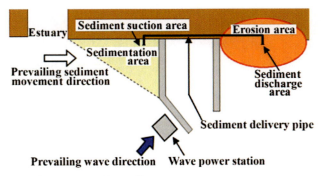

Fig. 8.23 Illustration of a coastal sediment management system using wave power.

8.3 Widespread Use of Wave Power

A reasonable lift of 4 m, a sediment transport drop of 2 m, a transport distance of 700 m, an easily available upper limit of 0.6 m for the inner diameter of the transport pipe, a standard friction loss coefficient of 0.02, a slurry specific gravity of 1.2 and a standard efficiency of 0.7 give a flow velocity of 1.27 m/s and a flow rate of 0.36 m³/s, and the suction and discharge power for one hour per unit will be 256.8 MW×hr.

If the amount of sediment to be moved per location is set at 90,000 m³, the amount of sediment to be moved in one coast will be 180,000 m³ (assuming an average value of 50,000 m³ to 300,000 m³ of problematic sediment in one coast). Under the above suction and discharge conditions, the time required to move 180,000 m³ of sediment will be 39.9 days.

The annual cumulative electricity required for suction and discharge for these 39.9 days is approximately 246 GW×year (≈256.8MWh×24hr×39.9day), which is less than the annual cumulative power generation capacity of the one wave power station (approximately 250 GW×year). However, if the power station is operated continuously for 39.9 days, a large suction hole is expected to appear. Therefore, the operation should be carried out intermittently, depending on the condition of the suction hole, etc, similar to the case of moving river sediment.

The construction cost of one pump suction/discharge unit and the 50-year operation/repair costs will be as follows in Japan in 2020. However, no land acquisition costs incurred as the facilities will be located on coast management land.

(i) Construction Cost of One Suction/discharge Unit

① Length: 700m (in one place),

② Unit cost of pipe installation: 500,000 yen (per meter),

③ Cost of suction/discharge pumps (6 units): 15,000,000 yen (per site),

④ Cost of suction/discharge mobiles and installation (two sets): 120,000,000 yen (per site),
   If the suction/discharge pumps are fixed, large holes are created at the suction points and mountains at the discharge points. To prevent this, electrically movable devices are installed.

⑤ Construction cost of control shed etc.: 20,000,000 yen (per site),

⑥ Construction cost of one suction/discharge unit: 1,010,000,000 yen {=(①×②+③+④+⑤)×2}.
   ②〜⑤ can be expected to be reduced if large orders can be placed.

(ii) Operation and Repair Costs over 50 Years

⑦ Unit cost of labour and miscellaneous utilities: 8,000,000 yen (per year and per site),

⑧ Unit repair cost: 20,900,000 yen (per year and per site, 2% of the construction cost for durable materials, 10% of costs for pumps and mobiles),

⑨ 50-year operation and repair costs: 2,890,000,000 yen {(⑦＋⑧)×50×2}.

8. ENVIRONMENTAL PROTECTION AND WAVE POWER GENERATION

From (⑥+⑨), the cost of one pump suction/discharge unit over 50 years is estimated to be approximately 3.90 billion yen.

On the other hand, if sediment is moved by the conventional method using dump trucks, the total cost (for 50 years) of moving 180,000 m³/year of sediment similar to river sediment movement is approximately 9.75 billion yen.

If a measure for coastal erosion prevention by using detached breakwaters is selected, in the case that the standard lengths of the subject coast and each detached breakwater are 2,000 m and 100 m, the necessary number of detached breakwaters is 13 based on the open length of 50 m between each detached breakwater, the total cost for 50 years is approximately 14.63 billion yen (= 450,000,000yen×(1+0.03×50year)×13) based on the standard construction cost of 450,000,000 yen and the maintenance ratio of 3% to each detached breakwater.

Comparing the implementation cost due to dump trucks and the total cost of detached breakwaters with the total cost of approximately 4.81 billion yen (= 0.91 billion yen + 3.90 billion yen) for one wave power generation facility and one pump suction/discharge unit, sediment management using wave power generation is sufficiently cheaper and more advantageous in terms of environmental protection.

In this proposal, since various kinds of unit prices change with a country or time, it is not necessary to focus on how high the total cost of a single project will be. It has been pointed out that there is a high possibility that integrated sediment management from rivers to coasts by hydropower and wave power will be useful.

List of References in Chapter 8

1) United Nations World Water Assessment Programme (WWAP): *The United Nations World Water Development Report 2017. Wastewater: The Untapped Resource*, Paris, UNESCO, 2017.

2) Malone, T.C. (Convener), Ambulker, A., Bebianno, M.J. (Co-lead member), Bontempi, P., Krom, M., Kuosa, H., Montoya, J., Newton, A., Ossey, Y., Yunes, J.S., Smith, W., Sonesten, L., Sylaios, G., Wang, J. (Lead member), and Yin, K: Chapter 10. Changes in Nutrient Inputs to the Marine Environment, *The Second World Ocean Assessment,* Vol. 2, United Nations, 2021, pp.77-100.

3) Galgani, F. (Convener: marine debris), Stöfen-O'Brien, A. (Convener: dumping), Ambulkar, A., Azzaro, M., Bebianno, M.J. (Lead member), Bondareff, J., Griffiths, H., Hassellov, M., Ioakeimidis, C., Jambeck, J., Keener, P., de Oliveira Lana, F., Makarenko, I., Rochman, C., Schuyler, Q., Sobral, P., Vu, T.C. (Co-lead member), Topouzelis, K., Vethaak, D., Vlahos, P., Wang, J. (Co-lead member), and Weis, J.: Chapter 12. Changes in Inputs and Distribution of Solid Waste, Other than Dredged Material, in the Marine Environment, T*he Second World Ocean Assessment*, Vol. 2, United Nations, 2021, pp.151-184.

4) Joint Group of Experts on the Scientific Aspects of Marine Environmental Protection (GESAMP; editors, Kershaw, P.J. and Rochman, C.M.): *Sources, Fate and Effects of Microplastics in the Marine Environment, Part 2 of a Global Assessment*, GESAMP Report and Studies Series, No. 93, IMO/FAO/UNESCO-IOC/UNIDO/WMO/IAEA/UN/UNEP/UNDP, 2016.

5) Joint Group of Experts on the Scientific Aspects of Marine Environmental Protection (GESAMP; editors, Kershaw, P., Turra, A. and Galgani, F.): *Guidelines for the Monitoring and Assessment of Plastic Litter and Microplastics in the Ocean*, GESAMP Report and Studies Series, No. 99, IMO/FAO/UNESCO-IOC/UNIDO/WMO/IAEA/UN/UN Environment/UNDP/ISA, 2019.

6) Lebreton, L., van der Zwet, J., Damsteeg, J., Slat, B., Andrady, A. and Reisser, J.: *River Plastic Emissions to the World's Oceans*. Nature Communications, Vol. 8, Article 15611, 2017. Doi: 10.1038/ncomms15611

7) Van Emmerik, T., Kieu-Le, T., Loozen, M., van Oeveren, K., Strady, E., Bui, X., Egger, M., Gasperi, J., Lebreton, L., Nguyen, P., Schwarz, A., Slat, B. and Tassin, B.: *A Methodology to Characterize Riverine Macroplastic Emission Into the Ocean*, Frontiers in Marine Science, Vol. 5, 2018. https://doi.org/10.3389/fmars.2018.00372

8) Schmidt, C., Krauth, T. and Wagner, S.: Export of Plastic Debris by Rivers into the Sea, *Environmental Science and Technology*, vol. 51, No. 21, 2017, pp.12246–12253. https://pubs.acs.org/doi/full/10.1021/acs.est.7b02368

9) Ministry of the Environment: *Environmental Quality Standards for Water*, Japanese government. https://www.env.go.jp/content/900454947.pdf

8. ENVIRONMENTAL PROTECTION AND WAVE POWER GENERATION

10） Abo, K. and Yamamoto, T.: Oligotrophication and Its Measures in the Seto Inland Sea, Japan, *Bull. Jap. Fish. Res. Edu. Agen.*, No. 49, 2019, pp.21－26.

11） Cole, S., Codling, I.D., Parr, W. and Zabel, T.: *Guidelines for Managing Water Quality Impacts within UK European Marine Sites*, prepared by WRc Swindon Frankland Road Blagrove Swindon Wiltshire SN5 8YF for the UK Marine SACs Project, 1999, 449p. http://ukmpa.marinebiodiversity.org/uk_sacs/pdfs/water_quality.pdf

12） Sasaki, J., Imai, M. and Isobe, M.: Prediction Model for Dissolved Oxygen Concentration in Inner Bays, *Journal of Coastal Engineering*, Vol. 44, 1997, pp.1091-1095. (in Japanese)

13） Sasaki, J., Ishii M., and Isobe M.: Development of High-Resolution Flow Model in Inner Bay and Long-Term Reproduction Calculation in Tokyo Bay, *Journal of Coastal Engineering*, Vol. 45, 1998, pp.406-410. (in Japanese)

14） Sasaki, J., Isobe, M. and Fujimoto H.: Development of a Simple Blue Tide Prediction Method in Tokyo Bay, *Journal of Coastal Engineering*, Vol. 46, 1999, pp.1006-1010. (in Japanese)

15） Shigematsu, T., Mizuta, K. and Endo, T.: A Study on Remediation Index of Sea Bottom Environment and Its Estimation Method by Field Investigation, *Journal of JSCE*, B2 (Coastal Engineering), Vol.66, No.1, 2010, pp.1031-1035. (in Japanese)

16） Ministry of the Environment: *Results of the 2019 Marine Litter Survey*, Japanese government. https://www.env.go.jp/press/108800.html

17） Van Sebille, E., Wilcox, C., Lebreton, L., Maximenko, N., Hardesty, B.D., van Franeker, J.A., Eriksen, M., Siegel, D., Galgani, F. and Law, K.L.: A Global Inventory of Small Floating Plastic Debris. *Environmental Research Letters*, 10, 124006, 2015. Doi:10.1088/1748-9326/10/12 /124006

18） Jambeck, J., Geyer, R., Wilcox, C., Siegler, T.R., Perryman, M., Andrady, A.L., Narayan, R. and Law, K.L.: Plastic Waste Inputs from Land into the Ocean, *Science*, Vol.347, No.6223, 2015, pp. 768-771. Doi:10.1126/science.1260352

19） Law, K. L.: Plastics in the Marine Environment, *Annual Review of marine Science*, Vol. 9, 2017, pp.205-229. https://doi.org/10.1146/annurev-marine-010816-060409

20） Yamamoto, Y. and Minami, N.: Research on Control of Odor due to Decomposition of Drifting-ashore Seaweed, *Journal of Civil Engineering in the Ocean*, JSCE, vol.25, 2009, pp.975-980. http://library.jsce.or.jp/jsce/open/00011/2009/25-0975.pdf (in Japanese)

21） Yamamoto, Y. and Nariyoshi, K: Influence of Temperature and Humidity to Odor Control Method of Drifting-ashore Seaweed, *Journal of Civil Engineering in the Ocean*, JSCE, vol.26, 2010, pp. 933-938. http://library.jsce.or.jp/jsce/open/00011/2010/26-0933.pdf (in Japanese).

22) IPCC: Sections In: *Climate Change 2023 Synthesis Report. Contribution of Working Groups I, II and III to the Sixth Assessment Report of the Intergovernmental Panel on Climate Change* [Core Writing Team, Lee, H. and Romero, J. (eds.)]. IPCC, Geneva, Switzerland, 2023, pp.35-115.

Doi: 10.59327/IPCC/AR6-9789291691647

23) Chen, J., Dai, A. The Atmosphere Has Become Increasingly Unstable During 1979–2020 Over the Northern Hemisphere. *Geophysical Research Letters*, Vol. 50, Issue 20, 2023, e2023GL106125.

24) Allen, J. T.: *Climate Change and Severe Thunderstorms*, In Oxford Research Encyclopedia of Climate Science, 2018. https://doi.org/10.1093/acrefore/9780190228620.013.62.

25) IPCC: *Climate Change 2013: The Physical Science Basis. Contribution of Working Group I to the Fifth Assessment Report of the Intergovernmental Panel on Climate Change* [Stocker, T.F., Qin, D., Plattner, G.-K., Tignor, M., Allen, S.K., Boschung, J., Nauels, A., Xia, Y., Bex, V. and Midgley, P.M. (eds.)], Cambridge University Press, Cambridge, United Kingdom and New York, NY, USA, 2013, 1535p.

26) Hotta, H., Washio, Y., Masuda, Y., Ishii S., Miyazaki, T. and Kudo, K.: Power Generation Operation Test by Coastal Stationary Wave Power Generating Units, *Proc. of the 32nd Coastal Engineering conference*, 1985, pp.702-706.
http://library.jsce.or.jp/jsce/open/00008/1985/32-0702.pdf (in Japanese)

27) Kihara, K., Hoskawa, Y., Oosawa, H., Shimosako, K., Masuda, K., Ikoma, T., Nagata, S. and Kanaya, Y.: Field Experiment on Wave Energy Convertor of a Pw-owc Type in the Port of Sakata, Japan, *Journal of Civil Engineering in the Ocean*, JSCE, Vol.71, No.2, pp. I_103-I_108, 2015.
https://www.jstage.jst.go.jp/article/jscejoe/71/2/71_I_103/_pdf/-char/ja (in Japanese)

28) Takahashi, S.: Hydrodynamic Characteristics of Wave-Power-Extracting Caisson Breakwater, *Proc. of 21st Conference on Coastal Engineering*, ASCE, 1988, pp.2489-2503.
https://doi.org/10.1061/9780872626874.18

29) Tamaki, H., Abe, M. and Agawa, T.: Study on the Energy Conversion Efficiency of a Movable Wave Rudder Type Wave Power Generator, *Journal of Civil Engineering in the Ocean*, JSCE, Vol. 3, 1987, pp.29-34. https://doi.org/10.2208/prooe.3.29 (in Japanese)

30) Kobayashi, H., Rheem, C. and Maruyama, K.: Field Demonstration of Pendulum Type Wave Power Generation Syatem (Wave Rudder), *Journal of JSCE*, B2 (Coastal Engineering), Vol.73, No.2, 2017, pp. I_1453-I_1458. https:// www.jstage.jst.go.jp/article/kaigan/73/2/73_I_1453/_pdf (in Japanese) or https://seasat.iis.u-tokyo.ac.jp/rheem/test_site.pdf (in English)

31) Gomyo, M. and Horisawa M.: Fundamental Study on Wave Power Generation Using Shaft-type Linear Motors, *Journal of Coastal Engineering*, Vol. 52, 2005, pp.1241-1245.
https://www.jstage.jst.go.jp/article/proce1989/52/0/52_0_1241/_pdf/-char/ja (in Japanese)

32) New Energy and Industrial Technology Development Organization (NEDO): *White Paper on Renewable Energy Technology, 2nd Edition, Chapter 6 Current Status and Roadmap of Wave Power Generation Technologies*, 2014. https://www.nedo.go.jp/library/ne_hakusyo_index.html (in Japanese)

8. ENVIRONMENTAL PROTECTION AND WAVE POWER GENERATION

33) Ishii S., Masuda, Y., Miyazaki, T., Kudo, K., Hotta, H., Washio, Y. and Tsuzuki, T.: Report on the Actual Sea Area Experiment with the Wave Power Generator "KAIMEI", *Journal of Civil Engineering in the Ocean,* Vol. 2, 1986, pp.31-36.
https://www.jstage.jst.go.jp/article/prooe1986/2/0/2_0_31/_pdf (in Japanese)

34) Hotta, H., Washio, Y., Yokozawa, H. and Miyazaki, T.: R&D on Wave Power Device "Mighty Whale", *Renewable Energy*, Elsevier, Vol.9, Issues 1-4, 1996, pp.1223-1226.
https://www.sciencedirect.com/science/article/abs/pii/0960148196884977

35) Funada, J., Murai, M. and Li, Q.: Research on How the Arrangement of Arrayed PA-WECs Influences to Their Electric Power Generation, *Journal of the Japan Society of Naval Architects and Ocean Engineers*, Vol.31, 2020, pp.59-71.
https://doi.org/10.2534/jjasnaoe.31.59 (in Japanese)

36) Ocean Power Technologies, Inc.: *PB3 PowerBuoy*, 2024.
https://oceanpowertechnologies.com/platform/opt-pb3-powerbuoy

37) Kanki, H.: *Development of a Highly Efficient Gyroscopic Wave Power Generation System*, Japan Science and Technology Agency, 2004.
https://jglobal.jst.go.jp/en/detail?JGLOBAL_ID=202104016914242108 (in Japanese)

38) Tanaka, H., Yodogawa, M. and Manabe, Y.: *Development of the Wave-power Generation Device by the Floating Beach Device (the Conical Floating-body) with Wave-absorbing and Water-flow Features*, KAKEN, 2008. https://kaken.nii.ac.jp/ja/grant/KAKENHI-PROJECT-18560813 (in Japanese)

39) Chiba, S., Waki, M., Hirakaw, Y., Masude, K. and Ikoma, T.: Consistent Ocean Wave Energy Harvesting Using Electroactive Polymer (Dielectric Elastomer) Artificial Muscle Generators, *Applied Energy*, ELSEVIER, Vol.104, 2013, pp.497-502.
https://doi.org/10.1016/j.apenergy.2012.10.052

40) Mutsuda, H., Kawakami, K., Kurokawa, T., Doi, Y. and Tanaka, Y.: A Technology of Electrical Energy Generated from Ocean Power Using Flexible Piezoelectric Device, *Proc. 29th ICOOAE*, ASME, Vol.3, 2010, pp.313-321. https://doi.org/10.1115/OMAE2010-20103

41) EMEC: *the European Marine Energy Centre LTD.* https://www.emec.org.uk

42) NEMOS: *NEMOS 2019 Wave Energy Converter Prototype.* https://nemos.de/wec-2019-media

43) Takahashi, S. and Adachi, T.: *Wave Power around Japan from a Viewpoint of Its Utilization*, Technical Note of the Port and Harbour Research Institute, Japanese Ministry of Transport, No.654, 1989, 23P. (in Japanese)

44) Katayama, H., Yoneyama, H. and Shimosako, K.: Evaluation of Wave Power Around Japan for Appropriate Place Selection of Wave Power Converter, *Journal of JSCE*, B3 (Ocean Development), Vol.70, No.2, 2014, pp. I_73-I_78.
https://www.jstage.jst.go.jp/article/jscejoe/70/2/70_I_73/_pdf/-char/ja (in Japan)

Appendix

MANUALS OF NUMERICAL PREDICTION MODELS

1. Numerical Model for Shoreline Change Due to Waves ------------------ 388
   The program and samples of input data (Shoreline.zip) can be downloaded from
   https://www.rikohtosho.co.jp/main/wp-content/uploads/Shoreline.zip
2. Numerical Model for Beach Change Due to High Waves ----------------- 410
   The program and samples of input data (Wave-topochange.zip) can be downloaded from
   https://www.rikohtosho.co.jp/main/wp-content/uploads/Wave-topochange.zip
3. Numerical Model for Coastal Topographic Change Due to Tsunamis ------ 434
   The program and samples of input data (Tsunami-topochange.zip) can be down loaded from
   https://www.rikohtosho.co.jp/main/wp-content/uploads/Tsunami-topochange.zip

----- About the numerical simulation programs included with this book -----
1) The programming language is mainly Fortran 90.
2) It can only be compiled on Windows.
3) The operating environment has been checked under Windows 11 and the Intel oneAPI2021 Base & HPC Toolkit (Intel's C++/Fortran compiler).
4) Please provide a Fortran compiler by yourself.
5) Please provide drawing software by yourself.

> It has been verified that the accompanying numerical simulation programs produce generally adequate results as long as the appropriate input data are used. However, it should be noted that the authors and publishers cannot be held responsible for all the results of calculations made using the accompanying programs.

Many programming languages can be used to develop computer programs used for scientific and technical calculations, such as FORTRAN (1957), BASIC (1964), C (1972), C++ (1983), Fortran90 (1991), Python (1991), Java (1995), JavaScript (1997) and Julia (2012), in order of age. Among these languages, C++ and Fortran90 can be recommended because large numerical simulations require high computational speed and high reliability of results. Moreover, since learning C++ can be quite difficult for engineers without programming experience, the numerical simulation programs that can be downloaded are written based on Fortran90.

Appendix MANUALS OF NUMERICAL PREDICTION MODELS

1. Numerical Model for Shoreline Change Due to Waves
1.1 Overview of the Numerical Model

Longshore sediment transport mainly occurs within the wave breaking zone. When the relationship between the longshore sediment transport rate and the wave power at wave breaking working per unit time is investigated using experimental and field observation data, an equation for calculating the longshore sediment transport rate can be obtained from the wave breaking parameters (wave height, period and direction). Using this equation to calculate the longshore sediment transport rate Q, the longshore distribution of longshore sediment transport in the wave zone can be obtained by determining the longshore distribution of wave breaking parameters from the offshore wave parameters. Substituting the gradient of longshore sediment transport rate (dQ/dx) into the sand continuity Eq. (A-1-1), the change in the offshore length y from the land side reference line to the shoreline can be obtained from Eq. (A-1-2).

$$\frac{\partial A}{\partial t} = D\frac{\partial y}{\partial t} = \frac{\partial Q}{\partial x} \qquad \text{(A-1-1)}$$

$$\therefore \Delta y = \frac{\Delta Q}{\Delta x}\frac{\Delta t}{D} \qquad \text{(A-1-2)}$$

where D is the water depth obtained from the regression analysis between the amount of erosion ΔA of the cross-section of the beach and the movement of the shoreline in the cross-shore direction (recession or advance) Δy, obtained from bathymetric survey data in the upper figure in **Fig. A.1.1**. This is called the **movement height** (a variant of the closure depth).

If the equations and diagrams based on the small amplitude wave theory or the numerical calculation method for solving a parabolic equation to consider only the incident waves in the wave motion are used to calculate the wave deformation (by shoaling, refraction, and diffraction) from the offshore wave parameters to the breaking depth, the time required for the wave deformation calculation can be sufficiently reduced. Therefore, by providing time series data of offshore wave parameters over several years to several decades, long-term prediction of shoreline changes over a large area (several kilometers to several tens of kilometers) becomes possible. This numerical prediction model is called the **one-line shoreline model** or **shoreline change model**.

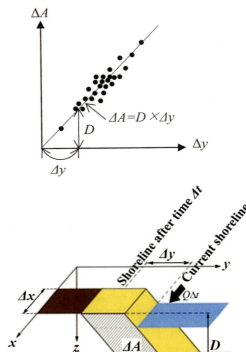

Fig. A.1.1
Concept of the one-line model
(= the shoreline change model)

The basic assumptions of this model are as follows:

388

1. Numerical Model for Shoreline Change Due to Waves

(a) Only longshore sediment transport due to waves is considered, while on-offshore sediment transport due to waves and sediment transport due to nearshore currents are ignored.

(b) The cross-shore profile of longshore sediment transport does not change in the longshore direction and the movement height is a constant value (it corresponds to the height from the beach cliff to the wave breaking depth).

1) Formula for Calculating Longshore Sediment Transport Rates

The longshore sediment transport rate Q [m³/s] is proportional to the power per unit width of a wave $EC_g \left(= 1/8 \times \rho g H_b^2 \times \sqrt{gh_b} \right)$, which according to Komar and Inman (1970) is expressed as follows.

$$
\begin{aligned}
Q &= \frac{K_1}{(\rho_s - \rho) g (1 - 0.4)} \times \left(\frac{1}{8} \rho g H_b^2 \sqrt{gh_b} \right) \left(\sin \theta_b \cos \theta_b \right) \\
&= \frac{K_1}{(\rho_s - \rho) g (1 - 0.4)} \times \left(\frac{1}{8} \rho g H_b^2 \sqrt{gh_b} \right) \left(\frac{1}{2} \sin(2\theta_b) \right)
\end{aligned}
\tag{A-1-3}
$$

where K_1 is the coefficient of longshore sediment transport (the value used by Komar et al. is 0.77, but is determined from verification simulations to match the measured shoreline change data for each coast); ρ_s is the density of sediment; ρ is the density of seawater; g is the gravitational acceleration; H_b is the wave breaking height (since the longshore sediment transport rate Q is proportional to the power per unit width of the mean wave energy flux EC_g, i.e. the squared of the mean wave height should be used); h_b is the wave breaking depth; and θ_b is the direction of the breaking waves. To obtain the volume of the sand layer, the net sand volume is divided by $(1 - 0.4)$ to account for the porosity of the sand layer (porosity 40% = 0.4).

If the change in wave height due to the diffraction effect of a breakwater cannot be ignored, the following formula by Ozasa and Brampton (1980) for calculating the longshore sediment transport rates can be used.

$$
\begin{aligned}
Q &= \frac{K_1}{(\rho_s - \rho) g (1 - 0.4)} \times \left(\frac{1}{8} \rho g H_b^2 \sqrt{gh_b} \right) \left(\frac{1}{2} \sin(2\theta_b) \right) \\
&- \frac{K_2}{(\rho_s - \rho) g (1 - 0.4)} \times \left(\frac{1}{8} \rho g H_b^2 \sqrt{gh_b} \right) \left(\frac{dH_b}{dx} \frac{\cos(\theta_b)}{\tan \alpha} \right)
\end{aligned}
\tag{A-1-4}
$$

where K_2 is the empirical coefficient (the value used by Ozasa and Brampton is $1.62 \times K_1$, but is determined from verification simulations to match the measured shoreline change data for each coast) and $\tan \alpha$ is the mean value of the on-offshore slope within the longshore sediment transport zone.

2) Movement Height

The movement height D can be obtained from the analysis of bathymetric data measured during a period of mild wave conditions every year so that the effect of on-offshore sediment transport is minimised (see the upper graph of **Fig. A.1.1**). Moreover, even if the existing bathymetric data are not available, it can be obtained by carrying out two bathymetry surveys, one in an area where shoreline

Appendix MANUALS OF NUMERICAL PREDICTION MODELS

retreat and advance due to longshore sediment transport takes place and the other in an area where it does not take place and by comparing these beach profiles.

If bathymetric measurements cannot be executed, then the significant limit depth h_c for sediment movement obtained using Eq. (A-1-5) proposed by Uda et al [2002] can be used as the movement height D in the case of general sandy coasts. This equation is the regression equation between the significant limit depth h_c and the incident significant wave height with a 95% non-exceedance probability H_{s95} (typically 1.5 m to 2.5 m in Japan), evaluated using measured wave data.

$$h_c = 3.64 \times H_{s95} \approx D \qquad \text{(A-1-5)}$$

3) Methods for Calculating Wave Propagation

When determining the longshore directional distribution of longshore sediment transport rates for the calculation of shoreline change, it is necessary to determine the longshore directional distribution of wave heights and directions from the wave data at the offshore boundary (assuming that the wave periods do not vary spatially).

If offshore observations are not available at the target coast, offshore wave data must be generated. If wave observations from a neighbouring coast that is assumed to have the same wave source area are available, the input wave data can be generated from refraction and diffraction calculations using those data. If wave observations from neighbouring coasts are not available, input wave data can be generated from wave forecasts using wind speed and direction data from weather information.

To obtain the wave heights and directions at the line of wave breaking from the input offshore wave data, wave deformation calculations due to shoaling, refraction and diffraction are required, and the wave heights after deformation can be obtained from Eq. (A-1-6). The breaking water depth h_b and wave height H_b can be determined by using the wave breaking criterion $H_b / h_b = \gamma$, where γ is the breaking limit index, usually taken as 0.78

$$H = K_s K_r K_d H_o \qquad \text{(A-1-6)}$$

where K_s is the shoaling coefficient, K_r is the refraction coefficient, K_d is the diffraction coefficient, and H_o is the offshore wave height.

The shoaling coefficient is determined using Eq. (A-1-7) based on the small amplitude wave theory.

$$K_s = \sqrt{C_{go}/C_g} \qquad \text{(A-1-7)}$$

where C_{go} is the group velocity of the offshore waves, C_g is the group velocity at the target position.

The refraction coefficient is determined using the Pade approximation Eq. (A-1-8) by Hunt (1979).

$$\left. \begin{array}{l} K_r = \omega\sqrt{F/(gh)}, \qquad Y = \omega^2 h/g, \\[2mm] F = Y + 1/\left(1 + 0.6522Y + 0.4622Y^2 + 0.0864Y^3 + 0.0675Y^4\right) \end{array} \right\} \qquad \text{(A-1-8)}$$

where $\omega = 2\pi/T$ is the angular frequency, g is the gravitational acceleration, and h is the water depth.

The wave direction θ_b at the breaking point due to wave refraction alone can be found from the following equation based on Snell's law.

$$\frac{\sin\theta_b}{L_b} = \frac{\sin\theta_o}{L_o} \tag{A-1-9}$$

where θ_o is the wave direction of the offshore wave, L_b is the wavelength at the breaking point and L_o is the wavelength of the offshore wave, and the wavelength L can be obtained using the dispersion relation equation as follows.

$$L = \frac{gT^2}{2\pi}\tanh\left(\frac{2\pi}{L}h\right) \Rightarrow L_o = \frac{gT^2}{2\pi}, \quad L_b = \sqrt{gh}T \tag{A-1-10}$$

The diffraction coefficient is obtained by the irregular wave diffraction calculation method based on the cumulative value of energy by wave direction $P_E(\theta)$ (expressed as a percentage), of Goda and Suzuki (1975). Eq. (A-1-11) is used for the semi-infinite length breakwater shown in **Fig. A.1.2.**

$$\left. \begin{array}{l} K_d = \sqrt{P_E(\theta)} \\[4pt] P_E(\theta) = 50\left[\tanh\left(\dfrac{S_{max}}{S_w}\theta\right)+1\right] \\[4pt] S_w = -0.000103\times(S_{max})^2 \\ \quad\quad +0.270\times(S_{max})+5.31 \end{array} \right\} \tag{A-1-11}$$

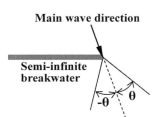

Fig. A.1.2
Illustration for diffraction calculation.

where S_{max} is the maximum value of the wave concentration parameter (10 for wind waves, 25 for swells with large wave steepness, and 70 for swells with small wave steepness), and θ is the angle from the main wave direction line to the point of interest, which takes negative values in the clockwise direction.

Eq. (A-1-12) is used for diffraction calculations at the opening between breakwaters shown in **Fig. A.1.3**.

$$K_d = \sqrt{P_E(\theta_1) - P_E(-\theta_2)} \tag{A-1-12}$$

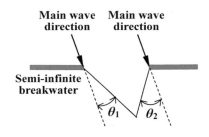

Fig. A.1.3
Illustration for diffraction calculation at the opening between breakwaters.

4) Methods for considering riverine sand supply and dredged volume

The sand supply from the river and the sand dredged volume can be considered as follows.

$$D\frac{dy}{dt} = \frac{dQ}{dx} - q_s + q_d \tag{A-1-13}$$

Appendix MANUALS OF NUMERICAL PREDICTION MODELS

where q_s is the sand supply [m³/s/m] between grid intervals dx in a river section and q_d is the sand dredged volume between grid intervals dx [m³/s/m].

5) Stability Condition When Calculating by the Difference Method

Substituting Eq. (A-1-4) for calculating longshore sediment transport rates into the sand continuity Eq. (A-1-1) gives an approximate diffusion equation for the shoreline distance y (length from a reference line to the shoreline).

$$\left. \begin{aligned} \frac{\partial y}{\partial t} &= \varepsilon \frac{\partial^2 y}{\partial x^2} \\ \varepsilon &= 2\left(H^2 C_g\right)_b K_1 \Big/ D + \left(H^2 C_g\right)_b K_2 \sin\left(\theta_b\right)\frac{\partial H_b}{\partial x}\Big/ D \end{aligned} \right\} \tag{A-1-14}$$

Therefore, from the stability condition for the diffusion equation, the following conditional equation must be satisfied when performing shoreline change calculations using the finite difference numerical method.

$$\Delta t \leq \frac{\left(\Delta x\right)^2}{2\left(\varepsilon\right)_{max}} \tag{A-1-15}$$

In this program, Eq. (A-1-16) is used instead of Eq. (A-1-15).

$$\Delta t = 0.2 \times \frac{\left(\Delta x\right)^2}{\left(\varepsilon\right)_{max}} \tag{A-1-16}$$

1. Numerical Model for Shoreline Change Due to Waves

1.2 Calculation Flow of the Program

The calculation flow of the program is shown in the diagram below.

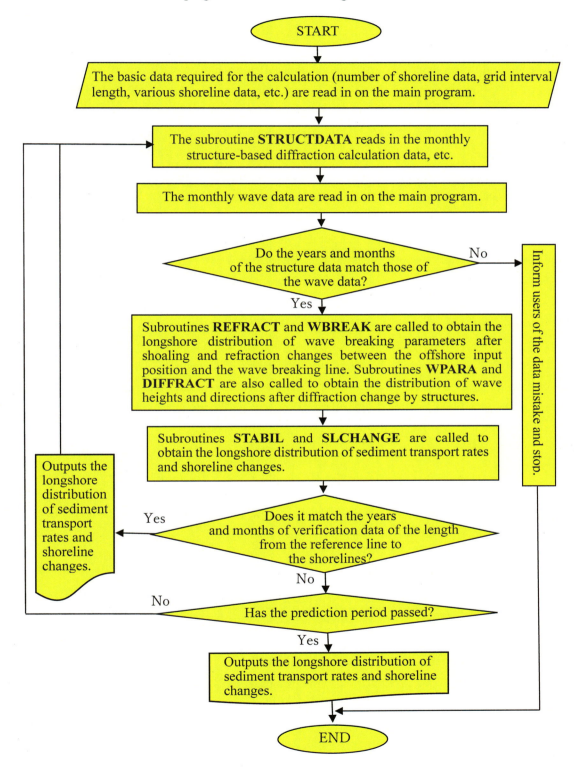

Fig. A.1.4 Calculation flow diagram for this program.

Appendix MANUALS OF NUMERICAL PREDICTION MODELS

1) Functions of the Main Program

The main program of the **ShorelineModel.For** reads basic data in and calls various subroutines to calculate the propagation of waves, longshore sediment transport rates, and shoreline change, and to output calculation results.

First, the basic calculation information (number of data on the length from the land side reference line to the shoreline, grid interval, coastal slope, movement height of sediment transport, longshore sediment transport coefficient, start year and end year of the calculation, boundary conditions, etc.), initial shoreline data, correction data of wave refraction coefficient and angle, sediment supply data from rivers, verification shoreline data (up to four times in the number of surveys), monthly wave data are read. In addition, the subroutine **STRUCTDATA** is called to read monthly structure information (data for calculating wave diffraction by structures, data for calculating sediment transit rates by groynes, and seawall position data).

Next, after confirming that the year and month of the structure data read monthly matches the year and month of the wave data, subroutines **REFRACT** and **WBREAK** are called to obtain the longshore distribution of wave breaking parameters after wave shoaling and refraction between the offshore boundary and the wave breaking line. Subroutines **WPARA** and **DIFFRACT** are then called to obtain the longshore distribution of wave heights and directions after diffraction changes due to structures. The subroutines **STABIL** and **SLCHANGE** are then called to obtain the longshore distribution of longshore sediment transport rates and shoreline distance changes, and the subroutine **OUTPUT** is called to output the longshore distribution of longshore sediment transport rates and shoreline changes only when the output date coincides with the year and month of shoreline verification data. Finally, when the calculation time reaches the pre-defined end of the calculation time, the subroutine **OUTPUT** is called to output the calculation results.

2) Function of Each Subroutine

(a) Subroutine SMOOTH

SMOOTH performs a moving average of the initial shoreline and the inclination angles between the shoreline grids to smooth these data.

(b) Subroutine STRUCTDATA

STRUCTDATA sets up data to calculate the diffraction of waves due to coastal structures (breakwaters, detached breakwaters, groynes, etc.), reads in data to calculate the sediment transit rates due to groyne-like structures, and reads the position data of seawalls as shoreline recession limits where seawalls are present.

These data must be generated on a monthly basis to assess shoreline changes associated with monthly differences in the length and number of structures installed.

(c) Subroutine RIVERSET

RIVERSET sets up the amount of sediment that is discharged from the rivers on each grid point in the estuarine areas within the computational domain.

1.Numerical Model for Shoreline Change Due to Waves

(d) Subroutine REFRACT

REFRACT calculates the shoaling coefficients from Eq. (A-1-7) based on the small-amplitude wave theory, the refraction coefficients from the Pade approximation Eq. (A-1-8) based on Hunt (1979), and the wave directions due to refraction from Snell's law.

(e) Subroutine WBREAK

When the ratio of the wave height to the water depth satisfies the wave breaking index (the wave height / the water depth = 0.78), **WBREAK** sets the wave height as the wave breaking height, and the water depth at that location as the wave breaking depth.

(f) Subroutine WPARA

WPARA calculates the wave concentration parameter.

(g) Subroutine DIFFRACT

DIFFRACT calculates the wave height and direction after the wave diffraction change based on Goda's irregular wave calculation method, using Eq. (A-1-11) for semi-infinite breakwaters and Eq. (A-1-12) for openings between breakwaters.

(h) Subroutine STABIL.

STABIL finds the suitable time step Δt to satisfy the stability criterion given in Eq. (A-1-16) for solving the diffusion equation.

(i) Subroutine SLCHANGE

SLCHANGE calculates the longshore distribution of the longshore sediment transport rates and determines the daily shoreline change distance from the sand continuity equation. The longshore sediment transport rates outside the two shoreline boundaries are set by calling the subroutine **BOUNDARY**. Sediment transport rates through groyne-like structures are calculated by calling the subroutine **DRIFTPASS**.

(j) Subroutine OUTPUT

OUTPUT provide output values such as the longshore distribution of longshore sediment transport rates, shoreline change lengths, and the length from the land side reference line to the shorelines.

Appendix MANUALS OF NUMERICAL PREDICTION MODELS

1.3 Program Execution Methods

To perform a numerical calculation, users must prepare an executable program made from the source program and input data, and then place the executable program and input data in the holder on a personal computer. Next, users should double-click the executable program or create a batch file in the same holder and double-click it to start the numerical simulations.

The file names of the target source program are **ShorelineModel.For** and **Shoreline.h**. The executable program **Shoreline.exe** can be created using the latest FORTRAN compilers such as Intel's. In **Shoreline.h**, the maximum number of grids in the longshore direction (NX) is set to 500 in order to save memory space on the computer. If users want to calculate over a larger area, change the NX in the parameter statement from 500 to the desired number and recompile. It should also be noted that NX should be carefully selected to achieve the required spatial resolution of shoreline change.

Input data files include **control.txt** (number of shoreline distance data, grid interval, coastal slope, movement height of sediment transport, longshore sediment transport coefficient, start year and end year of the calculation, boundary conditions, etc.), **shoreline.txt** (initial shoreline data, correction data of wave refraction coefficient and angle, sediment supply data from rivers, and validation shoreline data), **structure.txt** (monthly data for calculating wave diffraction by structures, monthly data for calculating sediment transit rates by groynes, and monthly data of seawall positions), and **waves.txt** (monthly data on the height, period and direction of waves). The method for creating these four input data files is described below.

1) Input Data Preparation Methods
(a) Coordinate System and Wave Direction Setting

The definition of the coordinate system, the length from the land side reference line to the shoreline $y(i)$ and input wave direction α used in this numerical model is shown in **Fig. A.1.5**.

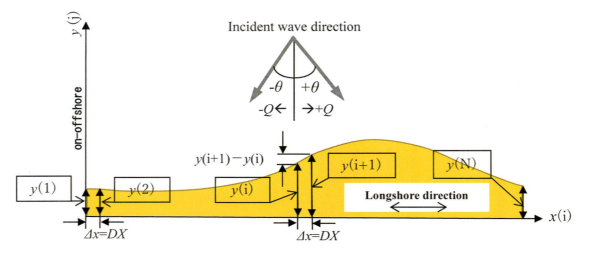

Fig. A.1.5 Coordinate system used in the numerical model.

1.Numerical Model for Shoreline Change Due to Waves

(b) Setting of the Input Data File **control.txt**

In this data file, the basic calculation information is prepared. The data reading program is as follows:

```
open(1,file='control.txt',status='old')
read(1,*) N                        !The number of total segments
read(1,*) DX                       !The spatial segment length (10.m - 25.m)
read(1,*) DRF                      !The water depth of a wave gauge (10.m - 40.m)
read(1,*) SLOPE                    !The beach - sea bottom slope (1./20. - 1./100.)
read(1,*) DCLOS                    !The closure depth (3.m - 15.m)
read(1,*) RK1                      !The longshore transport coefficient (0.77)
read(1,*) RK2                      !Ozasa Brampton's coefficient (1.62×0.77)
read(1,*) angle1                   !The arrangement angle1 (-)
read(1,*) angle2                   !The arrangement angle2 (-)
read(1,*) NENST                    !The year of the starting time
read(1,*) MONST                    !The month of the starting time
read(1,*) NENED                    !The year of the ending time
read(1,*) MONED                    !The month of the ending time
read(1,*) INDEXL, coefl, ratel
read(1,*) INDEXR, coefr, rater
close(1)
```

The actual data setting method and examples are shown below.

The relationship between the correction angle "angle1" and the correction value "angle2" is shown in Eq. (A-1-17), where "angle1" is the correction angle that converts the measured wave direction into the inclination angle used in wave deformation calculations θ relative to the y-axis in Fig. A.1.5. The inclination angle θ is positive when the longshore sediment transport rate Q is rightward.

$$\text{The inclination angle } \theta = \frac{\text{angle1 - the measured wave direction}}{\text{angle2}} \qquad \text{(A-1-17)}$$

Normally, the value '1' should be set for "angle2". However, when there is no wave direction data available and wind direction data has to be used instead to generate wave data, the existence range of wind direction data is often wider than that of wave direction data. In this case, the range of existence needs to be narrowed and adjusted by setting a value greater than '1' for "angle2". Furthermore, to exclude cases where the wave direction is from land, the program skips to the next wave data if the absolute value of the tilt angle θ is greater than 90 degrees.

Furthermore, after wave deformation calculations using the wave inclination angle θ to obtain the breaking wave angle θ_b, the inclination angle according to Eq. (A-1-18), which takes into account

Appendix MANUALS OF NUMERICAL PREDICTION MODELS

the shoreline inclination $\tan^{-1}[\{y(i+1)-y(i)\}/\Delta x]$ relative to the x axis in Fig. A.1.5, must be used to determine the longshore sediment transport rate.

$$\text{The inclination angle} = \theta_b - \tan^{-1}\left[\frac{y(i+1)-y(i)}{\Delta x}\right] \qquad (A\text{-}1\text{-}18)$$

Next, For the boundary condition on the left-hand side of Fig. A.1.5, users must set the leftmost longshore sediment transport rate $Q(1)$ as follows:

If INDEXL=1 is selected, a linear extrapolation of the increment [$Q(2)$ - $Q(3)$] is set to $Q(1)$.

If INDEXL=2 is selected, $Q(2)$ is set to $Q(1)$.

If INDEXL=3 is selected, the value obtained from the change over time equation [={coefl-ratel×(IYEAR-NENST)}×$Q(2)$] is set to $Q(1)$

where "coefl" and "ratel" are adjustment coefficients, IYEAR is the year of calculation and NENST is the year when the calculation starts, and are used when users want to reduce the longshore sediment transport rate on the left-hand side from year to year. If INDEXL is '1' or '2', "coefl" and "ratel" are not necessary and are neglected. If INDEXL is '3', set '1' to "coefl" and set an appropriate sediment transport reduction factor to "ratel", and the longshore sediment transport rate will be reduced year by year. If "coefl" and "ratel" are set to zero, the longshore sediment transport rate can be set to zero at the left-hand end.

The right-hand boundary condition in Fig. A.1.5 can also be set in the same way for the rightmost longshore sediment transport rate $Q(N+1)$.

If INDEXR=1 is selected, a linear extrapolation value is set to $Q(N+1)$.

If INDEXR=2 is selected, $Q(N)$ is set to $Q(N+1)$.

If INDEXR=3 is selected, the value obtained from the change over time equation [={(coefr-rater×(IYEAR-NENST)}×$Q(N)$] is set to $Q(N+1)$

where the treatment of "coefr" and "rater" is the same as that of coefl and ratel.

100	The total number N of shoreline distance data [Y(1) – Y(N) in Fig. A.1.5].
10	The grid interval length DX [unit: m] (this value is the interval length at which differences in the lengths of coastal structures can be evaluated).
40	The water depth DRF [unit: m] at the wave measurement position.
0.02	The coastal slope SLOPE [-] (this value must be set so that the ratio of the width of the wave breaking zone to the length of the groyne matches the local value).
5.4	The movement height of the longshore sediment transport DCLOS [unit: m].
0.1	The longshore sediment transport coefficient RK1 [-] (the optimum value from validation simulations must be set).
0.0	Ozasa-Brampton coefficient RK2 [-] (the optimum value from validation simulations must be set).
90.0	The correction angle "angle1" [unit: degrees] (this value is a correction angle to convert measured wave angles to the inclination from the y-axis in **Fig. A.1.5**).
5.0	The correction value "angle2" [-] (this value is a variation reduction factor in the case that wave directions are substituted with wind direction data).

1.Numerical Model for Shoreline Change Due to Waves

2000   The calculation start year NENST.

11   The calculation start month MONST.

2001   The calculation end year NENED.

1   The calculation end month MONED.

2 0.0 0.00   The selection number INDEXL for the left-hand boundary condition, adjustment factor "coefl" and "ratel".

2 0.0 0.00   The selection number INDEXR for the right-hand boundary condition, adjustment factor "coefr" and "rater".

(c) Setting of the Input Data File shoreline.txt

In this data file, the initial shoreline data, correction data of wave refraction coefficient and angle, sediment supply data from rivers, and validation shoreline data (maximum of four measurements) are prepared. The reading program is as follows:

```
open(2,file='shoreline.txt',status='old')
read(2,'(10F8.2)') (YORGN(I), I=1,N)      !The initial shoreline data (m)
read(2,'(10F8.2)') (OR10(I), I=1,NP)      !The refraction direction (-)
read(2,'(10F8.2)') (RKR10(I), I=1,NP)     !The refraction coefficient (-)
read(2,*) NR                              !The number of rivers
   if(NR .EQ. 0) go to 22
do I = 1,NR
read(2,*) NRST              !The start segment no. of discharge from a river
read(2,*) NRED             !The end segment no. of discharge from the river
read(2,*) QRDATA           !The sand discharge from the river (m³/year)
enddo
22 write(*,*) 'The number of rivers =', NR
read(2,*) NNC              !The number of verification data ( =< 4 )
if(NNC .EQ. 0) go to 33
do I = 1,NNC
read(2,'(2I6)') NYEAR(I), NMONTH(I)
read(2,'(10F8.2)') (YC(I,J), J=1,N)
enddo
33 close(2)
```

An actual example on the shoreline data of the number of N = 100 is shown below.

First, the data for the length YORGN [unit: m] from the reference line to the initial shoreline are set for each grid interval at the start of the simulation, 10 per line, for a total of N.

Next, the correction angle OR10 [unit: degrees] to the refraction wave direction and the correction coefficients RKR10 [-] to the refraction wave height are respectively set in the same way, 10 per line, for a total of N.

399

Appendix MANUALS OF NUMERICAL PREDICTION MODELS

The program calculates the refraction deformation only at two lines between the offshore boundary and the wave breaking line, which enables fast calculations over a large area and is therefore accurate enough if the coastline is almost straight. But, if the coastline is not straight, this method is not accurate enough and a correction to the refraction deformation calculation is required. Therefore, if the coastline is not remarkably straight, full-scale refraction calculations must be performed beforehand for the main wave directions and periods, and by averaging them, the correction angle OR10 for the refracted wave direction and the correction factor RKR10 for the refracted wave height are set to increase the accuracy of the refraction calculation. If no correction is required because the coastline is almost straight, '0.0' for the correction angle OR10 and '1.0' for the correction factor RKR10 must be set.

Furthermore, the sediment supply data from the rivers are set in the following order: the number of target rivers NR, the leftmost grid number NRST and the rightmost grid number NRED at the estuary of the first river, the sediment supply QRDATA [volume per year, m³/year] of the first river, the leftmost grid number NRST and the rightmost grid number NRED at the estuary of the second river, the sediment supply QRDATA of the second river [volume per year, m³/year], ... are set for the total number of target rivers.

Finally, the length data from the reference line to the shoreline used for the verification simulations are set in the following order: the number of shoreline surveys NNC, the first survey year and month NYEAR and NMONTH, the first data of the length from the reference line to the shoreline YC in 10 per line for N, the second survey year and month NYEAR and NMONTH, the second data of the length from the reference line to the shoreline YC in 10 per line for N, ... are set for the number of surveys. However, the maximum number of surveys is four.

100.00	100.00	100.00	100.00	100.00	100.00	100.00	100.00	100.00	100.00
100.00	100.00	100.00	100.00	100.00	100.00	100.00	100.00	100.00	100.00
100.00	100.00	100.00	100.00	100.00	100.00	100.00	100.00	100.00	100.00
100.00	100.00	100.00	100.00	100.00	100.00	100.00	100.00	100.00	100.00
100.00	100.00	100.00	100.00	100.00	100.00	100.00	100.00	100.00	100.00
100.00	100.00	100.00	100.00	100.00	100.00	100.00	100.00	100.00	100.00
100.00	100.00	100.00	100.00	100.00	100.00	100.00	100.00	100.00	100.00
100.00	100.00	100.00	100.00	100.00	100.00	100.00	100.00	100.00	100.00
100.00	100.00	100.00	100.00	100.00	100.00	100.00	100.00	100.00	100.00
100.00	100.00	100.00	100.00	100.00	100.00	100.00	100.00	100.00	100.00
0.00	0.00	0.00	0.00	0.00	0.00	0.00	0.00	0.00	0.00
0.00	0.00	0.00	0.00	0.00	0.00	0.00	0.00	0.00	0.00
0.00	0.00	0.00	0.00	0.00	0.00	0.00	0.00	0.00	0.00
0.00	0.00	0.00	0.00	0.00	0.00	0.00	0.00	0.00	0.00
0.00	0.00	0.00	0.00	0.00	0.00	0.00	0.00	0.00	0.00
0.00	0.00	0.00	0.00	0.00	0.00	0.00	0.00	0.00	0.00
0.00	0.00	0.00	0.00	0.00	0.00	0.00	0.00	0.00	0.00
0.00	0.00	0.00	0.00	0.00	0.00	0.00	0.00	0.00	0.00
0.00	0.00	0.00	0.00	0.00	0.00	0.00	0.00	0.00	0.00
0.00	0.00	0.00	0.00	0.00	0.00	0.00	0.00	0.00	0.00
1.00	1.00	1.00	1.00	1.00	1.00	1.00	1.00	1.00	1.00
1.00	1.00	1.00	1.00	1.00	1.00	1.00	1.00	1.00	1.00
1.00	1.00	1.00	1.00	1.00	1.00	1.00	1.00	1.00	1.00

1.Numerical Model for Shoreline Change Due to Waves

1.00	1.00	1.00	1.00	1.00	1.00	1.00	1.00	1.00	1.00
1.00	1.00	1.00	1.00	1.00	1.00	1.00	1.00	1.00	1.00
1.00	1.00	1.00	1.00	1.00	1.00	1.00	1.00	1.00	1.00
1.00	1.00	1.00	1.00	1.00	1.00	1.00	1.00	1.00	1.00
1.00	1.00	1.00	1.00	1.00	1.00	1.00	1.00	1.00	1.00
1.00	1.00	1.00	1.00	1.00	1.00	1.00	1.00	1.00	1.00
1.00	1.00	1.00	1.00	1.00	1.00	1.00	1.00	1.00	1.00

1   The Number of rivers whose sediment supply should be considered NR.

10   The leftmost grid number at the estuary of the first river NRST.

15   The rightmost grid number at the estuary of the first river NRED.

36500.0   The sediment supply volume per year QRDATA(m^3/year).

2   The number of shoreline surveys used for the verification simulations NNC (up to four times).

2000   11   The first survey year and month of the length from the reference line to the shoreline NYEAR and NMONTH.

100.00	100.00	100.00	100.00	100.00	100.00	100.00	100.00	100.00	100.00
100.00	100.00	100.00	100.00	100.00	100.00	100.00	100.00	100.00	100.00
100.00	100.00	100.00	100.00	100.00	100.00	100.00	100.00	100.00	100.00
100.00	100.00	100.00	100.00	100.00	100.00	100.00	100.00	100.00	100.00
100.00	100.00	100.00	100.00	100.00	100.00	100.00	100.00	100.00	100.00
100.00	100.00	100.00	100.00	100.00	100.00	100.00	100.00	100.00	100.00
100.00	100.00	100.00	100.00	100.00	100.00	100.00	100.00	100.00	100.00
100.00	100.00	100.00	100.00	100.00	100.00	100.00	100.00	100.00	100.00
100.00	100.00	100.00	100.00	100.00	100.00	100.00	100.00	100.00	100.00
100.00	100.00	100.00	100.00	100.00	100.00	100.00	100.00	100.00	100.00

2000   12   The second survey year and month of the length from the reference line to the shoreline NYEAR and NMONTH.

100.00	100.00	100.00	100.00	100.00	100.00	100.00	100.00	100.00	100.00
100.00	100.00	100.00	100.00	100.00	100.00	100.00	100.00	100.00	100.00
100.00	100.00	100.00	100.00	100.00	100.00	100.00	100.00	100.00	100.00
100.00	100.00	100.00	100.00	100.00	100.00	100.00	100.00	100.00	100.00
100.00	100.00	100.00	100.00	100.00	100.00	100.00	100.00	100.00	100.00
100.00	100.00	100.00	100.00	100.00	100.00	100.00	100.00	100.00	100.00
100.00	100.00	100.00	100.00	100.00	100.00	100.00	100.00	100.00	100.00
100.00	100.00	100.00	100.00	100.00	100.00	100.00	100.00	100.00	100.00
100.00	100.00	100.00	100.00	100.00	100.00	100.00	100.00	100.00	100.00
100.00	100.00	100.00	100.00	100.00	100.00	100.00	100.00	100.00	100.00

(d) Setting of the Input Data File **structure.txt**

In this data file, the monthly data for the calculations of wave diffraction by coastal structures (breakwaters, groynes, detached breakwaters, etc.), for the calculations of sediment transit rates by groyne-like structures, and for the locations of seawalls are prepared. The reading program is as follows:

```
    open(3, file='structure.txt', status='old')
555 read(3,'(3I5)', end=999) IYEAR, MONTH
    read(3,*) NJ                          !The number of diffraction points
```

Appendix MANUALS OF NUMERICAL PREDICTION MODELS

```
if(NJ.GT.0) then
do J=1,NJ
read(3,'(5I5,2F8.2)') IL(J),IR(J),IBREAK(J),INCDNC(J),IXJ(J),YJ(J),TKD(J)
enddo
end if
read(3,*) NG                            !The number of passing drift rates
if(NG.GT.0) then
do J=1,NG
read(3,'(I5,F8.2)') IG(J),YG(J)
enddo
end if
read(3,*) NS                            !The number of seawall segments
if(NS.GT.0) then
do J=1, NS
read(3,'(2I5)') IE,IW
read(3,'(10F8.2)') (YS(I),I=IE,IW)
enddo
end if
if (IYEAR.LT.NENED) go to 555
if ((IYEAR.EQ.NENED).and.(MONTH.LT.MONED)) go to 555
close(3)
```

The method to set the input data for each month and an example are shown as follows.

First, the target year and month IYEAR and MONTH are set.

Next, the total number of diffraction calculation points NJ is set. This number is usually twice the total number of breakwaters, groynes, detached breakwaters, etc. on the target coast since the left and right ends of these structures can become wave diffraction points. Moreover, the diffraction calculation information must be set from the structure whose diffraction points are the most offshore. In the case of **Fig. A.1.6**, the diffraction calculation information for the left-hand area '1' and the right-hand area '2' of the groyne (impermeable structure) is set first. Then, the the left-hand area '1' and the right-hand area '2' of the detached breakwater (transmissible structure) are set.

The information to be given for each diffraction point are as follows:

1) The grid point numbers for the leftmost IL and rightmost IR of the diffraction calculation range (as the calculation range, the equivalent cross-shore distance from the shoreline to the diffraction point should be taken on both sides. The if the distance offshore from the shoreline to the diffraction point of the groyne is 300 m and the grid interval length is 10 m then, '1' and '30' for the left side of the groyne, and '31' and '60' for the right side of the groyne are taken).

2) The number IBREAK of the diffraction calculation area ('1' for the left side of the structure, '2' for the right side).

1. Numerical Model for Shoreline Change Due to Waves

3) The identification number INCDNC of the target wave direction for diffraction calculations (② for incident waves from the right side, ① for incident waves from the left side).
4) The grid point number IXJ at which the target diffraction point is located ('30' if the target diffraction point is located on the left side of the groyne, '31' if on the right side).
5) The length YJ from the land side reference line to the diffraction point (400.0 m for the groyne).
6) The wave transmission coefficient of the target structure TKD ('0.0' for impermeable structures, '0.5' for normal block breakwaters).

The above is repeated for the total number of diffraction calculation points NJ.

Moreover, the total number of grid points NG at the locations of the groynes that should be considered for the sediment transport rates is set, and then the combination of the grid point number at of the location of each groyne IG and the length YG [unit: m] from the land side reference line to the tip of each groyne are set repeatedly for the total number of grid points at the groyne locations NG.

Finally, the number of seawall locations NS, which is the limit line of the shoreline recession, is set. Then, the leftmost and rightmost grid point numbers IE and IW are set for each seawall. In addition, the distance data YS [unit: m] from the land side reference line to the seawall at grid points between the left and right ends of each seawall are set. These combinations of IE, IW and YS are set repeatedly for the number of seawall locations NS.

The above data for the three types of structures to be set for each month are set repeatedly from the year and month when the simulation starts to the year and month when the simulation ends.

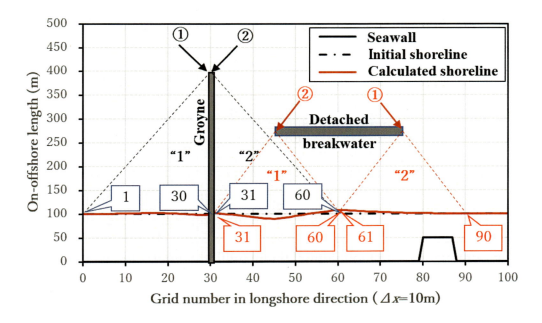

Fig. A.1.6 Illustration of the method for setting structure-related data.

Appendix MANUALS OF NUMERICAL PREDICTION MODELS

The following examples are given for the period from November 2000 to January 2001 based on **Fig. A.1.6**: in November 2000 there was only an impermeable groyne (a length of 400 m offshore, a wave transmission coefficient of 0.0), in December 2000 a detached breakwater (a detached length of 250 m, a wave transmission coefficient of 0.5) and half of a seawall were added, and in January 2001 the offshore length of the groyne was extended to 500 m.

```
2000    11
2
     1    30    1    2    30   400.00      0.0
    31    60    2    1    31   400.00      0.0
2
    30   400.00
    31   400.00
0
 2000    12
4
     1    30    1    2    30   400.00      0.0
    31    60    2    1    31   400.00      0.0
    31    60    1    2    45   250.00      0.5
    60    90    2    1    75   250.00      0.5
2
    30   400.00
    31   400.00
1
    80    87
    50.00    50.00    50.00    50.00    50.00    50.00    50.00    50.00
 2001     1
4
     1    30    1    2    30   500.00      0.0
    31    70    2    1    31   500.00      0.0
    31    60    1    2    45   250.00      0.5
    60    90    2    1    75   250.00      0.5
2
    30   500.00
    31   500.00
1
    80    94
    50.00    50.00    50.00    50.00    50.00    50.00    50.00    50.00    50.00    50.00
    50.00    50.00    50.00    50.00    50.00
```

(e) Setting of the Input Data File **waves.txt**

In this data file, the monthly wave height, period and direction data are set. The reading program is as follows:

```
      open(4,file='waves.txt',status='old')
  555 read(3,'(3I5)',END=999) IYEAR,MONTH
      CALL STRUCTDATA
C ===== LOOP DURING ONE MONTH ==================================
      read(4,'(3I6)',END=998) IYEARW,MONTHW,NDAYS
       if(IYEAR.NE.IYEARW) go to 997
```

404

1.Numerical Model for Shoreline Change Due to Waves

```
    if(MONTH.NE.MONTHW) go to 996
    do 1000 IDAY =1,NDAYS
    read(4,'(F4.2,F4.1,F4.0)') HRF,T,ZRF
    if(HRF.EQ.9.99 .OR. T.EQ.99.9 .OR. ZRF.EQ.999.0) go to 1000
          ⋮
    The shoreline changes for one day are calculated.
          ⋮
1000 continue
    if (IYEAR.LT.NENED) go to 555
    if ((IYEAR.EQ.NENED).and.(MONTH.LT.MONED)) go to 555
    close(3)
    close(4)
```

The method to set the wave height, period and direction data for each month and an example are shown as follows.

First, the target year and month IYEARW and MONTHW respectively and the number of days NDAYS are set.

Next, the wave height HRF [unit: m], wave period T [s] and wave direction ZRF [degrees, angle measured in the clockwise direction] of the energy-averaged wave representative for one day are set for the number of days. Here, '9.99 m' is set for the missing wave height data, '99.9 s' is set for the missing wave period data and '999 degrees' is set for the missing wave direction data, so that the program will skip sediment transport calculations when data are missing. An example is given below for the period November 2000 to January 2001.

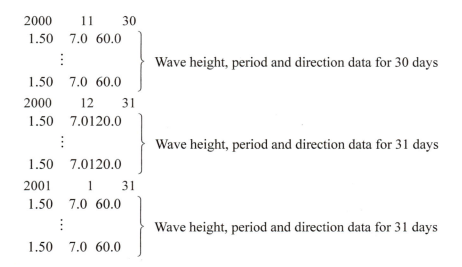

2) Output Method of Calculation Results

When the year and month of the sediment transport calculation coincide with the year and month of the length data from the reference line to the shoreline for verification, or when the end of the entire

Appendix MANUALS OF NUMERICAL PREDICTION MODELS

simulation period is reached, the output file **output.txt,** which contains the shoreline data and longshore sediment transport rates, is automatically created according to the following program.

```
      open(7, file=' output. txt', status=' unknown')
      write(7, 500) RK1
 500  format(1H ,'K1      =', F8.2)
      write(7, 510) RK2
 510  format(1H ,'K2      =', F8.2)
      write(7, 520) angle1
 520  format(1H ,'angle1 =', F8.2)
      write(7, 530) angle2
 530  format(1H ,'angle2 =', F8.2)
      write(7, 540) DX
 540  format(1H ,'DX(m) =', F8.2)
      write(7, 550) INDEXL, coefl, ratel
 550  format(1H ,' INDEXL=', I8,'    coef =', F10.3,'    rate =', F10.4)
      write(7, 560) INDEXR, coefr, rater
 560  format(1H ,' INDEXR=', I8,'    coef =', F10.3,'    rate =', F10.4)
      write(7, 570) IED-IST+1
 570  format(1H ,'Number=', I8)
      write(7, 580) IYEAR, MONTH
 580  format(1H ,'Year  =', I8,'    Month =', I4)
      write(7, 590)
 590  format(1H ,'   No.   Ywall(m)   Yorg(m)   Ycal(m)   Yreal(m)   Ycal-
     +Yreal   DY(m)   QNET(m³)   HB(m)   ZB(-)')
C
      do I = IST, IED
      write(7, 600)   I, YS(I), YORGN(I), Y(I), YC(NNC, I), Y(I)-YC(NNC, I), DY(I)
     +, QNET(I), HB(I), ZB(I)*RTD
 600  format(1H , I7, 6F10.2, F11.2, 2F8.2)
      enddo
C ------- LOCATIONS AND LENGTHS OF GROINS -------------------------------
      if(NG.NE.0) then
      write(7, 650) NG
 650  format(1H ,'* Locations and lengths of groins: No.=', I3)
      write(7,' (I8, F8.2)') (IG(J), YG(J), J=1, NG)
      end if
C ------- LOCATIONS AND LENGTHS OF DETACHED BREAKWATERS -----------------
      if(NJ.NE.0) then
      write(7, 700) NJ
 700  format(1H ,'* Locations and lengths of detached BWs: No.=', I3)
```

406

1.Numerical Model for Shoreline Change Due to Waves

```
write(7,'(I8,F8.2)') (IXJ(J),YJ(J),J=1,NJ)
end if
close(7)
```

The contents of the output are explained below.

Following are the outputs given by the program in sequence: the longshore sediment transport coefficients K1 and K2, the correction angle "angle1" and the correction value "angle2", the grid interval length DX, the left boundary condition (setting number INDEXL, adjustment factors "coefl" and "ratel") and the right boundary condition (setting number INDEXR, adjustment factors "coefr" and "rater"), the number of grid points to be output (IED -IST+1), the output year and month IYEAR and MONTH.

Next, the grid point number, the length data from the land side reference line to the seawall Ywall [unit: m], the initial data of the length from the reference line to the shoreline Yorg [m], the calculated data of the length from the reference line to the shoreline Ycal [m], the measured data of the length from the reference line to the shoreline Yreal [m], the difference data between the calculated and measured length from the reference line to the shoreline Ycal-Yreal [m], the difference data between the calculated and initial length from the reference line to the shoreline DY [m], the accumulated longshore sediment transport rate QNET [m^3, positive right direction], the wave breaking height HB [m] and the wave breaking direction ZB [degrees, positive right direction] are output for the number of grid points.

In addition, the total number of grid points NG at the locations of the groynes to estimate the sediment transit rates is output, and then the grid point number IG at the location of each groyne and the offshore length YG [m] from the land side reference line to the tip of the groyne are given repeatedly for the total number of NG.

Finally, the total number of diffraction calculation points NJ is output, and then the grid point number IXJ at the target diffraction point and the offshore length YJ [m] from the land side reference line to the diffraction point are given repeatedly for the total number of NJ.

```
K1     =     0.10
K2     =     0.00
angle1 =    90.00
angle2 =     5.00
DX(m) =     10.00
INDEXL=         2     coef =     0.000     rate =     0.0000
INDEXR=         2     coef =     0.000     rate =     0.0000
Number=       100
Year   =     2000     Month =   12
```

No.	Ywall(m)	Yorg(m)	Ycal(m)	Yreal(m)	Ycal-Yreal	DY(m)	QNET(m3)	HB(m)	ZB(-)
1	0.00	100.00	100.00	100.00	0.00	0.00	1012.92	0.00	0.00
2	0.00	100.00	100.12	100.00	0.12	0.12	1012.92	1.70	-1.05
3	0.00	100.00	100.23	100.00	0.23	0.23	1006.52	1.70	-1.06
4	0.00	100.00	100.35	100.00	0.35	0.35	993.84	1.70	-1.08
5	0.00	100.00	100.45	100.00	0.45	0.45	975.08	1.70	-1.10
6	0.00	100.00	100.55	100.00	0.55	0.55	950.60	1.70	-1.13
7	0.00	100.00	100.64	100.00	0.64	0.64	920.84	1.70	-1.16
8	0.00	100.00	100.72	100.00	0.72	0.72	886.24	1.70	-1.18
9	0.00	100.00	100.81	100.00	0.81	0.81	847.12	1.70	-1.12

Appendix MANUALS OF NUMERICAL PREDICTION MODELS

10	0.00	100.00	100.95	100.00	0.95	0.95	803.19	1.70	-0.95
				⋮					
91	0.00	100.00	100.13	100.00	0.13	0.13	-402.10	1.71	-2.59
92	0.00	100.00	100.10	100.00	0.10	0.10	-408.83	1.71	-2.57
93	0.00	100.00	100.07	100.00	0.07	0.07	-414.08	1.71	-2.55
94	0.00	100.00	100.06	100.00	0.06	0.06	-418.12	1.71	-2.54
95	0.00	100.00	100.04	100.00	0.04	0.04	-421.13	1.71	-2.53
96	0.00	100.00	100.03	100.00	0.03	0.03	-423.36	1.71	-2.52
97	0.00	100.00	100.02	100.00	0.02	0.02	-424.92	1.71	-2.51
98	0.00	100.00	100.01	100.00	0.01	0.01	-425.99	1.71	-2.50
99	0.00	100.00	100.01	100.00	0.01	0.01	-426.61	1.71	-2.50
100	0.00	100.00	100.00	100.00	0.00	0.00	-426.90	1.71	-2.50

* Locations and lengths of groins: No.= 2

    30   400.00

    31   400.00

* Locations and lengths of detached BWs: No.= 4

    30   400.00

    31   400.00

    45   250.00

    75   250.00

1.Numerical Model for Shoreline Change Due to Waves

1.4 References for the Shoreline Change Model

1）Komar, P.D. and Inman, D.L.: Longshore Sand Transport on Beaches, *Journal of Geophysical Research*, 75(30), 1970, pp.5914-5927.

2）Ozasa, H. and Brampton, A.H.: Models for Predicting the Shoreline Evaluation of Beaches Backed by Seawalls, *Report of the Port and Harbour Research Institute*, Vol.18, No.4, 1979, pp.77-104. (in Japanese)

3）Uda, T., Serizawa, M., Kumada, T., Karube, R. and Miura, M.: Relations among Wave Climate, Longshore Sand Transport Rate and Closure Depth, *Journal of Civil Engineering in the Ocean*, Vol.18, JSCE, 2002, pp.803-808. (in Japanese)

4）Hunt, J. N.: Direct solution of wave dispersion equation, *Journal of the Waterway Port Coastal and Ocean Division*, Vol.105, 1979, pp.457-459.

5）Goda, Y. and Suzuki, Y.: *Computation of Refraction and Diffraction of Sea Waves with Mitsuyasu' s Direction Spectrum*, Technical Note of the Port and Harbour Research Institute, No.230, 1975, 45p. (in Japanese)

Appendix MANUALS OF NUMERICAL PREDICTION MODELS

2. Numerical Model for Beach Change Due to High Waves
2.1 Overview of the Numerical Model

When high waves with large wave steepness strike the beach (spilling breakers), offshore turbulence energy transport dominates; on the other hand, when waves with small wave steepness (plunging breakers) break, the turbulence energy is transported in the onshore direction (Ting and Kirby, 1994). Consequently, with spilling breakers, offshore sediment transport dominates, and beach erosion occurs; with plunging breakers, onshore sediment transport dominates, and the beach tends to recover. Therefore, the generation, transport, and dissipation of turbulence energy due to wave breaking should be reproduced as accurately as possible in calculating fluid motion and sediment transport within the wave breaking zone. However, problems remain in the determination of wave breaking and the evaluation of wave energy loss in the wave breaking calculations using the Boussinesq model and other models. For example, the surface roller model of Schäffer et al. (1993) is prone to computational instability.

Vu et al. (2002) improved the Boussinesq and k-ε models to construct a horizontal 2D numerical model that can adequately reproduce the generation, transport, and dissipation of turbulent energy within the wave breaking zone and predict the transport of bed and suspended loads by waves and nearshore currents, as well as the topographic changes caused by them. The computational accuracy of their numerical model is at a sufficiently practical level, as shown in **Fig. A.2.1**, which is the result of a study by Yamamoto et al. (2012).

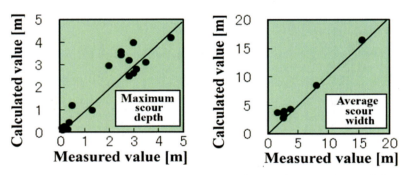

Fig. A.2.1 Comparison of measured results on frontal scouring of breakwaters and other structures (large-scale experiments by Port and Harbour Research Institute in Japan and field cases) with the simulation results using this numerical model (Yamamoto et al., 2012).

1) Hydrodynamic Models Including Beach Areas
(a) Water Motion Models in Shallow Water Areas

The governing equation for water motion is Boussinesq's equation, which can consider non-linearity, dispersion, and wave energy loss due to breaking waves. The equations of motion and continuity are as follows:

2. Numerical Model for Beach Change Due to High Waves

$$\frac{\partial q_x}{\partial t} + \frac{\partial}{\partial x}\left(\frac{q_x^2}{d}\right) + \frac{\partial}{\partial y}\left(\frac{q_x q_y}{d}\right) + gd\frac{\partial \eta}{\partial x} + \frac{h^3}{6}\left[\frac{\partial^3}{\partial x^2 \partial t}\left(\frac{q_x}{h}\right) + \frac{\partial^3}{\partial x \partial y \partial t}\left(\frac{q_y}{h}\right)\right]$$
$$-\frac{h^2}{2}\left(\frac{\partial^3 q_x}{\partial x^2 \partial t} + \frac{\partial^3 q_y}{\partial x \partial y \partial t}\right) - M_{bx} + \frac{f_b}{d^2}|q_x|q_x = 0 \tag{A-2-1}$$

$$\frac{\partial q_y}{\partial t} + \frac{\partial}{\partial x}\left(\frac{q_x q_y}{d}\right) + \frac{\partial}{\partial y}\left(\frac{q_y^2}{d}\right) + gd\frac{\partial \eta}{\partial y} + \frac{h^3}{6}\left[\frac{\partial^3}{\partial y^2 \partial t}\left(\frac{q_y}{h}\right) + \frac{\partial^3}{\partial x \partial y \partial t}\left(\frac{q_x}{h}\right)\right]$$
$$-\frac{h^2}{2}\left(\frac{\partial^3 q_y}{\partial y^2 \partial t} + \frac{\partial^3 q_x}{\partial x \partial y \partial t}\right) - M_{by} + \frac{f_b}{d^2}|q_y|q_y = 0 \tag{A-2-2}$$

$$\frac{\partial q_x}{\partial x} + \frac{\partial q_y}{\partial y} + \frac{\partial \eta}{\partial t} = 0 \tag{A-2-3}$$

where q_x and q_y are the flow rates per unit width defined by $q_x = \int u dz$ and $q_y = \int v dz$ using the velocity u in the x direction and v in the y direction, η is the water surface elevation, h is the mean water depth, d is the total water depth ($= h + \eta$), g is the acceleration of gravity, f_b is the sea bottom friction coefficient (Yamamoto et al., 1996), t is the time, x is the coordinate in the offshore direction, and y is the coordinate in the longshore direction. M_{bx} and M_{by} are the wave energy loss terms due to wave breaking and are expressed as follows.

$$M_{bx} = \frac{\partial}{\partial x}\left(f_D v_t \frac{\partial q_x}{\partial x}\right) + \frac{\partial}{\partial y}\left(f_D v_t \frac{\partial q_x}{\partial y}\right)$$
$$M_{by} = \frac{\partial}{\partial x}\left(f_D v_t \frac{\partial q_y}{\partial x}\right) + \frac{\partial}{\partial y}\left(f_D v_t \frac{\partial q_y}{\partial y}\right) \tag{A-2-4}$$

where v_t is the eddy kinematic viscosity coefficient and f_D is an empirical coefficient (usually 1.5 is good) that accounts for the fact that part of the wave energy lost during wave breaking becomes turbulent energy.

The horizontal 2D, depth-integrated equations for the generation and dissipation of turbulent energy in the wave breaking zone and the offshore transport of k and ε (the depth-integrated turbulence energy and its dissipation rate, respectively) are as follows.

$$\frac{\partial k}{\partial t} + \frac{\partial \tilde{u} k}{\partial x} + \frac{\partial \tilde{v} k}{\partial y} = P_r - \varepsilon + \frac{\partial}{\partial x}\left(\frac{v_t}{\sigma_t}\frac{\partial k}{\partial x}\right) + \frac{\partial}{\partial y}\left(\frac{v_t}{\sigma_t}\frac{\partial k}{\partial y}\right) \tag{A-2-5}$$

$$\frac{\partial \varepsilon}{\partial t} + \frac{\partial \tilde{u} \varepsilon}{\partial x} + \frac{\partial \tilde{v} \varepsilon}{\partial y} = \frac{\partial}{\partial x}\left(\frac{v_t}{\sigma_\varepsilon}\frac{\partial \varepsilon}{\partial x}\right) + \frac{\partial}{\partial y}\left(\frac{v_t}{\sigma_\varepsilon}\frac{\partial \varepsilon}{\partial y}\right) + \frac{\varepsilon}{k}(C_{1\varepsilon}P_r - C_{2\varepsilon}\varepsilon) \tag{A-2-6}$$

$$v_t = C_\varepsilon k^2 / \varepsilon \tag{A-2-7}$$

where $\tilde{u}$ and $\tilde{v}$ are the depth-averaged ensemble mean velocities in the x and y directions, respectively; $\sigma_t (= 1)$, $\sigma_\varepsilon (= 1.3)$, and $C_\varepsilon (= 0.09)$ (Jones and Launder, 1972) are the closure coefficients. $C_{1\varepsilon} (= 1.44)$ and $C_{2\varepsilon} (= 1.92)$ represent the generation and decay rate of the turbulent energy dissipation rate ε.

The generation of turbulent energy P_r consists of generation by bottom resistance, generation by shear stress, and generation by breaking waves, and is expressed by the following equation.

$$P_r = P_{rb} + P_{rs} + P_{rw} \tag{A-2-8}$$

where P_{rb}, P_{rs}, and P_{rw} are the generation of turbulent energy due to bottom resistance, shear stress, and wave breaking respectively.

Appendix MANUALS OF NUMERICAL PREDICTION MODELS

The generation of turbulent energy due to bottom resistance is considered small and ignored in this model.

The generation of turbulent energy due to shear stress is expressed by the following equation.

$$P_{rs} = v_t d\left[\left(\frac{\partial u}{\partial y}\right)^2 + \left(\frac{\partial v}{\partial x}\right)^2\right]$$

(A-2-9)

where u and v are the velocities in the x and y directions respectively.

The generation of turbulent energy due to wave breaking is assumed to be proportional to the kinetic energy in the surface roller and is expressed by the following equation.

$$P_{rw} = \alpha_p g C \delta \beta_d / d$$

(A-2-10)

where C is the wave celerity, α_p (= 0.33) and β_d (= 0.08) are constants, and δ is the thickness of the surface roller, estimated by the method of Schäffer et al. (1993). The wave breaking point is defined as the point where the water surface slope exceeds a certain limiting value. The surface roller then moves toward the shore while damping. When the surface slope in front of the surface roller falls below a certain limiting value, the surface roller disappears and the wave recovers. This method reproduces the formation, deformation, and movement of the surface roller from moment to moment, the loss of wave energy due to wave breaking, and the generation, transport, and dissipation of turbulent energy.

After a wave breaks, it moves up the slope, attenuating its energy due to large-scale eddies and bottom friction, and then begins to return. As this wave flows downstream, the next wave comes upstream. Thus, the shoreline constantly moving on the slope in the on-offshore direction. To calculate such run-up wave motion, in this model the shore-side condition in each grid mesh is setup in a manner similar to that of Hibberd and Peregrine (1979).

(b) Water Motion Models in Wave Run-up Areas

In wave run-up areas, the following equation is used to model the wetting and drying process on the beach:

$$\frac{\partial q_x}{\partial t} + \frac{1}{S}\frac{\partial}{\partial x}\left(\frac{Sq_x^2}{d}\right) + \frac{1}{S}\frac{\partial}{\partial y}\left(\frac{Sq_xq_y}{d}\right) + gd\frac{\partial \eta}{\partial x} - \frac{1}{S}\frac{\partial}{\partial x}\left[dv_t S\frac{\partial(q_x/d)}{\partial x}\right]$$
$$-\frac{1}{S}\frac{\partial}{\partial y}\left[dv_t S\frac{\partial(q_x/d)}{\partial y}\right] + \frac{f_b}{d^2}Qq_x = 0$$

(A-2-11)

$$\frac{\partial q_y}{\partial t} + \frac{1}{S}\frac{\partial}{\partial x}\left(\frac{Sq_yq_x}{d}\right) + \frac{1}{S}\frac{\partial}{\partial y}\left(\frac{Sq_y^2}{d}\right) + gd\frac{\partial \eta}{\partial y} - \frac{1}{S}\frac{\partial}{\partial x}\left[dv_t S\frac{\partial(q_y/d)}{\partial x}\right]$$
$$-\frac{1}{S}\frac{\partial}{\partial y}\left[dv_t S\frac{\partial(q_y/d)}{\partial y}\right] + \frac{f_b}{d^2}Qq_y = 0$$

(A-2-12)

$$\frac{\partial f_y q_x}{\partial x} + \frac{\partial f_x q_y}{\partial y} + \frac{\partial S\eta}{\partial t} = 0$$

(A-2-13)

where, S is the wetted area ratio (the ratio of the area where the flow exists to the grid area) in the computational grid, and f_x and f_y are the ratios of the wetted area of the computational grid parallel to

412

2. Numerical Model for Beach Change Due to High Waves

the x and y directions to the grid interval length.

(c) Water Motion Model in Permeable Structures

The water motion in a porous structure must reproduce the motion of the fluid through a cavity (a space other than the solid part). Therefore, the wet volume ratio (the ratio of the volume portion of the grid where the flow exists to the volume of the entire grid) is defined for each computational grid in the porous structure. Ignoring the dispersion of waves in the porous structure and integrating in space using the wet volume ratio, the continuity equation and the equations of motion in a porous structure are expressed as follows.

$$\frac{\partial q_x}{\partial t} + \frac{1}{K}\frac{\partial}{\partial x}\left(\frac{Kq_x^2}{d}\right) + \frac{1}{K}\frac{\partial}{\partial y}\left(\frac{Kq_x q_y}{d}\right) + gd\frac{\partial \eta}{\partial x} - \frac{1}{K}\frac{\partial}{\partial x}\left[dv_t K\frac{\partial (q_x/d)}{\partial x}\right]$$
$$-\frac{1}{K}\frac{\partial}{\partial y}\left[dv_t K\frac{\partial (q_x/d)}{\partial y}\right] + \frac{f_b}{d^2}Qq_x = 0 \tag{A-2-14}$$

$$\frac{\partial q_y}{\partial t} + \frac{1}{K}\frac{\partial}{\partial x}\left(\frac{Kq_y q_x}{d}\right) + \frac{1}{K}\frac{\partial}{\partial y}\left(\frac{Kq_y^2}{d}\right) + gd\frac{\partial \eta}{\partial y} - \frac{1}{K}\frac{\partial}{\partial x}\left[dv_t K\frac{\partial (q_y/d)}{\partial x}\right]$$
$$-\frac{1}{K}\frac{\partial}{\partial y}\left[dv_t K\frac{\partial (q_y/d)}{\partial y}\right] + \frac{f_b}{d^2}Qq_y = 0 \tag{A-2-15}$$

$$\frac{\partial Kq_x}{\partial x} + \frac{\partial Kq_y}{\partial y} + \frac{\partial S\eta}{\partial t} = 0 \tag{A-2-16}$$

where K is the wetted volume ratio. Using the volume portion where the flow exists in the computational grid as V and the volume of the entire grid as V_o, it is expressed by the following equation.

$$K = \frac{V}{V_0} \tag{A-2-17}$$

When the water level does not exceed the top of the structure, the wetted volume ratio in the grid is equal to the wetted area ratio, but when the water level exceeds the top of the structure, the wetted area ratio is 1.

2) Sediment Transport Model Including Beach Areas

(a) Calculation of Bed Load Sediment Transport

The bed load transport rate in a oscillating flow is calculated using the following equation by Ribberink (1998).

$$\Phi_b = \frac{q_b}{\sqrt{\Delta gd_{50}^3}} = \begin{cases} m\left[\left|\theta_s'(t)\right| - \theta_c\right]^{1.65}\dfrac{\theta_s'(t)}{\left|\theta_s'(t)\right|} & \left(\theta_s'(t) \geq \theta_c\right) \\ 0 & \left(\theta_s'(t) < \theta_c\right) \end{cases} \tag{A-2-18}$$

where Φ_b is the dimensionless bed load transport rate, q_b is the bed load transport rate per unit time through a unit width of the vertical section perpendicular to the transport direction [bed load rate,

413

Appendix MANUALS OF NUMERICAL PREDICTION MODELS

m³/m/s], d_{50} is the median grain size of the bed load, Δ is the relative density of the bed load [$= (\rho_s - \rho)/\rho$], ρ_s is the density of the bed load, ρ is the density of water, and m is the bed load transport coefficient determined from the validation simulations for each beach, which is about 11 for fine sand. $\theta_s'(t)$ is Shields number in the oscillating flow, expressed by the following equation.

$$\theta_s'(t) = \frac{0.5 f_w' \left| u_b(t) \right| u_b(t)}{(\rho_s - \rho) g d_{50}} \tag{A-2-19}$$

where $u_b(t)$ is the reciprocating velocity on the sea bottom boundary layer, f_w' is the friction coefficient expressed as follows.

$$f_w' = \begin{cases} e^{\left\{ 5.2(k_s / \hat{a})^{0.194} - 5.98 \right\}} & (k_s / \hat{a} < 0.63) \\ 0.3 & (k_s / \hat{a} \geq 0.63) \end{cases} \tag{A-2-20}$$

$\hat{a}$ is the velocity amplitude of the on-offshore direction near the sea bottom, k_s is the representative roughness length on the sea bottom expressed as follows.

$$k_s = \max \left\{ 3d_{50}, d_{50} \left[1 + 6 \left(\left| \overline{\theta_s} \right| - 1 \right) \right] \right\} \tag{A-2-21}$$

$\left| \overline{\theta_s} \right|$ is the absolute value of the Shields number defined using Eq. (A-2-22).

$$\left| \overline{\theta_s} \right| = \frac{\left| \tau_b(t) \right|}{(\rho_s - \rho) g d_{50}} \tag{A-2-22}$$

$$\left| \tau_b(t) \right| = 0.5 \rho f_w' u_b(t)^2 = 0.25 \rho f_w' \hat{U}^2 \tag{A-2-23}$$

$\hat{U}$ is the velocity amplitude of the oscillating flow.

θ_c is the transport limit Shields number, Eq. (A-2-24) from van Rijn (1993) is used.

$$\theta_c = \frac{\tau_{bc}}{(\rho_s - \rho) g d_{50}} = \begin{cases} 0.24 D_*^{-1} & 1 < D_* < 4 \\ 0.14 D_*^{-0.64} & 4 \leq D_* < 10 \\ 0.04 D_*^{-0.1} & 10 \leq D_* < 20 \\ 0.013 D_*^{0.29} & 20 \leq D_* < 150 \\ 0.055 & 150 \leq D_* \end{cases} \tag{A-2-24}$$

where D_* is the nondimensional value given in Eq. (A-2-25) using the kinematic viscosity coefficient v.

$$D_* = d_{50} \left(\frac{g \Delta}{v^2} \right)^{1/3} \tag{A-2-25}$$

In the case of an inclined bottom with slope angle β, the transport limit Shields number θ_{sc} given in Eq. (A-2-26) is used instead of θ_c, because the effect of gravity must be considered.

$$\theta_{sc} = \theta_c \frac{\sin(\phi + \beta)}{\sin \phi} \tag{A-2-26}$$

where ϕ is the repose angle.

2. Numerical Model for Beach Change Due to High Waves

(b) Calculation of Suspended Load Transport Rate

The transport of the suspended load is also important in the calculation of beach profile changes. The on-offshore transport of the suspended load due to turbulence in the wave zone is represented by the following equation.

$$\frac{\partial C_n}{\partial t} + \frac{\partial \tilde{u} C_n}{\partial x} + \frac{\partial \tilde{u} C_n}{\partial y} = \frac{\partial}{\partial x}\left(\nu_t \frac{\partial C_n}{\partial x} \right) + \frac{\partial}{\partial y}\left(\nu_t \frac{\partial C_n}{\partial y} \right) - C_s + C_{ut} \tag{A-2-27}$$

where C_n is the depth integrated suspended load concentration, C_s is the suspended load rate that settles to the bottom and C_{ut} is the suspended load rate that rolls up from the bottom.

The settling rate C_s and the roll-up rate C_{ut} are evaluated assuming a vertical distribution of the suspended load. If the vertical distribution of the suspended load concentration is $C_n(z)$, then C_s and C_{ut} are expressed as follows.

$$C_{ut} = -\nu_t \left. \frac{\partial C_n(z)}{\partial z} \right|_{z=z_a} \tag{A-2-28}$$

$$C_s = w_s C_n(w_s / 2) \tag{A-2-29}$$

On real coasts, sheet flow conditions prevail almost everywhere including inside and outside the wave breaking zone. Therefore, Eq. (A-2-30) from Soulsby (1997), which gives the vertical distribution of the suspended load in sheet flow conditions, is used.

$$C_n(z) = C_a (z / z_a)^{-b} \tag{A-2-30}$$

where b is called the Rouse number and is expressed as follows.

$$b = \frac{w_s}{\kappa u_*} \tag{A-2-31}$$

κ is the Kalman constant, u_* is the friction velocity, and w_s is the settling velocity of sand particles from van Rijn (1984), which can be found by the following equation.

$$w_s = \begin{cases} \dfrac{\nu D_*^3}{18 d_{50}} & (d_{50} \leq 100 \mu\mathrm{m}) \\[3mm] \dfrac{10\nu}{d_{50}}\left[\left(1 + 0.01 D_*^3 \right)^{0.5} - 1 \right] & (100\mu\mathrm{m} < d_{50} \leq 1000\mu\mathrm{m}) \\[3mm] \dfrac{1.1\nu D_*^{1.5}}{d_{50}} & (d_{50} > 1000\mu\mathrm{m}) \end{cases} \tag{A-2-32}$$

C_a and z_a are obtained from the proposal of Zyserman & Fredsøe (1994) as follows.

$$C_a = \frac{0.331(\theta_s - 0.045)^{1.75}}{1 + 0.720(\theta_s - 0.045)^{1.75}} \tag{A-2-33}$$

$$z_a = 2 d_{50} \tag{A-2-34}$$

(c) Calculation of Topographic Beach Area Change

The following conservation equation for drifting sand is used to calculate topographic change.

$$\frac{\partial \zeta}{\partial t} = -\frac{1}{1 - \varepsilon_s}\left(\frac{\partial q_{bx}}{\partial x} + \frac{\partial q_{by}}{\partial y} - C_s + C_{ut} \right) \tag{A-2-35}$$

Appendix MANUALS OF NUMERICAL PREDICTION MODELS

where ζ is the local bottom height, q_{bx} and q_{by} are the bed load transport in the x and y directions respectively, and ε_s is the porosity of the sand layer.

3) Difference Schemes and Numerical Simulation Method

The above partial differential equations are computed by numerically solving them in space and time. For spatial differences, a staggered grid is used. The advection terms in the equations of motion are solved on a central difference scheme, while the advection terms in the other equations are solved using an upwind difference scheme. The other terms are solved using a central difference scheme. The Crank-Nicolson scheme is used for time differencing.

When calculations are carried out using the Crank-Nicolson scheme, it is necessary to use other variable values at a new time to find the unknown variable of interest. For example, when solving the equations of motion to obtain the flow rates, the water level and eddy viscosity coefficient at the new time step must be used. This means that the calculations must be repeated until a convergence is reached. To facilitate the rapid convergence in the iteration process, the calculation time interval Δt is determined by the stability condition (C.F.L. condition) of the calculation.

4) Wave Input Method

At the free-transmitting boundary, the direction of wave propagation is considered and the radiating Sommerfeld boundary condition is adopted to give the water level variation at the boundary. At the offshore boundary, the given input waves must be allowed to enter the computational domain and the reflected waves from the computational domain must be allowed to exit the domain freely.

The water level elevation for irregular waves η is assumed to be expressed as a superposition of a number of component waves, as shown in the following equation.

$$\eta(x,y,t) = \sum_{n=1}^{\infty} \sum_{m=1}^{\infty} A_{nm} \cos\{k_n \cos\theta_m x + k_n \sin\theta_m y - \sigma_n t + \varphi_{nm}\} \qquad \text{(A-2-36)}$$

where A_{nm} is the single amplitude of the component wave, k_n is the wavenumber of the component wave, θ_m is the angle of propagation of the component wave, $\sigma_n = 2\pi \times f_n$ is the angular frequency of the component wave, f_n is the frequency of the component wave, and φ_{nm} is the initial phase of the component wave, which is randomly distributed between 0 and 2π.

If the number of component waves is limited to N frequencies and M directions, the equation becomes as follows.

$$\eta(x,y,t) = \sum_{n=1}^{N} \sum_{m=1}^{M} A_{nm} \cos\{k_n \cos\theta_m x + k_n \sin\theta_m y - \sigma_n t + \varphi_{nm}\} \qquad \text{(A-2-37)}$$

The amplitude of a component wave in an arbitrary interval (f, $f+\Delta f$; θ, $\theta+\Delta\theta$) is expressed as follows.

$$A_{nm} = \sqrt{2E_n(f)\Delta f \frac{G_m(f,\theta)}{2\pi}\Delta\theta} \qquad \text{(A-2-38)}$$

where $E_n(f)$ is the frequency spectral density function and $G_m(f,\theta)$ is the directional distribution function.

2. Numerical Model for Beach Change Due to High Waves

The wave celerity, group celerity, wavenumber, and angular frequency in the irregular wave calculations of this model are the peak wave celerity, peak group celerity, peak wavenumber, and peak angular frequency of a representative wave. The input wave is then obtained by using the following Bretschneider-Mitsuyasu type spectrum and calculated based on the energy equidivision method.

$$S(f) = 0.258 H_{1/3}^2 T_{1/3} \left(T_{1/3} f\right)^{-5} \exp\left[-1.03\left(T_{1/3} f\right)^{-4}\right] \tag{A-2-39}$$

Moreover, the approximation is made by disregarding some of the high-frequency components and distributing the energy of the disregarded component waves equally to the remaining component waves. The approximation accuracy of each component wave varies depending on how the representative frequency is taken. Although the approximation accuracy of the high-frequency components is not perfect, it was decided to disregard this region because the energy components corresponding to those frequencies are small. The representative angular frequency is the angular frequency corresponding to the significant wave period.

Furthermore, the directional spectral density function of the waves $E_1(f,\theta)$ is given by the product of the frequency spectral density function $E(f)$ and the directional distribution function $G(\theta,f)$, as follows.

$$E_1(f,\theta) = E(f)G(\theta,f) \tag{A-2-40}$$

where $G(\theta,f)$ is the following Mitsuyasu type directional distribution function.

$$G(\theta,f) = G_0 \cos^{2S}\left(\frac{\theta}{2}\right) \tag{A-2-41}$$

$$G_0 = \frac{1}{\pi} 2^{2S-1} \frac{\left[\Gamma(S+1)\right]^2}{\Gamma(2S+1)} \tag{A-2-42}$$

where $\Gamma(S+1)$ is the gamma function, and S is the wave energy concentration parameter given by Eq. (A-2-43).

$$S = \begin{cases} S_{max}\left(\dfrac{f}{f_p}\right)^5 & \text{in the case of } f \le f_p \\[4mm] S_{max}\left(\dfrac{f}{f_p}\right)^{-2.5} & \text{in the case of } f > f_p \end{cases} \tag{A-2-43}$$

S_{max} is a parameter that expresses the degree to which the incident direction of each frequency component wave is concentrated in the principal direction and is as follows.

$$S_{max} = \begin{cases} 10 & \text{in case of wind waves} \\ 25 & \text{for swells with short attenuation distances} \\ 75 & \text{for swells with long attenuation distances} \end{cases} \tag{A-2-44}$$

f_p is the peak frequency of the frequency spectrum and is defined as follows.

$$f_p = \frac{1}{1.05 T_{1/3}} \tag{A-2-45}$$

For the calculation of irregular waves, the double-summation method is used, and 1320 component waves (120 frequency components × 11 directional components) are superimposed on the input.

5) Processing of the Wave Runup Tip

At the wave runup tip, as there are both wetted areas with flow and non-wetted areas without flow in the grid, the wetted area ratio S in the grid is less than 1.

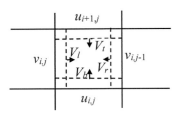

Fig. A.2.2 Illustration of how the runup tip is treated.

To estimate S, as shown in **Fig. A.2.2**, if the runup wave enters the grid from four directions, the velocities of propagation at the tip of the wave runup from each direction are expressed by the following equations.

$$V_t = \max\left[\sqrt{g(d_{ij} + d_{i+1j})} - u_{i+1,j}, 0\right], \quad V_b = \max\left[\sqrt{g(d_{ij} + d_{i-1j})} + u_{i,j}, 0\right]$$

$$V_l = \max\left[\sqrt{g(d_{ij} + d_{ij+1})} - v_{i,j}, 0\right], \quad V_r = \max\left[\sqrt{g(d_{ij} + d_{ij-1})} + v_{i,j-1}, 0\right]$$

If the values obtained by dividing the distances from the grid boundaries calculated using these propagation velocities by the grid interval length are expressed as ℓ_t, ℓ_b, ℓ_l, and ℓ_r respectively, S is obtained by the following equation.

$$S = 1 - \max\left[(1-\ell_t - \ell_b), 0\right] \times \max\left[(1-\ell_l - \ell_r), 0\right] \quad \text{(A-2-46)}$$

In addition, if the water depth in the grid mesh falls below 1 mm during the calculation, it is assumed that there is no more wetted area in the grid ($S = 0$).

2. Numerical Model for Beach Change Due to High Waves

2.2 Calculation Flow of the Program

The calculation flow of the program is shown in the diagram below.

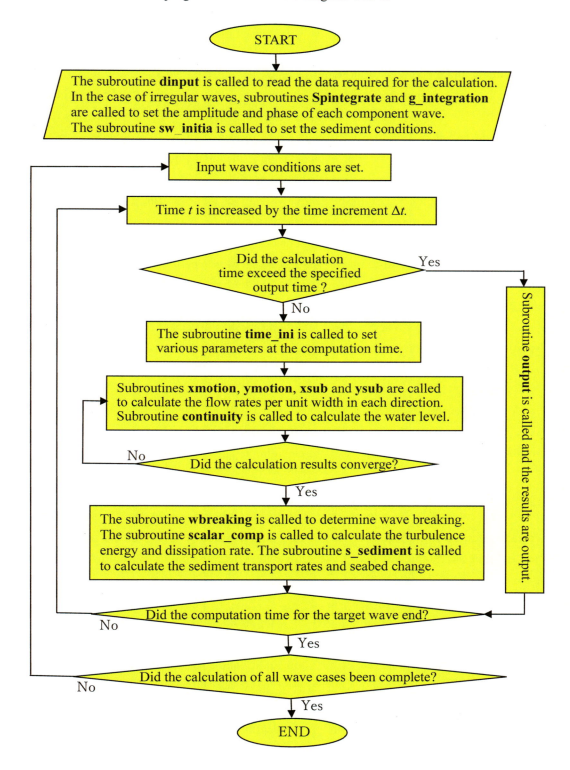

Fig. A.2.3 Calculation flow diagram of this program.

Appendix MANUALS OF NUMERICAL PREDICTION MODELS

1) Function of Main Program

The main program of **WAVETOPO.FOR** controls data reading, wave field calculation, topographic change calculation, and output of calculation results.

First, the subroutine **dinput** is called to read in various data required for the calculations (calculation range, grid interval length, bottom information, input wave information, structure information, and topography information).

Next, the subroutine **ini_set** is called to set the initial conditions and wave conditions for the calculation case and to start the calculation of wave fields and topographic changes. The subroutine **time_ini** is called to initialise each calculation time and to set the parameters required for the calculation. It then sets the water level change due to the input waves on the open boundary and enters an iterative loop. The subroutines **xmotion** and **ymotion** are then called to calculate the flow rates per unit width in the x and y directions respectively. The subroutine **continuity** is then called to update the parameters based on the newly obtained calculation results. It checks whether the calculations have converged and, if so, exits the iterative calculation loop. The subroutine **wbreaking** is then called to determine the wave breaking and to calculate the wave energy loss due to wave breaking. The subroutine **scalar_comp** is then called to calculate the turbulence energy and its dissipation rate. The subroutine **edvis_comp** is also called to calculate the eddy viscosity coefficient.

Finally, it checks whether the calculation time has exceeded the pre-set time, and if so, calls subroutine **output** to output the calculation results. These are repeated until the final calculation case.

2) Function of Each Subroutine

(a) Subroutine dinput

The subroutine **dinput** reads the data required for the calculation and calls the subroutine **breakwater_set** to set the coastal structure conditions in the computational domain.

(b) Subroutine breakwater_set

The subroutine **breakwater_set** sets the calculation conditions for the coastal structures (breakwaters, detached breakwaters, groynes, artificial reefs, etc.) within the computational domain and calls the subroutine **vol_comp** to calculate the effective volume ratio of each calculation grid in the computational domain.

(c) Subroutine vol_comp

The subroutine **vol_comp** calculates the effective volume ratio of each calculation grid in the computational domain to determine the permeability of the structures.

(d) Subroutine ini_set

The Subroutine **ini_set** sets the initial topographic conditions, sediment transport conditions for the computational domain, calls the subroutine **wave_set** to set the incident wave conditions, and further calls the subroutine **sw_initia** to set various parameters for sediment transport.

2. Numerical Model for Beach Change Due to High Waves

(e) Subroutine wave_set

The subroutine **wave_set** sets the wave energy spectrum and the amplitudes and initial phases of the component waves for each input wave condition, and calls the subroutine **gamma_int**, the **gamma function,** and the subroutine **sp_set** to set the wave spectrum It also calls the subroutine **wavkl** to determine the wavelength, wavenumber, wave celerity, and group celerity of each component wave at the offshore open boundary.

(f) Subroutine gamma_int

The subroutine **gamma_int** finds the gamma function value for each variable value.

(g) Subroutine sp_set

The subroutine **sp_set** sets the wave energy spectrum and calls the subroutine **spintegrate** to find the amplitude and period of each component wave.

(h) Subroutine spintegrate

The subroutine **spintegrate** performs numerical integration of the wave energy spectrum to calculate the amplitude and period of each component wave of equal energy.

(i) Subroutine g_integration

The subroutine **g_intergation** integrates the wave direction spectrum to find the incident direction of each directional component wave.

(j) Subroutine wavkl

The subroutine **wavkl** determines the wavelength, wavenumber, wave celerity, and group celerity of each component wave at the offshore open boundary using the linear dispersion relation.

(k) Subroutine sw_initia

The subroutine **sw_initia** sets the various parameters required for the calculation of sediment transport.

(l) Subroutine time_ini

The subroutine **time_ini** initialises the calculation time, sets various parameters required for the calculation, and calls the subroutine **vol_comp** to calculate the effective volume ratio of the permeable structure.

(m) Subroutine xmotion

The subroutine **xmotion** sets up the matrix needed to calculate the flow rate per unit width in the x direction and calls the subroutine **xsub** to calculate the flow rate per unit width in the x direction.

421

Appendix MANUALS OF NUMERICAL PREDICTION MODELS

(n) Subroutine xsub

The subroutine **xsub** calculates the flow rate per unit width in the x direction by backcalculating the matrix representing the equation of motion in the x direction set by the subroutine **xmotion**.

(o) Subroutine ymotion

The subroutine **ymotion** sets up the matrix needed to calculate the flow rate per unit width in the y direction and calls the subroutine **ysub** to calculate the flow rate per unit width in the y direction.

(p) Subroutine ysub

The subroutine **ysub** calculates the flow rate per unit width in the y direction by backcalculating the matrix representing the equation of motion in the y direction set by the subroutine **ymotion**.

(q) Subroutine continuity

The subroutine **continuity** updates the flow rate per unit width obtained from each iteration, calculates the water level, and sets other parameters such as velocities.

(r) Subroutine wbreaking

The subroutine **wbreaking** determines wave breaking and calculates the wave energy attenuation due to wave breaking.

(s) Subroutine scalar_comp

The subroutine **scalar_comp** calculates the turbulent energy and dissipation rate.

(t) Subroutine edvis_comp

The subroutine **edvis_comp** calculates the eddy kinematic viscosity coefficient using the turbulent energy and dissipation rate.

(u) Subroutine s_sediment

The subroutine **s_sediment** obtains topographic changes by calculating the transport of the bed load and the suspended load.

(v) Subroutine output

The subroutine **output** outputs hourly calculated values for the topography, topographic change, flow velocity, and wave height in the computational domain.

In the case of irregular waves, the subroutines **Spintegrate** and **g_integration** are called to set the amplitude and phase of each component wave. Moreover, the subroutine **sw_initia** is called to set the bottom condition.

422

2. Numerical Model for Beach Change Due to High Waves

2.3 Program Execution Methods

To perform a numerical calculation, users must prepare an executable program made from the source program and input data required for the calculation, and then place the executable program and input data in the folder on a personal computer. Next, users should double-click the executable program or create a batch file in the same folder and double-click it to start the numerical simulations.

The filenames of the source programs are **WAVETOPO.FOR** and **wavetopo.h**. The executable program **wavetopo.exe** can be created with modern FORTRAN compilers such as Microsoft Intel. However, in **wavetopo.h**, the maximum number of grids in the offshore direction (nx) is set to '500' and the maximum number of grids in the longshore direction (ny) is set to '600' in order not to use too much computer memory. If users wish to calculate over a larger area, they simply need to change nx and ny in the parameter statement to the desired values and recompile (but be aware that setting them to '1000×1000' or more may exceed the memory capacity of the PC or cause calculation times to take tens of days). To prevent the influence of lateral boundaries, which do not exist in the real sea area, 50 grids are automatically added outside each lateral boundary.

The input data consists of five files: (a) general calculation condition data such as input waves, (b) coastal structure data, (c) bathymetry data, (d) a file **DATAF.DAT** that communicates the names of these three data files to the PC, and (e) a random number file **RANDNUM.DAT**. This random number file is required to generate random input waves and can be simply placed in the same numerical calculation folder without changing its content.

1) Preparation Methods of Input Data
(a) Setting of Coordinates and Wave Direction

The coordinate system and input wave direction used in this numerical model are shown in **Fig. A.2.4**.

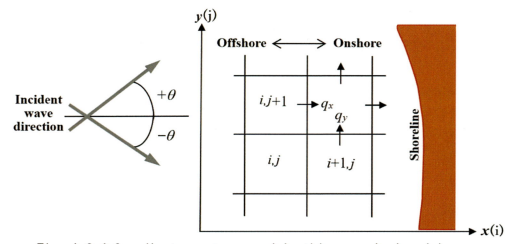

Fig. A.2.4 Coordinate systems used in this numerical model.

423

Appendix MANUALS OF NUMERICAL PREDICTION MODELS

(b) Setting of Input Data File DATAF.DAT

This input data file consists of the names of the following three data files.

(i) The file **fname1** consists of basic calculation condition data.

(ii) The file **fname2** consists of coastal structure data.

(iii) The file **fname3** consists of bathymetry data.

First, a file **DATAF.DAT** is created containing the names of these three data files. An example is shown below. These names can be changed for each calculation case for the sake of organization. "**wtcontrl.txt**", "**wtstruct.dat**", "**wtdepth.dat**".

The program to read the three data file names in this file is as follows:

```
fname='dataf.dat'
open(1,file=fname,status='old')
read(1,*) fname1,fname2,fname3
close(1)
```

(c) Data File fname1

The data file **fname1** for the basic calculation conditions must be created as follows (specific figures are examples):

91	Maximum number of grids in the x direction (imax, integer)
181	Maximum number of grids in the y direction (jmax, integer)
5.0	Grid interval length in the x direction (dx, real number, m)
5.0	Grid interval length in the y direction (dy, real number, m)
90	Number of grids in the output range of the calculation results from the centre of the y axis up and down (jpad, less than half of jmax)
0.4	Porosity of the sand layer, standard value = 0.4 (s_porosity, real number, dimensionless)
0.0002	Median grain size of sand (d_{50}, real number, m)
2.58	Specific gravity of sand, standard value = 2.58 (s_density, real, dimensionless)
2	Number of input wave conditions for topography change simulation (nccwave, integer)
'wave_condition 1'	Name of the first input wave dataset (chr, character)
21600.0	Action time of the first input wave (comp_time1, real number, s)
10.3	Significant period of the first input wave (wavet1, real number, s)
6.0	Significant wave height of the first input wave (waveh1, real number, m)
10.0	Principal angle of the first input wave (huong 1, real number, degrees) (refer to **Fig. A.2.4**)
10	Smax parameter of the first input wave (smax1, integer)
'wave_condition 2'	Name of the second input wave dataset (chr, character)
21600.0	Action time of the second input wave (comp_time1, real number, s)

424

2. Numerical Model for Beach Change Due to High Waves

8.4	Significant period of the second input wave (wavet1, real number, s)
4.0	Significant wave height of the second input wave (waveh1, real number, m)
-10.0	Principal angle of the second input wave (huong 1, real number, degrees) (refer to **Fig. A.2.4**)
10	Smax parameter of the second input wave (smax1, integer)

The program to read these data is as follows:

```
open(1,file=fname1,status='old')
read(1,*) imax      !Number of spatial meshes in the computation in x direction (integer)
read(1,*) jmax      !Number of spatial meshes in the computation in y direction (integer)
read(1,*) dx        !Spatial step in x direction (real, m)
read(1,*) dy        !Spatial step in y direction (real, m)
read(1,*) jpad      !Number of meshes from the center of the y axis for outputting (integer)
read(1,*) s_porosity     !Porosity of the sand, mean value = 0.4 (real, dimensionless)
read(1,*) d50       !Median grain size of the sand, in m (real, m)
read(1,*) s_density      !Relative density of sand (about 2.58) (real, dimensionless)
read(1,*) nccwave        !Number of wave conditions for the computation (integer)
do k=1,nccwave
read(1,*) chr            !Name of a set of wave data (character)
read(1,*) comp_time1(k)  !Time at the end of the computation in second (real, s)
read(1,*) wavet1(k)       !Wave period (real, s)
read(1,*) waveh1(k)      !Wave height (real, m)
read(1,*) huong 1(k)      !Offshore wave direction, in degree (real, degrees)
read(1,*) smax1(k)       !smax parameter used for the computation (integer)
enddo
close(1)
```

(d) Data File fname2

The coastal structures that can be included in "faname2" file are two types of non-submerged breakwaters (breakwaters, groynes, and detached breakwaters) and submerged breakwaters (artificial reefs). Breakwaters with complex planforms are decomposed into combinations of simple rectangular breakwaters as shown in **Fig. A.2.5**. Groynes and detached breakwaters are treated as simple rectangular breakwaters.

The method for setting up coastal structure data is described below.

chr	Name of breakwater data, usually "breakwater"
nbr	Number of simple breakwaters in the computational domain
nbrx1(i)	Coordinates in the x direction of the first vertex of the i th breakwater shown in **Fig. A.2.5** (grid number)
nbry1(i)	Coordinates in the y direction of the first vertex of the i th breakwater shown in **Fig. A.2.5** (grid number)
nbrx2(i)	Coordinates in the x direction of the second vertex of the same breakwater
nbry2(i)	Coordinates in the y direction of the second vertex of the same breakwater
nbrx3(i)	Coordinates in the x direction of the third vertex of the same breakwater
nbry3(i)	Coordinates in the y direction of the third vertex of the same breakwater

Appendix MANUALS OF NUMERICAL PREDICTION MODELS

nbrx4(i)	Coordinates in the *x* direction of the fourth vertex of the same breakwater
nbry4(i)	Coordinates in the *y* direction of the fourth vertex of the same breakwater
bheight(i)	Crown height of the simple breakwater above the tide level (m)
b_poro(i)	Porosity of the simple breakwater (0.0 for impermeable breakwater, 0.5 for permeable breakwater)
chr	Name of the artificial reef data, usually "artificial reef"
nrf	Number of artificial reefs in the computational domain
nrfx1(i)	Coordinates in the *x* direction of the first vertex of the *i* th artificial reef shown in **Fig. A.2.5** (grid number)
nrfy1(i)	Coordinates in the *y* direction of the first vertex of the *i* th artificial leaf shown in **Fig. A.2.5** (grid number)
nrfx2(i)	Coordinates in the *x* direction of the second vertex of the same artificial reef
nrfy2(i)	Coordinates in the *y* direction of the second vertex of the same artificial reef
nrfx3(i)	Coordinates in the *x* direction of the third vertex of the same artificial reef
nrfy3(i)	Coordinates in the *y* direction of the third vertex of the same artificial reef
nrfx4(i)	Coordinates in the *x* direction of the fourth vertex of the same artificial reef
nrfy4(i)	Coordinates in the *y* direction of the fourth vertex of the same artificial reef
rfheight(i)	Crown depth of the artificial reef below the tide level (m)
rf_poro(i)	Porosity of the artificial reef of interest

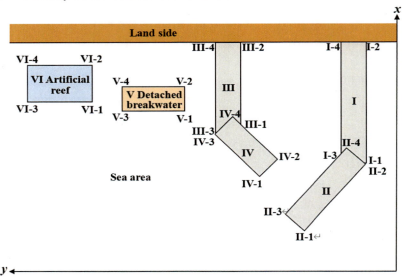

Fig. A.2.5 Illustration of how breakwater data is set.

An example of a coastal structure data file for the case shown in **Fig. A.2.5** is given below.

```
'breakwater'
5
     36    8   74    8   36   14   74   14   8.0   0.0
     16   25   35    8   22   29   40   12   8.0   0.0
```

2. Numerical Model for Beach Change Due to High Waves

46	40	74	40	46	46	74	46	8.0	0.0
31	35	36	30	44	46	49	42	8.0	0.0
52	54	60	54	52	70	60	70	8.0	0.0

'artificial_reef'

1

55	77	67	77	55	94	67	94	-1.0	0.5

The program for reading the coastal structure data is as follows:

```
open(unit=1,file=fname2,status='unknown')
read(1,*) chr
read(1,*) nbr                    ! number of breakwaters
if(nbr.eq.0) goto 41
do i=1,nbr
read(1,1) nbrx1(i),nbry1(i),nbrx2(i),nbry2(i),nbrx3(i),nbry3(i),nbrx4(i),nbry4(i)
  +     ,bheight(i),b_poro(i)              ! x, y coordinates of four points
enddo
41 continue
read(1,*) chr
read(1,*) nrf                    !number of artificial reef
if(nrf.eq.0) goto 42
do i=1,nrf
read(1,1) nrfx1(i),nrfy1(i),nrfx2(i),nrfy2(i),nrfx3(i),nrfy3(i),nrfx4(i),nrfy4(i)
  +     ,rfheight(i),rf_poro(i)            ! x, y coordinates of four points
enddo
1 format(8i5,2f6.2)
42 continue
```

The following points should be noted when setting up the coastal structure data.

(i) All coastal structures are represented as a combination of simple rectangular breakwaters. For example, in the case of **Fig. A.2.5**, there are four simple breakwaters (I-IV) constituting a harbour and a detached breakwater in the calculation area. These four simple breakwaters and the detached breakwater are represented by I, II, III, IV, and V respectively, and are set using the grid numbers (coordinates) in the x and y directions for each of the four vertices.

(ii) The input order of each simple breakwater is free, but the order of the coordinates (i,j) of the four vertices of each simple breakwater must be defined correctly. The order in which the vertices of each simple breakwater are set is as shown in **Fig. A.2.5**: right vertex on the offshore side of the breakwater (first), right vertex on the shore side of the breakwater (second), left vertex on the offshore side of the breakwater (third), left vertex on the shore side of the breakwater (fourth). For example, for the vertices of breakwater I, the order is I-1, I-2, I-3, I-4.

(iii) The simple breakwater should be a right-angled rectangle, but non-right angles are also acceptable. However, the coordinates of each vertex of the simple breakwater must satisfy the following conditions.

$$nbrx2 > nbrx1, \quad nbrx4 > nbrx3, \quad nbry3 > nbry1, \quad nbry4 > nbry2$$
$$nbrx2 - nbrx1 \geq nbrx3 - nbrx1, \quad nbry3 - nbry1 \geq nbry2 - nbry1$$

427

Appendix MANUALS OF NUMERICAL PREDICTION MODELS

$$\text{nbrx2 - nbrx1 = nbrx4 - nbrx3,} \qquad \text{nbry3 - nbry1 = nbry4 - nbry2}$$

Data for artificial reefs (submerged breakwaters) are prepared in the same way as for simple breakwaters.

(e) Data File fname3

This data file gives bathymetry data (water depth positive, ground height negative) within the computational domain. The coordinate origin of the computational domain is on the downside offshore of the sea area as shown in **Fig. A.2.4**. The x axis extends from the offshore to shore, and the y axis extends from right to left of the longshore direction as shown in **Fig. A.2.5**. The order of the bathymetry data is shown in **Fig. A.2.6**. The bathymetry data are first read in the shore direction from i = 1 on the x axis, at j = 1 on the y axis. This is repeated from j = 2 on the y axis upwards.

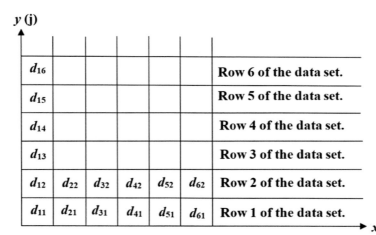

Fig. A.2.6 Illustration of how bathymetric data is placed.

The program for reading the bathymetry data is as follows:

```
open(unit=1,file=fname3,status='unknown')
do j=ky1,ky2
read(1,1) (d(i,j),i=kx1,kx2)
enddo
close(1)
1 format(100f8.2)
```

2) Output Method of Calculation Results

This program outputs the calculation results for each hour (3600 seconds) elapsed. The output items are the following five items, where the i in the data file name means the output after i time.

(i) Wave height distribution, data file name is "wavh_i.mdt".
(ii) x-directional component of flow velocity, data file name is "velu_i.mdt".
(iii) y-directional component of flow velocity, data file name is "velv_i.mdt".
(iv) Ground height distribution, data file name is "dept_i.mdt".

2. Numerical Model for Beach Change Due to High Waves

(v) Erosion depth and sedimentation height distribution, data file name is "ched_i.mdt".

The program outputting the calculation results is as follows:

```
        kre=kre+1
        open(1,file='dchrw',status='unknown')
        open(2,file='dchru',status='unknown')
        open(3,file='dchrv',status='unknown')
        open(4,file='dchrd',status='unknown')
        open(5,file='dchrdd',status='unknown')
        chrw='wavh'
        chru='velu'
        chrv='velv'
        chrd='dept'
        chrdd='ched'
        ext= '.mdt'
        write(1,21)chrw,kre,ext
        write(2,21)chru,kre,ext
        write(3,21)chrv,kre,ext
        write(4,21)chrd,kre,ext
        write(5,21)chrdd,kre,ext
     21 format(a4,i4,a4)
        close(1)
        close(2)
        close(3)
        close(4)
        close(5)
        open(1,file='dchrw',status='old')
        open(2,file='dchru',status='old')
        open(3,file='dchrv',status='old')
        open(4,file='dchrd',status='old')
        open(5,file='dchrdd',status='old')
        read(1,23) chrw
        read(2,23) chru
        read(3,23) chrv
        read(4,23) chrd
        read(5,23) chrdd
     23 format(a12)       .
        close(1)
        close(2)
        close(3)
        close(4)
        close(5)

        open(1,file=chrw,status='unknown')
        write(1,'(2i5,3f10.2)') ipcount,jpcount,+0.0,+0.0,dx
        if(kwave.eq.'regular') then
        do i=3,iprintmax
        write(1,'(10f8.3)') (wavha(i,j)/amax1(wavhc(i,j),1.), j=jprint1+1,jprint2)
        enddo
        else
        do i=3,iprintmax
        write(1,'(10f8.3)') (4.004*sqrt(wavh2(i,j)/amax1(h_count(i,j),1.)), j=jprint1+1,jprint2)
        enddo
        endif
        close(1)
        open(2,file=chru,status='unknown')
```

Appendix MANUALS OF NUMERICAL PREDICTION MODELS

```
write(2,'(2i5,3f10.2)') ipcount,jpcount,+0.0,+0.0,dx
do i=3,iprintmax
write(2,'(10f8.3)') (uaver(i,j)/tvelaver,j=jprint1+1,jprint2)
enddo
close(2)
open(3,file=chrv,status='unknown')
write(3,'(2i5,3f10.2)') ipcount,jpcount,+0.0,+0.0,dx
do i=3,iprintmax
write(3,'(10f8.3)') (vaver(i,j)/tvelaver,j=jprint1+1,jprint2)
enddo
close(3)
open(4,file=chrd,status='unknown')
write(4,'(2i5,3f10.2)') ipcount,jpcount,+0.0,+0.0,dx
do i=3,iprintmax
write(4,'(10f8.3)') (bed(i,j,2)*b_i(i,j)*b_rf(i,j)+bedst(i,j)-wlevel_average, j=jprint1+1,jprint2)
enddo
close(4)
open(5,file=chrdd,status='unknown')
write(5,'(2i5,3f10.2)') ipcount,jpcount,+0.0,+0.0,dx
do i=3,iprintmax
write(5,'(10f8.3)') ((bed(i,j,2)-bedini(i,j))*b_i(i,j)*b_rf(i,j)+bedst(i,j), j=jprint1+1,jprint2)
enddo
close(5)
```

Where "ipcount" and "jpcount" are the number of output grids in the x axis and y axis directions respectively.

The results of the simulation using the sample data provided with the accompanying numerical program are illustrated and presented below for reference.

Fig. A.2.7 shows the initial topographic data where a fishing port with breakwaters (porosity is 0.0) and a detached breakwater (porosity is 0.3) have been built on a sandy beach with a beach slope of 1/10 within the wave breaking zone. The distribution of significant wave heights, depth changes, and ground height and water depth after 6 hours from the start of the simulation for this case are shown in **Figs. A.2.8, A.2.9, and A.2.10** respectively.

Since waves do not enter from the lateral boundaries located at the upper and lower edges as shown in **Fig. A.2.8**, the topographic change near the two lateral boundaries is weaker than in the centre. Therefore, in **Figs. A.2.9 and A.2.10**, the areas between 170 m from the upper and lower edges, where the influence of the lateral boundary of the computational domain is strong have been cut off. Significant erosion occurred around the steep slopes near the shoreline. In addition, some erosion occurred due to the wave reflection in front of the breakwaters and the detached breakwater.

2. Numerical Model for Beach Change Due to High Waves

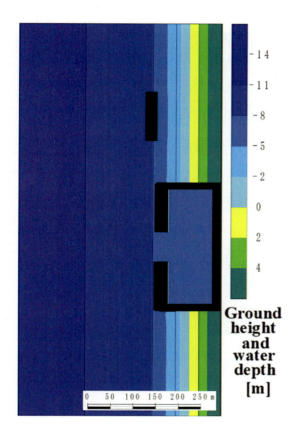

The black areas are a permeable breakwater (a detached breakwater) and breakwaters of a harbour.

Fig. A.2.7
Initial topography of the model coast.

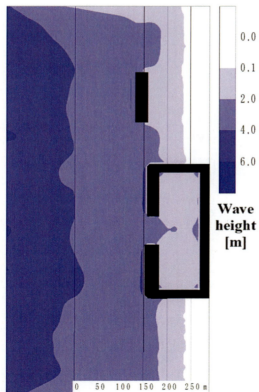

Fig. A.2.8
Distribution of significant wave heights after 6hrs from the start.

Appendix MANUALS OF NUMERICAL PREDICTION MODELS

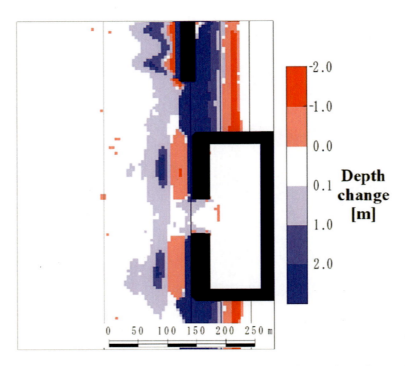

Fig. A.2.9 Distribution of depth changes after 6hrs from the start.

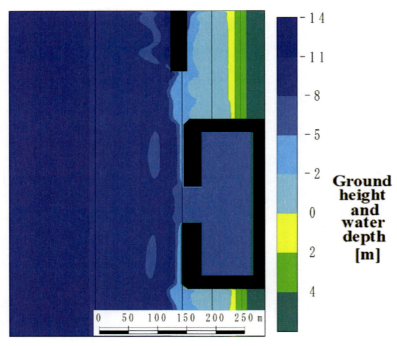

Fig. A.2.10 Topography after 6hrs from the start.

2. Numerical Model for Beach Change Due to High Waves

2.4 References for the Topographic Change Model for High waves

1) Ting, F.C.K. and Kirby, J.T.: Observation of undertow and turbulence in laboratory surf zone, *Coastal Engineering*, Vol.24, 1994, pp.51-80.

2) Schäffer, H.A., Madsen, P.A. and Deigaard, R.: A Boussinesq model for waves breaking in shallow water, *Coastal Engineering*, Vol.20, 1993, pp.185-202.

3) Ca, T. V., Yamamoto, Y. and Tanimoto, K.: Simulation of wave dynamics and scouring near coastal structures by a numerical model, *Proc. 28th Int. Conf. on Coastal Eng.*, ASCE, 2002, pp.1817-1829.

4) Yamamoto, Y., Charusrojtanadech, N. and Sirikaew, U.: Topographic Change Prediction of the Beach or the Seabed in the Front of a Coastal Structure, *Proc. 22nd Int. Offshore and Polar Eng. Conf.*, ISOPE, 2012, pp.1488-1495.

5) Yamamoto, Y., Yamaji, T. and Asano, G.: Importance of Wave Grouping in Wave Overtopping Calculations and Its Engineering Evaluation Method, *Journal of Coastal Engineering*, JSCE, Vol. 43, 1996, pp.741-744. (in Japanese)

6) Jones, W.P., and Launder, B.E.: The prediction of laminarization with a two-equation model of turbulence, *Int. J. Heat Mass Transfer.*, Vol.15, 1972, pp.301-314.

7) Hibberd, S., and Peregrine, H.D.: Surf and runup on beach: A uniform bore, *J. Fluid Mech.*, Vol.95, 1979, pp.323-345.

8) Ribberink, J.S.: Bed-load transport for steady flows and unsteady oscillatory flows, *Coastal Engineering*, Vol.34, 1998, pp.59-82.

9) van Rijn, L.C.: *Principles of sediment transport in rivers, estuarie and coastal seas*, Aqua Publications, Amsterdam, The Netherlands, 1993.

10) Soulsby, R.: *Dynamics of marine sands. A manual for practical application*, Thomas Telford, UK, 1997, 249p.

11) Van Rijn L.C.: Sediment transport: part I: bed load transport, *J. Hydraul. Div.*, *Proc. ASCE*, 110 (HY10), 1984, pp.1431-1456.

12) Van Rijn L.C.: Sediment transport: part II: suspended load transport, *J. Hydraul. Div.*, *Proc. ASCE*, 110 (HY11), 1984, pp.1613-1641.

13) Van Rijn L.C.: Sediment transport: part II: suspended load transport, *J. Hydraul. Div.*, *Proc. ASCE*, 110 (HY12), 1984, pp.1733-1754.

14) Zyserman J.A. and Fredsøe, J.: Data analysis of bed concentration of suspended sediment, *J. Hydrau. Engg.*, ASCE, 120, 1994, pp.1021-1042.

Appendix MANUALS OF NUMERICAL PREDICTION MODELS

3. Numerical Model for Coastal Topographic Change Due to Tsunamis

3.1 Overview of the Numerical Model

Following the study of Vu et al. (2010), this numerical model performs inundation calculations including inland areas by solving the two-dimensional non-linear long wave equations as the equations of motion for inundation. The input variable is the temporal changes in water level from an appropriate offshore depth. The model then can be used to calculate the topographic changes due to spatial variation of bed load transport rates, and settling and uptake rates of suspended sediment load.

Improving the accuracy of fault models that control tsunami generation is crucial for the progress of tsunami research. This will improve the reproduction of past tsunamis and the prediction of future tsunamis, thereby providing useful knowledge about the generation and propagation of tsunamis and the process of tsunami disaster generation. On the other hand, if the main purpose is to evaluate the effectiveness of tsunami countermeasures, improving the accuracy of tsunami propagation and run-up calculations is more important, and the accuracy of input waveform reproduction becomes less important. In addition, the number of cases for comparison can be increased by reducing the calculation range from a location relatively close to the coast, rather than from the offshore location where the fault occurs, to the shoreline. Therefore, a fault model is not incorporated into this numerical model.

Floods and storm surges are also long wave phenomena similar to tsunamis. These can be calculated using this numerical model by changing the water level input method to suit the respective phenomena.

The finite difference method is used for numerically solving the governing equation, and the equations of motion and continuity of inundation consider the inundated area within each calculation grid as a ratio to the grid area [see Yamamoto (2006)]. The method proposed in the Inundation Simulation Model Manual (draft) by Kuriki et al. (MLIT in Japan, 1996) is adopted for the roughness coefficients in the equations of motion. In addition, the Ribberink (1998) equation for a oscillatory flow is used to calculate the bed load transport. The equations that can calculate the suspended load rates due to sedimentation and roll-up are used to determine the suspended load transport. Soulsby's (1997) equation is adopted for the vertical distribution of suspended load transport.

The studies of Yamamoto et al. (2008, 2018) are used as references for applying this model to tsunami inundation problems. The studies of Vu et al. (2010), Naruyoshi et al. (2012), Yamamoto et al. (2019), and Ahmadi et al. (2020) are used for it's application to topographic change problems due to tsunamis.

1) Prediction Model for Inundation by Long Waves

(a) Basic Equations of Inundation Prediction Model

This model consists of mass and momentum conservation law for the inundation flow. However, as overland inundation may result in areas that cannot be inundated within a single grid, the basic equations for overland inundation are derived using the wetted area ratio S in a horizontal 2D control volume as follows.

3. Numerical Model for Coastal Topographic Change Due to Tsunamis

Continuity equation:

$$\frac{\partial f_y q_x}{\partial x} + \frac{\partial f_x q_y}{\partial y} + \frac{\partial S \eta}{\partial t} = 0 \tag{A-3-1}$$

Equation of motion in the x direction:

$$\frac{\partial q_x}{\partial t} + \frac{1}{S}\frac{\partial}{\partial x}\left(\frac{S q_x^2}{d}\right) + \frac{1}{S}\frac{\partial}{\partial y}\left(\frac{S q_x q_y}{d}\right) + gd\frac{\partial \eta}{\partial x} - \frac{1}{S}\frac{\partial}{\partial x}\left[dv_t S\frac{\partial(q_x/d)}{\partial x}\right]$$
$$-\frac{1}{S}\frac{\partial}{\partial y}\left[dv_t S\frac{\partial(q_x/d)}{\partial y}\right] + \frac{f_d}{d^2}Q q_x = 0 \tag{A-3-2}$$

Equation of motion in the y direction:

$$\frac{\partial q_y}{\partial t} + \frac{1}{S}\frac{\partial}{\partial x}\left(\frac{S q_y q_x}{d}\right) + \frac{1}{S}\frac{\partial}{\partial y}\left(\frac{S q_y^2}{d}\right) + gd\frac{\partial \eta}{\partial y} - \frac{1}{S}\frac{\partial}{\partial x}\left[dv_t S\frac{\partial(q_y/d)}{\partial x}\right]$$
$$-\frac{1}{S}\frac{\partial}{\partial y}\left[dv_t S\frac{\partial(q_y/d)}{\partial y}\right] + \frac{f_d}{d^2}Q q_y = 0 \tag{A-3-3}$$

where q_x and q_y are the flow rates per unit width defined by $q_x = \int u dz$ and $q_y = \int v dz$ using the velocities u and v in the x and y directions respectively (z is the depth coordinate), η is the water level elevation from the still water surface, f_x and f_y are the proportions of the flooded area on the sides parallel to the x and y directions of the computational grid respectively, S is the proportion of the flooded area in the computational grid (wetted area ratio), d is the total water depth (the depth from the still water surface $h + \eta$), g is the acceleration of gravity, v_t is the eddy kinematic viscosity coefficient for inundation (assumed to be equivalent to the water depth as the size of the eddy) and f_d is the resistance coefficient due to houses, structures and trees. Q is the vector composite of the flow components in both directions ($= \sqrt{q_x^2 + q_y^2}$).

When calculating inundation on land, the coefficients f_x, f_y, and S are introduced into the basic equations to allow for non-inundated areas within the control volume in the upstream region. However, since the continuity equation considers the equilibrium between the inflow and outflow of the control volume in the x and y directions and the water level fluctuations over the entire volume area, f_y is considered in the first term, f_x in the second term and S in the third term. The equations of motion, on the other hand, only consider S, since the motion concerning the entire mass in the control volume is considered.

(b) Treatment of Resistance Coefficients due to Ground Surface Constituents

This model considers resistance to flow due to buildings and trees, based on the water depth and the roughness in the equations of motion. The resistance coefficients due to buildings and trees are set according to "Inundation Simulation Model Manual (draft)" by Kuriki et al. (the Public Works Research Institute, Ministry of Land, Infrastructure and Transport in Japan, 1996).

$$f_d = \alpha\frac{gn^2}{d^{1/3}} \tag{A-3-4}$$

Appendix MANUALS OF NUMERICAL PREDICTION MODELS

where α is the correction factor (basically $\alpha = 1$, but as f_d is abnormally large when depth d is close to zero, a reasonable result can be obtained either by setting $f_d = 0.01$ on the seabed and $\alpha = 0.01$ on land, or by setting $\alpha = 1$ and fixing it to 0.1 when $f_d > 0.1$) and n is the roughness coefficient.

In urban areas, especially since the building density is only reflected in the area ratio, a method was adopted to set the roughness coefficient for the building zone according to the building occupancy ratio (= percentage of building area in the grid) and to combine it with the roughness coefficients for other land uses. In other words, a weighted average is used to obtain the roughness coefficient n_0 for areas other than buildings, and the floodplain roughness coefficient n is obtained from the building occupancy ratio O_s and the inundation depth d, using the following equation.

$$n^2 = n_0^2 + 0.020\frac{O_s}{100 - O_s}d^{4/3} \tag{A-3-5}$$

$$n_0^2 = \frac{n_1^2 A_1 + n_2^2 A_2 + n_3^2 A_3}{A_1 + A_2 + A_3} \tag{A-3-6}$$

$$n_1 = 0.06, \quad n_2 = 0.027, \quad n_3 = 0.05$$

where the subscripts 1, 2, and 3 in the area A and the roughness coefficient n refer to agricultural land, roads, and other land uses in each grid. Agricultural land includes rice paddies, fields, forests, orchards, and bamboo forests. National roads and major regional roads are considered as roads. Road areas also include footpaths along roads. Other land uses include wastelands, grasslands, wetlands, and salt pans.

Various facilities within the flood zone can significantly impact the propagation of flood waters, flood extent, and inundation depths. These impacts should be appropriately modelled by considering the impact magnitude and the complexity of the model.

Roads built on mounds and embankments are treated as a kind of wall. These ssignificantly influence the propagation of flood waters and should therefore be included in the calculation conditions as much as possible.

In addition, the overtopping from river embankment formula can be used to calculate the overtopping flow from coastal embankments, and the formula proposed by Yamamoto et al. (2008) is also available for tsunami-specific overtopping flows. Furthermore, if the overtopping flow rate due to high waves during a storm surge is used as an input condition, the Goda diagram for overtopping rates can be used.

2) Prediction Model for Topographic Change
(a) Calculation of Bed Load Transport Rates

The Ribberink (1998) equation obtained from oscillatory flow experiments is used to calculate the bed load transport rates.

$$\Phi_b = \frac{q_b}{\sqrt{\Delta g d_{50}^3}} = \begin{cases} C_B\left[\left|\theta_s'(t)\right| - \theta_{sc}\right]^{1.65}\dfrac{\theta_s'(t)}{\left|\theta_s'(t)\right|} & \left(\theta_s'(t) \geq \theta_{sc}\right) \\ 0 & \left(\theta_s'(t) < \theta_{sc}\right) \end{cases} \tag{A-3-7}$$

where Φ_b is the dimensionless bed load transport rates, q_b is the bed load transport rates per unit width and per unit time (m³/m/s), d_{50} is the median grain size of the bed load, Δ is the relative density of the

3. Numerical Model for Coastal Topographic Change Due to Tsunamis

bed load ($= (\rho_s - \rho)/\rho$), ρ_s is the density of the bed load, and ρ is the density of seawater. C_B is the bed load transport coefficient (the standard value is 11 for fine sand, but varies according to the median grain size, the uniformity coefficient, and the dry density of the local beach sediment).

For the exponent '1.65' in Eq. (A-3-7), '1.5' is considered correct from the concept of power, but perhaps due to a bias in the experimental measurements, of the values calculated with '1.65' gives the best fit with the measured value.

$\theta'_s(t)$ is the Shields number in a oscillatory flow and is expressed by the following equation.

$$\theta'_s(t) = \frac{0.5\rho f'_w |u_b(t)| u_b(t)}{(\rho_s - \rho) g d_{50}} \tag{A-3-8}$$

where $u_b(t)$ is the velocity in the offshore direction above the seabed boundary layer, and f'_w is the friction coefficient, expressed by the following equation.

$$f'_w = \begin{cases} e^{\{5.2(k_s/\hat{a})^{0.194} - 5.98\}} & (k_s/\hat{a} < 0.63) \\ 0.3 & (k_s/\hat{a} \geq 0.63) \end{cases} \tag{A-3-9}$$

where $\hat{a}$ is the velocity amplitude in the offshore direction near the seabed, and k_s is the representative seabed roughness length expressed by the following equation.

$$k_s = \max\left\{3d_{50}, d_{50}\left[1 + 6\left(\overline{|\theta_s|} - 1\right)\right]\right\} \tag{A-3-10}$$

where $\overline{|\theta_s|}$ is the time-averaged absolute value of the Shields number expressed by Eq. (A-3-11).

$$\overline{|\theta_s|} = \frac{\overline{|\tau_b(t)|}}{(\rho_s - \rho) g d_{50}} \tag{A-3-11}$$

$$\overline{|\tau_b(t)|} = 0.5\rho f'_w \overline{u_b(t)^2} = 0.25\rho f'_w \hat{U}^2 \tag{A-3-12}$$

where $\overline{|\tau_b(t)|}$ is the time-averaged absolute value of the shear stress on the seabed, and $\hat{U}$ is the velocity amplitude of the oscillatory flow.

The critical Shields number θ_{sc} on the seabed of the slope angle β is expressed by the following equation where the effect of gravity is considered.

$$\theta_{sc} = \theta_c \frac{\sin(\phi + \beta)}{\sin\phi} \tag{A-3-13}$$

where ϕ is the angle of repose of the bed load; θ_c is the critical Shields number on the horizontal seabed, using the following equation from van Rijn (1993).

$$\theta_c = \frac{\tau_{bc}}{(\rho_s - \rho) g d_{50}} = \begin{cases} 0.24 D_*^{-1} & 1 < D_* < 4 \\ 0.14 D_*^{-0.64} & 4 \leq D_* < 10 \\ 0.04 D_*^{-0.1} & 10 \leq D_* < 20 \\ 0.013 D_*^{0.29} & 20 \leq D_* < 150 \\ 0.055 & 150 \leq D_* \end{cases} \tag{A-3-14}$$

where D_* is a dimensionless value with a kinematic viscosity coefficient ν, expressed by the following equation.

$$D_* = d_{50}\left(\frac{g\Delta}{\nu^2}\right)^{1/3} \tag{A-3-15}$$

437

Appendix MANUALS OF NUMERICAL PREDICTION MODELS

Instead of determining the bed load transport coefficient C_B from a validation simulation using measured topographic change data, it can be determined from Eq. (A-3-16) by obtaining C_b, C_1, and C_2 from **Fig. A.3.1** and **Fig. A.3.2**, following the work of Yamamoto et al. (2019) and Ahmadi et al. (2020).

$$C_B = C_b \times C_1 \times C_2 \quad (A\text{-}3\text{-}16)$$

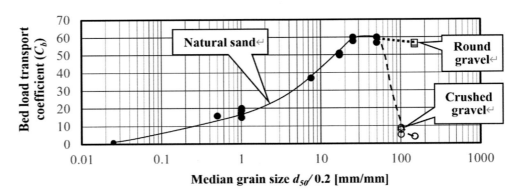

Fig. A.3.1 Influence of the median grain size to the bed load transport coefficient. (uniformity coefficients U = 1.5 - 3, dry densities $\rho_s \fallingdotseq 1.5$ g/cm³, refer to Ahmadi et al. (2020)).

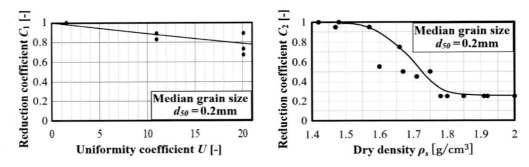

Fig. A.3.2 Relationship of the uniformity coefficient and the dry density to the bed load reduction coefficient C_1 and C_2 (d_{50} = 0.2 mm, $\rho_s \fallingdotseq 1.5$ g/cm³, refer to Ahmadi et al. (2020)).

(b) Calculation of Suspended Load Transport

In the case of large uniformity coefficients of the experiments in **Fig. A.3.2**, silt components are included (the proportion of silt components is around 10%). So, the suspended load transport is also important for the modeling of topographic change due to tsunami. The on-offshore transport of the suspended load due to a turbulent flow can be expressed by the following equation.

$$\frac{\partial C_n}{\partial t} + \frac{\partial \tilde{u} C_n}{\partial x} + \frac{\partial \tilde{u} C_n}{\partial y} = \frac{\partial}{\partial x}\left(v_t \frac{\partial C_n}{\partial x}\right) + \frac{\partial}{\partial y}\left(v_t \frac{\partial C_n}{\partial y}\right) - C_s + C_{ut} \quad (A\text{-}3\text{-}17)$$

where C_n is the depth-integrated suspended load concentration, $\tilde{u}$ is the depth-averaged ensemble mean velocity, v_t is the eddy kinematic viscosity coefficient, C_s is the suspended load rate that settles to the bottom, and C_{ut} is the suspended load rate that rolls up from the bottom. Here, the settling rate

3. Numerical Model for Coastal Topographic Change Due to Tsunamis

C_s and the roll-up rate C_{ut} are evaluated with the same treatment as that of (2) in 2) of **Section 2.1** in the **Appendix**.

(c) Calculation of Topographic Inundation Area Change

The following conservation equation for drifting sand is used for the calculation of topographic change in which inundation areas are included.

$$\frac{\partial \zeta}{\partial t} = -\frac{1}{1-\varepsilon_s}\left(\frac{\partial q_{bx}}{\partial x} + \frac{\partial q_{by}}{\partial y} - C_s + C_{ut}\right) \quad \text{(A-3-18)}$$

where ζ is the local ground height, q_{bx} and q_{by} are the bed load transport rates in the x and y directions respectively, and ε_s is the porosity of the sand layer.

3) Difference Schemes and Processing of the Wave Runup Tip

The partial differential equations of this model are solved numerically in space and time. For spatial differencing, a staggered grid is used. The advection terms in the equations of motion are solved on a central difference scheme, while the advection terms in the other equations are solved using an upwind scheme. The other terms are solved using a central difference scheme. The Crank-Nicolson scheme is used in the time differencing. Since the Crank-Nicolson scheme is an implicit method, iterative calculations are required for every time interval Δt. Moreover, the time interval Δt is determined by the stability condition (C.F.L. condition).

The wave runup tip is processed using the same method as that in 5) of **Section 2.1** in the **Appendix**. If the water depth in the grid falls below 1 mm during the calculation, it is assumed that the wetted area ratio in the grid $S = 0$.

The coordinate systems and the difference grids are shown in **Fig. A.3.3**.

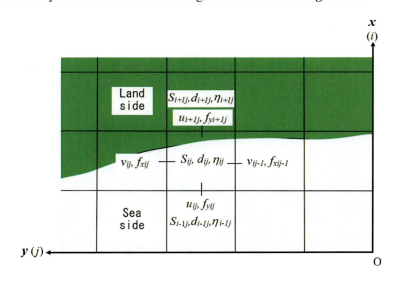

Fig. A.3.3 Coordinate systems and difference grids.

Appendix MANUALS OF NUMERICAL PREDICTION MODELS

3.2 Calculation Flow of the Program

The calculation program **ts_runupscour5.for** consists of a main program and several subroutines. The specification of the arrays used in the main program and subroutines is summarised in the header file **runupscour5.h** (current x-directional grid count 1100, y-directional grid count 1600).

1) Function of Main Program

The main program of **ts_runupscour5.for** controls input data reading, flow field and topographic change calculation, and outputting calculation results.

First, the subroutine **dinput** is called to read in various data required for the calculations (calculation time, an output time interval, water level elevation data to be input from the offshore boundary, data on ground height and water depth, building occupancy data, bottom sediment data, etc.).

Next, the subroutine **ini_set** is called to set the initial conditions and to start the calculation of flow fields and topographic changes. The subroutine **sfrac_comp** is then called to calculate the wetted area ratio in each grid. The subroutine **time_ini** is then called to set initial conditions at the time step of interest. The subroutine **edvis_comp** is also called to calculate the eddy viscosity coefficient.

Then, the calculation proceeds to an iterative loop. First, the subroutine **bc_incoming** is called to set the water level elevation at the offshore boundary. The subroutine **solveqx** is called to calculate the flow rates per unit width in the x direction and the subroutine **solveqy** is called to calculate the flow rates per unit width in the y direction. The subroutine **iter_update** is then called to store the calculation results of the water level and the flow rates per unit width. It then checks for convergence and decides whether to exit the calculation loop.

If a topographic change calculation is desired, the subroutine **s_sediment** is called to perform the topographic change calculation.

When the time for outputting the calculation results has been reached, the output subroutine **outputmdt** (**outputplt** for TECHPLOT is hidden) is called to output the results of the intermediate calculations. Furthermore, if the calculation end time is reached, the calculation is terminated.

2) Function of Each Subroutine
(a) Subroutine **dinput**

The subroutine **dinput** reads the conditions required for inundation and topographic change calculations. Here, the data required for the calculations are read from the "fname", "fname1" to "fname7" files.

The basic condition data are read in from "fname".

The ground height or water depth [m] for all grids are read in from "fname1".

The square of Manning's roughness coefficients for all grids are read in from "fname2".

The building occupancy ratios [%] for all grids are read in from "fname3".

The data to determine whether to carry out topographic change calculations for all grids (0 if the calculations are carried out, 1 if not) are read in from "fname5".

440

3. Numerical Model for Coastal Topographic Change Due to Tsunamis

Bed load transport coefficients for all grids are read from "fname6".

The sediment median grain sizes [mm/10] for all grids are read from "fname7".

The subroutine **sw_initia** is also called to set up the necessary parameters for the topographic change calculations.

(b) Subroutine `sw_initia`

This subroutine sets the various parameters required for the sediment transport rate calculation.

(c) Subroutine `ini_set`

This subroutine sets the initial conditions. The subroutine **vol_comp** is then called to determine the percentage of the fluid volume occupied by the fluid excluding the solid part in each initial computational grid, and the percentages of the flooded parts (flow present) of the sides parallel to the x and y directions in each grid. The subroutine **sfrac_comp** is then called to determine the movement of the leading surface of the flood wave in each grid and the wetted area ratio. The subroutine **time_ini** is also called to initialise the calculation for the new time.

(d) Subroutine **vol_comp**

This subroutine calculates the percentage of fluid volume occupied by the fluid excluding the non-submerged area on each grid.

(e) Subroutine **sfrac_comp**

This subroutine calculates the movement of the leading surface of the flood wave in each grid and the wetted area ratio.

(f) Subroutine **time_ini**

This subroutine initialises the calculation for the new time.

(g) Subroutine **edvis_comp**

This subroutine calculates the eddy kinematic viscosity coefficient in the computational domain.

(h) Subroutine **bc_incoming**

This subroutine inputs the tsunami water level elevation at each time step from the offshore boundary.

(i) Subroutine **solveqx**

This subroutine sets up the required data for the equation of motion in the x direction and the array of calculated results. The subroutine **thomasx** is called to back-calculate the array and obtain the flow rates per unit width in the x direction.

(j) Subroutine **thomasx**

Appendix MANUALS OF NUMERICAL PREDICTION MODELS

This subroutine back-calculates an array representing the difference equation of motion in the x direction using the Thomas Algorithm (refer to Tridiagonal matrix algorithm - Wikipedia, http s://en.wikipedia.org/wiki/Tridiagonal_matrix_algorithm).

(k) Subroutine **solveqy**

This subroutine sets up an array of required data and calculates results for the equation of motion in the y direction. The subroutine **thomasy** is then called to back-calculate the array and obtain the flow rates per unit width in the y direction.

(l) Subroutine **thomasy**

This subroutine back-calculates an array representing the difference equation of motion in the y direction using the Thomas Algorithm.

(m) Subroutine **iter_update**

This Subroutine stores the results of the current iterative calculation and prepares for the next iteration.

(n) Subroutine **s_sediment**

This subroutine calculates the bed load transport rates, the suspended load transport rates, and the topographic change.

(o) Subroutine **outputmdt**

This subroutine outputs the calculation results for each time (sec) set in the output interval of the input data file **runupscour.ctl**. The outputs are the following four items, where "i" in the data file name means the output after time i.

(i) The output of the change in ground height [m] is stored in the data file **dhed_i.mdt**.

(ii) The output of the ground height [m] is stored in the data file **bed_i.mdt**.

(iii) The output of the flow velocity in the x direction [m/s] is stored in the data file **u_i.mdt**.

If users want to output the flow velocity in the y direction, they only need to copy the FORTRAN executable statements of this subroutine relating to the output of the flow velocity in the x direction and change the output file name to 'output¥v' and the velocity array name to v(i,j,2).

(iv) The output of the water level elevation [m] is stored in the data file **eta_i.mdt**.

442

3. Numerical Model for Coastal Topographic Change Due to Tsunamis

3.3 Program Execution Methods

1) Executive Method of the Program

To execute a numerical simulation, an executable program made from two source programs, a set of input data containing the necessary data such as calculation conditions, and a sub-folder **output** (empty folder) for storing the output files of calculation results must be prepared.

Then, after the executable program, input data group, and empty sub-folder **output** are placed in a folder, double-click the executable program to start the simulation, and the calculation results are stored in the sub-folder **output**. Alternatively, if users create a batch file that contains the name of the executable program and the command "pause" in the same folder and double-click, the simulation starts and calculation results are stored in the sub-folder. If the program terminates abnormally, an error message is given, which is useful for investigating the cause of the problem. This batch file is also useful for automatically executing multiple simulations.

The source programs of interest are the computation program **ts_runupscour5.for** and the header file **runupscour5.h**, which controls the array, and are used to create an executable program using modern FORTRAN compilers such as Intel's. However, in **runupscour5.h**, the maximum number of grids in the offshore direction (nx) is set to '1100' and the maximum number of grids in the longshore direction (ny) to '1600', in order not to waste computer memory space. If users desire simulations in a wider calculation range, the nx and ny in the parameter statement must be changed from '1100×1600' to the desired values. When users change the number of grids in **runupscour5.h** or the contents of **ts_runupscour5.for**, they must compile the programe using the new values and create a new executable program (since there are existing executable programs **0.01gn2d.exe** and **1.0gn2d.exe**, it is better to give the new program a different name). In this case, it is necessary to move the new executable program to the folder where all input data files and **output** are placed.

There are two compilation methods: "debug mode" and "release mode". If there are no mistakes in the source program, compiling in "release mode" will speed up the calculation.

2) Preparation Methods of Input Data

(a) Grid Data for Water Depth and Ground Height

Grided data for water depth and ground height should be generated from the offshore area (the water depth is sufficiently deeper than twice the maximum tsunami height and can be regarded as almost uniform in the longshore direction), to the inundation limit area on the land side (the ground height is sufficiently higher than twice the maximum tsunami height). In this numerical simulation program, as shown in **Fig. A.3.3**, the coordinate origin is placed on the offshore side, the x axis is in the off-onshore direction, and the y axis is in the longshore direction. The grid interval is set between 5 m and 20 m to consider differences in the thickness of embankments and the density of houses. The reference height should be the mean tide level at the time of the calculation.

(b) Other Grided Data.

Data files on ground surface roughness, building occupancy ratios, range information for topographic change calculation, bed load transport coefficients, and median grain sizes of the bottom

Appendix MANUALS OF NUMERICAL PREDICTION MODELS

sediment must be prepared in the same format as the data file of ground height and water depth (the grid interval length, the number of grids in the x and y directions must be the same) as follows.

(i) Ground surface roughness: The square of Manning's roughness is set to assess surface roughness due to vegetation, etc. '0.06×0.06' is appropriate for arable land, '0.027×0.027' for roads, and '0.05×0.05' for wastelands.

(ii) Building occupancy ratios: The ratios of the non-flooded areas by buildings and trees to each grid area are set to assess the surface roughness. Around 70% is appropriate for densely built-up areas, 30% for private houses, 5% for forests, and 1% for wastelands.

(iii) Range information for topographic change calculation: The information indicating whether a topographic change in a grid should be calculated. '1' is set for the grid where topographic change need not or should not be calculated, and '0' for the grid where it should be calculated.

(iv) Bed load transport coefficients: values between '0' and '60' are set for calculation grids. For beaches constantly exposed to wave action, it is relatively easy to set a representative value due to the progressive sorting of the beach sand. However, for megatsunamis, it is unreasonable to set a representative value for the entire inundation area, consisting of various bed materials. It is more reasonable to set two representative values for the following two areas, the first in the lowland area and the second in the upland area. The negative influence of this setting is small because it is not necessary to consider multiple oscillating flows over several hours, as is the case with wind-driven high waves.

(v) Median grain sizes of the bottom sediment: For each grid, the grain size value is set by multiplying the value in milimeters by '10' to reduce the number of digits in the input data. In other words, '0.05' is set for 0.005 mm, and '2.0' is set for 0.2 mm. The value is then reset to '1/10' when the data is read.

(c) Data for Calculation Control

Users must create a data file consisting of the basic conditions to control the calculations (e.g. the upper limit for the calculation time interval Δt, the total calculation time, the output interval, time series data of water level elevation as input data from the offshore boundary, names of grid data files to be read, etc.).

Here, the time series data of the water level elevation are given at the offshore boundary (the grid number $i = 1$ in the x direction, the grid number $j = 1$ to ny in the y direction) in order to perform inundation calculations including in shallow water areas. This can be generated from the tide records from stations close to the target coast or determined from the existing numerical calculation results.

Further, it is easy to modify the numerical simulation program so that the overflow rates by existing formulae from the coastal dike/seawall or natural coastal cliffs are input as flow rates per unit width and the inundation can be calculated only on the landward side.

3. Numerical Model for Coastal Topographic Change Due to Tsunamis

3) Specific Example of Input Data Settings

The red boxes in **Fig. A.3.4** are executable programs (either **0.01gn2d.exe** or **1.0gn2d.exe** is acceptable, the former calculates the resistance coefficients on land at $0.01gn^2/d^{1/3}$, the latter calculates the same coefficients in $gn^2/d^{1/3}$ and fixes it at '0.1' if it is greater than '0.1'), input data files (7 types, but in this example the distribution data of the squared values of the ground surface roughness coefficients, **rough2.mdt**, has been ignored), and the sub-folder **output** that stores the output data files. These must be prepared in the same folder for the execution of simulations. Here, **checkwrite.dat** and **depth.mdt** are check files that the computer automatically creates during computation.

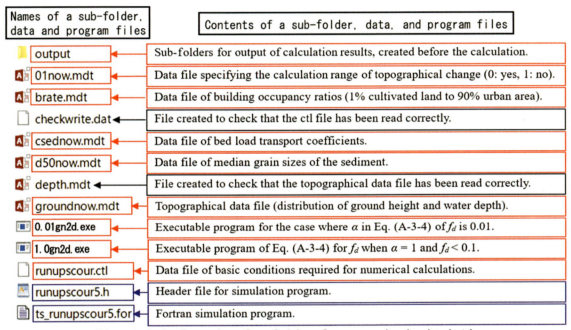

Fig. A.3.4 Example of a folder for numerical simulations.

(a) Sub-folder **output**

This sub-folder must be always prepared in the name **output** before the simulations are started to store the files of simulation resullts (output data) such as **bed__.mdt** [m] for the ground height and water depth, **dbed__.mdt** [m] for the change in ground height, **eta__.mdt** [m] for the inundation depth, and **u__.mdt [m/s]** for the flow velocity in the off-onshore direction.

(b) Data of Ground Height and Water Depth

An example of this data file is **groundnow2.mdt** (see **Fig. A.3.5**), where the number of grids in the x (off-onshore) and y (longshore) directions (1085×1501 in the example), the coordinates of the origin in the x and y directions (usually 0.0, 0.0), and the grid interval length (usually Δx = 5 to 20 m) are set on the first line. The ground height and water depth (positive values above the water surface and negative values below the water surface are standard, unit: m) are set in the second and subsequent lines. The sample data provided for the accompanying numerical program is prepared in a similar

Appendix MANUALS OF NUMERICAL PREDICTION MODELS

format to **groundnow.mdt** with a grid interval length of 25 m (number of grids 521×721), as the calculation time is very long for **groundnow2.mdt** with a grid interval length of 12 m.

The format to be read in the simulation program is as follows:

```
read(1,*) imax,jmax,xor,yor,dx
do i = 1, imax
  read(1,'(10f8.2)') (bed(i,j,2),j=1,jmax)
enddo
```

(c) Data of Building Occupancy Ratio

The building occupancy ratio is the area percentage (1 - 99%) occupied by buildings and trees in each grid cell and is used to evaluate the roughness in the equations of motion. An example of this data file is **brate.mdt** (see **Fig. A.3.6**). The same information as the data file of the ground height and water depth is set in the first line for checking the contents, but this line is skipped when running the calculation, and then the building occupancy ratio [%] is set after the second line.

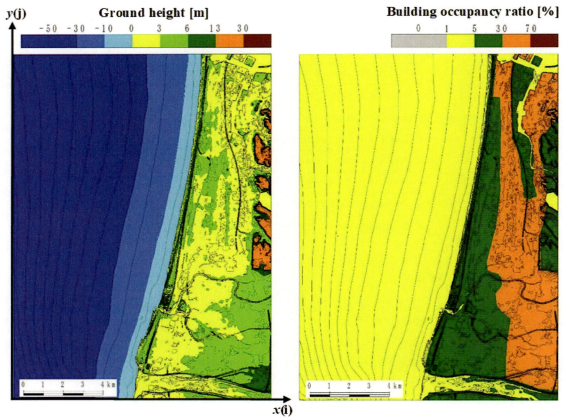

Fig. A.3.5 Example of topographic data (ground height and water depth).

Fig. A.3.6 Example of building occupancy ratios.

The format to be read in the simulation program is as follows:

```
read(1,*)
do i = 1, imax
```

3. Numerical Model for Coastal Topographic Change Due to Tsunamis

```
read(1,'(10f8.2)') (bldr(i,j),j=1,jmax)
enddo
```

(d) Data Specifying Calculation Range

An example of this data file is **01now.mdt** (see **Fig. A.3.7**). The first line is skipped, and then '0' and '1' data ('0' if the topographic change is calculated, '1' if not, no units) are set after the second line, corresponding to the data of the ground height and water depth. The reading format is as follows:

```
read(1,*)
do i = 1, imax
read(1,'(10f8.2)') (s_l(i,j),j=1,jmax)
enddo
```

(e) Data of Bed Load Transport Coefficient

The bed load transport coefficient C_B can be obtained from a validation simulation using measured data or from Eq. (A-3-16) using coefficients C_b, C_1, and C_2 obtained from **Fig. A.3.1** and **Fig. A.3.2**. An example of this data file is **csednow.mdt** (see **Fig. A.3.8**).

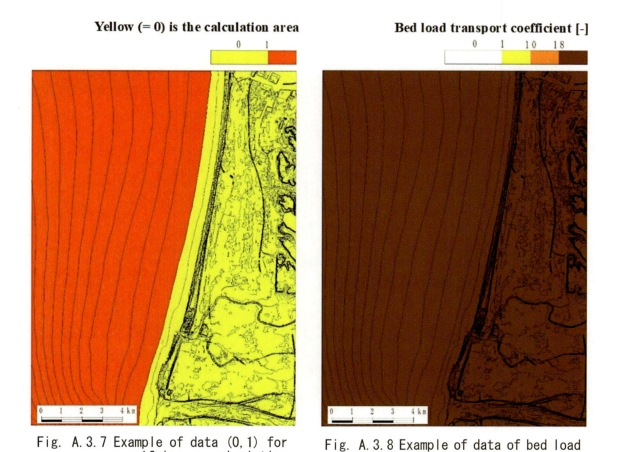

Fig. A.3.7 Example of data (0, 1) for specifying a calculation range.

Fig. A.3.8 Example of data of bed load transport coefficients.

The first line is skipped, and then the bed load transport coefficients (1 - 60, no units) are set after the second line. The reading format is as follows:

Appendix MANUALS OF NUMERICAL PREDICTION MODELS

```
read(1,*)
do i = 1, imax
read(1,'(10f8.2)') (sedc(i,j),j=1,jmax)
enddo
```

(f) Data of Median Grain Size

An example of this data file is **d50now.mdt** (see **Fig. A.3.9**). The first line is skipped, and then the median grain sizes of the bottom sediment (the median grain size value in mm multiplied by '10') are set after the second line. For example, '0.05' is set for 0.005 mm, or '2' is set for 0.2 mm). The value is reset to '1/10' when the data is read. The read format is as follows:

```
read(1,*)
do i = 1, imax
read(1,'(10f8.5)') (d50(i,j),j=1,jmax)
enddo
```

(g) Data for controlling Simulations

An example of data file for controlling numerical simulations is **runupscour.ctl**, the contents of which are listed below.

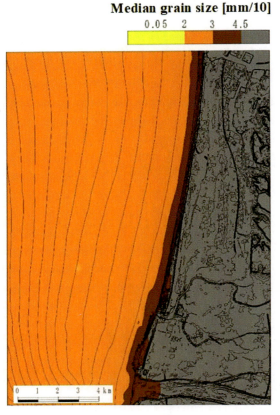

Fig. A.3.9 Example of distribution of median grain sizes [mm/10].

```
0.5  1980.  60.0      dt-upper limit(sec), duration(sec), output interval(sec)
8.00  6.00    11      water height(m) at j=1 & max, the discreted number of an incident wave
0.0 0.15 0.3 1.0 0.53 0.15 0.15 0.24 0.05 -0.10 -0.14    time-series of the incident wave
groundnow.mdt         ground height data (m)
0                     iground, if the seaward of ground height data is minus, iground=0
0    0.0025           iksw, square of roughness coefficient in the case of iksw=0
rough2.mdt            data of the quare of roughness coefficient that are used in the case of iksw=1
1    1.0              irsw, mean ratio (%) of building area in a mesh in the case of irsw=0
brate.mdt             ratio (%) of building area in each mesh that is used in the case of irsw=1|
'yes'                 ! sediment transport option, yes or no
01now.mdt             ! '0' is a mesh where topographical change is calculated, '1' is a mesh where isn't
1    11.0             ! ksedc, mean value of sediment rate coefficient in the case of ksedc=0
csednow.mdt           ! sediment rate coefficients that are used in the case of ksedc=1
0.0002                ! mean value of median grain size (m) in the case of ksedc=0
d50now.mdt            ! median grain sizes (mm/10) that are used in the case of ksedc=1
0.4                   ! mean porosity of the sand layer
2.58                  ! relative density of sand
```

(i) '0.5sec' is the upper limit of the time interval Δt. To obtain stable calculation results, should be set it at 0.5sec or less.

(ii) '1980sec' is the total calculation time of the target simulation case and should be set according to the required time.

3. Numerical Model for Coastal Topographic Change Due to Tsunamis

(iii) '60sec' is the output interval and should be set so that it does not exceed the memory capacity while considering the purpose of use. Normally, the interval is between 60 and 300 seconds.

(iv) '8.00m' and '6.00m' are the maximum water level elevation of the input tsunami at the offshore boundary (i,j) = (1,1) and (i,j) = (1, maximum value), and are interpolated and distributed on the offshore boundary.

(v) '11' is the number of divisions of the input tsunami water level elevation change over time, which can be increased up to a maximum value of '20'.

(vi) '0.0, 0.15, 0.3, 1.0, ... ' are the temporal variation data using the ratio of each temporal water level elevation to the maximum water level elevation at the offshore boundary of the input tsunami from the start to the end of the simulation, the total number of which must match the above number of divisions (currently '11'). In addition, the total time of this data on change over time must match the '1980sec' above. The reason for normalisation by the maximum water level elevation is that the pattern of water level elevation change over time can be seen not to change much on an open coast 100 km away, but the water level elevation itself should be different considerably if the offshore depth below the still water surface is not same. Also, the method of multiplying the time series data of normalised water level elevation at an observation station by the maximum water level elevation at the offshore boundary of the target coast for simulation saves labour in the preparation of the water level elevation data.

(vii) **groundnow.mdt** is a data file containing the ground height and water depth. The number of grid points in the off-onshore (x) and longshore (y) directions, the origin coordinates (usually x, y = 0.0, 0.0), and the grid interval length (usually 5 - 20 m) are read in on the first line. The ground height and water depth [m] are read in. If a data file with a different name is used, it must be rewritten to that name.

(viii) "iground" is set to '0' when the ground height and water depth data is positive above the still water surface and negative below the still water surface. In the opposite case, it is set to '1'.

(iv) "iksw" is set to '0' when a detailed distribution of surface roughness coefficients is not desired. However, the square of the surface roughness coefficient representative of the target coast (standard value is '0.0025') must always be set to the right of this '0'.

(x) If "iksw" is '0', **rough2.mdt** is skipped without being used, and its contents can be empty. But, it must not be deleted to prevent the order of the data from being shifted. On the other hand, if "iksw" is '1', because the information of this data file (distribution of the squared values of the surface roughness coefficients) is used, the data file for the distribution of the squared values of the surface coefficients must be created in the same format as the data file for the median grain size of the bottom sediment (no need to multiply by 10 as in the median grain size data.) and placed in the folder containing all the input data files.

(xi) "irsw" is set to '0' when the distribution of building occupancy ratios is not considered. However, the representative building occupancy ratio on the target coast must always be set to the right of this '0'. Currently '1.0' for wastelands is set as the representative building occupancy ratio, but is ignored if "irsw" is set to '1'.

Appendix MANUALS OF NUMERICAL PREDICTION MODELS

(xii) **brate.mdt** is skipped without being used if "irsw" is '0', so its contents can be empty, but it must not be deleted to prevent the order of the data from being shifted. On the other hand, if "irsw" is '1' as it is now, this data file (distribution of building occupancy ratios) is used, so it must be created in advance and placed in the folder containing all the input data files.

(xiii) **"yes"** is set if topographic change calculations are desired. If not desired, set **"no"** to skip the following data and not perform the topographic change calculation.

(xiv) **01now.mdt** is a data file that specifies the range in which topographic change calculations are performed within the entire calculation area. The calculations are performed on grids with a value of '0.0 ' and are not performed on grids with a value of '1.0'. (assuming an area where topographic changes can be ignored, e.g. where the area is covered with concrete or asphalt).

(xv) "ksedc" is set to '0' when the detailed distribution of the bed load transport coefficients and median grain sizes are not considered. However, the representative bed load transport coefficients for the target coast must always be set to the right of this '0'. Currently '11' for fine sand is set as the representative bed load transport coefficient, but is ignored if "ksedc" is set to '1'.

(xvi) **csednow.mdt** is skipped without being used if "ksedc" is '0', so its contents can be empty. But it must not be deleted to prevent the order of the data from being shifted. On the other hand, if "ksedc" is '1' as it is now, this data file (distribution of the bed load transport coefficients) is read, so it must be created in advance and placed in the folder containing all the input data files.

(xvii) '0.0002m' is the representative median grain size of the sediment and is used when "ksedc" is '0'. If "ksedc" is '1', it is skipped. However, it must not be deleted to prevent the order of the data from being shifted.

(xviii) **d50now.mdt** is not used if "ksedc" is '0' and is skipped, so its contents can be empty. But it must not be deleted to prevent the order of the data from being shifted. On the other hand, if "ksedc" is '1' as it is now, this data file (distribution for the median grain sizes) is read, so it must be created in advance and placed in the folder containing all the input data files.

(xix) '0.4' is the porosity of the sand layer.

(xx) '2.58' is the density of the bottom sediment.

However, the example data files prepared for this accompanying numerical program were produced with a grid interval length of 25 m (number of grids 521×721), because the data used in the actual calculations with a grid interval length of 12 m (e.g. groundnow2.mdt) have taken a very long time to calculate.

450

3. Numerical Model for Coastal Topographic Change Due to Tsunamis

4) Specific Examples of Calculation Results

The contents of all output files in the sub-folder **output** are as follows: on the first line, the number of grid points along the x and y axis, the x and y coordinates of the origin [m], and the grid interval length [m]; on the second and subsequent lines, the calculated values for $i = 1$ with $j = 1$ to the maximum value, then for $i = 2$ with $j = 1$ to the maximum value, and so on. The calculated values [m] or [m/s] are output in this order.

As an example of tsunami simulation, **Fig. A.3.10** and **Fig. A.3.11** show the inundation area and the distribution of velocities in the off-onshore direction 22 minutes after the start of the simulation using **groundnow2.mdt** as the ground height and water depth data, respectively. **Fig. A.3.12** shows the topographic change at 33 minutes after the tsunami run-up was completed.

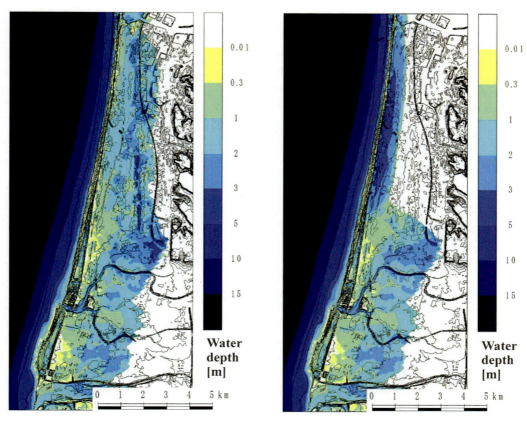

Fig. A.3.10 Examples of water depths and inundation depths (the left is the case of $f_d = 0.01gn^2/d^{1/3}$, the right is the case of $f_d = gn^2/d^{1/3} < 0.1$).

451

Appendix MANUALS OF NUMERICAL PREDICTION MODELS

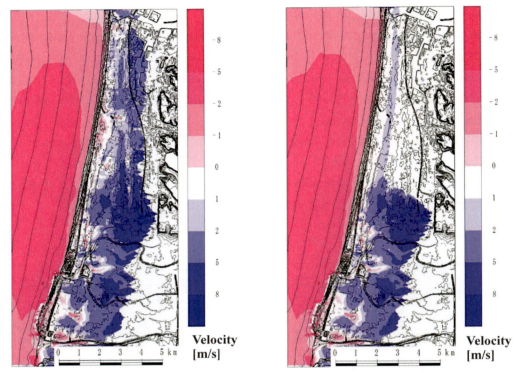

Fig. A.3.11 Examples of velocities in the off-onshore direction (the left is the case of $f_d = 0.01gn^2/d^{1/3}$, the right is the case of $f_d = gn^2/d^{1/3} < 0.1$).

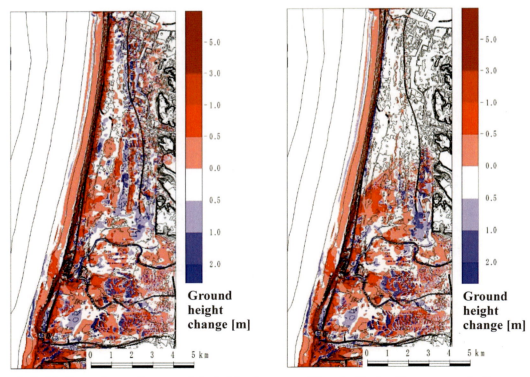

Fig. A.3.12 Examples of ground height changes (the left is the case of $f_d = 0.01gn^2/d^{1/3}$, the right is the case of $f_d = gn^2/d^{1/3} < 0.1$).

3. Numerical Model for Coastal Topographic Change Due to Tsunamis

3.4 References for the Topographic Change Model for Tsunamis

1) Vu, C.T., Yamamoto, Y. and Charusrojthanadech, N.: Improvement of Prediction Methods of Coastal Scour and Erosion due to Tsunami Back-flow, *Proc. 20th International Offshore and Polar Engineering Conference*, 2010, pp.1053-1060.

2) Yamamoto, Y.: Design Process of Coastal Facilities for Disaster Prevention, *Proc. Schl. Eng. Tokai Univ.*, Ser. E, Vol. 31, 2006, pp.11-19.

3) Kuriki, M., Suetsugu, T., Unno, H., Tanaka, Y., and Kobayashi, H.: *Flood Inundation Simulation Model Manual* (*Draft*), Technical Note of Public Works Research Institute, MLIT, Japan, No. 3400, 1996, 197p. (in Japanese)

4) Ribberink, J.S.: Bed-load Transport for Steady Flows and Unsteady Oscillatory Flows, *Coastal Engineering*, Vol.34, 1998, pp.59-82.

5) Soulsby, R.: *Dynamics of Marine Sands. A Manual for Practical Application*, Thomas Telford, UK, 1997, 249p.

6) Yamamoto, Y., Wuttichan, W. and Arikawa, T.: Improvement of Prediction Method of Coastal Damage due to Tsunami, *Journal of Coastal Engineering*, Vol.55, JSCE, 2008, pp.301-305. (in Japanese)

7) Nariyoshi, K., Yamamoto, Y. and Ishii, S.: Improvement of Prediction Models of the Scour of a Seawall and the Topographic Change of a Wide Coastal Area due to Tsunami, *Journal of JSCE*, A1 (Structural & Earthquake Engineering), Vol.68, No.4, 2012, pp. I_1179-I_1185. (in Japanese)

8) Yamamoto, Y., Hayakawa, M. and Ahmadi, S.M.: Simulation Methods for Flood Damage and Scour Damage on Land by Huge Tsunami, *Journal of JSCE*, B2 (Coastal Engineering), Vol.74, No.2, 2018, pp. I_193-I_198. (in Japanese)

9) Yamamoto, Y., Hayakawa, M. and Ahmadi, S.M.: Proposal of a Rational Prediction Method of Scour on Land by a Huge Tsunami, *Journal of JSCE*, B2 (Coastal Engineering), Vol.75, No.2, 2019, pp. I_697-I_702. (in Japanese)

10) Ahmadi, S.M., Yamamoto, Y. and Vu, T.C.: Rational Evaluation Methods of Topographic Change and Building Destruction in the Inundation Area by a Huge Tsunami, *Journal of Marine Science and Engineering*, Vol.8, 762, 2020. Doi:10.3390/jmse8100762

11) van Rijn, L.C.: *Principles of sediment transport in rivers, estuarie and coastal seas*, Aqua Publications, Amsterdam, The Netherlands, 1993.

SYMBOL LIST

A_e	Dean's shape factor for equilibrium beach section
A_{mp}	the amplitude of waves
B	the opening width between breakwaters, the crown width, the parameter
B_{SWL}	the width of the breakwater at the still water surface
B_b	the front shoulder width below the still water surface
C	the phase velocity, wave celerity $(= L/T = \omega/\kappa)$, the additional mass coefficient, the depth-integrated suspended load concentration
$C(z)$	the vertical distribution of suspended load concentration
C_D	the drag coefficient
C_L	the lift coefficient
C_M	the inertia force coefficient, the mass coefficient
C_{MA}	the apparent mass coefficient
C_a	the suspended load concentration at position a
C_b	the wave celerity at the wave breaking position
C_g	the wave group velocity
C_{gb}	the group velocity at wave breaking position
C_{go}	the group velocity offshore
C_o	the wave celerity of offshore waves, the reference concentration at the bottom
C_s	the amount of suspended load settling to the bottom
C_{ut}	the amount of suspended load being rolled up from the bottom
C_ε	the closure coefficient $(=0.09)$
$C_{1\varepsilon}$	the generation rate of the turbulence energy dissipation rate ε $(= 1.44)$
$C_{2\varepsilon}$	the decay rate of the turbulence energy dissipation rate ε $(= 1.92)$
D	the diameter, the distance, the representative dimension of a stone and a block, the slip amount of the fault, the cross-sectional diameter of the drifting object, the representative size of the impact cross-section, the number of main mangrove trunks per unit area, the movement height (the height obtained from the regression analysis of the eroded area ΔA of the cross-sectional area A of the sandy beach and the recession length Δy)
D_{eff}	the effective height of the deformed concrete block
E	the total average energy per unit area due to the wave in one cycle, the longitudinal modulus of elasticity of a log, the representative stiffness of the impacting and impacted objects,

455

	the non-local sink term due to evaporation
$E(f_i)$	the total average energy per unit area of the i-th component wave
E_i	Young's modulus
E_k	the average kinetic energy over a wavelength
E_p	the average potential energy over a wavelength
E_s	the wave energy per unit area and unit time
ΣE_y	the accumulated kinetic energy
$\overline{E}$	the mean energy of irregular waves
F	the fetch or the VOF function,　the tsunami force per unit width on the ground, the fluid force that moves the sediment particle,　the wave energy transport
$F(\ )$	the integral kernel function
F_D	the drag force
F_G	the caisson weight reduced Buoyancy
F_H	the total horizontal wave force
F_L	the lift force
$F_R(x)$	the distribution functions of the resistance force R
F_S	the horizontal salt diffusion term
F_T	the horizontal thermal diffusion term
F_U	the total lift force
F_X	the turbulent momentum flux in the X direction
F_Y	the turbulent momentum flux in the Y direction
F_s	the wave energy transport in the stable state
$F_s(x)$	the distribution function of the external force S
F_{hmax}	the maximum horizontal wave force
F_i	the impact force [N]
F_{ik}	the impact force [kN]
F_{min}	the minimum fetch
F_r	the Froude number
F_u	the horizontal diffusion term in the x direction
F_v	the horizontal diffusion term in the y direction
F_w	the horizontal diffusion term in the z direction
$\Sigma F_y \Delta t$	the accumulated impulse
$\vec{F}_{flow,i}$	the fluid force vector acting on element i
$\vec{F}_{int,i}$	the interaction force vector between the elements
$G(f, \theta)$	the directional distribution function
H	the wave height

H_I	the incident wave height,　the tsunami height in the sea area (= tsunami wave height)
H_L	the wave height of long- period gravity waves
$H_{L1/3}$	the significant wave height of long- period gravity waves
H_{Lm}	the mean wave height of long- period gravity waves
H_R	the reflected wave height
H_b	the breaking wave height
H_{bc}	the breaking wave height with flow
H_i	the incident wave height,　the tsunami height at the shore side
H_{ib}	the thickness of the water flow on the crown of the structure
H_{in}	the incident wave height at the position of the breakwater
H_n	the amplitude of each partial tide
H_m	the mean wave height
H_{max}	the maximum wave height,　the maximum tsunami height at the shoreline
H_{mean}	the average tsunami height at the shoreline
H_o	the offshore wave height,　the tsunami height at the offshore side
$H_o{}'$	the equivalent offshore wave height
H_{rms}	the root mean square of wave height
H_s	the significant wave height,　the wave height on the shoreline
H_0	the height from the reference plane to the mean sea level during the observation period
$H_{1/3}$	the highest one-third wave height,　the significant wave height
$H_{1/10}$	the highest one-tenth wave height
$H_{2\%}$	the wave height in excess of the upper 2%
$\overline{H}$	the mean wave height
I_i	the inertia tensor of element i
I_{TSU}	the importance factor for a tsunami
$J_0(\ )$	the zero-order Bessel function
K	the coefficient of longshore sediment transport rate, the coefficient of on-offshore sediment transport rate
K_D	the dimensionless quantity related to stability of covering materials, the horizontal diffusion coefficient
K_L	the dimensionless quantity representing the stability of covering stones and blocks
K_R	the reflection coefficient
K_T	the transmission coefficient
K_d	the diffraction coefficient
$K_d(f,\theta)$	the diffraction coefficient of the component (regular) waves
K_e	the wave attenuation coefficient

K_f	the wave height reduction factor (reduction coefficient) due to bottom friction
K_h	the vertical diffusion coefficient
K_{loss}	the wave energy loss rate
K_m	the vertical eddy kinematic viscosity
K_r	the refraction coefficient
$K_r(f,\theta)$	the refraction coefficient of component (regular) waves
K_s	the shoaling coefficient
$K_s(f)$	the shoaling coefficient of component (regular) waves
K_θ	the wave runup reduction ratio determined from the wave incident angle θ
K_1	the complete elliptic integral of the first kind
K_2	the complete elliptic integral of the second kind
L	the wavelength, the length of the fault, the length of the drifting object, the dimension of length
L_o	the offshore wavelength
L_{om}	the mean wavelength offshore
L_{op}	the offshore wavelength of the peak frequency
L_p	the wavelength of the peak frequency, the horizontal length from the leading edge of the crown of the structure to the maximum scour depth
M	the required mass of each covering material for stabilization, the magnitude of an earthquake (this magnitude is calculated from the maximum amplitude of the seismic waveform)
M_G	the moment of the caisson weight reduced Buoyancy around the shoreward heel of the target caisson
M_H	the moment of the horizontal wave force around the shoreward heel of the target caisson
M_U	the moment of the vertical wave force around the shoreward heel of the target caisson
M_b	the energy loss term due to wave breaking, the required mass of a concrete block
M_s	the required mass of a stone
M_{xx}	the momentum flux in the x direction through a surface perpendicular to the x axis
M_{xy}	the momentum flux in the y direction on a surface perpendicular to the x axis
M_{yx}	the momentum flux in the x direction on a surface perpendicular to the y axis
M_{yy}	the momentum flux in the y direction through a surface perpendicular to the y axis
M_w	the moment magnitude (this is calculated from the fault moment by multiplying the fault area by the displacement)
N	the wave action density (the directional spectrum represented by the wavenumber vector)
$N(\sigma,\theta)$	the wave action density (the directional spectrum divided by the relative frequency)
$N_0(\ )$	the zero-order Neumann function

N_s	the Stability Number (the dimensionless quantity that expresses the degree of stability)
N_w	the incident number of storm waves
P	the water pressure,　the wave pressure,　the non-local source term of precipitation
$P(r)$	the atmospheric pressure at a distance r from the typhoon centre
$P(z)$	the water pressure at a vertical coordinate z
$P_D(x)$	the damage probability
P_X	the gradient hydrostatic pressure in X direction
P_Y	the gradient hydrostatic pressure in Y direction
P_a	the one atmospheric pressure (= 1013hpa)
P_c	the atmospheric pressure at the typhoon centre (= $P_a - \triangle p$)
$P_d(z)$	the hydrodynamic pressure at a vertical coordinate z
P_{ob}	the excess pore water pressure during the back flow on the surface of the backfill layer in the structure
$P_{ob\max}$	the maximum excess pore water pressure (the maximum value of dynamic pore water pressure) at the lowest edge of front face of the dike or the seawall during back flow
$P_{d\max}$	the maximum hydrodynamic pressure
P_f	the probability of failure
P_{fa}	the allowable probability of failure
P_r	the turbulent energy generation
P_{rb}	the generation of turbulence energy due to bottom resistance
P_{rs}	the generation of turbulence energy due to shear stress
P_{rw}	the generation of turbulence energy due to breaking waves
p_u	the maximum lift pressure
P_w	the amount of electricity,　the amount of power generated per unit width
P_∞	the atmospheric pressure of the ambient air ($\fallingdotseq$ 1013hpa)
Q	the longshore sediment transport rate,　the flow rate of the water supply
R	the radius of the earth,　the wave runup height,　the tsunami run-up height from the frontal ground,　the radius of maximum wind speed, the strength,　the resistance force, the maximum run-up height × 1.3 for calculating the impact force due to the tsunami
R_L	the runup height of long- period gravity waves
$R_{L1/3}$	the significant runup height of long- period gravity waves
R_{Lm}	the mean runup height of long- period gravity waves
R_{Lmax}	the maximum runup height of long- period gravity waves
R_e	the Reynolds number
R_f	the reliability

R_{max}	the maximum runup height of waves,　the dynamic response ratio to the impact load
R_{mean}	the mean runup height of waves
R_x	the resistance term in the x direction
R_y	the resistance term in the y direction
R_z	the resistance term in the z direction
$R_{1/3}$	the highest one-third runup height waves,　the significant runup height of waves
$R_{1/10}$	the highest one-tenth runup height of waves
$R_{2\%}$	the upper 2% excess runup height of waves
S	the parameter representing the concentration of directional energy distribution of waves, the ratio of the inundated area to the total area of the control volume,　the salinity, the external force
$S(f)$	the frequency energy spectral density function, the frequency spectrum (the wave energy per unit frequency corresponding to the frequency)
$S(f,\theta)$	the directional wave spectrum ($= S(f) \times G\,(f,\,\theta)$)
S_E	the function of energy supply and dissipation
S_T	the omnidirectional spectrum of waves
S_{bf}	the energy dissipation due to seabed friction
S_{ds}	the energy dissipation due to wave breaking
S_{in}	the energy supply from wind to the waves
S_{max}	the maximum value of the concentration parameter
S_{nl}	the energy transport due to non-linear interaction between wave components
S_p	the terms for generating the input waves in the continuity equation
S_u	the terms for generating the input waves in the equation of motion
S_v	the terms for generating the input waves in the equation of motion
S_w	the terms for generating the input waves in the equation of motion
S_{wc}	the energy dissipation due to whitecaps
S_{wb}	the energy dissipation due to shallow water wave breaking
S_{xx}	the radiation stress obtained by averaging the momentum flux in the x direction crossing a surface perpendicular to the x axis over one wave period
S_{xy}	the radiation stress obtained by averaging the momentum flux in the y direction on a surface perpendicular to the x axis over one wave period
S_{yx}	the radiation stress obtained by averaging the momentum flux in the x direction on a surface perpendicular to the y axis over one wave period
S_{yy}	the radiation stress obtained by averaging the momentum flux in the y direction crossing a surface perpendicular to the y axis over one wave period

T	the wave period,　the water temperature
T_I	the wave period of incident waves
T_L	the wave period of long- period gravity waves
$T_{L1/3}$	the significant wave period of long- period gravity waves
T_{Lm}	the mean wave period of long- period gravity waves
T_T	the wave period of transmitted waves
T_m	the mean wave period
T_{max}	the maximum wave period
T_o	the offshore wave period
T_p	the peak wave period
T_s	the significant wave period
T_{term}	the observed period of water surface elevation
T_1	the resonance period of the secondary undulation
T_2	the resonance period of the seiche
$T_{1/3}$	the highest one-third wave period,　the significant wave period
$T_{1/10}$	the highest one-tenth wave period
$\vec{T}_{flow,i}$	the torque due to fluid force acting on element i
$\overline{T}$	the mean wave period
U	the flow velocity in the x direction,　the dimensionless velocity,　the depth averaged u, the maximum flow velocity at the top of the covering work,　the uniformity coefficient
$U(z)$	the wind velocity at a height z
U_{gr}	the wind speed of the gradient wind
U_m	the mass transport velocity
U_{max}	the maximum wind speed
U_n	the amplitude of each partial current
U_r	the Ursell number
U_0	the velocity of the constant current component of the tidal current
$U_{0.2}$	the mean wind speed 0.20 ft above the mean water surface
U_{30}	the mean wind speed 30 ft above the water surface
U_{10}	the mean wind speed 10 m above the water surface
$U_{19.5}$	the mean wind speed 19.5 m above the water surface
$\vec{U}$	the flow velocity vector
V	the flow velocity in the y direction,　the impact velocity,　the depth averaged v
$V(r)$	the typhoon wind velocity at r
$V(x)$	the velocity of the longshore current at a distance x offshore from the shoreline
V_o	the velocity of the drift current on the sea surface

V_{max}	the maximum flow velocity at the lowest edge of front face of the coastal structure during back flow
V_x	the drifting speed of the small vessel
$\overline{V}$	the mean value of the offshore distribution of longshore current velocity
W	the average power (= generating power) per unit width due to the wave in one cycle, the dead weight, the weight of bottom sediment in water, the small vessel weight, the width of the fault
W_f	the power of the frictional stress
W_b	the dissipation term due to wave breaking
X	the distance along the breakwater from the diffraction start point
X	the longitude coordinate in spherical coordinates
Y	the distance in shore direction from the diffraction start point of the breakwater
Y	the latitude coordinate in spherical coordinates
Y_r	the ratio of the occurrence interval of rip currents to the wave breaking zone width x_b
Z	the dimensionless water surface elevation, the limit state function, the failure function
a	the horizontal acceleration of the water particles, the half radius of the impact surface, the coefficient
a_i	the wave amplitude of the i-th component wave
a_m	the horizontal amplitude of the water particle at the seabed surface
a_{max}	the amplitude of the horizontal acceleration
a_x	the horizontal acceleration of the water particles
a_z	the vertical acceleration of the water particles
b	the wave line spacing, the width of a block, the width of a breakwater, the coefficient, the Rouse number
b_i	the bay width at the shore side
b_o	the wave line spacing offshore, the bay width at the offshore side
cn	Jacobi's elliptic function
d	the water depth (= $h + \eta$), the water depth on the rubble foundation, the representative grain size, the inner diameter of the pipeline
d_{50}	the median grain size
$d_{i,j}$	the backfill thickness from the lowest edge of the front face of the structure at the points i and j
d_{max}	the total backfill thickness from the lowest edge of front face to the top of the structure
d_o	the standard sand grain size (= 0.25mm = 0.00025m)
d_s	the backfill thickness from the lowest edge of the front face of the structure which is adjusted to match the natural beach stabilising slope.

462

d_t	the thickness of the sand layer in front of the coastal structure (= the height from the ground surface in front of this structure to the lowest edge of the front face of this structure)
$\overline{d}$	the mean water depth
dF	the wave force acting on a small height dz
dF_D	the drag force acting on a small height dz
dF_I	the inertia force acting on a small height dz
dF_{max}	the maximum value of the wave force acting on a small height dz
d_B	the water depth of the platform below the water surface of the structure
dn	Jacobi's elliptic function
$\overrightarrow{e_w}$	the unit vector of wave direction at a wave breaking position
$\overrightarrow{e_n}$	the unit vector of tangential direction of the wave crest line
f	the wave frequency,　the friction loss coefficient,
	the coefficient of fluid force at the suction layer ($\fallingdotseq 1$)
f'	the friction coefficient as the function of sediment particle size
$f_R(x)$	the probability density functions of the resistance force R
f_b	the sea-bottom friction coefficient
f_{cl}	the Coriolis coefficient
f_d	the resistance coefficient due to structures and trees
f_i	the wave frequency of the i-th component wave
f_n	the modification factor of the amplitude of each partial tide
f_p	the peak wave frequency
f_s	the friction coefficient at sea level
$f_s(x)$	the probability density functions of the external force S
f_w	the sea-bottom friction coefficient ($= 2 \times f_b$)
f_w'	the friction coefficient in the oscillatory flow field
f_x	the length ratio of the inundated part to the length of one side parallel to the x directions in the control volume
f_y	the length ratio of the inundated part to the length of one side parallel to the y directions in the control volume
f_{xy}	the spatial distribution function of the amplitude of the water surface elevation η
$f_Z(x)$	the normal probability density function of the limit state function Z
f_*	the dimensionless frequency
f_{p*}	the dimensionless peak frequency
$\overrightarrow{f_{ls}}$	the interaction vector between the liquid and solid phases
$\overrightarrow{f_{col}}$	the collision force vector between solid phase particles.
$\overline{f}$	the mean wave frequency

g	the acceleration of gravity
h	the water depth below the still water surface, the water head, the water depth at the front of the dike or the seawall
h_I	the tsunami height on the ground (= inundation depth)
h_b	the wave breaking depth below the still water surface
h_{bc}	the wave breaking depth with flow
h_{bmax}	the peak thickness of the water flow on the crown during tsunami back flow
h_c	the crown height above the still water surface, the critical depth of sediment movement
h_e	the net head
h_f	the inundation depth in front of the target wall on the ground
h_i	the water depth below the still water surface at the shore side
h_{max}	the height of the uppermost point where the tsunami pressure acts the peak thickness of the water flow on the crown during tsunami run-up
Δh_{max}	the maximum scour depth due to high waves or megatsunamis
h_o	the water depth below the still water surface at the offshore side, the height from the still water surface to the top of pillar roots
h_r	the water depth from the still water surface to the crown of the artificial reef, the peak water depth in return flows on the slope
k	the Karman constant (= 0.41), the turbulence energy integrated in the depth direction, the effective axial stiffness between the drifting object and the drifted object, the smaller between the effective stiffness of the drift and the transverse stiffness of the target structure
k_c	the hydraulic conductivity within the sand and gravel layer
k_d	the correction factor
k_e	the equivalent roughness
l	the length of the pipeline, the mean length of the harbor or the lake in the direction of interest
m_d	the mass of the drifting object
m_i	the i-th moment, the mass of the impacting object 1 and the impacted object 2, the mass of element i
n	the number of data, the ratio of the wave group velocity to the wave celerity
p	the water pressure, the wave pressure, the total pressure
p_a	the atmospheric pressure
p_{max}	the maximum tsunami pressure
p_s	the hydrostatic water pressure
p_{sm}	the maximum hydrostatic water pressure

p_u	the maximum value of uplift pressure
p_1	the maximum wave pressure per unit length (width) during the push wave
p_2	the wave pressure per unit length (width) at the seabed surface during the push wave
p_1'	the minimum wave pressure per unit length (width) during the backwash
p_2'	the wave pressure per unit length (width) at the seabed surface during the backwash
$p(H)$	the Rayleigh distribution
$p(>H)$	the probability of occurrence of a wave height greater than a certain wave height H
$p(z)$	the tsunami force per unit area
$p(\eta)$	the probability density function of water surface elevation
Δp	the difference between one atmospheric pressure (1013 hPa) and the lowest pressure at the tropical cyclone centre
q	the flow rate per unit width, the average wave overtopping rate per second of unit width [m³/m/s], the nonhydrostatic pressure of water, the suction (outflow) rate of backfill materials including porosity per unit time for unit width, the peak flow rate per unit width of falling water
$q(x)$	the on-offshore sediment transport rate
q_T	the average wave overtopping rate per wave of unit width [m³/m/T]
q_b	the bed load transport rate, the on-offshore sediment transport rate at the wave breaking position
q_{bc}	the bed load transport rate by the flow
q_{bw}	the bed load transport rate by waves
q_{bx}	the bed load transport rate in the x direction
q_{by}	the bed load transport rate in the y direction
q_{in}	the local sources of water per unit of volume
q_m	the mass flow rate of wind-blown sand
q_{out}	the local sinks of water per unit of volume
q_p	the on-offshore sediment transport rate at the wave plunge position
q_r	the regular wave overtopping rate per wave of unit width [m³/m/T]
q_x	the flow rate per unit width in the x direction, the sediment transport rate in the x direction, the suction (outflow) rate of backfill materials including porosity per unit time for unit width in the x direction
q_y	the flow rate per unit width in the y direction, the on-offshore sediment transport rate per unit width and unit time, the sediment transport rate in the y direction, the suction (outflow) rate of backfill materials including porosity per unit time for unit

	width in the y direction
q_z	the on-offshore sediment transport rate at the run-up start position
$\vec{q}$	the vector-displayed sediment transport rate
r	Euler's constant ($= 0.5772$), the curvature radius of the isobaric line, the linear distance from the typhoon centre to the target position, the unit volume weight of the drifting object
r_o	the distance from the typhoon centre to the position of maximum wind speed
$\vec{r_i}$	the relative position vector between the elements
sn	Jacobi's elliptic function
t	the time, the duration
Δt	the time interval
t_b	the time delay
t_d	the action time of the impact force
t_{min}	the minimum duration
t_r	the arrival time, the time taken for the water flow to reach the frontal ground from the crown
$\tan\alpha$	the mean beach slope, the surface slope of the coastal structure
$\tan\alpha_e$	the mean beach slope in equilibrium (final) with respect to the target high wave
$\tan\alpha_0$	the mean beach slope immediately before the onset of the target high wave
$\tan\alpha_f$	the foreshore slope
$\tan\beta$	the mean seabed slope, the initial sea bottom slope
$\tan\beta_c$	the equilibrium seabed slope at which sediment transport ceases
$\tan\beta_w$	the seabed slope in the wave direction
u	the horizontal velocity in the x direction
ub	the water particle horizontal velocity at the still water surface at the wave breaking position, the horizontal velocity at the seabed surface, the flow velocity during the back flow on the surface of the backfill layer in the structure
$u_b(t)$	the oscillatory flow velocity over the seabed boundary layer
u_{bmax}	the amplitude of the bottom velocity of the water particles during wave breaking
u_{max}	the amplitude of the horizontal velocity of the water particles, the maximum velocity of the fluid transporting the drifting object
u_r	the velocity of the return flow at the slope
u_x	the flow velocity in the x direction
u_y	the flow velocity in the y direction, the vertical velocity of the water flow jumping from the crown
u_*	the friction velocity

u_{*_c}	the critical friction velocity
$\vec{u}$	the velocity vector
$\vec{u}_i$	the velocity vector of element i
$\tilde{u}$	the ensemble mean velocity in the x direction averaged across the water column, the depth-averaged velocity (averaged over one cycle for oscillatory flow)
$\hat{u}_b$	the amplitude of the horizontal velocity of the water particles at the seabed surface
v	the horizontal velocity in the y direction, the propagation velocity of tsunamis
$\tilde{v}$	the ensemble mean velocity in the y direction averaged across the water column
w	the vertical velocity in the z direction
w_s	the settling velocity of the bottom sediment
w_x	the wind speed in the x direction
w_y	the wind speed in the y direction
x	the horizontal coordinate, the external force parameters such as inundation depth and flow velocity
x_b	the wave breaking position
x_i	the horizontal distance from the front face of the structure to the point i
x_{max}	the total horizontal length from the front face to the back end of the structure
x_o	the average horizontal position of the water particles
x_p	the wave plunge position
x_r	the run-up end position
x_z	the run-up start position
y	the horizontal coordinate
Δy	the shoreline recession
z	the vertical coordinate (up and down from the still water surface), the crown height
z_f	the height from the still water surface to the top of the revetment
z_o	the average vertical position of the water particles, the roughness length
$\Gamma()$	the gamma function
Φ	the damage function
A	the coefficient, the parameter, the surface slope angle, the mean slope angle from the wave breaking depth to the wave run-up height, the strike angle when the water flow strikes the frontal ground
α_c	the critical surface slope angle of the coastal structure
α_p	the predominant refraction wave direction
β	the coefficient, the parameter, the mean surface slope angle, the reliability index
γ	the parameter, the wave amplitude – water depth ratio, the wave height - water depth ratio for wave breaking limits

γ_H	the partial safety factor for wave height
γ_s	the specific gravity of sediment, the specific gravity of the covering work
γ_p	the energy attenuation effect due to plasticity
γ_v	the porosity
γ_x	the area permeability in the x direction
γ_y	the area permeability in the y direction
γ_z	the area permeability in the z direction, the partial safety factor for the resistance
δ	the thickness of the surface roller, the thickness of the frictional boundary layer, the friction depth (the depth to which the flow direction is opposite to the wind direction), the dip (the angle of the fault slope) of the fault
$\delta(\ )$	the delta function
ε	the dissipation rate of turbulence energy, the constant eddy viscosity in the vertical direction, the coefficient of sediment transport rate dependent on the local seabed slope
ε_I	the phase angle of the incident wave
ε_R	the phase angle of the reflected wave
ε_i	the phase angle of the i-th component wave
ε_s	the porosity of the sand layer
ζ	the vertical travel distance ($= z - z_o$) of a water particle, the local bottom height
$\overrightarrow{\nabla \zeta}$	the slope vector of the terrain
η	the water surface elevation, the tide level, the power generation efficiency
η'	the height from the still water surface to the top of the wave pressure
η_I	the water surface elevation of incident waves
η_M	the highest high storm tide
η_R	the water surface elevation of reflected waves
η_{max}	the maximum tidal height during a storm surge, the height to the top of tsunami pressure distribution
η_0	the mean water surface elevation, the elevation in water level due to low pressure
η_{rms}	the root mean square of water surface elevation
η_s	the surge of seawater level due to wave setup or surf beat, the mean water surface elevation above the shoreline
$\bar{\eta}$	the average water surface elevation from the still water surface
θ	the wave direction angle
θ'	the angle at which the wave pressure is greatest in the incident angle range $\theta \pm 15°$
θ_I	the incident wave angle
θ_R	the reflection angle

θ_T	the refraction angle
θ_b	the incident angle of the breaking wave from the normal to the shoreline
θ_c	the critical Shields number, the dimensionless critical suction resistance during back flow, the dimensionless liquefaction pressure during back flow
θ_o	the offshore wave direction angle
θ_s	the Shields number, the dimensionless suction (outflow) force during back flow
$\theta_s'(t)$	the Shields number in oscillatory flow
θ_w	the wind direction, the angle from x-direction to the direction of breaking waves
$\bar{\theta}$	the mean wave direction angle
κ	the wave number ($= 2\pi/L$), Kalman's constant ($= 0.4$)
κ_p	the peak wavenumber of the JONSWAP spectrum
$\tilde{\kappa}$	the mean wave number
$\vec{\kappa}$	the wavenumber vector
Λ	the parameter, the coefficient
λ_{nl}	the parameter
μ	the viscosity coefficient of seawater, the parameter, the mean value
μ_z	the mean value of the limit state function Z
v	the kinematic viscosity coefficient
v_e	the sum of the molecular and eddy kinematic viscosity coefficients
v_i	Poisson's ratio
v_t	the eddy kinematic viscosity coefficient
v_v	the vertical eddy kinematic viscosity coefficient
ξ	the horizontal travel distance ($= x - x_o$) of a water particle, the Iribarren number
ξ_m	the wave breaking similarity parameter
ξ_{mc}	the threshold of wave breaking type
ξ_o	the Iribarren number for offshore waves, the surf similarity parameter
ρ	the density of seawater
ρ^*	the density considering the buoyancy force of seawater
ρ_a	the air density
ρ_i	the density of element i
ρ_s	the density of sand (usually around 2.5), the density of the sand layer (usually around 1.8)
ρ_w	the reference density of seawater
σ	the relative frequency (angular frequency in a coordinate system moving with the flow) the standard deviation
$\sigma_H{}'$	the variational coefficient of wave data

σ_f	the yield stress of the objecting object
σ_η	the standard deviation of the water surface elevation
σ_t	the closure coefficient ($=1$)
σ_z	the standard deviation of the limit state function Z
σ_ε	the closure coefficient ($=1.3$)
$\bar{\sigma}$	the mean relative frequency
τ	the shear stress, the maximum frictional stress on the bottom under the coexistence of flow and waves
τ_b	the frictional stress (shear stress) at the seabed
τ_{cr}	the movement-limiting frictional stress on the bottom material
τ_f	the maximum suction (outflow) force per unit area during back flow
τ_m	the horizontal diffusion force per unit area
τ_r	the effective suction resistance per unit area during back flow, the effective liquefaction pressure during the backflow
τ_w	the wave-induced stress
$\hat{\tau}_b$	the amplitude of the frictional stress at the seabed
ϕ	the velocity potential, the angle between the reef surface at the point of maximum flow velocity on the artificial reef and the horizontal plane, the internal friction angle of sediment, Hallermeier's non-dimensional number
φ	the latitude, the strike (the horizontal angle from the north to the fault plane) of the fault
χ	the parameter
ω	the angular frequency ($= 2\pi/T$), the angular velocity due to the rotation of the earth
$\bar{\omega}$	the angular velocity vector of element i
$\tilde{\omega}$	the mean angular frequency
Δ	the propagation distance from the epicentre to the target coast

TECHNICAL TERM LIST

allowable stress design /329

allowable wave overtopping rates /297

angular frequency /2

armour units /137

artificial reef(s) /141, 285

asperity /193

astronomical tides /167

back beach /210

backfill material suction /290

backshore /210

bathymetric change models based on waves and nearshore currents /228, 236

beach /210

beach face /210

beat waves /3

bed load /220

bed load transport rates /221

berm /211

biodegradability /354

biodegradable plastics /355

biomass plastics /355

biomass power /361

bioplastics /355

block mound breakwater /66

blue tides /342

bore /80

breaking wave pressure /125

breakwater /286

Bretschneider-Mitsuyasu frequency spectrum /11

broken wave pressure /125

Bruun rule /218

CADMAS SURF /88, 200, 239, 295

cardinal datum level (C.D.L.) /169

Chemical Oxygen Demand (COD) /343

Cnoidal wave theory /54,57

coastline /210

coastal currents /152

coastal dike /103, 282

cold currents /165

complete movement limit depth /221

composite breakwater /68

composite type /124

conservation law of drifting sand /226

continental shelf waves /181

continuity equation of drifting sand /226

Coriolis force /41, 163

Cost-Benefit Ratio (B/C) /318

Coupled Discrete Model / 28

Coupled Hybrid Model / 28

crest /66

critical depth for sediment movement /220

critical Shields number /213

cross-shore sediment transport /229

crown /66

current-rips /166

cyclone /175

cyclone storm /175

damage limit overtopping rates /297

dangerous semicircle /21

Decoupled Propagation Method /28

deep sea waves /45

Delft3D /239

density of sediment /214

471

design water depth /293

design wave height /293

detached breakwater(s) /66, 246, 284

detached breakwater works /300

diffraction coefficient /62

dip /192

directional distribution function /13

directional spectrum /13

Discrete Interaction Approximation (DIA) /31

dispersion relation equations /44

dispersion relation formula /44

Dissolved oxygen (DO) /343

Distinct Element Method (DEM) /241

drift currents /152, 162

drifting sand /220

drift phenomenon /56

dry density /213

dynamical theory of tides /168

ebb tide /168

edge waves /158

Ekman spiral /163

Ekman transport /163

Electroactive Polymer Artificial Muscle (EPAM) /363, 371

energy loss rate /65

epicentral earthquake /184

equilibrium theory of tides /168, 169

equivalent offshore wave height /65

Eulerian method /40

fetch /10

finite amplitude wave theory /41

Finite Volume Community Ocean Model (FVCOM) /351

flood tide /168

flowing sand /220

foreshore /210

Framework for Aquatic Biogeochemical Models (FABM) /351

freak wave /289

frequency domain analysis method /4

frequency energy spectral density function /9

frequency spectrum /8, 9

friction depth /162

front scour /290

general movement limit depth /221

geostrophic wind /20

geothermal electric power /361

global warming /311

gradient wind / 20

gravity type /124

gravity waves /2

Greenhouse Gas Inventory /360

Green's low /188

group velocity /49

groynes /246, 284

groyne works /299

Hallermeier's non-dimensional number /218

harmonic analysis of tides /169

headland defence works /303

Healy's method /69

highest one-tenth wave /6

highest one-tenth wave height /6

highest one-tenth wave period /6

highest one-third wave /6

highest one-third wave height /6

highest one-third wave period /6

highest wave /6

high tide /168

high tide shoreline /210

horizontal two-dimensional seabed change models /237

Hudson's formula /138

hurricane /175

hydroelectric power /361

ideal fluid /40

impact breaking wave pressure /125

improved virtual slope angle /106

Incompressible Smoothed Particle Hydrodynamics method (ISPH method) /242

initial movement limit depth /221

infragravity /158

infragravity waves /2, 96

inland earthquake /184

Intergovernmental Panel on Climate Change (IPCC) /311

intermediate sea waves /45

intertropical convergence zone /174

intraplate earthquake /184

Iribarren number /67

KAIMEI /368

Kyoto Protocol /360

Lagrangian method /40

landing site /287

legged type /124

limit depth for sediment movement /220

limit state design /330

limit state function /330

littoral drift /229

long-period gravity waves /2, 102

longshore bar /211

longshore currents /155

longshore drift /229

longshore sediment transport /229

longshore sediment transport rate(s) /228, 231

long waves /45

low tide /168

low tide shoreline /210

Lumped Triad Approximation (LTA) /33

magnitude (M) /183

major tsunami warning /326

mass transport /56

mass transport velocity /56

maximum period /7

maximum wave height /7

median grain size /213

meteorological anomaly /309

meteorological deviation /309

Meteorological Research Institute model (MRI) /28

meteorological tides /167

microplastics /354

Mighty Whale /368, 369

mild slope equation /86

minimum duration time /10

Mitsuyasu-type directional distribution function /14

Mitsuyasu type II frequency spectrum /10

model for predicting beach changes based on Bagnold's concept /235

modified JONSWAP frequency spectrum /12

moment magnitude (M_w) /183

movement height /230, 232

Moving Particle Semi-implicit method (MPS method) /241

MRI-II /28

473

MRI-III   /28

multiline models   /228, 233

multiple defenses   /317

navigable semicircle   /21

neap tide   /168

nearshore current(s)   /80, 152

nearshore devices   /363

non-critical areas   /311

normal beach   /211

normal seashore   /211

numerical wave analysis method   /86

ocean currents   /152

ocean-trench earthquakes   /182

offshore breakwater(s)   /246, 284

offshore devices   /363

offshore waves   /45

offshore wave height   /59

offshore zone   /211

one line model   /228, 230

on-offshore sediment transport   /229, 232

on-offshore sediment transport rate   /228

outer-rise earthquakes   /182, 184

outflow (suction)   /255

parabolic wave equation   /87

parameter method   /19

partial standing waves   /52

partial tides   /169

particle implemented simulator for physical and engineering research (PARISPHERE) /242

Particle method   /241

Paris Agreement   /360

peak frequency   /10

perfect fluid   /40

Performance-based design   /329

phase velocity   /2

photovoltaic energy   /361

pier   /287

Pierson-Moskovitz frequency spectrum /11

piezoelectric element   /363

plate-boundary earthquakes   /182

plunging breaker   /72

PNJ method   /27

Potential of hydrogen (PH)   /343

PowerBuoy   /369

probability of failure   /330

quays   /288

quay wall   /287

radiation stress   /100, 152

rake angle   /192

red tides   /342

reflection coefficient   /65

refraction coefficient   /60

relative frequency   /31

relative water depth   /53

relative wave height   /53

reliability-based design   /330

reliability index (β)   /332

renewable energy   /360

resilient facilities   /319

rip currents   /155

ripples   /2, 16

roaring waves   /3

rolling   /220

root mean square of wave heights   /6

saltation   /220

sand bypass method   /304

sand discharge systems from dammed reservoirs /303

sand recycling method   /304

Saville's virtual slope   /105

474

SBEACH model /228, 234

seawall /103, 282

secondary undulation /180

sediment transport /220

seiche /180

semi-diurnal tide /168

shallow sea waves /45

Shields law of similarity /214

Shields number /222

shoaling coefficient /59

shoreline change model /228, 230

shoreward currents /155

significant wave /6

significant wave height /6

significant wave method /19

significant wave period /6

Simulating WAve Nearshore model (SWAN) /19, 28

sliding /220

sloping type /124

slurry transportation /304

small amplitude wave theory /41

SMB method /19

Smoothed Particle Hydrodynamics method (SPH method) /241

soil cohesion /213

solitary wave /58

soliton fission /195

specific gravity of sediment /213

spectral analysis method /4, 7

spectral method /19

spilling breaker /73

spring tide /168

stability number /141

standing wave pressure /125

standing waves /51

step /211

STOC /194

Stokes wave theory /54

storm surge(s) /2, 174

stormy beach /211

stormy seashore /211

Stream Function Method (SFM) /54

strike /192

subduction zone earthquakes /182

surface movement limit depth /221

surface roughness /213

surface tension /41

surf beat /96

surf similarity parameter /73, 158

surging breaker /73

suspended load /220

suspended load transport /225

swell (swells) /2, 16

the highest high-water level (HHWL) /293

the highest high storm tide (η_M) /293

the mean monthly-highest water level (MWL) /293

Theory of Bound Long Waves (BLW) /98

Theory of Time-varying Breakpoint-forced Long Waves (BFLW) /98

thermohaline currents /152

three-dimensional two-way coupled fluid-structure-sediment interaction model (FSSM) /239

tidal waves /2

tidal current(s) /152, 167

tidal force /167, 168

tides /167

tide generating force /167

time-domain analysis method /4

tombolo /300

475

training jetty /286

training wall /286

transmission coefficient /65

trenches /185

tropical cyclone /175

tropical storm /175, 176

trough /211

troughs /185

tsunami(s) /2, 45, 174, 182

tsunami advisory /326

tsunami debris impact forces /200

tsunami warning /326

typhoon /176

undertow /155

uniformity coefficient /213

upright type /124

Ursell number /53, 218, 233

Van der Meer's formula /139

variational coefficient of wave data (σ_H') /333

velocity potential /41

virtual slope angle /105

Wallops frequency spectrum /12

warm currents /165

wash load /220

wave absorbing breakwater /66, 283

wave absorbing structure /283

wave dissipating breakwater /283

wave dissipating structure /283

wave action density /17, 31

wave-by-wave analysis method /4

wave celerity /2

wave diffraction /62

wave forecasting and hindcasting /18

wave groupiness /4

wave height /2

wave height-to-depth ratio /53

wavelength /2

WAve Model (WAM) /19, 28

wave number /2

wave period /2

wave period-averaged momentum equation /160

wave power /362

wave power generation methods /363

wave ray method /86

wave refraction /59

wave setdown /154

wave setup /155

wave shoaling /59

wave steepness /53

WAVEWATCH-III /28

westward intensification /165

wharves /288

whitecaps /18

Wilson method /19

wind-blown sand /269, 290

wind duration /10

wind power /361

wind waves /2, 16

XBeach /237

zero down-crossing method /4

zero up-crossing method /4

2011 off the Pacific coast of Tohoku Earthquake /183